MOLECULARLY IMPRINTED SENSORS

MOLECULARLY IMPRINTED SENSORS:

Overview and Applications

DR. SONGJUN LI

School of Materials Science & Engineering, Jiangsu University, Zhenjiang 212013, China

DR. YI GE, DR. SERGEY A. PILETSKY,
DR. JOSEPH LUNEC

Cranfield Health, Vincent Building, Cranfield University, Cranfield, Bedfordshire, MK43 0AL, UK

ELSEVIER

AMSTERDAM • BOSTON • HEIDELBERG • LONDON • NEW YORK • OXFORD
PARIS • SAN DIEGO • SAN FRANCISCO • SINGAPORE • SYDNEY • TOKYO

Elsevier
The Boulevard, Langford Lane, Kidlington, Oxford OX5 1GB, UK
Radarweg 29, PO Box 211, 1000 AE Amsterdam, The Netherlands

First edition 2012

British Library Cataloguing in Publication Data
A catalogue record for this book is available from the British Library

Library of Congress Cataloging-in-Publication Data
A catalog record for this book is available from the Library of Congress

ISBN: 978-0-444-56331-6

Printed and bound in Great Britain

12 13 14 15 16 10 9 8 7 6 5 4 3 2 1

Contents

Contributors

Behrouz Akbari-adergani Food & Drug Laboratory Research Center, Food & Drug Department, Ministry of Health and Medical Education, Tehran, Iran

Daniel Antuña-Jiménez Departamento de Química Física y Analítica, Universidad de Oviedo, Julián Clavería, Oviedo, Spain

M. Carmen Blanco-López Departamento de Química Física y Analítica, Universidad de Oviedo, Julián Clavería, Oviedo, Spain

Shunsheng Cao School of Materials Science and Engineering, Jiangsu University, China. Cranfield Health, Cranfield University, Cranfield, Bedfordshire, UK

Juanrong Chen School of Environment, Jiangsu University, China

Iva Chianella Cranfield Health, Vincent Building, Cranfield University, Cranfield, MK43 0AL, Bedfordshire, UK

Thomas J. Cleij Hasselt University, Institute for Materials Research (IMO), Agoralaan, Belgium. Maastricht University, Maastricht Science Programme, The Netherlands

Goretti Díaz-Díaz Departamento de Química Física y Analítica, Universidad de Oviedo, Julián Clavería, Oviedo, Spain

Franz L. Dickert Department of Analytical Chemistry, University of Vienna, Austria

Hanne Diliën Hasselt University, Institute for Materials Research (IMO), Agoralaan, Belgium

Frederik Horemans Hasselt University, Institute for Materials Research (IMO), Agoralaan, Belgium

Naseer Iqbal University of Vienna, Department of Analytical Chemistry, Vienna, Austria

Mehran Javanbakht Department of Chemistry, Amirkabir University of Technology, Tehran, Iran

Marzena Kaniewska Department of Chemistry, University of Warsaw, Pasteura 1, 02-093 Warsaw, Poland

Konstantinos Karras Cranfield Health, Vincent Building, Cranfield University, Cranfield, MK43 0AL, UK

Dhana Lakshmi Cranfield Health, Vincent Building, Cranfield University, Cranfield, MK43 0AL, UK

Jianping Li College of Chemistry and Bioengineering, Guilin University of Technology, Guangxi, China

Peter A Lieberzeit University of Vienna, Department of Analytical Chemistry, Vienna, Austria

Yibin Liu Beijing National Laboratory for Molecular Sciences, MOE Key Laboratory of Bioorganic, Chemistry and Molecular Engineering, College of Chemistry and Molecular Engineering, Peking University, Beijing, China

M. Jesús Lobo-Castañón Departamento de Química Física y Analítica, Universidad de Oviedo, Julián Clavería, Oviedo, Spain

Fang Long School of Materials Science and Engineering, Jiangsu University, China

Silvia Marson School of Applied Science, Cranfield University, UK

Zi-Hui Meng School of Chemical and Environmental Engineering, Beijing Institute of Technology, Beijing, China

Arturo J. Miranda-Ordieres Departamento de Química Física y Analítica, Universidad de Oviedo, Julián Clavería, Oviedo, Spain

Adnan Mujahid Department of Analytical Chemistry, University of Vienna, Austria. Institute of Chemistry, University of the Punjab, Pakistan

Tsutomu Nagaoka Department of Applied Chemistry, Graduate School of Engineering, Osaka Prefecture University, 1-2 Gakuen-cho, Naka-ku, Sakai, Japan

Sergey Piletsky Cranfield Health, Cranfield University, Cranfield, Bedfordshire, UK

Stas Piletsky Cranfield Health, Cranfield University, Cranfield, Bedfordshire, UK

Weicheng Sheng School of Materials Science and Engineering, Jiangsu University, China

Hiroshi Shiigi Department of Applied Chemistry, Graduate School of Engineering, Osaka Prefecture University, 1-2 Gakuen-cho, Naka-ku, Sakai, Japan

Jin Tan Tianjin Key Laboratory of Food Biotechnology, College of Biotechnology and Food Science, Tianjin University of Commerce, China

Shiho Tokonami Nanoscience and Nanotechnology Research Center, Research Organization for the 21st Century, Osaka Prefecture University, 1-2 Gakuen-cho, Naka-ku, Sakai, Japan

Marek Trojanowicz Department of Chemistry, University of Warsaw, Pasteura 1, 02-093 Warsaw, Poland

Paulino Tuñón-Blanco Departamento de Química Física y Analítica, Universidad de Oviedo, Julián Clavería, Oviedo, Spain

Patrick Wagner Hasselt University, Institute for Materials Research (IMO), Agoralaan, Belgium. IMEC, Division IMOMEC, Belgium

Shanshan Wang Beijing National Laboratory for Molecular Sciences, MOE Key Laboratory of Bioorganic, Chemistry and Molecular Engineering, College of Chemistry and Molecular Engineering, Peking University, Beijing, China

Ge Wei College of Chemistry and Bioengineering, Guilin University of Technology, Guangxi, China

Michael J. Whitcombe Cranfield Health, Vincent Building, Cranfield University, Cranfield, MK43 0AL, UK

Weiwei Wu School of Materials Science and Engineering, Jiangsu University, China

Xiu-Ping Yan Research Center for Analytical Sciences, College of Chemistry, and State Key Laboratory of Medicinal Chemical Biology, Nankai University, Tianjin, China

Yun Zhang College of Chemistry and Bioengineering, Guilin University of Technology, Guangxi, China

Zhiyuan Zhao School of Materials Science and Engineering, Jiangsu University, China

Meiping Zhao Beijing National Laboratory for Molecular Sciences, MOE Key Laboratory of Bioorganic, Chemistry and Molecular Engineering, College of Chemistry and Molecular Engineering, Peking University, Beijing, China

Dongqin Zhao Beijing National Laboratory for Molecular Sciences, MOE Key Laboratory of Bioorganic, Chemistry and Molecular Engineering, College of Chemistry and Molecular Engineering, Peking University, Beijing, China

Xiaocui Zhu Beijing National Laboratory for Molecular Sciences, MOE Key Laboratory of Bioorganic, Chemistry and Molecular Engineering, College of Chemistry and Molecular Engineering, Peking University, Beijing, China

Foreword

I welcome the timely publication of this book on molecularly imprinted sensors and their applications. The impact of sensors and sensing systems is enormous and covers multifarious applications ranging from environmental monitoring and protection to pharmaceutical separation and analysis, and from defense and security to medicine and health care. It encompasses everything from electronics and materials to biotechnology and nanotechnology. The emerging challenges associated with public exposure to pollution and hazardous substances have fueled an urgent need for novel sensors and detection technologies, and scientists in the field have been struggling to meet these challenges. A large body of knowledge in sensors and sensing systems is now available, but we still cannot straightforwardly create reliable, stable, reproducible, highly specific, and sensitive sensors for many applications. In reality, there are also significant challenges in the fabrication of these sensors, which is usually complicated and needs careful and time-consuming processing. Thus, novel technologies and sensing materials are in huge demand. Sometimes known as plastic antibody technology, the latest developments in molecular imprinting may provide a solution to at least some of these problems.

This book comprises 14 chapters covering the field of molecularly imprinted sensors (MIPs). The emphasis is on highlighting rapid, specific, sensitive, inexpensive, in-field, online, and/or real-time detection by MIPs. In the opening chapter, Tuñón-Blanco et al. furnish us with a critical review on molecularly imprinted electrochemical sensors, covering their past, present, and future prospects. A comprehensive and critical analysis of chemical sensors, grouped around the different types of transduction, is presented. The relevance of the type of transduction in defining the requirements for the design of MIPs and the different strategies for coupling MIPs to potentiometric, conductometric, impedimetric, voltammetric, and amperometric sensors are highlighted. Chapter 2, by Li et al., focuses on the use of MIPs as the sensitive materials and recognition elements in chemical sensors. The classification of molecular imprinting techniques and methods of preparing MIPs for developing sensitive films for sensors is summarized and key milestones are reviewed. The recognition mechanism of molecularly imprinted sensors to template molecules is described based on dynamics, thermodynamics, equilibrium adsorption, steric action, and molecular clusters. In Chapter 3, Cao et al. introduce the development of molecularly imprinted polymer-based sensors with applications in environmental monitoring. In order to help readers better understand the development of MIPs, the synthetic strategies of MIP-based sensors are also introduced. The next chapter, by Nagaoka et al., introduces a novel approach to molecularly imprinting polymers by using conducting polymers and related materials. These authors have successfully created molecular recognition properties in polypyrrole through overoxidation and used this technique for sensing applications. In Chapter 5, Cleij et al. present a straightforward and versatile MIP-based sensor platform, which is compatible with both impedimetric and microgravimetric detection. The optimization of MIP synthesis for sensor purposes has been demonstrated for a variety of target molecules. The researchers have further demonstrated that well-functioning MIPs can be obtained with different synthetic MIP protocols, depending on the target and measuring environment of interest. In the following two chapters, Yan et al. present the use of molecularly imprinted materials in the fabrication of fluorescent sensor arrays, and Trojanowicz et al. describe chiral sensors based on MIPs. Challenges remain in improving the separation efficiency in a single operation unit for detection in the case of chiral separation. The use of quartz crystal microbalances is addressed by Iqbal and Lieberzeit, who summarize different imprinting strategies in Chapter 9. Zhao et al. review recent developments in luminescent optical sensors based on nanoscale MIPs in Chapter 10. Javanbakht et al. introduce, in Chapter 11, molecularly imprinted polymer-based potentiometric sensors for determining drugs in pharmaceutical, biological, and environmental samples. Chianella et al. consider conductive polymers for plastic electronics in Chapter 12, while Meng describes molecularly imprinted sol-gel sensors in Chapter 13. The combination of Sol-Gel techniques with molecular imprinting can facilitate the mass transfer of target within the imprinted matrix. In the final chapter, Piletsky et al. provide an overview of the technology, with an emphasis on the performance of MIPs in optical, acoustic, and electrochemical sensors. Specific attention is paid to publications of a practical nature. Turning exciting science into practical products is a long and hard path, but it is clear that MIPs are coming of age. Serious attempts are now being made to realize their huge potential benefits in the form of tangible products that will lead to wealth generation and improvement in the quality of life.

It would take several extensive volumes to cover all the details involved in molecularly imprinted sensors. Thus, this book can only hope to provide an overview and to highlight some of the most important recent research. The editors have made an admirable attempt to include the most extensively studied areas, which should be of interest to a broad range of investigators and researchers.

I would like to thank both Elsevier and the leading editor Professor Songjun Li for their invitation to write this Foreword and I want to congratulate all the contributors for making this interesting book possible. Thanks also should be expressed to Jiangsu University and the National Science Foundation of China for supporting this work under the frameworks of distinguished professorship (start-up funding No. 12JDG001) and NSFC No. 21073068. It is my hope that this book will provide valuable information for a broad range of researchers.

By

Professor Anthony P.F. Turner, PhD, DSC, FRSC

Professor of Biosensors & Biolectronics, Editor-in-Chief, Biosensors & Bioelectronics *and Head of the Biosensors & Biolectronics Centre, IFM, Linköping University, S-58183, Linköping, Sweden.*

Tony's Biography

Professor Anthony (Tony) Turner's name is synonymous with the field of biosensors, and in 2010, he moved to Sweden to head a new *Centre for Biosensors and Bioelectronics* at Linköping University. His previous 35-year academic career in the UK culminated in the positions of Principal of Cranfield University at Silsoe and Distinguished Professor of Biotechnology. He retains a part-time position at Cranfield as Innovations Director, with responsibility for leveraging Cranfield University's IP via spinouts and licensing, and is *Professor Emeritus* there. He was elected a Fellow of the Royal Society of Chemistry in 1996 and invited to Fellowship of the Institute of Biology in 1999 and the Institute of Physics in 2006. He was awarded a Higher Doctorate for his exceptional contribution to biosensors by the University of Kent in 2001 and an Honorary DSc by the University of Bedfordshire in 2008, where he served as a Governor for six years. He was admitted to the United States' National Academy of Engineering in 2006, for his work on glucose sensors, environmental monitors, and synthetic recognition molecules. Professor Turner has over 200 refereed journal publications in the field of biosensors and biomimetic sensors and an *h*–index of 48. He co-founded and has edited the principal journal in the field, *Biosensors & Bioelectronics*, since 1985 and edited the first textbook on Biosensors in 1987. In 1990, he founded the World Congress on Biosensors for Elsevier and has chaired it since then. He led the team that pioneered the world's most successful type of biosensor—the mediated amperometric enzyme electrode for glucose for home use-and has been a board member of a number of start-ups. In addition to advising companies and governments worldwide in the general area of analytical biotechnology, he has served as an expert witness in patent litigations and arbitrations on three continents. He played a key role in coordinating research activities in medical and environmental sensors in the European Union and has led numerous research programs, concerted actions, and thematic networks since 1988.

Preface

The impact of sensors and sensing systems is enormous and covers multifarious applications, including environmental monitoring, pharmaceutical separation and analysis, defense and security, as well as medicine and health care. It encompasses everything from electronics and biomaterials to synthetic and surface chemistry and nanotechnology. The emerging challenges associated with public exposure to pollution and hazardous substances have fueled an urgent need for novel sensors and sensing systems to support toxicogenomic studies and protect populations. The emerging threats such as avian flu pandemic, as well as potential terrorist attacks, bring a sense of urgency to the development of novel rapid diagnostic tools for national and international organizations.

A large body of knowledge in sensors and sensing systems are now available; however, a very small percentage of it directly addresses practical needs in creating commercially successful, reliable, stable, and reproducible sensors. There is also a significant challenge in the preparation of the robust sensors for a broad range of practical applications. This demand stimulates progress aimed at development of novel sensing materials and detection principles. One of the most promising groups of materials that mimic biorecognition functions and provide an alternative to relatively unstable antibodies and receptors is molecularly imprinted polymers.

Inspired by the mechanisms of antibody/antigen recognition, molecular imprinting provides a straightforward route for creating binding sites for desired analyte molecules (template), comparable to those of natural antibodies. During the fabrication process, the template and functional monomers are first allowed to form a self-assembled complex where the functional monomers are regularly positioned around the template. Polymerization is then performed to fix this self-organized architecture in place, followed by removal of the imprinted template from the polymeric networks, which thereby leaves behind binding sites complementary to the template. In this way, the match between the template and binding sites constitutes an induced molecular memory, which makes the prepared imprinted polymers capable of recognizing the imprint species. In comparison with antibodies, molecularly imprinted polymers have the merits of easy preparation, reusability, and robustness for chemical and physical stresses. The molecularly imprinted sensors can be used under harsh conditions, including elevated temperature and pressure, presence of metal ion, and organic solvent. Furthermore, molecularly imprinted sensors can be fabricated using standard microchip fabrication protocols such as photolithography, and they would have an extended lifetime. Molecularly imprinted sensors could benefit from the new development of nanosized imprinted materials ("plastic antibodies"). The nanosized imprinted materials can significantly improve binding kinetics, which represents a stumbling block in the development of sensors for real-time analysis, and could find straightforward application in microfluidic devices (e.g., lateral flow assays). To some extent, the real potential of molecularly imprinted sensors is still unclear, and this emerging technology requires careful and thorough analysis. By providing a relatively comprehensive profile on the current knowledge and technologies associated with molecular imprinting, we hope to furnish an insight into new opportunities for science and analytical applications in the diagnostic field. Thus, this book can be used not only as a textbook for advanced undergraduate and graduate students, but also as a reference for researchers and scientists in molecular imprinting, sensors, electronics, materials, biotechnology, nanotechnology, and the like.

Several books with many chapters are probably not enough to cover all of the new developments made in molecular imprinting. Therefore the contributions presented here are selected with the aim of addressing the most interesting issues associated with the particular niche of developing successful MIPs for diagnostic applications. Because of the multidisciplinary nature of this subject, a large number of experts with different backgrounds have been invited to contribute to this book. We are extremely grateful for their efforts.

We would also like to thank Elsevier for the editing endeavor, whose careful editorial work has made publication of this book feasible. Thanks also should be expressed to the Jiangsu University and the National Science Foundation of China for supporting this work under the frameworks of distinguished professorship (start-up funding No. 12JDG001) and NSFC No. 21073068. It is our hope that this book will provide a valuable reference for a broad range of researchers.

Edited by

Songjun Li, PhD (Email: Lsjchem@ujs.edu.cn)
Yi Ge, PhD (Email: y.ge@cranfield.ac.uk)
Sergey A. Piletsky, PhD (Email: s.piletsky@cranfield.ac.uk)
Joseph Lunec, PhD (Email: j.lunec@cranfield.ac.uk)

Editors' Biographies

Dr. Songjun Li is a Distinguished Professor of Functional Polymers in Jiangsu University and is currently the president of the Chinese Advanced Materials Society. He was chairman of the 1st International Congress on Advanced Materials. Dr. Li has published over 60 papers in peer-reviewed journals and has edited 5 books for prestigious publishers including Wiley-VCH, Elsevier, Bentham Science, Nova Science, and Research Signpost. He has been the Principal Investigator for one EU Marie Curie FP7-IIF project and two Chinese National Science Funding projects. He is also currently editor-in-chief of the *Journal of the Chinese Advanced Materials Society,* and sits on the editorial boards of *American Journal of Environmental Sciences, Chemical Sensors, Journal of Public Health and Epidemiology,* the *Open Electrochemistry Journal,* and *Journal of Computational Biology and Bioinformatics Research.*

Dr. Li was appointed an associate professor by Central China Normal University, in the wake of his PhD degree awarded by the Chinese Academy of Sciences in 2005. As a postdoctoral associate, he joined the University of Wisconsin-Milwaukee (USA) in 2008, followed by his research under the support of Marie Curie Actions in Cranfield University (UK), where he worked with the world-renowned scientists Professor Anthony P.F. Turner and Professor Sergey A. Piletsky during 2009–2011. He joined Jiangsu University as distinguished professor in December 2012, where he is leading the research group of "Molecular Imprinting and Catalytic Nanoreactors" in the School of Materials Science and Engineering.

Dr. Yi Ge currently is the Course Director and a Lecturer in Nanomedicine in Cranfield Health at Cranfield University. He obtained his bachelor's degree (with First Class Honors) in Biopharmaceutics and went on to an MPhil degree in Pharmaceutical Chemistry at Aston University. Afterward, he moved to the University of Sheffield for a PhD in Chemistry. He was later employed as a research scientist in a UK pharmaceutical company and was then a postdoctoral research associate at Imperial College London, before joining Cranfield University in the group of advanced sensor and smart material in 2006. He is a member of the Royal Society of Chemistry and has served as an expert reviewer for the Engineering and Physical Sciences Research Council and Biotechnology and Biological Sciences Research Council. His activity in the field of nanotechnology was recognized by the Institute of Nanotechnology, and he was admitted as a Professional Fellow of the Institute. He has been appointed as a Visiting Professor at the University of Jinan (China) and the Associate Editor for Advanced Materials Letters. One of his research interests and activities centers on novel nanomaterials and advanced sensor and sensing technology, where he has extensive experience in the design, synthesis, characterization, and integration of nanosized materials for various sensing and/or biological applications.

Dr. Sergey A. Piletsky is Professor of Bioorganic and Polymer Chemistry at Cranfield University. He received his MS (Chemistry) from Kiev State University in 1985 and his PhD in Bioorganic Chemistry from the Institute of Bioorganic and Oil Chemistry, Kiev, in 1991. After working in Tokyo, Kalmar, Regensburg, and Berlin universities, he joined Cranfield University in 1998. In 2002 he became Professor of Bioorganic and Polymer Chemistry and Head of Cranfield Biotechnology Centre. His professional interests include molecularly imprinted polymers, conjugated polymers, sensor and membrane technology, and the chemistry of dormant radicals. Professor Piletsky has received a number of awards, including JSPS, DFG, and Leverhulme Fellowships, the award of the President of Ukraine, and the Royal Society Wolfson Research Merit Award.

Dr. Joseph Lunec is a graduate in Chemistry from Queen Mary College, University of London, and has a PhD and DSc in Chemical Pathology from the University of London. He is a Fellow of the Royal College of Pathologists and periodically a government adviser on medical toxicology issues. Prior to joining Cranfield Health, Professor Lunec was Professor of Molecular Toxicology, Director of Genomics Centre, Kings College, London; Director of Medical Toxicology Laboratories, St Thomas's Hospital NHS Trust; and Deputy Director of the Medical Toxicology Unit. His research interests are in three main areas: genotoxicity, atherosclerosis, and inflammatory joint disease. He has lectured on Master's courses in Toxicology and Molecular Pathology. Professor Lunec has also supervised numerous PhD students and acted as external examiner throughout the UK and overseas.

CHAPTER

1

Molecularly Imprinted Electrochemical Sensors: Past, Present, and Future

Daniel Antuña-Jiménez, Goretti Díaz-Díaz, M. Carmen Blanco-López, M. Jesús Lobo-Castañón, Arturo J. Miranda-Ordieres, Paulino Tuñón-Blanco

Departamento de Química Física y Analítica, Universidad de Oviedo, Julián Clavería, Oviedo, Spain

OUTLINE

Molecularly Imprinted Sensors.
DOI: 10.1016/B978-0-444-56331-6.00001-3

1. INTRODUCTION

Electrochemical sensors are devices that give information about the composition of a system in real time by coupling a chemically selective layer (the recognition element) to an electrochemical transducer. In this way, the chemical energy of the selective interaction between the chemical species and the sensor is transduced into an analytically useful signal. Due to the simplicity of the procedures and instrumentation required, they are the largest and the oldest group of chemical sensors. They attract great interest nowadays because they are easy to miniaturize and integrate into automatic systems, without compromising analytical characteristics.

Different families of electrochemical sensors can be recognized depending on the electrical magnitude used for transduction of the recognition event: potentiometric (change of membrane potential); conductometric (change of conductance); impedimetric (change of impedance); and voltammetric or amperometric (change of current for an electrochemical reaction with the applied voltage in the first case, or with time at a fixed applied potential in the latter).

The first step when a chemical sensor is used as an analytical tool is the selective recognition of the species of interest for a given application (the analyte) [1]. In the real world, the concept of "interference-free samples" does not exist. Selectivity should therefore be the most important feature under study during the development of a new sensor. Although there is no apparent relation with other analytical characteristics, it is known that when there is a high concentration of an interfering compound, the linear range is reduced, and the limits of detection are higher. Therefore, highly selective recognition elements are required for the development of sensors.

Other features of interest at sensor design involve short response times, as well as a high mechanical strength and chemical resistance to changes of operating media. Nevertheless, many sensors have been developed under critical conditions, when the high selectivity of biological receptors such as antibodies or enzymes compensate for the weak mechanical and chemical strength of the device.

Molecularly imprinted polymers (MIPs) were conceived as artificial tailor-made receptors for molecular recognition. They combine the robust mechanical and chemical strength characteristic of these cross-linked materials with selectivity for the target analyte that mimics that of the natural receptors (in the most successful cases).

The work of the research group headed by K. Mosbach about twenty years ago showed the great potential of analytical methods with regard to these materials. MIPs were immediately applied to the separation field (solid-phase extraction, chromatography), with excellent results, even when the separation of chiral species was undertaken. However, the use of MIPs as artificial receptors in sensors is undergoing slower development. One reason may lie in the intrinsic principles governing the construction of a chemical sensor: whereas a chromatographic column can be considered as a series of theoretical plates (equilibrium stages), the sensor response is governed by the interaction between the recognition element (i.e., the MIP) and the analyte, either a surface or a bulk interaction, in a single step. On the other hand, the design of a chemical sensor is more demanding because the recognition element needs to be coupled to the transducer.

In the early 1990s, Mosbach and collaborators pointed out the usefulness of MIPs for chemical sensor design. The first studies of MIP-based chemical sensors used some form of electrochemical transduction, such as capacitive [2] or amperometric [3]. Initial studies on optical sensors were also published [4]. It was clear from these early works that the different transductions had different requirements for coupling the MIP to the transducer. It was also observed that the nonconducting nature of the acrylic polymers, most commonly used in the imprinting process, presented a problem for the design of devices based on dynamic electrodic techniques, in which electrolysis occurs during measurements. In the effort to modify the features of these materials without compromising their chemical recognition ability, composites with electrically conductive polymers were synthesized [5].

This chapter reviews the work on the development of electrochemical sensors based on MIPs. The discussion is grouped around the different types of transduction, which involve different design requirements and different forms of coupling the MIP with the transducer.

2. POTENTIOMETRIC SENSORS

Potentiometric sensors form one of the most important groups of chemical sensors and may be one of the lowest cost analytical devices available today. The construction of potentiometric sensors relies on preparing a membrane that is permselective for a particular ion, and thereby develops a membrane potential in contact with a solution of the analyte ion. The electrodes built up with those membranes are called ion selective electrodes (ISEs). This principle has been applied for decades in constructing chemical sensors that are as robust as the well-known glass electrode for pH measurements.

The interaction between the membrane and the analyte ion produces a potential change across the interface of the sample and the membrane phase that depends on the activity of the target ion in solution. The experimental setup used to measure the membrane potential is schematically represented in Fig. 1.1. The measurement is carried out under equilibrium

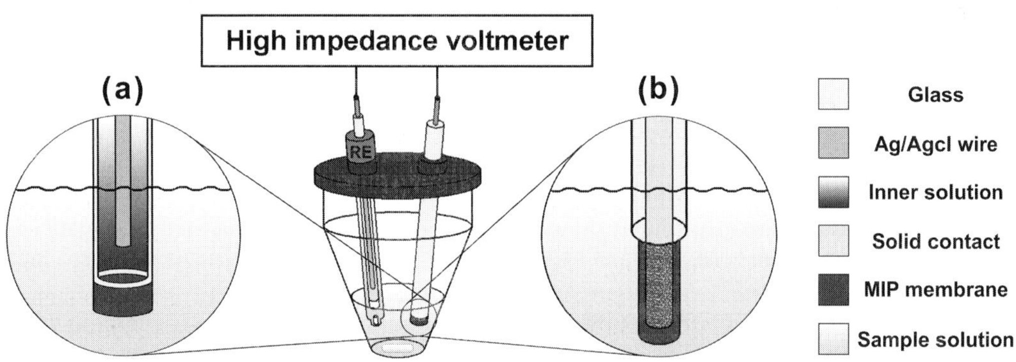

FIGURE 1.1 Potentiometric measurement setup. (a) with an inner reference electrode ISE, (b) coated-wire ISE. RE: reference electrode.

conditions (null intensity current) and corresponds to the potential of the cell constructed with the ISE and an external reference electrode. To achieve that, a potentiometer with high-input impedance is required. The semiempirical Nicolskii-Eisenman equation, which establishes a linear relationship between the cell potential and the logarithm of the target ion activity, is the response function of these sensors.

A key component of the permselective membrane is a lipophilic complexing agent that selectively binds to the target analyte ion. The selectivity of the response strongly depends on the selectivity of this interaction. Thus the search for selective synthetic receptors that could be immobilized at the membranes has been a permanent research field of interest. Molecularly imprinted polymers satisfy the requirements to act as one of these agents, usually called ionophores or ion carriers.

A look over the literature shows that potentiometric sensors using MIPs have undergone intense development in recent years—to such an extent that they represent the most established principle among the electrochemical family for MIP sensor design. A recent review surveys the advances in the field of MIP-based potentiometric sensors [6]. Their success might rely on the matching of the polymer features and the requirement for the design; that is, the most usual MIPs, acrylic polymers, are nonconductive and hydrophobic, and they can become part of user-friendly robust membranes of the type usually termed polymeric membranes.

The sensing element described can be implemented in different ways, and the potentiometric sensors using MIPs as artificial receptors follow some of the most popular designs that have been used for the development of potentiometric sensors [7,8]. These can be classified as faradaic coupled devices (with a conventional inner reference electrode) and capacitive coupled devices (with solid inner contacts, coated wire ISEs, and ion-sensitive field-effect transistors (ISFETs)).

2.1. Faradaic Coupled Devices

This is the classical design in which the permselective membrane is placed separating two compartments with solutions with different activity for a given ion (the sample and the internal electrolyte solution). Thus, these ISEs contain an internal solution with a fixed activity of the analyte ion and an inner reference electrode as shown in Fig. 1.1a. The molecular recognition interaction between the selective ionophore and the analyte causes the extraction of the target ion into the membrane. This process occurs at both the outer and the inner membrane surface, giving rise to the membrane potential at the equilibrium. Although ionic conductivity through the membrane is needed, no significant transport of the analyte ion should occur within the membrane, in such a way that the response is mainly dictated by a localized surface phenomenon. In fact, recent breakthroughs in lowering the limits of detection of these sensors are related to reduction of ion fluxes, which are never completely eliminated [9].

Most of the membranes incorporating MIPs have been prepared following recipes similar to those of conventional ISEs. These devices have been called polymer inclusion membranes [6]. As noted above, the MIP particles play the role of the ionophore, but to make a useful membrane a matrix is also needed. The unique properties of poly(vinyl chloride) (PVC) have made it the standard matrix in other ISEs as well. In addition, a plasticizer is required.

Typically, the MIP particles are dispersed in a PVC matrix (~30 wt. %), where the plasticizer (solvent) is added (~66%). The solvent sets the dielectric constant and provides suitable mechanical resistance. Plasticizers of different polarity can be employed; the most common in MIP-based sensors is the polar solvent 2-nitrophenyl octyl ether (*o*-NPOE), although other solvents such as di-n-octyl phthalate (DOP) and dibutyl sebacate (DBS) have also been used. The membranes usually require the additional incorporation of lipophilic ions or ionic additives to ensure permselectivity and, at the same time, to decrease the membrane resistance [10].

Most often, the proportions of these components in the membrane are little altered when conventional ionophores are replaced by MIPs. For instance, the weight of the MIP is between 6 and 23%, it is directly related to the sensitivity of the device, and it is limited by the solubility in the membrane components. In respect to the percentage of PVC, the proportion of 1:2 with respect to the solvent is usually kept.

Recent reports on conventional MIP-based potentiometric sensors [11—22] show that the response of the sensor depends on different features of the membrane: type and amount of the plasticizer, amount and binding characteristics of the MIP, and presence of ionic additives. All the aspects of the membrane preparation are optimized in order to obtain better analytical performance (working concentration range and limit of detection). On the other hand, since the membranes respond to ions, the pKa of the species needs to be considered in order to choose the appropriate pH for measurement. The suitable value would be that for stability of one of the ionic forms of the analyte, avoiding secondary simultaneous equilibria in which the ionic species of interest could participate. Sometimes the cleaning of the sample in a preliminary stage requires operating at a given pH value that should be changed to induce the ionic form of the analyte required for detection with ISEs. For instance, a flow system to separate proteins and cations that could interfere with the determination of melamine in milk samples was developed [13]. Melamine passes through the system in a neutral form, and then a change of pH allows its detection as a cationic form with an ISE.

As could be expected, the ISE is sensitive to the MIP preparation route. The performance of membranes incorporating traditional bulk polymers, where subsequent grinding and sieving are needed, is frequently compared with that of the MIPs prepared by suspension polymerization or multistep-swelling routes that give rise to spherical monodispersed polymeric particles [17]. Surprisingly, bulk polymers showed better analytical performance.

As it was shown, plasticized PVC is the most commonly used membrane matrix as it exhibits good compatibility with the MIP particles, easy handling, and chemical inertness. However, the slow leaching of the different components and especially the relatively high ion mobility may be restrictive for some applications. Thus, following the recent trends in the design of mass-controlled ISEs for trace analysis trying to reduce the primary ion leaching from the ion-selective membrane, which by contaminating the sample solution at the membrane/solution interface worsens the attainable limit of detection (LOD), monolithic membranes are prepared in situ by direct casting of the MIP precursor solution on a surface [21,22]. After polymerization, the membrane formed is placed at the end of a tube, with an internal solution in contact with the inner reference electrode. The performance of both types of membranes is summarized in Table 1.1.

Potentiometric sensors based on this design have been constructed for pesticides and biomedical substrates that are mainly of an organic nature, normally under their cationic

TABLE 1.1 Analytical Characteristics of MIP-ISEs with an Inner Reference Electrode (Faradaic Coupled Devices)

Analyte	Membrane composition (%wt.)	Measuring range / M	LOD / M	mV/decade	Response time	Operational life	Ref.
Atrazine	PVC:MIP:DOP:NaTPB 25:13:55:7	1×10^{-7}–1×10^{-2}	5×10^{-7}	+56	2 min	3 months >30 times	[11]
Levamisole	PVC:MIP:DBS 29.3:9.8:61.0	2.5×10^{-6}–1×10^{-1}	1×10^{-6}	+57	15 s	4 months	[12]
Melamine	PVC:MIP:NPOE:NaTFBP 31.3:6.1:61.1:1.5	5×10^{-6}– 1×10^{-2}	n/a	+ 54.1	16 s	>2 months	[13]
Promethazine	PVC:MIP:NPOE 15:7.5:77.5	5×10^{-7}– 1×10^{-1}	1×10^{-7}	+31.2	50 s	2 months >10 times	[14]
Dy (III)	PVC:MIP:NPOE:NaTPB 22:22:50:6	8×10^{-6}– 1×10^{-1}	2×10^{-6}	+21.7	10 s	3 months >20 times	[15]
UO_2^{2+}	Id. Id.	2×10^{-8}–1×10^{-6} 1×10^{-6}–1×10^{-2}	2×10^{-8}	+15 +29	2 min	3 months >30 times	[16]
Pinacoyl methyl phosphonate	PVC:MIP:*o*-NPOE 23:23:54	4×10^{-8}–1×10^{-3}	4×10^{-8}	n/a	5 min	2 months >15 times	[17]
Methylphosphonic acid	Id.	5×10^{-8}– 1×10^{-1}	5×10^{-8}	59.0 ± 0.5	5 min	2 months	[18]
Phorate	Id.	1×10^{-9}–1×10^{-6}	1×10^{-9}	n/a	5 min	3 months >40 times	[19]
Chlorpyrifos	PVC:MIP:*o*-NPOE:TDMAC 30:7.6:60:1.5	2×10^{-9}–5×10^{-8}	9.6×10^{-10}	dE/dt Transient response	2 min	n/a	[20]
Atrazine	Monolithic	3×10^{-5}–1×10^{-3}	3×10^{-5}	25	2 s	>2 months	[21]
Diethyl chlorophosphate	Monolithic with *o*-NPOE	1×10^{-6}–1×10^{-2}	1×10^{-6}	36.5	2 min	3 months	[22]

Membrane components: *o*-NPOE:-*o*-nitrophenyloctyl; DOP: dioctylphtalate; ODS-SAM: octadecysiloxane self-assembled monolayer; TDMAC: tridodecylmethylammonium-chloride.
n/a: not available.

form. Murray et al. [23] was the first group to prepare potentiometric sensors for metals such as uranium and lead incorporating MIP particles dispersed in PVC (Imprinted Polymer Inclusion Membrane). Prasad et al. obtained dysprosium [15] and uranil cation [16] MIP sensors based on the same principle with better selective coefficient over alkali and alkali earth and transition metals, by adding a nonpolarizable ligand such as 5,7-dichloroquinoline-8-ol during the preparation of the polymer. This additive remains trapped in the polymeric matrix and assists the rebinding process. The limits of detection were lowered even more, down to 1×10^{-8} M when other complexing agents such as ethylenediaminetetraacetic acid (EDTA) were added.

Only a few examples of MIP-based potentiometric sensors have been reported for anionic [18] or neutral species [19,22]. In the case of neutral molecules, the potential response can be explained taking into account the partial shrinking of the MIP upon rebinding of the template, which increases the permeability of the membrane. This fact is combined with the hydrolysis of the analyte at the working pH [22]. Just recently a new principle for nonionic potentiometric sensing devices has been reported [20]. The device is based on a polymer inclusion membrane whose composition is similar to that of the Ion Selective Electrodes: MIP particles (7.6 wt.%), the anion exchanger tridodecylmethylammonium chloride (1.5 wt.%), *o*-NPOE (60.6 wt.%), and PVC (30.3%). The novelty is that the MIP (uniform-sized) is used as a synthetic carrier that selectively binds neutral species, but the analytical signal used is the change of potential occurring in a second step, in the presence of a charged compound (indicator ion) with a structure similar to that of the neutral analyte. This idea was applied to the determination of chlorphyrifos, an organophosphorus pesticide. The principle is similar to that of a displacement assay, since after binding of the analyte (template) there are less binding sites available for the indicator ion; therefore the potential response to this ion decreases when the membrane is in contact with the analyte. A transient response is used as a linear relationship between the initial slope of the potential change with time and the concentration of analyte bound to the membrane can be established, and is the basis of the quantification. This strategy could be the starting point for developing a new generation of potentiometric sensors for neutral species determination at trace levels. The sensor developed in this work was highly selective and sensitive, but with a narrow linear range. Unlike other membranes developed for ionic species, in this work the membranes prepared with MIP synthesized by precipitation polymerization showed better performance than those constructed with conventional bulk polymerization routes.

2.2. Capacitive Coupled Devices

Potentiometric sensors involving ISEs described in the previous section can be considered to follow a "symmetrical arrangement". This arrangement results in a stable potential at both the inner boundary of the membrane and the interface of the inner reference electrode. This makes sensors very stable, though not easily miniaturized. Another possible arrangement discards the use of this inner solution and is based on an asymmetrical setup in which one side of the ion selective membrane is in direct contact with a solid phase, while the other side is exposed to the (aqueous) sample. The internal solid contact can be a conductor, a semiconductor, or even an insulator. Devices based on this principle that have been developed with MIPs can be divided into two categories: those parallel to the *coated wire* electrodes

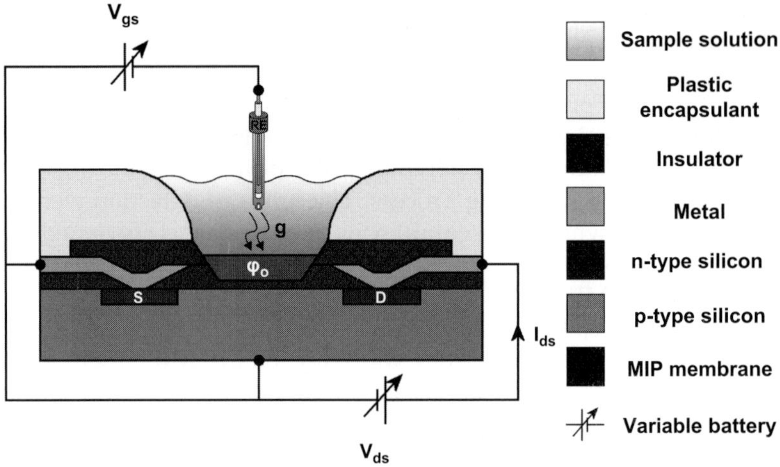

FIGURE 1.2 ISFET device scheme. g: gate, d: drain, s: source, V_{gs}: gate-source voltage, V_{ds}: drain-source voltage, φ_o: surface potential, I_{ds}: drain-source current, RE: reference electrode.

design (Fig. 1.1b) and those that integrate the polymeric membrane with a field-effect transistor device (Fig. 1.2).

In order to develop a solid-contact potentiometric sensor, the proper design of the solid/membrane interface is critical to ensure that a stable electrode potential is obtained. In general, a fast and reversible ion-to-electron transduction system is required to avoid a slow drift of the electrode potential.

One of the first strategies used to develop MIP-based solid-contact potentiometric devices was the preparation of a conducting polymer that contains recognition sites in the form of doped ions. In this way an easy integration of the ion-recognition sites and the ion-to-electron transducer event is achieved. Following this approach, a solid-state NO_3^--ISE was obtained by electrosynthesizing polypyrrole doped with nitrate on a glassy carbon electrode [24].

Alternatively, the ion-selective membrane containing the MIP as ionophore can be cast over an electrode surface, either a noble metal (Pt, Au) or carbon material (glassy carbon or graphite), reproducing the principles governing the preparation of *coated wire electrodes*. In this case, the ionic charge transfer occurring at the membrane during the binding of the analyte to the membrane receptor sites produces an excess of surface charge. This creates an electric field that acts on the conducting electrons of solid contact. A potential change at the electrode is developed, and the interface between the membrane and the electrode acts as a condenser. They rely on the blocked interface between an ionic conductor (the membrane) and the electronic conductor, and thus they show limited stability. The electrode is coated with a conventional ion-selective membrane in which the acrylic MIP, prepared by bulk polymerization, is incorporated as ionophore together with a plasticizer in a PVC matrix according to the traditional recipe (Table 1.2). By this approach, glassy carbon and graphite electrodes were used as solid contact in plasticized PVC membrane-based ISEs for the determination of various organic compounds such as antibiotics, β-blockers and pesticides in

TABLE 1.2 Characteristics of MIP-ISEs with Solid Contact

Analyte	Membrane composition (solid contact)	Measuring range / M	LOD / M	mV/decade	Response time	Operational life	Ref.
Nitrate	Polypyrrol	5×10^{-5}–5×10^{-1}	2×10^{-5}	−56 ± 1	<24 s	55 days	[24]
Oxytetracycline	PVC:MIP:*o*-NPOE / 35.1:3.5:61.4 (GCE)	from 8×10^{-6}	3.5×10^{-6}	55.5 ± 3.6	<15 s	>2 weeks	[25]
Chlormequat	PVC:MIP:BEHS / 34.5:1.8:63.6 (GCE)	6.2×10^{-6}–1.0×10^{-2}	4.1×10^{-6}	53.2 ± 0.5	<10 s	n/a	[26]
Metoprolol	PVC:MIP+DOP / 20.2:13.4:66.4 (GE)	2×10^{-7}–8×10^{-3}	1.26×10^{-7}	55.4 ± 1	14 s	>6 months	[27]
Ciprofloxacin	PVC:MIP:*o*-NPOE / 35.4:2.6:61.9 (GE)	2×10^{-5}–n/a	1×10^{-5}	50.0 ± 0.1	15 s	>7 weeks	[28]
Amoxicillin	Id.	4.86×10^{-5}–n/a	2.9×10^{-5}	60.7 ± 2.7	20 s	>4 weeks	[29]
Chlortetracycline	Id.	from 4.6×10^{-5}	4.1×10^{-5}	62 ± 5	15 s	>7 weeks	[30]
Trimethoprim	PVC:MIP:*o*-NPOE / 33:1:66 (GE)	from 4×10^{-7}	3×10^{-7}	59.7 ± 2.5	<15 s	>5 weeks	[31]
Quinine	PVC:MIP:CPB or TDMAC / 33:1:66 (Au)	4×10^{-6}–1×10^{-2}	1.2×10^{-6}	61.3 ± 1.1	10 s	3 months	[32]
Serotonin	MIP (in situ) (Pt/plasma polymer)	1×10^{-6}–1×10^{-2}	1×10^{-10}	24.7	12 s	200 days >50 times	[33]
Cetizirine	MIP (precipitation) (CPE)	1×10^{-6}–1×10^{-2}	7×10^{-7}	28 ± 0.9	20 s	>5 months	[34]
Hydroxyzine		1×10^{-6}–1×10^{-1}	7×10^{-7}	29.4 ± 1.0	15 s	>5 months	[35]
Methylphosphonic acid	ODS-SAM (ITO)	5×10^{-5}–6.2×10^{-1}	5×10^{-5}	>59.2	50 s	>210 times	[36]
Dipicolinic acid	Id.	1.5×10^{-6}–1.94×10^{-2}	1.5×10^{-6}	>59	25 s	>550 times	[37]
Chiral amino acids	Id.	5×10^{-6}–1.2×10^{-2}	5×10^{-5}	180	160 s	>200 times	[38]

Membrane components: *o*-NPOE: *o*-nitrophenyloctyl; BEHS: b(bis(2-ethylhexyl) sebacate); DOP: dioctylphtalate; CPB: tetrakis (4-chlorophenyl borate); TDMAC: tridodecylmethylammoniumchloride; ODS-SAM: octadecysiloxane self-assembled monolayer.
Solid contact: GCE: glassy carbon electrode; GE: graphite electrode; Au: gold electrode; CPE: carbon paste electrode.
n/a: not available.

pharmaceuticals, clinical samples, or food stocks [25–31]. Following the same approach, a quinine sensor was prepared over a gold electrode and was used in a wall-jet flow injection system for determination of quinine at tonic water [32]. The analytical characteristics of these sensors are summarized in Table 1.2.

A similar approach was followed at the design of a potentiometric sensor for serotonin, but in this case a uniform pinhole-free ultrathin layer of a polymer obtained by plasma polymerization acts as an interface between the sensing element and the transducer. The plasma-polymer with appropriate conductivity is deposited over a planar Pt electrode, and then, using a method of swelling and thermal radical polymerization, the serotonin-imprinted polymer was prepared in situ over the modified Pt. Excellent sensitivity and selectivity to serotonin were thus obtained with very short response time [33].

Another way of obtaining a stable response can be, in principle, the simple mixing of the ion-selective membrane with carbon paste (graphite powder and paraffin as binder). In this case, the contact with the permselective membrane has a very large area, and thus the interface is relatively stable. ISEs constructed just mixing the molecularly imprinted polymer and the carbon paste were applied to potentiometric sensing of cetirizine [34] and hydroxyzine [35] in pharmaceuticals and biological fluids, with excellent analytical characteristics (Table 1.2)

Another approach to solid-contact ISE involves the use of self-assembled monolayers (SAMs). Their lipophilicity hinders the formation of an aqueous layer between the membrane and the conducting transducer, thus providing a more effective coupling of ionic and electronic conductivity. In addition, they allow surface imprinting; when assembled in the presence of a template molecule, and after its removal, footprint cavities are created in the monolayer matrix in such a way that the rebinding of the template to the sensor surface causes a change in its potential that is potentiometrically monitored. An octadecylsiloxane monolayer was covalently bound on indium tin oxide electrodes (ITO) in the presence of different templates giving rise to sensors for methylphosphonic acid [36], dipicolinic acid [37], or even chiral amino acids [38]. In all cases the sensor response was associated with local change in pH that takes place upon analyte rebinding. A similar approach was used for the selective detection of different proteins using SAMs of alkanethiols with hydroxyl terminal groups as the imprinted surface [39,40].

Potentiometric sensors based on field-effect devices forms another group of solid-contact potentiometic devices: the ion-sensitive field effect transistors (ISFETs). In this case the ion-selective membrane is deposited directly over the input dielectric of the field-effect transistor. The structure of an ISFET is thus the same as that of a normal metal-oxide-semiconductor field-effect transistor (MOSFET) except for the transistor gate as shown in Fig. 1.2. Commonly, two n-type zones are incorporated in a substrate of p-doped silicon. The substrate is covered by an insulating layer, and finally the ion-selective membrane is applied on top of the gate. The electrical path is closed with a reference electrode, and both are placed in the sample solution. At a certain gate potential, and upon the application of the appropriate potential between the source and drain of the ISFET, a constant current flows between the source and drain. The interaction of the analyte with the ion-selective membrane alters the gate potential and thus the current between the source and drain changes. The ISFET output could be this change in the drain current or most usually, operating in the constant current mode, the feedback voltage that is applied from an external voltage in order to compensate the change in the gate potential and thereby maintain the drain current constant.

The group of I. Willner has pioneered the development of ISFETs with MIPs as recognition elements. For this purpose they initially used TiO_2 sol-gel as inorganic polymeric structure for molecular imprinting [41−44]. pH sensitive ISFETs, commercially available, were used in these studies, in such a way that deprotonation of the hydroxylated Ti-OH imprinted interface controls the charge, and thereby the potential of the gate. Local changes in pH induced by the analyte rebinding were used for determination of different acidic organic compounds such as chloroaromatic acids [41,44], benzylphosphonic acids, and thiophenols [43]. Strict control of the adhesion to the semiconductor substrate and thickness of the film are required. These imprinted devices showed reasonable sensitivities, even revealing in some cases chiroselectivity [42]. This concept was extended to the determination of triazine herbicides [45] using imprinted acrylamide-methacrylate copolymers prepared in situ over the gate of Al_2O_3 ISFET. Likewise, acrylamide and acrylamidephenylboronic acid were used as functional monomers for the semicovalent imprinting of mononucleotides [44] and the cofactors β-nicotinamide adenine dinucleotide (NAD) and its phosphorylated form (NADP) [46]. The functional monomer 1-allyl-2-thiourea was also used for in situ preparation of a noncovalent MIP for the potentiometric determination of cyclic adenosine monophosphate, (cAMP) [47] with limited sensitivity but very good selectivity. These types of sensors attract great interest due to the ease of miniaturization and compatibility with mass-production processes. The combination of these devices with microelectromechanical systems (MEMS) allowed the fabrication of a platform that incorporates both sensor and readout circuits [48] for the determination of creatinine.

In general, it can be said that the major developments of MIP-based potentiometric sensors occurred approximately during the last five years . The field has reached some level of maturity, since practically all the recent works include analysis of complex samples such as urine, serum, or environmental samples. The linear ranges covered at least two or three orders of magnitude, and the limits of detection are suitable for the analytical applications proposed. In some cases the LOD achieved are very competitive with those of other methods of analysis, and the sensors developed can be used for trace analysis. In addition, they exhibited excellent selectivity as the typical selectivity factors ($K^{pot}{}_{A,X}$) reported over interferences related to the analytical problem are usually in the range of −2 to −5 (logarithmic scale). On the other hand, the response time of these sensors is normally < 20 s, and the lifetime is 3 to 6 months.

Only scarce works related to the preparation of MIP-based potentiometric sensors for neutral molecules have been reported. One of the strategies that could be useful for determination of these species involves the catalytic conversion of the neutral analyte giving rise to charged species, which can be monitored by a change of potential at an ion-selective polymeric membrane. This application would mimic the design of many enzymatic potentiometric sensors. The requirement therefore is that the MIP used as recognition element exhibits catalytic properties, acting as an artificial enzyme. The potentiometric sensors constructed with these polymeric artificial enzymes would benefit from the good mechanical strength of the MIPs and would be less prone than the biosensors to deteriorate with changes in chemical conditions. Although this is a promising area of research, no MIPs with catalytic activity have yet been applied to the development of potentiometric sensors.

3. CONDUCTOMETRIC AND IMPEDIMETRIC SENSORS BASED ON MIPs

Conductance and impedance measurements are nonselective in nature, and only their coupling to a selective recognition element, such as MIPs, may convert them to relatively selective methods. Measurements are performed by imposing an alternating voltage between two electrodes, two Pt electrodes for conductometric devices, and one reference electrode and the sensor in the case of impedimetric ones, and obtaining the alternating current (AC) as response. Although in some cases the potentiostatic three-electrode system is used for impedimetric measurements, an electrochemical system with two electrodes is sufficient because only a negligible current will actually flow through the reference electrode. The experimental setup used in each case is schematically represented in Figs. 1.3 and 1.4.

As a result of this measurement, the so-called impedance (AC resistance) is obtained. When a sinusoidal voltage is applied, there is a phase difference between the applied voltage and the response current, and both capacitive and resistive factors tend to impede the flow of current. The impedance (Z) is usually described by a complex number $Z = R + j\,X_C$ whose real and imaginary parts represent their components for a phase delay of 0° and 90°, respectively. The resistive factor, R, critically depends on electrically conducting defects in the electrochemical cell. In principle, a change in conductance, the reciprocal of resistance, of the electrochemical cell can be used to follow any reaction that produces a change in the number of ion, its charge, or mobility. On the contrary, the capacitive reactance, X_C, arises from

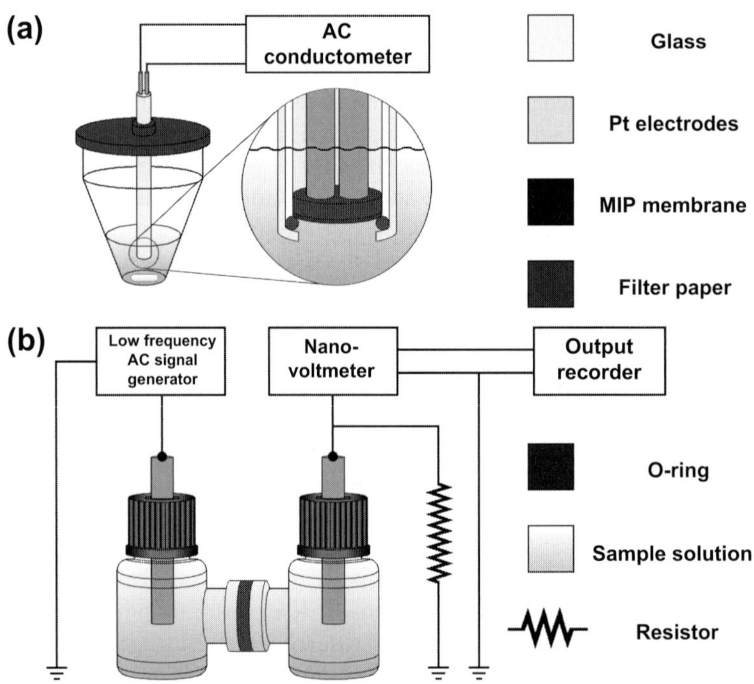

FIGURE 1.3 Different instrumental setups used in conductometric devices. (a) probe type and (b) separated cells.

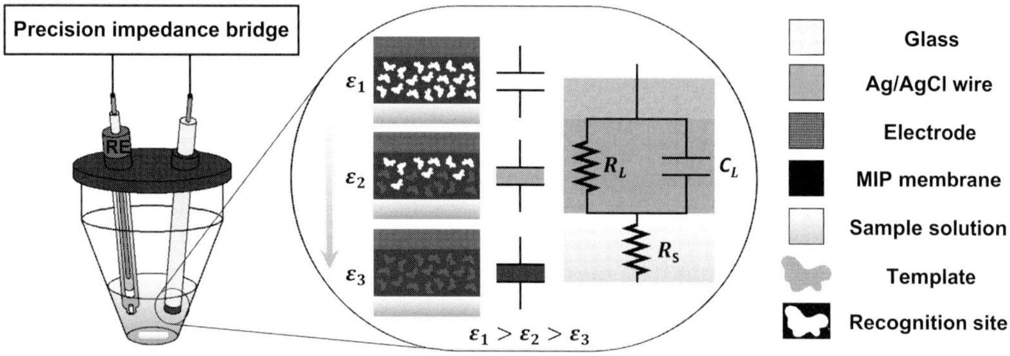

FIGURE 1.4 Scheme of the electrochemical cell used with capacitive sensors. The inset illustrates how the signal is obtained and the equivalent circuit of the cell. ε: dielectric constant; R_L: MIP layer resistance; C_L: MIP layer capacitance; R_S: electrolyte resistance; RE: reference electrode.

charging the double layer at the electrode surface. If the capacitive contribution predominates, the device is a capacitive sensor. An electrically blocked electrode is used in this case. Capacitance describes the dielectric properties of the electrically insulating layer, which includes the recognition element, sandwiched between two conductive phases—the electrode and the electrolyte. Any change at the dielectric constant of the recognition layer induced by the selective interaction with the analyte can thus be monitored.

3.1. Conductometric Devices

Only a limited number of MIP coupled conductometric devices [49–56] have been reported, mainly in the early development of MIP-based chemical sensors. In these devices, the MIP is incorporated into a membrane placed in contact with the sample; the resistance of the membrane changes when interacting with the analyte, and the percentage change in conductance is measured as signal. Two different instrumental configurations have been proposed for measuring the analytical signal (Fig. 1.3).

The setup proposed by Mosbach et al. [49] uses particulated acrylic MIPs synthesized by a classical bulk polymerization protocol and is illustrated in Fig. 1.3a. The membrane incorporating the MIP is prepared by pressing the polymer particles on a polished bare surface. The disk was in contact with two platinum wires and held on a filter paper. A discriminating output for benzyltriphenylphosphonium ions (the template species) was achieved with MIP membranes with respect to those prepared with nonimprinted polymers. The increased conductance observed after exposing the MIP sensor to the template is explained in terms of enrichment of the analyte and counterions in the polymer particles. As the analyte ion must overcome the diffusion barrier imposed by the filter paper, very long response times are obtained (hours).

In the alternative instrumental setup proposed by Piletsky et al. (Fig. 1.3b), the MIP membrane is placed separating the two compartments of a conventional electrochemical cell in which two Pt electrodes are immersed. An initial difficulty was the preparation of suitable membranes. Early studies have used glass filters [50–52] as polymerization support.

The filter was impregnated with the pre-polymerization mixture, and radical polymerization was performed within the pores. In order to reduce the long response times (about 30 min), thinner membranes are proposed. Thin and flexible MIP membranes can be obtained in a photopolymerization process using methacrylate-oligourethane copolymers. The polymerization mixture was cast between two quartz plates to obtain 60–120-nm-thick membranes [53,54]. An increase in the membrane conductance is induced by the interaction with the template. In this way sensing of atrazine was possible, down to a detection limit of 5 nM, and other related herbicides in the 1–100 nM concentration did not interfere. The response time was reduced to 6–15 min depending on the membrane thickness.

A very similar instrumental setup was described for the sensitive screening of haloacetic acids in drinking water by means of MIP-based conductometric measurements [55]. The sensor could be used in the range 25–1000 μg L^{-1} (1.5×10^{-7} to 6×10^{-6} M for trichoroacetic acids), and the detection limit was 0.2–5 μg L^{-1} ($7 \times 10^{-10} - 4 \times 10^{-8}$ M), depending on the compound. The MIP was incorporated as powder in a membrane made with PVC and dibutyl phthalate as plasticizer, similar to those constructed for ISEs. Two electrodes were placed close to the membrane surface and the other two away. The outer electrodes were used to apply a constant current and those closer to the membrane to measure the voltage drop across the membrane. In this case, an opposite behavior is observed after template rebinding. The membrane conductivity decreases (impedance increases) as the template concentration increases, following a hyperbolic curve. Although the response mechanism was not investigated, it is proposed that the binding of the analytes to MIP could take place through the formation of ion pairs between the haloacetic acids and vinylpyridium groups in the recognition sites, causing a resistance increase. Alternatively, a reduction in the PVC chain mobility induced by the interaction between the template and the MIP embedded in the membrane could account for the observed behavior.

The main drawback of the conductometric devices is their poor selectivity, especially when measurements are performed in buffers with a high ionic strength. A reference sensor, containing nonimprinted polymer (NIP) is always used. A true coupling between the recognition element and the transducer could be achieved by using interdigitated electrodes, which would enable differential measurement between a MIP-modified surface and a reference (NIP-modified) one. Thus, the conductometric design for the screening of haloacetic acids in water [55] was extended to a lab-on-a-chip device [56]. Screen-printed interdigitated gold electrodes were covered by MIP thin-film layers and were used in a flow-through arrangement for online sensing of haloacetic acids in chlorinated water. The integrated miniaturized device allows minimizing the sample volume (2.5 mL) and reducing the assay time to 2 min.

3.2. Capacitive Devices

In the case of capacitive measurements, the detection must be performed in a regime where the sensor exhibits near-ideal capacitor behavior (phase angle value close to 90°). For that reason the polymer layer that incorporates the MIP must be electrically insulating. Additionally, ultrathin films are preferred in order to improve the diffusion of the analyte and thus the sensitivity of the sensor. The instrumental setup used for these measurements is shown in Fig. 1.4, which also shows the equivalent circuit used to model the sensor interface.

Electropolymerization permits obtaining ultrathin polymer layers (between 10 and 50 nm thickness), although it is difficult to ensure the good insulating properties that are so important for a capacitive sensor. This problem has been overcome by filling the defects in the polymer layers created over gold electrodes with alkanethiols, resulting in insulating self-assembled monolayers. The first work regarding this strategy reported a capacitive sensor for phenylalanine synthesized by cyclic voltammetry from a phenol solution [57] onto a gold electrode previously modified with mercaptophenol to guarantee film adherence. The corresponding MIP layer was filled with octanethiol. Likewise, a sensor for glucose was developed. In this case, *o*-phenylendiamine [58] was used as the polymerization monomer and 1-dodecanethiol as the alkanethiol that provides the insulating properties to the MIP layer. Additionally, a sensor against glutathione was synthesized with the same monomer over a monolayer of 2-mercaptoethane [59]. Following the same approach, other recent examples of capacitive sensors can be found, such as those for cholesterol [60] and glutamic acid [61], which have been synthesized from the monomers 2-mercaptobenzylimidazole and *o*-phenylendiamine-dopamine, respectively. The last one allows the enantioselective recognition of glutamic acid [61]. The insulating properties of electropolymerized films can be improved controlling the scan rate during the growth process. Thus, by just using smaller scan rates (5 mV s^{-1}), it was observed that no additional thiol-SAM sublayer is needed to ensure the near-ideal capacitive behavior of the sensing interface. This approach was used for the determination of tegafur (the pro-drug of the chemotherapy compound 5-fluorouracil) [62], nicotine [63], and theophylline [64]. In all cases, the analyte is directly detected by a decrease in the electrical capacitance of the interface upon template rebinding, which is related to a decrease in the dielectric constant of the MIP film as is illustrated in Fig. 1.4. Analytical characteristics are summarized in Table 1.3. In general, the calibration curves are nonlinear or linear in a very narrow range, and long response times are obtained. The stability of the response is also limited, as can be expected for MIPs prepared with a low cross-linking degree.

An alternative approach to form highly cross-linked acrylic MIPs on the surface of a capacitive transducer was developed. This methodology is based on the formation of a thin film (around 10 nm) of an acrylic polymer directly on the blocked transducer surface [65]. The substrate used was either polypropylene or gold electrodes modified with a self-assembled monolayer of an insulating compound, such as mercaptohexadecane, to give hydrophobic character and block electrochemical surface reactions. A MIP against the herbicide desmetryn was prepared over these surfaces by grafting photopolymerization. The protocol involved the immobilization of benzophenone, a photoinitiator, and after washing, the polymerization was carried out under UV light without porogen. The capacitance of the sensor decreased as the concentration of desmetryn increased due to a decrease of the dielectric constant of the polymer layer upon analyte rebinding [66]. The selectivity toward other herbicides was good, although the sensor was used only at the mM range. Using the same approach, a capacitive sensor for the determination of creatinine was developed. In this case, the sensor could be used to measure creatinine at the micromolar range, but with a short linear range [66,67]. The output was indifferent to the addition of creatine, urea, or glucose. These devices give more reproducible responses as compared with the output of molecularly imprinted sensors designed using electropoymerized films.

It was demonstrated that capacitive measurement is an effective way for sensing the formation of MIP-analyte complexes on electrode surfaces. Although these devices have

TABLE 1.3 Characteristics of Capacitive Sensors Using MIPs as Recognition Element Constructed Over Gold Surfaces

Analyte	Membrane composition (formation)/thickness	Measuring range / M	LOD / M	Response time	Operational life	Ref.
Phenylalanine	PP (E) + octanethiol / ~16 nm	3×10^{-6}–4.8×10^{-5}	3×10^{-6}	60 min	Limited (single use)	[57]
Glucose	PPD (E) + 1-dodecanethiol / ~95 nm	1×10^{-4}–2×10^{-2}	5×10^{-5}	~20 min	10 h	[58]
Glutathione	PPD (E) + 1-dodecanethiol / 90 nm	2.5×10^{-5}–3×10^{-4}	1.25×10^{-6}	~20 min	n/a	[59]
Cholesterol	PMBI (E) + 1-dodecanethiol / ultrathin	Up to 3×10^{-5}	4.2×10^{-7}	15 min	1 month	[60]
l-glutamic acid	Co–PPD–PD (E) + 1-dodecanethiol / <1 nm	1.7×10^{-5}–2.5×10^{-4}	4.7×10^{-6}	~5 min	>30 times	[61]
Tegafur	PAP (E) / ~65 nm	Up to 8×10^{-5}	~10^{-5}	7 min	Limited	[62]
Nicotine	PD (E) / ~40 nm	1×10^{-6}–2.5×10^{-5}	5×10^{-7}	10 min	3 months	[63]
Theophylline	PP(E) / 70 nm	Up to 1.5×10^{-5}	1×10^{-6}	10 min	n/a	[64]
Desmetryn	Acrylic (GP) + 1-hexadecanethiol / ~10 nm	1×10^{-4}–1×10^{-3}	1×10^{-4}	5 min	6 months	[65]
Creatinine		8×10^{-5}–6×10^{-4}	1×10^{-5}	2 min	6 months	[66]

Membrane formation: E: electropolymerization; GP: grafting photopolymerization.
Membrane components: PP: polyphenol; PPD: poly(*o*-phenylendiamine); PMBI: poly-2-mercaptobenzilimidazole; PD: polydopamine; PAP: poly(m-aminophenol).
n/a: not available.

not been used for the detection and quantification of target molecules in complex matrixes, they are compatible with microfabrication techniques. Thus, following modern approaches in biosensing, interdigitated capacitance sensors (IDCs) for the determination of amino acids [68] and protein allergens [69] have been recently developed, demonstrating that IDC is an effective approach to miniaturize MIP-based sensors. This paves the way for inexpensive production of integrated microdetection systems for multiple applications.

4. VOLTAMMETRIC AND AMPEROMETRIC SENSORS

This group of electrochemical sensors involves the measurement of current flow as the primary transduction phenomenon and thus belongs to the group of faradaic coupled devices. Measurements are performed by applying a potential (variable in voltammetric and fixed in amperometric devices) between the sensor and a reference electrode in a three-electrode configuration electrochemical cell. The current that flows through the counter electrode and the sensor is continuously monitored and used as analytical signal. Figure 1.5 shows the instrumental setup used in this case.

The presence of electroactive species, in the electrolyte solution is thus required for this kind of measurements. Thereby these sensors are in principle useful for the determination of electroactive species, although indirect approaches have also been described in which an electroactive probe (different from the analyte) is used in solution to obtain the electrochemical response.

Voltammetric and amperometric detection requires that the MIP used as recognition element allows diffusion of the electrochemically active species to the electrode surface where electron transfer takes place and outward diffusion of the electrochemically generated products. Ideally, the recognition layer should have a high electrical conductance. Thereby, MIPs with intrinsic conductivity are the ideal basis for the design of these sensors, and different methods to provide conductivity to the materials usually employed for MIP synthesis have been investigated.

FIGURE 1.5 Instrumental setup for (a) voltammetric and (b) amperometric sensors. RE: reference electrode; CE: counter electrode.

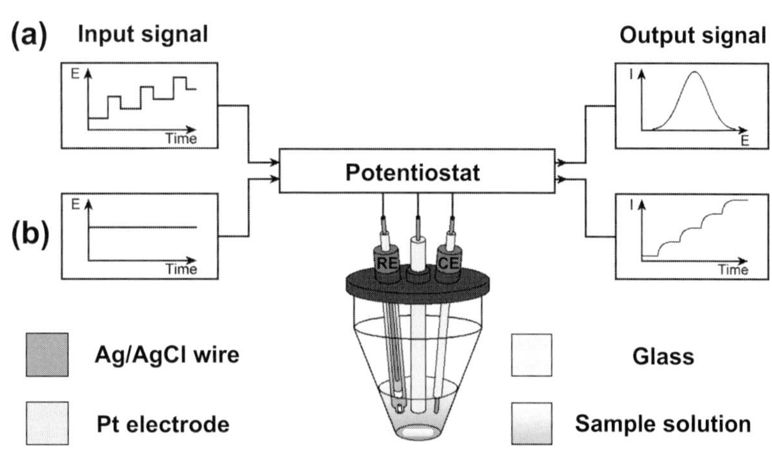

An important aspect at the design of a faradaic sensor is the integration between the recognition element and the transducer. When a MIP is used as the recognition element, the synthesis of MIP films at the transducer surface in a single step allows an easy and low time-consuming sensor preparation. For the construction of these sensors, the MIPs can be synthesized ex situ, that is, by preparing the MIP in a previous step, followed by a subsequent integration stage with the transducer (drop coating, composites with different support matrix) or in situ, that is, directly over the electrode surface (electropolymerization, spin-coating, grafting, self-assembled monolayers). Voltammetric and amperometric MIP-based sensors are examined in the following sections using as classification criteria the different protocols that have been reported for integrating MIP-recognition layers on different conductive supports.

4.1. Ex situ Affinity Sensors

Initial approaches for the preparation of electrochemical sensors with MIPs as the recognition element consists of two stages: first, the synthesis of the MIP, and second, the integration of the MIP with the transducer. Conventional polymerization techniques (bulk, precipitation) could be used for preparation of the MIP. The integration step could be as simple as placing a drop of a suspension of the MIP on the electrode surface, or entrapping the MIP into a different matrix, such as gels, membranes, electrogenerated polymers, or gold and carbon (nano)materials.

The first attempt at MIP entrapment was carried out with gels as supports [3]. This work described an agarose-based sensor for morphine determination that involved two steps. First, morphine bound selectively to the MIP, and second, a nonelectroactive compound (codeine) completely displaced morphine from the MIP binding sites. The release of the morphine originated a current that can be used to quantify the analyte by an amperometric method. There are other matrices in which the synthesized MIP could be entrapped. For example, T. Alizadeh [70] employed poly(epichlorohydrin) to entrap the particles of a paraoxon-MIP, which were drop coated on the surface of glassy carbon and graphite electrodes. Polyvinyl chloride (PVC) was also used to integrate rifamycin-MIP particles with a glassy carbon electrode [71]. In order to improve the conductivity of the composite, carbon black can be added to the PVC-MIP suspension before drop coating onto the electrode surface. Sol-gels could also be used as an entrapment matrix. A suspension of folic acid-imprinted particles, previously synthesized by bulk methodology (10–15 μm sieved fraction), in tetraethyl orthosilicate-based (TEOS-based) sol-gel was spin-coated onto the surface of a pencil graphite electrode. Graphite powder has been added to the above-mentioned suspension to improve conductivity. The sensor as prepared was able to detect this vitamin in aqueous, serum, and multivitamin samples [72].

Looking for an intimate integration between the MIP and the transducer, electrogenerated polymer films were used as the means of MIP entrapment. Electropolymerization is a straightforward means of obtaining polymer films with a certain thickness by controlling the number of cycles or the current that is applied to the electrode. An electrosynthesized polymer could be used as the matrix to support MIP particles and establish their contact to the electrodic surface. For this purpose, MIP particles were suspended in a solution of the monomer, and then polymerization was carried out. MIP particles should have micrometer size in order to prepare homogeneous MIP suspensions and consequently achieve a successful integration between the MIP and the transducer. Therefore precipitation polymerization was usually employed for synthesis of the MIP. This integration strategy has been used to develop an amperometric sensor for morphine onto an ITO electrode, electropolymerizing a substituted thiophene monomer at a constant potential [73]. Polypyrrole was also used to integrate (-)ephedrine imprinted beads with a glassy carbon electrode [74]. The entrapment matrix is obtained after electropolymerization of pyrrol in a suspension of MIP and concentrations of (-)ephedrine above 5×10^{-4} M were detected. Table 1.4 summarizes the analytical characteristics of the sensors developed using as integration strategy the entrapment in a proper material.

Following the goal of providing conductivity to the MIP itself before its integration with the transducer, a wide variety of composites have been developed. Table 1.5 gives an overview of the sensors prepared using this integration strategy. The early works used graphite particles in order to prepare carbon pastes that were modified with the MIPs. For preparation of this type of composite, MIPs with small particle size are required in order to obtain a homogeneous distribution of binding sites through the composite [75]. Consequently, when polymer particles proceed from crushing, grinding, and sieving steps after bulk polymerization, a traditional way to isolate the small fraction has been the preparation of polymer suspensions and the collection of the solids in the supernatants [75]. Nevertheless, with new polymerization routes, such as precipitation polymerization [76–80], this is not required, although the supernatant is collected in order to get finer and smaller MIP particles. Once the MIP particles were obtained, they were mixed with graphite and a binder such as paraffin oil or n-eicosane and the paste was packed carefully into the well of the working electrode.

Following this trend, a conductive imprinted composite MIP was synthesized incorporating carbon powder (1–2 μm) in the polymerization mixture. This mixture was placed into glass capillaries, where polymerization took place, giving rise to a monolithic MIP-carbon composite in the form of a fiber. During polymerization carbon particles were separated from the polymer and grouped in a zone that gives conductivity to the monolith. The imprinted fibers were then partially inserted in a glass capillary filled with agarose, which was fixed inside a micropipette tip for better handling and electrically connected to Cu wire to be used as a working electrode (fiber 3 mm long) in voltammetric measurements. This design was used for the development of microsensors for folic acid determination in aqueous, blood, serum, and pharmaceutical sample [81]. This strategy does not lead to a true integration of the MIP and the conductive material due to the large size of the graphite particles resulting in their segregation during the polymerization process.

Preparation of hybrid materials with MIPs and carbon nanomaterials has been a significant step forward in the development of composites between acrylic MIPs and conducting matrices. As a polymerizable core, polymers grown onto carbon substrates present some advantages as compared to their macroscopic counterparts: greater surface-to-volume ratio, accessibility to the maximum number of recognition sites, and lower diffusion times of the template. Because of the complexity and large time consumption of the synthesis of these nanocomposites, simple integration strategies are currently being used. Therefore, once the hybrid material is prepared, a drop coating onto the transducer is performed, as is discussed below.

Despite their promising advantages, the vast majority of the work reported so far focuses on the morphological characterization of these new composite nanomaterials; only a few

TABLE 1.4 Analytical Characteristics of ex situ Affinity Sensors Prepared by Entrapment Approach

Analyte	Entrapment matrix	Sensor (electrode, electrochemical technique)	Measuring range / M	LOD / M	Operational lifetime	Ref.
Morphine	MIP (bulk) in an agarose gel	Amperometric (GCE)	3.5×10^{-7}–3.5×10^{-5}	1.8×10^{-7}	>4 weeks	[3]
Paraoxon	MIP (bulk) in a poly(epichlorhydrine) matrix	Voltammetric (GCE, GE, CV)	3.8×10^{-9}–7.5×10^{-7}	1.0×10^{-9}	n/a	[70]
Folic acid	MIP (bulk) into an sol-gel matrix	Voltammetric (PGE, DPCSV)	1.6×10^{-8}–3.5×10^{-7}	4.8×10^{-9}	n/a	[72]
Morphine	MIP (precipitation) into an electrosynthesized polymer	Amperometric (ITO)	1×10^{-7}–2×10^{-6}	3×10^{-7}	n/a	[73]
Ephedrine	MIP (bulk) into an electrosynthesized polymer	Voltammetric (GCE, CV)	5×10^{-4}–3×10^{-3}	5×10^{-4}	3 days	[74]

Electrode: GCE: glassy carbon electrode; GE: graphite electrode; PGE: pencil graphite electrode; ITO: indium tin oxide electrode.
Electrochemical technique: CV: cyclic voltammetry; DPCSV: differential pulse cathodic stripping voltammetry.
n/a: not available.

TABLE 1.5 Analytical Characteristics of ex situ Affinity Sensors Prepared Using Carbon (nano)materials-MIP Composites

Analyte	Sensor preparation strategy	Sensor (electrode, electrochemical technique)	Measuring range / M	LOD / M	Operational lifetime	Ref.
Metronidazole	MIP (bulk), graphite and n-eicosane as binder	Voltammetric (CPE, DPV)	3.3×10^{-10}–1.5×10^{-8}	2.1×10^{-10}	10 days 20 times	[75]
Parathion	MIP (precipitation), graphite and n-eicosane as binder	Voltammetric (CPE, DPV)	1.7×10^{-9}–9×10^{-7}	2×10^{-10}	n/a	[76]
p-nitrophenol	Id.	Voltammetric (CPE, DPV)	8×10^{-9}–5×10^{-6}	3×10^{-9}	n/a	[77]
2,4,6-trinitrotoluene	Id.	Voltammetric (CPE, SWV)	5×10^{-9}–1×10^{-6}	1.5×10^{-9}	n/a	[78]
Caffeine	Id.	Voltammetric (CPE, DPV)	6×10^{-8}–2.5×10^{-5}	1.5×10^{-8}	n/a	[79]
Promethazine	Id.	Voltammetric (CPE, SWV)	7×10^{-9}–4×10^{-7}	3.2×10^{-9}	n/a	[80]
Folic acid	MIP synthesized in presence of carbon powder	Voltammetric (composite fiber, DPCSV)	1.3×10^{-9}–9.8×10^{-9}	4.5×10^{-10}	n/a	[81]
Uric acid	MIP (precipitation) onto the surface of acrylated MWCNTs	Amperometric (GCE)	8×10^{-5}–5×10^{-4}	2.2×10^{-5}	n/a	[84]
Dopamine	Id.	Amperometric (GCE)	5.0×10^{-7}–2.0×10^{-4}	5.0×10^{-7}	n/a	[85]
Insulin	Id.	Voltammetric (PGE, DPASV)	5.8×10^{-11}–5.6×10^{-9}	1.9×10^{-11}	25 days	[86]
Tetracycline	MIP (precipitation) onto the surface of MWCNTs-GNPs	Voltammetric (GCE, CV)	2.3×10^{-7}–9×10^{-5}	9×10^{-8}	n/a	[87]
Dopamine	MIP (precipitation) onto the surface of graphene	Amperometric (GCE)	1.0×10^{-7}–8.3×10^{-4}	1.0×10^{-7}	3 weeks >30 times	[88]

Sensor preparation strategy: MWCNTs: multi-walled carbon nanotubes; MWCNTs-GNPs: multi-walled carbon nanotubes modified with gold nanoparticles.
Electrode: GCE: glassy carbon electrode; GE: graphite electrode; CPE: carbon paste electrode; PGE: pencil graphite electrode.
Electrochemical technique: CV: cyclic voltammetry; DPV: differential pulse voltammetry; SWV: square wave voltammetry; DPASV: differential pulse anodic stripping voltammetry.
n/a: not available.

publications describe its use in the development of analytical devices (Table 1.5). Both carbon nanotubes (CNTs) and graphene have been used as conductive core to prepare acrylic-imprinted core-shell nanomaterials.

Graphene is a single aromatic sheet of sp^2-bonded carbons, and CNTs are graphene sheets rolled up either as single-layer cylinders (single-walled carbon nanotubes, (SWCNTs)) or as an onion-like multilayer structure (multi-walled carbon nanotubes, (MWCNTs)). There are two main different approaches used for grafting polymers onto the surface of carbon nanotubes [82] or graphene sheets [83]: covalent attachment and noncovalent adsorption that may be followed by a covalent functionalization of the attached functional groups.

Functionalized CNTs could be obtained by means of covalent attachment. The first common step is a chemical oxidation with strong acids of pristine CNTs obtaining mainly carboxylic acid groups onto the surface of shorter CNTs. Second, activation of carboxylic acid is achieved with thionyl chloride [84] or a greener method such as carbodiimide reaction [85]. Finally, allylalcohol [84] or allylamine [85] takes part in a nucleophilic attack to the activated groups, resulting in acrylated nanotubes suitable for subsequent polymerization. This strategy produces MIP nanolayers with an average thickness of about 15–20 nm that can be modulated up to 30 nm by increasing the concentration of the polymerization monomers. Although this approach [85] does not solve the problem of parallel uncontrolled polymerization leaving ungrafted polymers, this methodology was used to develop MIP-MWCNTs for dopamine [85] and uric acid [84] as target molecules in amperometric sensors. These devices are constructed by drop-coating the electrodic material (usually glassy carbon, GC) with aliquots of the nanocomposite material which remains attached to the electrode by means of the adhesive force among the nanocomposite and the bare GC [84] or by the aid of agarose gelling [85]. Even though the electrode surface was not homogeneously covered by the MIP-MWCNTs, which leads to a sensitivity loss, at the same time, selectivity was improved when compared with GC as itself. In order to obtain a homogeneous coating over the surface of the electrode, another approach has been developed by the group of Prasad et al. [86]. A suspension of a 10–15-nm-thick MIP layer grafted onto MWCNTs was dropped onto the surface of a pencil graphite electrode (diameter: 0.5 mm) and spun for 30 s at 3000 rpm. Thereby, MIP nanocomposite is attached onto the surface of a pseudo-microelectrode by means of adhesion forces, allowing the determination of insulin by differential pulse anodic stripping voltammetric (DPASV) at trace level (1.86×10^{-11} M) in pharmaceutical and blood serum samples.

The attachment of gold nanoparticles (GNPs) to CNTs is a new way of obtaining hybrid materials with the GNPs catalytic properties that can be used as core for MIP synthesis. To explore this possibility, an imprinted polymer was synthesized over the surface of MWCNTs decorated with GNPs [87]. No further derivatization was performed over the decorated nanotubes (5–8 nm of diameter), and polymerization was carried onto their surface giving rise to MIP layers with thickness in the range of 12 to 22 nm. Taking advantage of the electrocatalytic oxidation of tetracycline over GNPs, a sensor against this template was obtained by means of drop-coating over a glassy carbon electrode. Although electron transfer is partially blocked by the molecularly imprinted polymer layer, very good analytical characteristics for tetracycline determination are obtained (LOD of 0.04 mg L^{-1} or 9×10^{-8} M). This work opens a new route for new transduction approaches involving GNPs–CNT nanostructures as conductive substrate for the development of molecular imprinting electrochemical sensor.

As far as we know, only an electrochemical sensor was achieved by using reduced graphene oxide as conductive substrate [88]. With a noncovalent approach, Congo red was used to modify graphene sheets, which improves the substrate solubility in the polymerization solvent and allows polymer chains growth onto the double bond structure of the adsorbed colorant. In this way a pre-polymerization complex is obtained over the surface of graphene sheet by means of simply noncovalent interactions, and subsequent radical polymerization gives rise to the MIP-graphene composite. By simply drop-coating this material onto a glassy carbon electrode, an amperometric electrochemical sensor against dopamine with good selectivity toward epinephrine and ascorbic acid was obtained with a detection limit of 1×10^{-7} M, five times lower than using CNTs as carbon substrate [85]. Results obtained in this first work encourage the use of this nanomaterial as electroactive substrate to achieve conductive transduction in future molecularly imprinted electrochemical sensors.

With a different goal, magnetic nanoparticles derivatized with silica were used as core in the synthesis of MIPs, using a sol-gel approach [89]. The magnetic core of these particles facilitated the removal of the template from the imprinted material during the processing steps prior to construction of the sensor and subsequent immobilization on the transducer surface simply with an external magnet. In this way a suspension of magnetic core-shell imprinted microspheres against phenylephrine was dropped onto the surface of a glassy carbon electrode in order to develop an amperometric sensor that allows phenyleprine determination in the range 1×10^{-6} to 1×10^{-4} M with a limit of detection of 1×10^{-7} M.

4.2. In situ Affinity Sensors

Other approaches for the integration between MIPs and electrochemical transducers are based on the concept of synthesizing the recognition element directly onto the surface of the electrode. By this way the electron transfer process that gives rise to the analytical signal is favored. The single-step methodology allows an easy and low time-consuming sensor preparation. Below we describe different approaches to generate MIPs directly on the electrode surface.

Initial attempts at integration of molecular imprinted polymers with electrochemical transducers relied on the porosity of the polymers, which were prepared as films on the surface of conventional carbon electrodes. This fact allowed a detection system based on a selective extraction of the analyte into the MIP followed by diffusion through the pores toward the electrode surface. For this purpose, a drop of the polymerization mixture was deposited on the surface of the electrode. Then polymerization took place, giving a thick imprinted film (in the μm range) onto the electrode surface [90,91]. In order to obtain thinner layers, spin coating was performed before polymerization. This strategy was carried out to prepare voltammetric sensors for vanillylmandelic acid [71], homovanillic acid [92], and phenylephrine [93]. Methacrylic acid has been used as the functional monomer and ethylene glycol dimethacrylate as the cross-linking agent in the synthesis of acryilic MIPs, which acted as the recognition element in the above-mentioned sensors. A major drawback of this strategy is the difficulty in controlling the film formation process, which leads to poor reproducibility of the sensors (Table 1.6).

Sol-gel technology was also used to prepare molecularly imprinted polymers with the aim of achieving the integration with the transducer. The major advantage of sol-gel materials is their ability to form thin films by drop/spin coating. However, these materials have low

TABLE 1.6 Analytical Characteristics of in situ Affinity Sensors Prepared by Drop-Coating and Spin-Coating Approaches

Analyte	Sensor preparation strategy	Sensor (electrode, electrochemical technique)	Measuring range / M	LOD / M	Operational lifetime	Ref.
Atrazine	Drop coating + polymerization (acrylic)	Voltammetric (Au, CV)	1×10^{-6}– 1×10^{-5}	n/a	3 days	[90]
Atrazine cyanuric acid	Id.	Amperometric (GCE, GE)	n/a	9×10^{-7} 2×10^{-5}	n/a	[91]
Vanillylmandelic acid	Spin coating + polymerization (acrylic)	Voltammetric (GCE, DPV)	1×10^{-4}– 1.5×10^{-3}	1×10^{-4}	n/a	[71]
Homovanillic acid	Id.	Voltammetric (GCE, DPV)	5×10^{-8}– 1×10^{-5}	7×10^{-9}	n/a	[92]
Phenylephrine	Id.	Voltammetric (GCE, DPV)	n/a	n/a	n/a	[93]
Phi-NO_2	Spin coating (sol-gel)	Voltammetric (GCE, LSV)	$1 \times 0\ 10^{-8}$– 2.0×10^{-5}	1.0×10^{-9}	30 days	[96]
Parathion	Id.	Voltammetric (GCE, DPV)	5.0×10^{-9}– 1.0×10^{-4}	1.0×10^{-9}	30 days 50 times	[97]
Creatinine	Id.	Voltammetric (GE, DPCSV)	1.09×10^{-5}– 8.8×10^{-4}	2.9×10^{-6}	40 times	[98]
Uric acid	Id.	Voltammetric (DPCSV)	8.7×10^{-5}– 1.1×10^{-3}	2.4×10^{-5}	43 times	[99]
Barbituric acid	Id.	Voltammetric (GE, DPCSV)	3.9×10^{-5}– 7.8×10^{-4}	1.2×10^{-5}	n/a	[101]
Rifamycin SV	Drop coating	Voltammetric (GCE, DPV)	2.5×10^{-7}– 6.3×10^{-6}	2.5×10^{-7}	25 times	[102]
Uric acid	Id.	Amperometric (ITO)	Up to 1.125×10^{-3}	3×10^{-7}	40 times	[103]

Analytes: Phi-NO_2: O,O-dimethyl-(2,4-dichlorophenoxyacetoxyl)(3-;nitrophenyl)methinephosphonate.
Electrode: Au: gold electrode; GCE: glassy carbon electrode; GE: graphite electrode.
Electrochemical technique: CV: cyclic voltammetry; DPV: differential pulse voltammetry; LSV: linear sweep voltammetry; DPCSV: differential pulse cathodic stripping voltammetry.
n/a: not available.

stability and reproducibility. Disadvantages of sol-gel material are a consequence of their inherent nature, which causes their evolution over time and consequently the loss of the structure at the imprinted cavities [94,95].

The older and easier methodology to use imprinted sol-gel materials is their integration with electrochemical transducers by conventional strategies such as drop/spin coating (Table 1.6). In some cases, the sol was prepared and the template was added to this sol in a subsequent step. Macrocyclic compounds like *p-tert*-butylcalix[6]arene [96] or *p-tert*-butylcalix[6]1,4-crown-4 [97] were added to the polymerization mixture in order to create adequate hydrophobic cavities for the noncovalent binding of the templates.

Prasad et al. have developed some voltammetric sensors by drop-coating the imprinted sol-gel over the surface of graphite electrodes. First, the MIP adduct was prepared combining the template, and the monomers melamine and chloranile. Then, this adduct was mixed with the sol (TEOS) and graphite, and a semisolid viscous liquid was obtained. Finally, a drop of this liquid was deposited over the surface of the electrode, and subsequently spin-coated and then left dry overnight. This strategy has been used to develop sensors against serum or urine metabolites [98–100] and drugs [101]. These devices exhibited good analytical performance and enough sensitivity to determine trace levels of these compounds in both aqueous and biological samples.

Polyphosphazenes can also be used as supporting material for molecular imprinting technique because these linear polymers adopt a tridimensional structure when drying in the presence of a template molecule. With this simple approach, a rifamycin SV voltammetric sensor was described [102], with detectability in the submicromolar range. The main drawback of the design is its limited stability, related to structural damages of the imprinted recognition cavities due to the non-cross-linked nature of the polymer. A similar approach was recently described for construction of an amperometric sensor for determination of uric acid [103]. In this case an amine-imide-type conducting polymer is synthesised from N,N-bis(4-aminophenyl)-N′,N′-diphenyl-1,4-phenylenediamine and 3,3′,4,4′-benzophenotetra carboxylic dianhydride. A mixture of the polymer and template was drop-coated on ITO electrode and subjected to a program of temperature to remove the solvent and obtain a uniform MIP film. The sensitivity of this MIP sensor depends on film thickness, and under optimal conditions it allows uric acid determination from 3×10^{-7} M.

Focusing on improving the intimate contact among the bare electrode and the imprinted material as recognition element and ensure stable adhesion, surface grafting methodologies have been developed (Table 1.7). Bifunctional compounds are used for this purpose. They are capable of binding the electrode (by means of covalent or gold-sulphur interactions) and of serving as an anchor point for a subsequent polymerization.

Self-assembled monolayers (SAMs) with allyl mercaptane is a straightforward means of grafting vinyl groups onto the surface of gold electrodes, and in a subsequent step a MIP could be anchored to these groups. Unfortunately, there is also little control of the thickness of the MIP layer, which limits the diffusion of the analyte toward the recognition sites. However, there are some examples in the literature of amperometric sensors using this strategy. First, a sensor against bilirubin was described [104], using a 114-nm-thick MIP film synthesized onto the surface of a mercaptane-modified gold electrode. The sensor exhibited sensitivity enough to determine this metabolite in adult human serum, and a portable design was also developed [105]. Second, a catalytic amperometric sensor against dopamine was developed. The SAM was made by attaching thioglycolic acid on a gold electrode, and then an electron transfer mediator (quercetin) was covalently bound to the SAM. Finally, the MIP was photopolymerized over this structure after a previous spin coating of the pre-polymerization mixture that controlled the thickness of the imprinted layer [106].

In order to perform this strategy over carbon electrodes, a condition has to be reached: the working electrode matrix needs to contain acrylated groups that should be activated prior to

TABLE 1.7 Analytical Characteristics of in situ Affinity Sensors Prepared by Layer-by-Layer, Surface Grafting and SAMs Approaches

Analyte	Sensor preparation strategy	Sensor (electrode, electrochemical technique)	Measuring range / M	LOD / M	Operational lifetime	Ref.
Bilirubin	SAMs + polymerization (acrylic)	Amperometric (Au)	$0–1.7 \times 10^{-5}$	n/a	n/a	[104,105]
Dopamine	Id.	Amperometric (Au)	$0–1.16 \times 10^{-4}$	5.4×10^{-3}	n/a	[106]
Dopamine	Surface grafting + spin coating (acrylic)	Voltammetric (home-made electrode, DPASV)	$6.5 \times 10^{-9}–5.5 \times 10^{-7}$	1×10^{-9}	4 weeks	[107]
Dopamine	Surface grafting + spin coating (sol-gel)	Voltammetric (ITO, CV)	$2 \times 10^{-6}–0.8 \times 10^{-3}$	2×10^{-6}	14 days	[108]
Imipramine	Surface grafting + dip coating (sol-gel)	Voltammetric (ITO, DPV)	$5.0 \times 10^{-6}–1.0 \times 10^{-3}$	1.0×10^{-9}	n/a	[109]
l-histidine	Surface grafting + electrodeposition (sol-gel)	Voltammetric (ITO, DPV)	$2 \times 10^{-6}–1 \times 10^{-3}$	5.8×10^{-9}	n/a	[110]
Tyramine	Layer-by-layer (sol-gel)	Amperometric (Au)	$1.08 \times 10^{-7}–1 \times 10^{-5}$	5.7×10^{-8}	2 weeks	[111]
Benzylpenicillin	Layer-by-layer + electrodeposition (sol-gel)	Voltammetric (CE, DPV)	$5 \times 10^{-8}–1 \times 10^{-3}$	1.55×10^{-9}	30 days	[112]
Streptomycin	Id.	Voltammetric (Au, DPV)	$1.0 \times 10^{-6}–1.0 \times 10^{-3}$	1.5×10^{-9}	60 min	[113]
Bisphenol A	Id.	Amperometric (Au)	$1.13 \times 10^{-7}–8.21 \times 10^{-3}$	3.6×10^{-9}	5 times	[114]
Clindamycin	Id.	Voltammetric (Au, SWV)	$5.0 \times 10^{-7}–8.0 \times 10^{-5}$	2.44×10^{-8}	n/a	[115]
Thymidine	Id.	Amperometric (CE)	$2 \times 10^{-6}–2.2 \times 10^{-5}$	1.6×10^{-9}	n/a	[116]
l-phenylalanine	Id.	Amperometric (CE)	$5.0 \times 10^{-7}–1.0 \times 10^{-4}$	1.0×10^{-9}	1 month	[117]

Electrode: Au: gold electrode; ITO: indium tin oxide electrode; CE: carbon electrode.
Electrochemical technique: DPV: differential pulse voltammetry; DPASV: differential pulse anodic stripping voltammetry; CV: cyclic voltammetry; SWV: square wave voltammetry.

MIP formation. In this way, a home-made carbon electrode was prepared in a glass tube with different carbon material mixtures (MWCNT, graphite, or a mixture of both) and an acrylated sol (3-trimethoxysylilpropylmethacrylate). An iniferter was adsorbed onto the surface of the carbon electrode and was photochemically activated prior to incorporation of the MIP. A drop of the acrylic pre-polymerization mixture was spin-coated, and then polymerization took place. Using this strategy three dopamine sensors were developed; the one prepared with only MWCNT exhibited the best analytical performance [107].

For grafting imprinted sol-gel films, the oxidized surface of ITO electrodes could be used as substrate. For example, an ITO electrode was modified with aminopropyltriethoxysilane, which provided amino groups that interacted with the template in the binding sites and improved the template recognition. The precursor of the imprinted sol-gel was deposited by spin coating, and the sensor was dried overnight [108]. Moreover, an ITO electrode previously modified with 3-mercaptopropyltrimethoxysilane was prepared for the detection of imipramine. Commercial gold nanoparticles were attached by gold-sulfur bonds to the surface of the electrode, and afterward the precursor of the imprinted sol-gel was dip-coated. Gold nanoparticles improved the conductivity of the sensor, and an imipramine concentration of 1×10^{-9} M could be detected with this device [109]. Finally, electrodeposition has also been used in combination with the surface modification of the electrode surface by a grafting approach. MWCNTs were covalently bonded to an ITO electrode, and then the electrodeposition of the imprinted sol-gel took place by cyclic voltammetry [110]. The sensitivity of this sensor was suitable for determining nanomolar concentrations of L-histidine in biological fluids.

Multilayer systems, which are composed of different conducting materials, have been obtained by a layer-by-layer deposition and represent a straightforward strategy to provide conductivity to the recognition element. The procedure consisted of simple drop coating followed by a drying step. In the last step, the MIP layer was prepared in situ over the modified electrode. Huang et al. [111] modified a glassy carbon electrode with MWCNTs decorated with GNPs and chitosan, and the tyramine precursor of the imprinted sol-gel was subsequently drop-coated. The amperometric sensor developed this way was able to detect tyramine concentrations up to 5.7×10^{-10} M in food samples. Electrodeposition was also used in combination with the layer-by-layer approach, which leads to the deposition of a wide variety of coatings onto the surface of the electrode. MWCNTs [112–116] and GNPs [114] were used for this purpose, and the subsequent electrodeposition of the imprinted sol-gel was carried out by cyclic voltammetry. Following this strategy, amperometric and voltammetric sensors with outstanding sensitivity (limits of detection in the nanomolar range) were developed, and they were compiled in Table 1.7. Finally, a polyaniline film was electrogenerated onto the surface of a carbon electrode in order to improve conductivity. MWCNTs and β-cyclodextrin were also used to enhance the surface area of the electrode and favor the interaction with the template, respectively. With this approach, an L-phenylalanine amperometric sensor was developed showing increased amperometric currents when compared to a sensor prepared in absence of polyaniline [117].

Even though previously described methodologies guarantee adherence among the recognition element and the electrode, there is no control over the thickness of the MIP layer. In order to solve these problems, electropolymerization is an advantageous technique for preparation of MIP-based electrochemical sensors [118–127]. Molecular imprinting approaches for the electrosynthesis of conductive polymers include galvanostatic, potentiostatic, and cyclic voltammetric methods. These techniques allow depositing a uniform, molecularly imprinted layer with good adherence to the transducer surface, and controlling the thickness of the conductive polymer film by varying the electrodeposition conditions. In the majority of cases, the electrosynthesized imprinted layer was generated by cyclic voltammetry in a solution containing the monomer and the template molecule. Pyrrole, aniline, *o*-phenylenediamine, and *o*-aminophenol are monomers widely used for the electrosynthesis of MIPs onto graphite, glassy carbon, and gold electrodes. Alternatively, monomers that form a self-assembly monolayer on gold surfaces such as *p*- and *o*-aminothiophenol have been used [125–127]. Table 1.8 depicts voltammetric and amperometric sensors prepared this way. A voltammetric sensor for the determination of oxytetracycline should be mentioned due to its low LOD (6.49×10^{-10}), which is a consequence of an enzymatic amplification. Oxytetracycline-labeled peroxidase and the analyte itself competed for the binding sites of the MIP, and the oxytetracycline was determined according to the competition reaction by means of the voltammetric measurement of the peroxidase activity [124].

Apart from selectivity, sensitivity should also be taken into account in the development of sensors. The sensitivity of an imprinted sensor is related to the amount of effective imprinted sites on the sensor surface. Although the amount of the imprinted sites increases with the imprinted film thickness, thick MIP layers lead to slow diffusion of the analyte to the recognition sites, and consequently to inefficient coupling between the binding sites and transducer. To further increase the amount of effective imprinted sites on the sensor surface, the simplest method consists in enhancing the surface area of the electrode, which could be carried out by depositing gold nanoparticles (GNPs) at the surface of the electrode. A *p*-aminothiophenol self-assembled monolayer was prepared over the surface of a gold electrode. Then the template was adsorbed, and finally electropolymerization of *p*-aminothiophenol in the presence of the template (acetylsalicylic acid) and in situ generated gold nanoparticles was carried out. This sensor was able to determine low acetylsalicylic acid concentrations (LOD 3×10^{-10} M) in biological fluids [126]. Likewise, GNPs were electrochemically deposited onto the surface of a glassy carbon electrode at constant potential from a $HAuCl_4$ solution. *p*-Aminophenol was assembled to these gold nanoparticles, and then MIP electrosynthesis took place by cyclic voltammetry [127]. With this design, the sensitivity for determining the organophosphate pesticide chlorpyrifos was improved about two orders of magnitude with respect to a *p*-aminothiophenol imprinted polymer synthesized in absence of GNPs.

4.3. Catalytic Sensors

The sensors mentioned above rely on affinity MIP, and the analytical signal was based on the binding/release of the analyte (template) to/from the MIP recognition sites. Nevertheless, in recent years acrylic molecularly imprinted polymers with catalytic activity have been developed [128]. These materials are able to catalyze reactions with the same regio- and stereoselectivity of natural enzymes with the additional advantages of high selectivity, thermal and chemical stability, and low cost. To create catalytic activity into a MIP, a simple strategy is to include a compound that resembles the catalytic center of the mimicked enzyme

TABLE 1.8 Analytical Characteristics of in situ Affinity Sensors Prepared by Electropolymerization Approach

Analyte	Monomer	Sensor (electrode, electrochemical technique)	Measuring range / M	LOD / M	Operational lifetime	Ref.
Paracetamol	pyrrole	Voltammetric (PGE, DPV)	5×10^{-6}–5×10^{-4}	7.9×10^{-7}	10–15 times	[118]
Sulfamethoxazole	Id.	Voltammetric (PGE, DPV)	2.5×10^{-5}–7.5×10^{-4}	3.59×10^{-7}	3 months 50 times	[119]
4,6-dinitro-*o*-cresol	aniline *o*-phenylenediamine	Voltammetric (GE, CV, DPV)	8×10^{-7}–1×10^{-4}	2.01×10^{-7}	30–35 times	[120]
Dopamine	*o*-aminophenol	Voltammetric (Au, CV, DPV)	2×10^{-8}–0.25×10^{-6}	1.98×10^{-9}	1 month	[121]
Fluorescein, rhodamine, 2,4-D	*o*-phenylenediamine	Voltammetric (GE, SWV)	n/a	n/a	n/a	[122]
Triclosan	Id.	Amperometric (GCE)	2.0×10^{-7}–3.0×10^{-6}	8.0×10^{-8}	n/a	[123]
Oxytetracycline	Id.	Voltammetric (Au, DPV)	Up to 1×10^{-7}	6.49×10^{-10}	1 month	[124]
MTMC	*o*-aminothiophenol	Amperometric (Au)	5×10^{-8}–3.5×10^{-7}	1.34×10^{-8}	3 months 10 times	[125]
Acetylsalicylic acid	*p*-aminothiophenol	Voltammetric (Au, DPV)	1×10^{-9}–1×10^{-7}	3×10^{-10}	10 days	[126]
Chloropyrifos	Id.	Voltammetric (GCE , CV)	5.0×10^{-7}–1.0×10^{-5}	3.3×10^{-7}	n/a	[127]

Analyte: 2,4-D: 2,4-dichlorophenoxyacetic acid; MTMC: *m*-tolyl methylcarbamate.
Electrode: Au: gold electrode; PGE: pencil graphite electrode; GE: graphite electrode; GCE: glassy carbon electrode.
Electrochemical technique: DPV: differential pulse voltammetry; CV: cyclic voltammetry; SWV: square wave voltammetry.
n/a: not available.

into the pre-polymerization mixture. During polymerization, the functional monomers are arranged around the template, which is adequately oriented into the catalytic center. Consequently, imprinted catalytic binding sites are generated into the MIP. These catalytic MIPs could be used as recognition elements in enzymatic-like sensors design. In these devices the monitored analytical signal proceeds from the redox behavior of the product, resulting in the catalytic reaction.

There are few publications, however, regarding this matter. Peroxidase catalytic activity was mainly mimicked, and metalloporphyrins have been widely used since they constitute the catalytic center of natural peroxidases. First, a catalytic molecularly imprinted amperometric sensor was prepared by electropolymerization of Co(III) tetrakis(aminophenyl) porphyrin, which acted as the functional monomer [129]. The pesticide 4-(2,4-dichlorophenoxy)butyric acid could be oxidized at the catalytic imprinted centers, and concentrations above 4×10^{-5} M could be detected.

Kubota et al. have prepared a catalytic MIP against *p*-aminophenol that introduce hemin (Fe(III) protoporphyrin IX) in order to form the imprinted catalytic centers. The MIP particles prepared by precipitation polymerization were dispersed in Nafion® and drop-coated onto the surface of a glassy carbon electrode. Reduction peak of the oxidized product obtained in the biomimetic peroxidase oxidation was used as an analytical signal to perform an amperometric sensor that shows good selectivity [130].

A similar approach has been used to develop a catalytic sensor for 2,4,6-trichlorophenol [131]. In this case, the catalytic MIP was synthesized as microgel (particles between 55 and 80 nm of diameter) by precipitation polymerization using hemin as one of the functional monomers. Then the MIP particles were suspended with CNTs in DMF and DMF/H_2O 1:1 followed by drop coating the mixture over a screen-printed electrode. The template is catalytically transformed to 2,6-dichloro-1,4-benzoquinone, which is voltammetrically determined. It should be noted that the catalytic conversion of the analyte will be transduced into a measurable response only when an efficient electric pathway has been established between the electrode surface and the catalytic center. Thus, the inherent insulating properties of the acrylic recognition material limit the performance of the sensor even when using CNTs to facilitate the integration.

A reliable solution to this problem requires providing conductivity to the polymeric network, and it has been recently achieved with the synthesis of a novel bifunctional monomer, N-phenylethylene diamine methacylamide (NPEDMA), which contains both aniline and methacrylamide functionalities and thus allows the preparation of MIP films with intrinsic conductivity [132]. A sensor against catechol was prepared onto a glass surface covered with a gold film by using this monomer. The aniline has been electropolymerized on the surface of the gold electrode, and the molecular imprinted polymer has grown in situ by grafting polymerization over this film from a solution of the monomer urocanic acid ethyl ester and $CuCl_2$. In this way the catalytic center was generated with the metallic center and 4-6 imidazole-containing residues where two copper ions were coordinated, recreating the active site of tyrosinase. This strategy satisfied both the recognition and the electronic path in the sensor development. The amperometric sensor against cathecol exhibited good analytical performance. The sensitivity of this device has been improved one order of magnitude by preparing polyaniline fibers instead of the conventional film on the electrode surface [133].

5. CONCLUSIONS

This chapter summarizes the major advances over the last decades in the developing area of electrochemical sensors using MIPs as recognition elements. The different sensors described clearly show the relevance of the type of transduction in defining the requirements for design of MIPs and the proper way of coupling to the transducer.

Although we are still a long way from developing MIP-based chemical sensors that fulfill all the requirements of portability, reliability, accuracy, ruggedness, long-term stability, and ease of use that ideally characterize the commercial sensors, the most recent advances described here allow envisaging a new generation of sensors that have improved analytical characteristics and are closer to commercialization. With this aim, improved detection limits, stability, and response times must be achieved for most of the applications.

Important impacts are emanating from novel hybrid nanomaterials; composites of MIPs and conducting nanomaterials such as carbon nanotubes, graphene, and metal nanoparticles represent a new strategy for providing conductivity to the imprinted polymers and thus facilitate the electrical contact between the imprinted polymer and the transducer. From an analytical point of view, use of this kind of hybrid materials is in its infancy, and many future applications are envisaged.

The use of MIPs for mimicking natural enzymes is another challenging goal that is in an early phase of development. Looking at the success of biosensor design using enzymes, one can expect that these synthetic catalysts will find a way to launch the MIP-based electrochemical sensor development, since they offer new strategies for detection and the possibility of achieving better analytical performance than affinity receptors. The viability of the concept has been proven; however, in order to obtain reliable sensors based on these recognition elements numerous challenges still have to be solved. An understanding of the molecular mechanism of these synthetic enzymes could lead to tailor-made catalytic systems, which would certainly contribute to major advances in this field.

Finally, the use of micro- and nanofabrication technologies in combination with the new imprinted materials will allow the preparation of micrometer-scale arrays of MIPs and the integration of these sensors in lab-on-chip systems, which will be the basis for different devices for real-time and online monitoring.

Based on recent advances in the field, discussed here, we can conclude that this kind of sensor has great potential for numerous applications in clinical, environmental, and industrial analysis. This is a very active area of research, and new and exciting developments will be achieved in the near future.

Acknowledgments

D. Antuña-Jiménez and G. Díaz-Díaz are grateful to FICYT (*Fundación para el Fomento en Asturias de la Investigación Científica Aplicada y la Tecnología*, Principado de Asturias Government) for a PhD Grant (Ref. BP09059) and for a postdoctoral contract (COF 09-20 Ref. E-33-2009-0183337), respectively. The Spanish government's Ministerio de Ciencia y Tecnología (CTQ2008-02429/BQU) and the European Social Fund (ESF) are also acknowledged.

References

[1] J. Janata, Principles of Chemical Sensors, Springer, U.S., 2009.
[2] E. Hedborg, F. Winquist, I. Lundström, L.I. Andersson, K. Mosbach, Sens. Actuators, A 37-38 (1993) 796.

[3] D. Kriz, K. Mosbach, Anal. Chim. Acta 300 (1995) 71.
[4] D. Kriz, O. Ramstroem, A. Svensson, K. Mosbach, Anal. Chem. 67 (1995) 2142.
[5] D. Kriz, L.I. Andersson, M. Khayyam, B. Danielson, P.O. Larsson, K. Mosbach, Biomimetics 3 (1995) 81.
[6] T. Prasada Rao, R. Kala, Talanta 76 (2008) 485.
[7] E. Bakker, P. Bühlmann, E. Pretsch, Chem. Rev. 97 (1997) 3083.
[8] P. Bühlmann, E. Pretsch, E. Bakker, Chem. Rev. 98 (1998) 1593.
[9] E. Bakker, E. Pretsch, TrAC, Trends Anal. Chem. 24 (2005) 199.
[10] U. Schaller, E. Bakker, U.E. Spichiger, E. Pretsch, Anal. Chem. 66 (1994) 391.
[11] K. Prasad, K.P. Prathish, J.M. Gladis, G.R.K. Naidu, T.P. Rao, Sens. Actuators, B 123 (2007) 65.
[12] S. Sadeghi, F. Fathi, J. Abbasifar, Sens. Actuators, B 122 (2007) 158.
[13] R. Liang, R. Zhang, W. Qin, Sens. Actuators, B 141 (2009) 544.
[14] T. Alizadeh, M. Akhoundian, Electrochim. Acta 55 (2010) 3477.
[15] K. Prasad, R. Kala, T. Prasada Rao, G.R.K. Naidu, Anal. Chim. Acta 566 (2006) 69.
[16] P. Metilda, K. Prasad, R. Kala, J.M. Gladis, T.P. Rao, G.R.K. Naidu, Anal. Chim. Acta 582 (2007) 147.
[17] V. Vishnuvardhan, K.P. Prathish, G.R.K. Naidu, T. Prasada Rao, Electrochim. Acta 52 (2007) 6922.
[18] K.P. Prathish, K. Prasad, T.P. Rao, M.V.S. Suryanarayana, Talanta 71 (2007) 1976.
[19] K. Prasad, K.P. Prathish, J.M. Gladis, G.R. Naidu, T.P. Rao, Electroanalysis 19 (2007) 1195.
[20] R.-N. Liang, D.-A. Song, R.-M. Zhang, W. Qin, Angew. Chem., Int. Ed. 49 (2010) 2556.
[21] G. D'Agostino, G. Alberti, R. Biesuz, M. Pesavento, Biosens. Bioelectron. 22 (2006) 145.
[22] K.P. Prathish, V. Vishnuvardhan, T.P. Rao, Electroanalysis 21 (2009) 1048.
[23] X. Zeng, G.M. Murray, Sep. Sci. Technol. 31 (1996) 2403.
[24] R.S. Hutchins, L.G. Bachas, Anal. Chem. 67 (1995) 1654.
[25] F.T.C. Moreira, A.H. Kamel, J.R.L. Guerreiro, M.G.F. Sales, Biosens. Bioelectron. 26 (2010) 566.
[26] A.H. Kamel, F.T.C. Moreira, S.A.A. Almeida, M.G.F. Sales, Electroanalysis 20 (2008) 194.
[27] M.S. Tehrani, M.T. Vardini, P.A. Azar, S.W. Husain, Int. J. Electrochem. Sci. 5 (2010) 88.
[28] H.M.V. Oliveira, F.T.C. Moreira, M.G.F. Sales, Electrochim. Acta 56 (2011) 2017.
[29] J. Guerreiro, M. Sales, F. Moreira, T. Rebelo, Eur. Food Res. Technol. 232 (2011) 39.
[30] J.R.L. Guerreiro, V. Freitas, M.G.F. Sales, Microchem. J. 97 (2011) 173.
[31] T.S.C.R. Rebelo, S.A.A. Almeida, J.R.L. Guerreiro, M.C.B.S.M. Montenegro, M.G.F. Sales, Microchem. J. 98 (2011) 21.
[32] A.H. Kamel, H.E.M. Sayour, Electroanalysis 21 (2009) 2701.
[33] T. Kitade, K. Kitamura, T. Konishi, S. Takegami, T. Okuno, M. Ishikawa, M. Wakabayashi, K. Nishikawa, Y. Muramatsu, Anal. Chem. 76 (2004) 6802.
[34] M. Javanbakht, S. Eynollahi Fard, M. Abdouss, A. Mohammadi, M. Reza Ganjali, P. Norouzi, L. Safaraliee, Electroanalysis 20 (2008) 2023.
[35] M. Javanbakht, S.E. Fard, A. Mohammadi, M. Abdouss, M.R. Ganjali, P. Norouzi, L. Safaraliee, Anal. Chim. Acta 612 (2008) 65.
[36] Y. Zhou, B. Yu, E. Shiu, K. Levon, Anal. Chem. 76 (2004) 2689.
[37] Y. Zhou, B. Yu, K. Levon, Biosens. Bioelectron. 20 (2005) 1851.
[38] Y. Zhou, B. Yu, K. Levon, Chem. Mater. 15 (2003) 2774.
[39] Y. Wang, Y. Zhou, J. Sokolov, B. Rigas, K. Levon, M. Rafailovich, Biosens. Bioelectron. 24 (2008) 162.
[40] Y. Wang, Z. Zhang, V. Jain, J. Yi, S. Mueller, J. Sokolov, Z. Liu, K. Levon, B. Rigas, M.H. Rafailovich, Sens. Actuators, B 146 (2010) 381.
[41] M. Lahav, A.B. Kharitonov, O. Katz, T. Kunitake, I. Willner, Anal. Chem. 73 (2001) 720.
[42] M. Lahav, A.B. Kharitonov, I. Willner, Chem.—Eur. J. 7 (2001) 3992.
[43] S.P. Pogorelova, A.B. Kharitonov, I. Willner, C.N. Sukenik, H. Pizem, T. Bayer, Anal. Chim. Acta 504 (2004) 113.
[44] M. Zayats, M. Lahav, A.B. Kharitonov, I. Willner, Tetrahedron 58 (2002) 815.
[45] S.P. Pogorelova, T. Bourenko, A.B. Kharitonov, I. Willner, Analyst 127 (2002) 1484.
[46] S.P. Pogorelova, M. Zayats, T. Bourenko, A.B. Kharitonov, O. Lioubashevski, E. Katz, I. Willner, Anal. Chem. 75 (2002) 509.
[47] A. Kugimiya, K. Kohara, Mater. Sci. Eng., C 29 (2009) 959.
[48] H.-H. Tsai, C.-F. Lin, Y.-Z. Juang, I.L. Wang, Y.-C. Lin, R.-L. Wang, H.-Y. Lin, Sens. Actuators, B 144 (2010) 407.
[49] D. Kriz, M. Kempe, K. Mosbach, Sens. Actuators, B 33 (1996) 178.

[50] S.A. Piletsky, Y.P. Parhometz, N.V. Lavryk, T.L. Panasyuk, A.V. El'skaya, Sens. Actuators, B 19 (1994) 629.
[51] S.A. Piletsky, E.V. Piletskaya, A.V. Elgersma, K. Yano, I. Karube, Y.P. Parhometz, A.V. El'skaya, Biosens. Bioelectron. 10 (1995) 959.
[52] S.A. Piletsky, E.V. Piletskaya, T.L. Panasyuk, A.V. El'skaya, R. Levi, I. Karube, G. Wulff, Macromolecules 31 (1998) 2137.
[53] T.A. Sergeyeva, S.A. Piletsky, T.L. Panasyuk, A.V. El'skaya, A.A. Brovko, E.A. Slinchenko, L.M. Sergeeva, Analyst 124 (1999) 331.
[54] T.A. Sergeyeva, S.A. Piletsky, A.A. Brovko, E.A. Slinchenko, L.M. Sergeeva, A.V. El'skaya, Anal. Chim. Acta 392 (1999) 105.
[55] R. Suedee, T. Srichana, C. Sangpagai, C. Tunthana, P. Vanichapichat, Anal. Chim. Acta 504 (2004) 89.
[56] R. Suedee, W. Intakong, F.L. Dickert, Anal. Chim. Acta 569 (2006) 66.
[57] T.L. Panasyuk, V.M. Mirsky, S.A. Piletsky, O.S. Wolfbeis, Anal. Chem. 71 (1999) 4609.
[58] Z.L. Cheng, E.K. Wang, X.R. Yang, Biosens. Bioelectron. 16 (2001) 179.
[59] L. Yang, W. Wei, J. Xia, H. Tao, P. Yang, Electroanalysis 17 (2005) 969.
[60] A. Aghaei, M.R.M. Hosseini, M. Najafi, Electrochim. Acta 55 (2010) 1503.
[61] R. Ouyang, J. Lei, H. Ju, Y. Xue, Adv. Funct. Mater. 17 (2007) 3223.
[62] H. Liao, Z. Zhang, H. Li, L. Nie, S. Yao, Electrochim. Acta 49 (2004) 4101.
[63] K. Liu, W.Z. Wei, J.X. Zeng, X.Y. Liu, Y.P. Gao, Anal. Bioanal. Chem. 385 (2006) 724.
[64] Z. Wang, J. Kang, X. Liu, Y. Ma, Int. J. Polym. Anal. Charact. 12 (2007) 131.
[65] T. Panasyuk-Delaney, V.M. Mirsky, M. Ulbricht, O.S. Wolfbeis, Anal. Chim. Acta 435 (2001) 157.
[66] T.L. Delaney, D. Zimin, M. Rahm, D. Weiss, O.S. Wolfbeis, V.M. Mirsky, Anal. Chem. 79 (2007) 3220.
[67] T. Panasyuk-Delaney, V.M. Mirsky, O.S. Wolfbeis, Electroanalysis 14 (2002) 221.
[68] J.J. BelBruno, G. Zhang, U.J. Gibson, Sens. Actuators, B 155 (2011) 915.
[69] C. Sontimuang, R. Suedee, F. Dickert, Anal. Biochem. 410 (2011) 224.
[70] T. Alizadeh, Thin Solid Films 518 (2010) 6099.
[71] M.C. Blanco-López, S. Gutiérrez-Fernández, M.J. Lobo-Castañón, A.J. Miranda-Ordieres, P. Tuñón-Blanco, Anal. Bioanal. Chem. 378 (2004) 1922.
[72] B.B. Prasad, R. Madhuri, M.P. Tiwari, P.S. Sharma, Sens. Actuators, B 146 (2010) 321.
[73] K.-C. Ho, W.-M. Yeh, T.-S. Tung, J.-Y. Liao, Anal. Chim. Acta 542 (2005) 90.
[74] E. Mazzotta, R.A. Picca, C. Malitesta, S.A. Piletsky, E.V. Piletska, Biosens. Bioelectron. 23 (2008) 1152.
[75] M.B. Gholivand, M. Torkashvand, Talanta 84 (2011) 905.
[76] T. Alizadeh, Electroanalysis 21 (2009) 1490.
[77] T. Alizadeh, M.R. Ganjali, P. Norouzi, M. Zare, A. Zeraatkar, Talanta 79 (2009) 1197.
[78] T. Alizadeh, M. Zare, M.R. Ganjali, P. Norouzi, B. Tavana, Biosens. Bioelectron. 25 (2010) 1166.
[79] T. Alizadeh, M.R. Ganjali, M. Zare, P. Norouzi, Electrochim. Acta 55 (2010) 1568.
[80] T. Alizadeh, M. Akhoundian, Electrochim. Acta 55 (2010) 5867.
[81] B.B. Prasad, R. Madhuri, M.P. Tiwari, P.S. Sharma, Biosens. Bioelectron. 25 (2010) 2140.
[82] N. Karousis, N. Tagmatarchis, D. Tasis, Chem. Rev. 110 (2010) 5366.
[83] Y.W. Zhu, S. Murali, W.W. Cai, X.S. Li, J.W. Suk, J.R. Potts, R.S. Ruoff, Adv. Mater. 22 (2010) 3906.
[84] P.Y. Chen, P.C. Nien, C.W. Hu, K.C. Ho, Sens. Actuators, B 146 (2010) 466.
[85] X.W. Kan, Y. Zhao, Z.R. Geng, Z.L. Wang, J.J. Zhu, J. Phys. Chem. C 112 (2008) 4849.
[86] B.B. Prasad, R. Madhuri, M.P. Tiwari, P.S. Sharma, Electrochim. Acta 55 (2010) 9146.
[87] H. Wang, H. Zhao, X. Quan, S. Chen, Electroanalysis 23 (2011) 1863.
[88] Y. Mao, Y. Bao, S. Gan, F. Li, L. Niu, Biosens. Bioelectron. 28 (2011) 291.
[89] G. Jin, W. Li, S. Yu, Y. Peng, J. Kong, Analyst 133 (2008) 1367.
[90] R. Shoji, T. Takeuchi, I. Kubo, Anal. Chem. 75 (2003) 4882.
[91] M. Pesavento, G. D'Agostino, R. Biesuz, G. Alberti, Electroanalysis 21 (2009) 604.
[92] Y. Diñeiro, M.I. Menéndez, M.C. Blanco-López, M.J. Lobo-Castañón, A.J. Miranda-Ordieres, P. Tuñón-Blanco, Anal. Chem. 77 (2005) 6741.
[93] L. Yao, Y. Tang, W. Zeng, Z. Huang, Anal. Sci. 25 (2009) 1089.
[94] A. Mujahid, P.A. Lieberzeit, F.L. Dickert, Materials 3 (2010) 2196.
[95] M.E. Díaz-García, R.B. Laíño, Microchim. Acta 149 (2005) 19.
[96] C. Li, C. Wang, C. Wang, S. Hu, Anal. Chim. Acta 545 (2005) 122.

[97] C. Li, C. Wang, B. Guan, Y. Zhang, S. Hu, Sens. Actuators, B 107 (2005) 411.
[98] A.K. Patel, P.S. Sharma, B.B. Prasad, Electroanalysis 20 (2008) 2102.
[99] A.K. Patel, P.S. Sharma, B.B. Prasad, Mater. Sci. Eng. C 29 (2009) 1545.
[100] A.K. Patel, P.S. Sharma, B.B. Prasad, Thin Solid Films 518 (2010) 2847.
[101] A.K. Patel, P.S. Sharma, B.B. Prasad, Int. J. Pharm. 371 (2009) 47.
[102] S. Gutierrez-Fernandez, M.J. Lobo-Castañón, A.J. Miranda-Ordieres, P. Tuñón-Blanco, G.A. Carriedo, F.J. Garcia-Alonso, J.I. Fidalgo, Electroanalysis 13 (2001) 1399.
[103] P.-Y. Chen, R. Vittal, P.-C. Nien, G.-S. Liou, K.-C. Ho, Talanta 80 (2010) 1145.
[104] M.-J. Syu, T.-C. Chiu, C.-Y. Lai, Y.-S. Chang, Biosens. Bioelectron. 22 (2006) 550.
[105] C.-Y. Huang, M.-J. Syu, Y.-S. Chang, C.-H. Chang, T.-C. Chou, B.-D. Liu, Biosens. Bioelectron. 22 (2007) 1694.
[106] P.-Y. Chen, P.-C. Nien, C.-T. Wu, T.-H. Wu, C.-W. Lin, K.-C. Ho, Anal. Chim. Acta 643 (2009) 38.
[107] B.B. Prasad, D. Kumar, R. Madhuri, M.P. Tiwari, Electrochim. Acta. 56 (2011) 7202.
[108] N. Gao, Z. Xu, F. Wang, S. Dong, Electroanalysis 19 (2007) 1655.
[109] X. Xu, G. Zhou, H. Li, Q. Liu, S. Zhang, J. Kong, Talanta 78 (2009) 26.
[110] Z. Zhang, Y. Hu, H. Zhang, L. Luo, S. Yao, Biosens. Bioelectron. 26 (2010) 696.
[111] J. Huang, X. Xing, X. Zhang, X. He, Q. Lin, W. Lian, H. Zhu, Food Res. Int. 44 (2011) 276.
[112] Y. Hu, J. Li, Z. Zhang, H. Zhang, L. Luo, S. Yao, Anal. Chim. Acta 698 (2011) 61.
[113] Y. Hu, Z. Zhang, H. Zhang, L. Luo, S. Yao, J. Solid State Electrochem. 16 (2012) 857.
[114] J. Huang, X. Zhang, Q. Lin, X. He, X. Xing, H. Huai, W. Lian, H. Zhu, Food Control 22 (2011) 786.
[115] Z. Zhang, Y. Hu, H. Zhang, S. Yao, J. Colloid Interface Sci. 344 (2010) 158.
[116] Z.-h. Zhang, Y.-f. Hu, H.-b. Zhang, L.-j. Luo, S.-z. Yao, J. Electroanal. Chem. 644 (2010) 7.
[117] Y.-f. Hu, Z.-h. Zhang, H.-b. Zhang, L.-j. Luo, S.-z. Yao, Talanta 84 (2011) 305.
[118] L. Özcan, Y. Sahin, Sens. Actuators, B 127 (2007) 362.
[119] S.P. Ozkorucuklu, Y. Sahin, G. Alsancak, Sensors 8 (2008) 8463.
[120] A. Gomez-Caballero, N. Unceta, M.A. Goicolea, R.J. Barrio, Sens. Actuators, B 130 (2008) 713.
[121] J. Li, J. Zhao, X. Wei, Sens. Actuators, B 140 (2009) 663.
[122] H.H. Weetall, K.R. Rogers, Talanta 62 (2004) 329.
[123] Y. Liu, Q.J. Song, L. Wang, Microchem. J. 91 (2009) 222.
[124] J. Li, F. Jiang, X. Wei, Anal. Chem. 82 (2010) 6074.
[125] M.-F. Pan, G.-Z. Fang, B. Liu, K. Qian, S. Wang, Anal. Chim. Acta 690 (2011) 175.
[126] Z. Wang, H. Li, J. Chen, Z. Xue, B. Wu, X. Lu, Talanta 85 (2011) 1672.
[127] C.G. Xie, H.F. Li, S.Q. Li, J. Wu, Z.P. Zhang, Anal. Chem. 82 (2010) 241.
[128] G. Wulff, Chem. Rev. 102 (2002) 1.
[129] E. Mazzotta, C. Malitesta, Sens. Actuators, B 148 (2010) 186.
[130] J.d.R.M. Neto, W.d.J.R. Santos, P.R. Lima, S.M.C.N. Tanaka, A.A. Tanaka, L.T. Kubota, Sens. Actuators, B 152 (2011) 220.
[131] G. Díaz-Díaz, M.C. Blanco-López, M.J. Lobo-Castañón, A.J. Miranda-Ordieres, P. Tuñón-Blanco, Electroanalysis 23 (2011) 201.
[132] D. Lakshmi, A. Bossi, M.J. Whitcombe, I. Chianella, S.A. Fowler, S. Subrahmanyam, E.V. Piletska, S.A. Piletsky, Anal. Chem. 81 (2009) 3576.
[133] F. Berti, S. Todros, D. Lakshmi, M.J. Whitcombe, I. Chianella, M. Ferroni, S.A. Piletsky, A.P.F. Turner, G. Marrazza, Biosens. Bioelectron. 26 (2010) 497.

CHAPTER

2

Molecularly Imprinted Polymers as Recognition Elements in Sensors

Jianping Li, Ge Wei, Yun Zhang

College of Chemistry and Bioengineering, Guilin University of Technology, Guangxi, China

OUTLINE

Molecularly Imprinted Sensors.
DOI: 10.1016/B978-0-444-56331-6.00002-5

1. INTRODUCTION

Molecularly imprinted sensors are prepared by the molecularly imprinted technique and used for electrochemical, optical, and other detections [1–5]. The molecularly imprinted electrochemical sensor (MIECS) has the advantages of high selectivity, high sensitivity, low detect limit, ease of miniaturization and automation, and lower cost in use due to the combination of the molecularly imprinted technique and the electrochemical detection [6–8]. Molecularly imprinted polymer (MIP) is the sensitive membrane of MIECS. An electrochemical response is obtained when the MIP membrane is recognized and combined with the target molecules (i.e., template molecule or imprinted molecule); then the electro-signal can be recorded, and the concentration of target molecules can be determined. MIECS is commonly classified as capacitive, potentiometric, and amperometric sensors in accordance with the determination process of target molecules. The detection of target molecules by MIECS is based on electrochemical signal changes, including current, capacitance, and potential of molecular imprinting sensitive film before and after the specific recognition of target molecules. The advantage of the capacitance and potentiometric sensors is that no other additional reagents or markers should be added besides the analyte; the sensor is also simple in operation with a low price cost. The amperometric MIP sensor is the most widely used sensor among MIECS. It can be used to detect the electroactive target molecules by the electrode reaction of target molecules, or detect the nonelectroactive target molecules indirectly [9–11]. In the process of detecting nonelectroactive substances, the measurements were performed by the use of probes such as potassium ferricyanate. Furthermore, the indirect detection of non-electroactive analytes may also be realized based on the competition reactions between the target molecules and their electroactive structural analogs [12,13].

2. THE MOLECULARLY IMPRINTED TECHNIQUE AND MOLECULARLY IMPRINTED POLYMER

The molecularly imprinted polymer is the functional sensing element of the molecularly imprinted sensors. The sensors can selectively recognize the template molecule with specificity based on the cavities in the spatial structure and binding sites in molecularly imprinted polymer membrane, perfectly match to a specific target molecule. In the synthesis process, the template molecules interact with monomer to form a multi-action spot, and the polymerization process will be remembered. When the template molecules are removed, the cavities that match with the spatial configuration of the template are obtained, and then they will selectively identify the template molecules and their analogue compounds.

The process for preparing molecularly imprinted membrane includes three steps:

Associating: The template molecules, cross-linker, and functional monomer with functional groups are chosen so as to form a complex by covalent or noncovalent binding [14].
Polymerization: Cross-linked copolymers entrapped with template molecules were obtained by the reactions of photopolymerization, thermal polymerization, electropolymerization, and the like, so that the functional groups of the monomer in

polymers are fixed in spatial arrangement and spatial orientation in order to get a three-dimensional porous polymer network with recognition ability [15].
Elution: Template molecules are eluted out and dissociated from polymers by appropriate eluent, and the cavities with three-dimensional structure that intentionally matched with the template molecular size and shape are formed for specific recognition of imprinted molecules containing functional groups [16–18].

For molecular imprinted sensors, the molecular imprinted polymers should have the following properties [19]:

(1) Appropriate rigidity: the cavities keep their original shape and size after the template molecules are removed from polymer.
(2) Suitable flexibility: The equilibrium of combination reaction between the template molecules and the cavities can be quickly reached.
(3) Favorable mechanical stability: The sensors are stable in renewal after each use.

2.1. Techniques for Preparing Molecularly Imprinted Polymer Membrane

According to the interactions between the template molecule and functional groups of monomer, techniques for preparing molecularly imprinted polymer membrane can be divided into two types: the covalent and the noncovalent.

(1) Covalent bond method (pre-assembly method)

This method was proposed by Wulff and his colleagues from Heinrich-Heine University, Germany, in the early 1970s [20,21]. According to this method, monomer-template complexes are constructed by reversible covalent bonding. First, template molecules covalently react with monomer to form borate esters, Schiff base, or imine acetal derivatives, and then produce cross-linking polymers by polymerization. Then the imprinted molecules are removed by breaking the covalent bond in the presence of eluent. Small molecule compounds are often used as functional monomer. The commonly used functional monomers are boric acid and diols containing ethylene, also a mixture of silane containing acid esters and the like, in which boric acid ester is the most representative one. However, the method is rarely used due to a very limited selection of chemical reactions [22,23].

(2) Noncovalent method (self-assembly method)

The noncovalent method was proposed by Professor Mosbacht and his colleagues in the late 1980s, and it is the most effective and most commonly used method for preparing MIPs [24]. The prearranged self-assembly of the template molecules and monomers is obtained after the weak interaction between template molecule and functional groups of monomer by electrostatic forces (including ion-dipole and dipole-dipole), hydrogen bonding, hydrophobic interaction, charge transfer, metal chelate, and even van der Waals. Then the arrangements are preserved after polymerization. Among the noncovalent interactions, the most important one is the ion-dipole, and the most commonly used one is the hydrogen bond. Therefore, it is easy to remove the template molecules by the simple method of solvent extraction. But compared with the covalent method, the noncovalent method's selectivity is relatively lower.

The main difference between the covalent and noncovalent method involves the different mechanisms used in the combination of monomer and template molecules. The molecular recognition properties of the polymer are determined mainly by the interaction between guest molecules and the role of the template molecules. For both methods, each has strong points in the preparation of molecularly imprinted sensors; it should be selected according to specific circumstances. But the two methods are not mutually exclusive. For example, Whitcombe [25] skillfully combined both methods to prepare MIPs. With the aid of 4-vinyl-phenyl carbonate as a functional monomer, Whitcombe synthesized molecular imprinting complexes by the covalent method. Then the complexes were hydrolyzed with alkaline; the covalent bonds between functional monomers and molecules were broken; and the process gave rise to noncovalent molecular recognition sites on MIPs which could identify the template molecules by phenyl on recognition sites via the hydrogen bond.

2.2. The Synthesis of Molecular Imprinted Polymer Film

The following reagents need to be adopted in preparing the molecularly imprinted membrane: the template molecule, functional monomer, cross-linker, initiator, and solvent.

(1) Template molecule

Various substances can act as the template molecules in the molecular imprinting technique. For example, carbohydrates, amino acids and their derivatives, organic amines, vitamins, proteins, and other molecules have been successfully used as template molecules in the preparation of MIPs. Typically, these molecules contain highly polar groups (such as carboxyl and amino). It is easy to prepare high-performance MIPs because strong polar groups and the functional monomer could form more stable molecular complexes. The molecular imprinted polymers with high selectivity and affinity can be produced from the imprinted molecules that can form hydrogen bonds with the functional monomer, due to the outstanding advantages of hydrogen bonds in terms of its directionality, saturation and strength. Macromolecules, supramolecular cells, and even metal ions are also used as the template molecule.

(2) Functional monomer

Functional monomer can provide specific functional groups in polymerization or copolymerization. In various polymerization reactions, the most widely used one is free radical polymerization.

During the polymerization process, the molar ratio of the imprinted molecule and functional monomer has a great effect on the generation of the specific cavities. Generally, with the ratio of function monomer to the template molecule increasing, one can make the self-assembly of template molecule to functional monomer easy and fully completed. On the one hand it will do harm to the polymerization for the high ratio of proportion of functional monomer, since the excess of functional monomer leads to an increase in nonselective binding sites generated by nonassembled functional monomer residues in the polymer. On the other hand, excess functional monomer may cause self-aggregation, which results in reducing selective binding sites. The molar ratios of template molecule to functional monomer are generally controlled as 1:4. In addition, functional groups of the imprinted molecules and the properties of solvents should also be considered during the preparation.

(3) Cross-linker

Cross-linkers bridge the linear molecules so as to bond with each other among molecules to establish a network, and they also promote or regulate the polymer chain formation. The cross-linker in the synthesis of molecular imprinted polymers acts in three aspects: (a) they control the morphology of MIPs; (b) they fix imprinted recognition sites; and (c) they affect the mechanical stability of the MIP array.

The amount of cross-linker directly affects the number and the cross-linking degree of active functional monomers in the unit mass of imprinted polymer, whereas the number of active functional monomer and cross-linking degree directly influences the selectivity of MIPs and the binding capacity. In conclusion, the ratio of functional monomer to cross-linker has a great impact on the properties of molecularly imprinted polymers.

When the amount of cross-linker is less, the cavity configuration of MIPs cannot maintain itself in a stable state due to insufficient cross-linking. As a result, it reduces the molecular imprinted polymer's recognition ability. However, an excess of the cross-linker decreases the number of functional monomer in unit mass and then reduces the number of the recognition sites. Especially for the noncovalent molecular imprinting, the cross-linker can reduce the swelling of MIPs in different solvents, improve selectivity, and enhance the ability to distinguish the structural analogues. It also makes the analyte quickly diffuse into the recognition sites by increasing the polymer chain gap. Generally, the cross-linking degree of MIPs should be higher than 80%.

(4) Initiator

The initiation methods used in preparing MIPs are mainly divided into radical polymerization, radiation polymerization, and electrochemical polymerization methods.

The radical polymerization initiator is known as the free radical initiator used in polymerization. It is the compounds tending to thermally decompose into free radicals (i.e., primary radical) that can be used for initiation of vinyl radical polymerization and copolymerization, and also for unsaturated polyester curing cross-linking and polymer cross-linking reaction.

According to the molecular structure, the initiators generally can be divided into the peroxide initiator such as benzoyl peroxide (oil soluble), potassium persulfate (water soluble); azo compound initiator, such as azobisisobutyronitrile (oil soluble), V-50 initiator (water soluble); redox initiator, such as hydrogen peroxide–ferrous salt (water soluble), and so on.

Azobisisobutyronitrile and azobisisobutyronitrile G nitrile are generally used as initiators in radical polymerization. The initiation methods commonly used are thermal initiation (60 °C) and photo-initiation, usually at 0 °C. The adoption of the initiation method largely depends on the characteristics of the template molecules. In noncovalent molecular imprinting, polymers synthesized by photo-initiation at low temperature have more excellent properties than that prepared by thermal initiation at high temperature, which was proven by O'Shannessy [26]. Compared with thermal initiation, the photo-initiator has the following advantages: the complexes of template molecule-functional monomer are more stable; unstable compounds can also be used as template molecules; and higher selectivity can be obtained by changing the physical properties of the polymer. However, Milojkovi [27] had tried to initiate polymerization reactions by γ-ray, but the thermal stability and selectivity of the polymer are unsatisfactory.

In addition, Panasyuk [28] prepared imprinted polymers by electrochemical polymerization for the detection of phenylalanine.

(5) Solvent

Solvent can provide the porous structure polymer and promote the bonding of guest molecules. Solvent molecules enter the inside of polymer during polymerization, and then are removed in the elution process. The space originally occupied by solvent molecules become cavities and are left inside the polymer. The polymer obtained in the polymerization with the lack of solvent will become dense, which is not available to the diffusion of template molecules during binding and elution.

For noncovalent MIPs, there is significant influences for the solvent on the interaction between template molecules and functional monomer. The solvent should not only have a high solubility on the imprinted molecule, but also promote the interaction between template molecules and functional monomers. In general, polar solvent might reduce the combination of imprinting molecules and function monomers, in particular, and affect the formation of hydrogen bonds, which makes the recognition performance of MIPs becoming worse. So, solvents with a low dielectric constant are often used such as benzene, toluene, xylene, chloroform, and methylene chloride.

2.3. Imprinted Membrane Polymerization

The methods of MIP polymerization are divided into bulk polymerization [29–31], in situ polymerization [32], precipitation polymerization [33], suspension polymerization [34], two-step swelling and polymerization [35], surface imprinting [36,37], and sol-gel methods [38–40]. Among them, bulk polymerization, surface imprinting, and sol-gel methods are often used in preparing electrochemical sensors.

When polymerized with the bulk polymerization, the functional monomer, molecular imprinting, cross-linker, and initiator are dissolved in the solvent on a percentage. After degassing and deoxygenation, the monolithic polymers are obtained. The monolithic polymers are smashed and screened to be small polymer particles that could be used to prepare a molecularly imprinted carbon paste electrode. Bulk polymerization is simple, but the specific cavities in the polymers may suffer damage during the smash processor, and the sizes of the particles are not uniform. Furthermore, it is difficult to wash out the template molecules due to the large diffusion resistance in the internal polymers.

For the surface imprinting method, the template molecules and functional monomers react in an organic solvent to form polymers, the surface of the polymers is modified to get MIPs and the imprinted cavities are on the surface of the particles. In order to meet the application of the sensors, the surface imprinting method is further developed as electropolymerization, in situ photopolymerization, and the sandwich technique.

Molecularly imprinted sol-gel film technology combines the molecular imprinting and sol-gel techniques. The sol-gel technique is usually used to prepare solid-containing oxides or other compounds of inorganic or metal alkoxide by the procedures of solution, sol-gel cure, and then heat-treated. The molecularly imprinted sol-gel technology is the method used to prepare a rigid material with the inorganic network structure, into which the template molecules were introduced via the sol-gel process. After the template molecules are removed, they can show good affinity to the template molecules.

3. MECHANISM OF MOLECULAR IMPRINTING

3.1. Kinetic Theory of Molecular Imprinting

Whitcombe et al. [41] successfully prepared the imprinted polymers with two analyte-specific binding sites by using the molecular self-assembly method. In this study, thermodynamic analysis had been carried out to investigate the chemical equilibrium processes of these polymers in their preparation and applications. As shown in this study, the equilibrium processes could be described by the following equation based on certain assumptions:

$$[MTM] = K^2 \cdot [T] \cdot [M]^2 \tag{2.1}$$

where K is the association constant; [T] is the concentration of template molecule; [M] is the concentration of functional monomer; and [MTM] is the concentration of the desired template monomer complex.

The assumptions are described as follows:

(1) The strength of both interactions between two functional monomers and one template molecule is identical.
(2) The two association events between two functional monomers and one template molecule are independent.

This equation enables one to calculate the equilibrium concentrations of TM and MTM when the binding equilibrium is reached.

Further assumptions based on equation (2.1) are as follows:

(1) The two-point binding force for the template molecules is very strong (in comparison with the one-site binding format), so that excellent selectivity can still be achieved even if the MTM concentration is very low.
(2) No strong nonspecific interactions take place between template molecules and polymer chains.
(3) The association constant K is the same in the preparation and application.

So, one can get the following equation:

$$N = N_S + N_{NS} = P\{K^2 \cdot [L] \cdot [MTM] + K \cdot [L] \cdot ([M_0] - 2[MTM])\} \tag{2.2}$$

where N, N_S, and N_{NS} are total, specific, and nonspecifically binding, respectively, L is the free ligand concentration, and P is a polymer concentration factor (in relation to the concentrations of the template molecules and the polymer).

It is also assumed that:

(1) The $[M_0]$ and $[T_0]$ that formed the polymer are the original concentrations of the monomers and template molecules, respectively.

(2) The polymer was formed at a $[M_0]/[T_0]$ ratio of 2.
(3) 50% of the polymerization mixture consists of porogen.
(4) The distribution ratios of MTM, MT, and M (the monomers that did not bind to the template molecules) sites are the same as their equilibrium ratios in the polymerization mixture.
(5) All the template molecules in the cavities can be removed completely after the polymerization, and all the vacant sites can be re-occupied by free template molecules.

Based on these assumptions, the selectivity of the prepared MIPs can be evaluated using the equation (2.2), where the K value can be calculated with a host–guest chemistry software system, and other related values can be obtained from experiments such as chromatography, spectroscopy, and NRM.

Whitcombe et al. [41] have used equations (2.1) and (2.2) to evaluate several studies reported by Mosboch and co-workers, and the calculated results are in good agreement with the experimental data.

3.2. Thermodynamic Theory of Molecular Imprinting

Jencks's groups have pioneered in the field of thermodynamics of molecular imprinting [42,43]. Williams et al. suggested that the trend of molecular imprinting reactions can be characterized with the following equation [44]:

$$\Delta G = \Delta G_{(trans+rot)} + \Delta G_{rotors} + \Delta G_H + \Sigma\Delta G_i + \Delta H_{conform} + \Delta G_{vdw} \quad (2.3)$$

where ΔG is the change in Gibbs free energy for the formation of the monomer-template complex; $\Delta G_{(trans+rot)}$ is the change in translational and rotational Gibbs free energy; ΔG_{rotors} is the change in Gibbs free energy caused by restricted rotational motion of template molecules; ΔG_H is the change in Gibbs free energy caused by hydrophobic interactions; $\Sigma\Delta G_i$ is the total change in Gibbs free energy of the whole groups participated in the reactions; $\Delta H_{conform}$ is the strain energy in the complex in relation with the introduction of unfavorable bond lengths, bond angles, dihedral angles, and so on; and ΔG_{vdw} is the change in Gibbs free energy caused by van der Waals forces.

This is a relatively complete formula. It takes into account not only the specificity of groups but also the spatial selectivity. Nevertheless, this formula is intricate, and in some cases some changes in Gibbs free energy are negligible. So, Nicholls et al. [45] proposed a simplified formula expressed as:

$$\Delta G = \Delta G_{(trans+rot)} + \Delta G_{rotors} + \Delta G_H + \Delta G_{vdw} + \Sigma\Delta G_i \quad (2.4)$$

Such simplification is based on the following assumptions:

(1) No conformation constraints in the formation of monomer-template complex.
(2) No effects of van der Waals forces.
(3) The polymerization and identification processes are carried out in the same solvent, and no change in conformation occurs during the recognition process.
(4) In the highly cross-linked MIP, the groups' motions are restricted.

In the case of these experiments conducted on nonpolar organic solvents, where there is no hydrophobic force, the formula (2.4) can be further simplified as:

$$\Delta G = \Delta G_{(trans+rot)} + \Delta G_{rotors} + \Delta G_{vdw} + \Sigma\Delta G_i \quad (2.5)$$

These simplified formulas may help to simplify the calculation process, but at the same time its scope of application is reduced. In some cases, the experimental and calculated data even deviate rather substantially. In the meantime, these factors that had been omitted in the simplified formula would play important roles.

3.3. The Equilibrium Adsorption Theory of Molecular Imprinting

The equilibrium adsorption theory of molecular imprinting makes use of the adsorption isotherms of MIPs and imprinted molecules to gain a variety of chemical information, including their binding energy, the type of binding sites, and distribution [46]. The three most commonly used equilibrium isotherm models for curve fitting are the Langmuir, Bi-Langmuir, and Freundlich equations [47]. The adsorption models recently developed include the mass-action model and the steric mass-action model.

3.3.1. *The Model of Adsorption Equilibrium Isotherm*

The adsorption equilibrium relationship of imprinted molecules on the MIPs is defined as the relationship between the equilibrium concentration q of solute adsorbed and the concentration c of free solute in bulk solution. In theory, q as a function of c and temperature can be expressed as:

$$q = f(c, T)$$

Since physical adsorption generally occurs at a specific temperature, q is just a function of c, and their resulting relationship is plotted on a curve that is known as the adsorption isotherm. In the Henry adsorption equilibrium, q is a linear function of c:

$$q = mc$$

where m is the partition coefficient. This formula works well for low solute concentrations, but it is not available at high solute concentrations because of the nonlinear adsorption equilibrium. The Freundlich empirical equation has been alternatively used to describe such nonlinear equilibrium:

$$q = kc^{1/n}$$

where k and n ($1 < n < 10$) are constants. When the adsorbent shows large adsorption ability to the solute, n may be greater than 10. Because q has little effect on c, such adsorption is nearly irreversible, resulting in a rectangular isotherm.

(1) The discrete model (Langmuir model)

The Langmuir model is the simplest and most commonly used model to describe the adsorption processes. It is the first theoretical adsorption model and currently is commonly used to investigate the thermodynamics of molecular imprinting. The Langmuir model, which is originally derived from the gas adsorption process from a kinetic point of view, is based on the following assumptions: (a) molecules only bind to specific sites on a solid surface; (b) each adsorption site can only adsorb one molecule, and all the molecules adsorbed at the solid surface finally form a monolayer; (c) all the adsorption sites have equal energy; and (d) no interactions occur between the molecules adsorbed. Based on this monolayer adsorption theory, the Langmuir equilibrium equations can be expressed as:

$$q = \frac{q_m c}{k_d + c} \text{ or } q = \frac{q_m K_b c}{1 + K_b c}$$

where q_m is the saturated adsorption amount; k_d is the equilibrium dissociation constant; and K_b is the binding constant.

If n solute molecules adsorb at one active site, a general formula can be derived from the above one as:

$$q = \frac{q_m K_b c^n}{1 + K_b c^n}$$

As for the monomer adsorption of n components, the equilibrium equation should be expressed as follows:

$$q = \frac{q_m K_{bj} c_i}{1 + \sum_j^n K_{bj} c_j}$$

Using the Langmuir model to describe the adsorption process is based on this assumption that the whole surface of MIPs possesses uniform adsorption energy. However, different areas of the MIPs actually have nonuniform surfaces with diverse adsorption energy. Therefore, the Langmuir adsorption model used for non-uniform surfaces that only correspond to one energy level, and their overall adsorption isotherm should be the statistic sum of adsorption isotherms of all energy levels.

(2) Continuous model (Freundlich model)

The Langmuir adsorption model is only applicable when the system has one type of binding sites. In fact, diverse factors in MIPs influence the formation of binding sites, thus resulting in a variety of binding methods. The affinity distribution curve, which describes the relationship between the number of binding sites (N) and binding constant (K), should be continuous. In other words, each K value determined corresponds to a specific number of binding sites. The study of Shimizu et al. [48] suggested that the binding properties of MIPs could be more accurately described using the Freundlich model, which was favorable for the fitting of isotherm curve of subsaturated binding sites (namely, the attenuation part of the index).

Freundlich summarized a series of isotherm equations to derive an empirical formula:

$$\frac{X}{m} = Kp^{\frac{1}{n}}$$

where n was a constant; X was the amount of solute adsorbed; m was the amount of solid adsorbent; and p was the gas equilibrium pressure of solute. If X/m and p are represented by the solute concentrations in the stationary phase and the mobile phase, respectively, this empirical formula above can be expressed as:

$$c_s = Kc_m^{\frac{1}{n}}$$

After a logarithmic operation, one can get:

$$\lg c_s = \lg K + \frac{1}{n} \lg c_m$$

Shimizu et al. [49] reported a comparative study of 12 types of previously reported noncovalent MIPs prepared with different imprinted molecules, binding forms, and polymerization conditions. They found that 11 types of polymers could be simulated well using the Freundlich model.

In a study by Sellergren and co-workers [50], the adsorption isotherm was separately fitted to several different models including Langmuir, Bi-Langmuir, and Freundlich. The results implied that the choice of adsorption model largely depended on the types of functional monomers and imprinted molecules, polymerization conditions, composition of mobile phase, and pH. In a study of Baggianni et al. [51], seven models were used to fit the adsorption isotherms for investigating the adsorption of three kinds of MIPs that were prepared separately by using imprinted molecules with different concentrations. Their study demonstrated that substrate molecules covalently bounded to MIPs to form crystal nuclei. The resultant MIPs presented a far larger adsorption capability compared with those prepared in the absence of crystal nuclei, thus verifying the hypothesis of recognition cluster theory of molecular imprinting originally proposed by the authors.

3.3.2. *Steric Mass-action Model*

In 1999, Brooks and Cramer [52] reported a novel adsorption equilibrium model to describe protein ion exchange adsorption, that was, the steric mass-action model (SMAM). SMAM is a nonmechanistic model based on the law of mass action. Because this model can exactly describe the real protein ion exchange adsorption processes, it has been successfully applied to a series of protein adsorption studies in various ion exchange systems by combining with appropriate mass-transfer equations [53].

Su and co-workers [54] developed a novel model for surface active site adsorption of heavy metal ions on the basis of the affinity mechanism and SMAM of proteins. Its application to the adsorption equilibrium study of heavy metal ions and mycelium was also reported. In their study, Ni^{2+} was chosen as the imprinted ion to prepare a series of mycelium surface molecular imprinting adsorbents with different binding site concentrations. Model parameters were quantitatively determined by investigating the effects of various concentrations of H^+ and Ni^{2+} on the equilibrium parameters. This novel model was based on the following assumptions:

(1) Adsorption equilibrium exists between the adsorbent surface and N types of ions, and the binding constant of each type of ion is K_{ai}.
(2) Both the solid adsorbent and the liquid phase are in thermodynamic states, where concentrations can be used instead of activities.

(3) During the adsorption process, one metal ion can bind several active sites, and such multisite binding can be expressed as the number of featured binding sites, n_i.

(4) The concentration of imprinted cavity on the adsorbent surface is equal to the imprinting ion concentration, and all the ions used for imprinting can be thoroughly eluted.

(5) At relatively high H^+ concentrations, the decrease in adsorption amount of metal ions is due to the binding of H^+ to part of active sites.

(6) The residual functional groups on the adsorbent surface are still able to chelate metal ions. Some fissures in the adsorbent surface as well as certain groups of the mycelium and the cross-linker can also act as active sites for metal ion chelation. All of these active sites are collectively known as nonimprinted molecules. Their active site number is inversely proportional to the concentration of imprinted cavities in the adsorbent surface with a scale factor of σ. Their adsorption to H^+ is also in accordance with this proportional relationship.

(7) This model does not take into account the effects of nonideal behaviors such as the changes in three-dimensional structures caused by the adsorption.

For the aqueous solution containing N types of heavy metal ions, one can use N equilibrium equations to describe the stoichiometric equilibrium between each type of heavy metal ion and the available active sites on mycelium surface:

$$c_{pi} + n_i \overset{\frown}{L}_a \rightleftharpoons Q_i \quad i = 1, 2, 3, \ldots, N$$

where c_{pi} is the concentration of heavy metal ion in solution (mmol/L); Q_i is the concentration of metal ion-active site complex (mmol/g); and $\hat{L}_a$ is the concentration of available active sites including the imprinted and nonimprinted active sites (mmol/g).

The number of total nonimprinted active sites is: $\overline{L}_n = \sigma_i \bullet L_n$
where L_n is the number of imprinted cavities in the absorbent surface, namely, the imprinted ion concentration (mmol/g).

The number of available imprinted active sites is: $\overline{L}_a = \sigma_i \bullet L_n$
where L_a is the number of available imprinted ion concentration (mmol/g).

The concentration of active sites available on the adsorbent surface can thus be expressed as:

$$\hat{L}_a = (1 + \sigma_i)L_a$$

The equilibrium constant K_{ai} is defined as:

$$K_{ai} = \frac{Q_i}{c_{pi} \bullet [(1 + \sigma_i)L_a]^{ni}} \quad i = 1, 2, 3, \ldots, N$$

When there are only heavy metal ions existing in aqueous solution, the active sites on the adsorbent surface just bind to these metal ions. At relatively high H^+ concentrations, however, part of the active sites may bind to H^+, which in turn weakens the binding between active sites and heavy metal ions. In this case, by assuming that only these active sites that do not bind to H^+ are able to adsorb metal ions, the effects of H^+ on the adsorption can be characterized by the equilibrium relationship between binding sites and H^+.

3.4. Steric Action Theory of Molecular Imprinting Recognition

Template molecules (imprinted molecules) combine with functional monomer to form multiple action sites, and such actions will be recorded during the polymerization processes. The removal of template molecules from the resultant polymers allows for the formation of imprinted cavities that possess multiple binding sites for template molecules and match their spatial configuration as well. These cavities will show the selective recognition ability to the template molecules and their analogues. Because S-type sample molecules match up well with their imprinted cavities in space structures, strong interactions will take place between their chemical functional groups and functional groups of the cavities in more favorable spatial locations. In contrast, the imprinted cavities prepared from S-type molecules show an unsatisfactory space-matching ability to R-type sample molecules; only weak interactions occur between their functional groups and these molecules in less favorable spatial locations. Thus it can be seen that the spatial configuration shows a large effect on recognition abilities of molecular imprints.

The recognition ability of MIPs to sample molecules is attributed to their matching in chemical groups and spatial configuration. That is, the selectivity of MIPs is highly dependent on the number and intensity of interactions between template molecules and imprinted groups, in addition to the shape and rigidity of the template molecules. The chiral recognition mechanism of MIPs can be more reasonably described using the two-site binding theory originally reported by Wulff et al. in the a study of imprinted cavities [55].

Whether MIPs are used for catalysis or separation (or other purposes), it is very important for the substrates selected to correctly enter the imprinted cavities. For a cavity with two binding sites, either of the following two cases will occur when a template molecule enters into the cavity and bind to its functional groups (as shown in Fig. 2.1):

(1) The template molecule first binds to one active site and further binds to another site together via its moving. This is the most ideal situation.

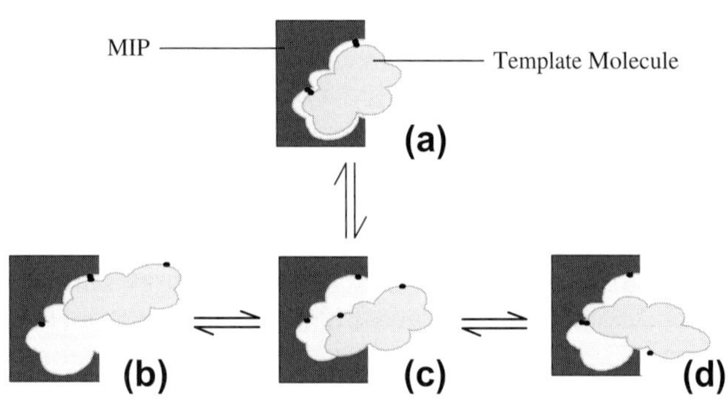

FIGURE 2.1 A schematic diagram for two-site binding: (a) the two-site binding mode; (b) the first one-site binding mode; (c) entering of template molecule into the cavity; and (d) the second one-site binding mode.

(2) After the binding of one active site, another site cannot be further bound by the template molecule, since it lacks an efficient movement due to the stereo-hindrance effect (such as b and d shown in the Fig. 2.1).

It is prerequisited for this template molecule to break the bound bonds before forming the correct two-site binding. In fact, the binding constant of two-site binding is greater than that of single-site binding. So, most substrate molecules are able to enter the imprinted cavities in the right way and bind to their functional groups. The reason may be that the shapes of gaps and cavities between polymer chains enable the preliminary orientation and screening of substrate molecules. Generally, the rate of two-site binding is relatively slow. Wulff and co-workers have further found that compared with the spatial shapes of cavities, the spatial orientation and type of imprinted sites in cavities have larger effects on the selectivity of MIPs [56].

3.5. Molecular Cluster Theory

The main recognition theory of MIPs currently focuses on the specific recognition mechanism of nanocavities generated in the polymerization of template molecules or functional monomers. According to the lock-key principle, the shape of these nanocavities is complementary to that of the template. The noncovalent interactions that govern the molecular recognition mechanism are established between the binding sites and a single, isolated template molecule. As regards these polymers prepared via noncovalent imprinting technology, this lock-key binding principle should be intended as effectively operating in a complex system constituted by a multiplicity of binding sites, with a wide and perhaps continuous distribution of affinities for the template, but not in a system possessing a single, well-defined class of binding sites. Some experimental results show that the molecular recognition behavior of the imprinted binding sites is more complex than that traditionally considered, presumably because it involves template–template interactions in addition to distinct classes of binding site.

In 2004, Baggiani et al. [51] originally proposed the hypothesis of molecular clusters. That is, template molecules first covalently bind to the binding sites to generate one or more crystal nuclei that will promote the subsequent formation of molecular clusters as a result of the increased cooperative-like template–template molecular interactions. The cavities in the MIPs are finally filled up with the formed molecular clusters, thus finishing the specific recognition of chiral imprinted molecules. The authors prepared three types of MIPs by using 2,4,5-trichlorophenoxyacetic acid (2,4,5-T) and 2-(2,4,5-trichlorophenoxyacetoxy)-ethylmethacrylate (2,4,5-TEMA) as template molecules at different volume ratios. A series of experimental parameters were calculated, such as affinity constants, binding site density, and binding capacity. The adsorption isotherms of polymers prepared were additionally fitted by using several isotherm models including Bi-Langmuir, Freundlich-Jovanovic, and Freundlich-Langmuir. Experimental results confirmed that substrate molecules covalently linked to MIPs to form crystal nuclei. The resulting MIPs showed a far larger adsorption capability in comparison with those prepared in the absence of crystal nuclei. Therefore, the number of substrate molecules that could be adsorbed would increase with the increase of the adsorption sites of MIPs, by assuming that the crystal nuclei formed had no effects on the adsorption.

4. FACTORS INFLUENCING THE SELECTIVITY OF MOLECULARLY IMPRINTED SENSORS AND THE ROUTES FOR SELECTIVITY IMPROVEMENT

In general, the recognition ability of MIPs toward template molecules is governed mainly by the following three factors.

4.1. Selectivity of Imprinting Reactions

Imprinting reactions are selective reactions between the functional groups of monomers in MIPs and that of template molecules. The selectivity of imprinting reactions plays a decisive role in the selectivity of molecularly imprinted sensors.

Imprinting reactions involve reactions in which reversible covalent bonds as well as noncovalent bonds are formed, including hydrogen bond, ionic bond, metal-ligand bond, P-P forces, hydrophobic forces and van der Waals forces. Thus, all the factors influencing the formation of reversible covalent bonds and noncovalent bonds will influence the recognition of MIPs toward template molecules.

(1) Inhibitors of functional groups

Inhibitors are substances that can react with functional groups. After the interactions, the template molecules are not able to bind to these functional groups because the inhibitors have made them lose their activities in addition to occupying the cavity space.

(2) The change in spatial orientation of functional groups

Some substrates or conditions such as solvent and temperature are capable of causing the change in spatial orientation of functional groups. Either the reaction for covalent bond formation or the reaction for noncovalent bond formation has a certain direction. If the spatial orientation of functional groups does not match that of the template molecules, their binding strength and rate will be highly affected, or even the binding reactions cannot take place [57].

(3) Electrostatic repulsion and steric hindrance effects

MIPs are highly cross-linked copolymers whose polymer chains and other groups existing in these chains could impede the reactions between template molecules and functional groups and in turn make the MIPs reduce or even loss their selectivity and affinity [58].

(4) Effects of solvents

Some properties of solvents have a large effect on imprinting reactions, such as solvent polarity, dielectric constant, protonation, and complexation. These factors influence both the selectivity and affinity of MIPs and substrates via the molecular interactions [59–60]. The solvent acting as the inhibitor of functional groups shows its effects on MIPs. It also influences the reaction activity of substrate and functional groups through weak interactions such as the formation of hydrogen bonds. In particular, weak interactions generally govern the formation of molecular self-assembly. It is very difficult for the substrate molecules to react with the polymer functional groups if hydrogen bonds or other types of bonds are once formed between them and the solvent. So, the preparation of self-assembled MIPs is generally carried out in nonpolar organic solvents.

In recent years, breakthrough advances in molecular imprinting technology in which water served as the solvent have been achieved [61–63]. In these studies, although the MIPs could still not be prepared in aqueous solution, they were capable of recognizing and binding substrate molecules in the aqueous phase. This result at least indicated that molecular imprinting reactions could indeed occur in an aqueous phase.

A series of experiments have demonstrated that the number of active sites and functional groups on MIPs has a large effect on the selectivity and affinity of these polymers. Generally, the substrate selectivity and affinity of MIPs is increased with increasing active sites [64–65]. And the number of the functional groups on MIPs decisively depends on the number of functional groups on template molecules.

Although the important effect of the specific binding site number on the selectivity and affinity of MIPs has been well demonstrated by a large number of experimental facts, recently Mosbach et al. originally proposed a new theory on imprinting mechanisms in a study of MIP affinity for peptides [66]. This theory holds that the highly ordered structures that are formed from functional monomers combined with the ion bonds or (and) hydrophobic interactions play a major role in the selectivity and affinity of MIPs. However, it currently seems that this theory suffers from the main problem associated with the lack of sufficient experimental evidences. The authors only investigated the imprinting behaviors of biomacromolecules. Small molecules display many chemical and physical properties that are quite different from those of biomacromolecules.

4.2. Matching Between the Spatial Structures of Cavities in MIPs and the Configuration and Conformation of Template Molecules

The perfect matching between the spatial structures of cavities in MIPs and the configuration and conformation of template molecules will allow the functional groups of MIPs to sufficiently approach and specifically bind to those template molecules. The structures and shapes of the cavities in MIPs are not completely rigid and inflexible. In fact, in a solvent there exists a swelling phenomenon of MIPs which are prepared via either molecular self-assembly or molecular preorganization. The swelling of MIPs will allow for changes in sizes and shapes of cavities, which in turn changes their selectivity and affinity to some extent [67]. Researchers experimentally found that the changes in conformation of MIPs were due to the interactions between the solvent and the polymer chains, and they could be directly determined by measuring the change in the optical rotation before and after the interactions of the polymers with template molecules [65–70].

Different effects of polymer swelling take place on MIPs prepared by using different methods. For the MIPs that are prepared via molecular preorganization, they will swell in a solvent after the elution of template molecules. When the cavities in MIPs are reoccupied by template molecules, the volume of the polymers will decrease even to their original value as a result of the interactions between the function groups and the template [67]. This indicates that the interaction force between the preorganized MIPs and the template molecules is stronger than that between the solvent and the polymer chains. On the other hand, the volume decrease phenomenon does not occur on these self-assembled MIPs. This implies that the interaction force between the polymers and the template molecules is weaker than or at most equal to that between the solvent and the polymer chains. The volume decrease phenomenon can be considered as one of the direct evidences that the solvent shows a larger effect on the molecular self-assembly than that shows on the molecular preorganization. In general, the specific interactions between functional groups and the template molecules play a main role in selectivity and affinity of the polymers for substrate molecules.

In the enantiomeric recognition of MIPs, there exists a phenomenon that the MIPs which are prepared using a substrate as the template molecule can be used for recognizing its enantiomers and vice versa. This phenomenon is termed cross-selectivity. For example, if a polymer is imprinted with D-fructose or D-galactose, naturally, it will bind to the corresponding template molecules in the substrate solution. However, the D-fructose-based MIPs prefer to be able to bind to L-galactose in a substrate mixture, while those prepared from a D-galactose template prefer to be able to bind to L-fructose in a substrate mixture [71]. Such a phenomenon lies in the fact that the cross-reaction molecules (e.g., D-fructose and L-galactose, D-galactose and L-fructose) have the same arrangement of functional groups but different extension directions of the hydroxymethyl.

This phenomenon indicates that the selectivity and affinity of the MIPs are mainly dependent on the specific binding of template molecules to these polymers via the interactions between their functional groups. However, the size and shape of cavities will play a decisive role in the recognition of MIPs for the substrates that have the same arrangement of functional groups. Overall, both functional groups and polymer cavities govern the selectivity and affinity of the MIPs; neither of the two can be dispensed with.

4.3. Kinetic Processes Involved in the Recognition of MIPs for Substrate Molecules

As mentioned above in "3.4. Steric Action Theory of Molecular Imprinting Recognition," the binding constant of two-site or multisite binding is greater than that of single-site binding. Thus, most substrate molecules are capable of entering the imprinted cavities in the right way and in turn bind to the functional groups. This may be due to the fact that the shapes of gaps and cavities between polymer chains may allow for the preliminary orientation and screening of substrate molecules. Nevertheless, the rate of two-site or multisite binding is relatively slow.

A large number of experimental evidences demonstrate that the selectivity of MIPs for substrate can be significantly improved by increasing reaction temperature, presumably for the following three main reasons [72]:

(1) The rate of two-site or multisite binding increases with increasing temperature.
(2) At relatively high temperatures, the single-site-bound substrate would dissociate from cavity surface and further be recognized and bounded by functional groups in a more stable two-site binding form.
(3) The polymers will swell at high temperatures, and these cavities, which are not available at room temperature, can be exploited for substrate recognition.

As regards the same imprinted polymer, different cavities have different selectivities for substrate and thus make various degrees of contributions to the polymer selectivity [73].

This is owing to the different degrees of matching between the substrate and the functional groups of different cavities in polymers.

Many experimental results show that, although the MIPs have high specificities for corresponding template molecules, these functional groups on the imprinted polymers, which do not bind to the template molecules during the polymerization processes of functional monomers, are able to bind to other foreign molecules. This kind of binding is often referred to as nonspecific binding, which can also occur between the functional groups inside the imprinted cavities and the foreign molecules. The nonspecific binding generally leads to a decrease in the selectivity of MIPs for template molecules [74]. In a study reported by McNiven and co-workers [75], chemical reagents such as iodoform were used to treat the MIPs to not only protect the specific binding sites but also inhibit the nonspecific binding reactions, and a high selectivity was thus achieved.

Acknowledgments

This study was financially supported by the National Natural Science Foundation of China (21165007).

References

[1] K. Haupt, K. Mosbach, Molecularly imprinted polymers and their use in biomimetic sensors, Chem. Rev. 100 (2000) 2495–2504.
[2] A. Merkoçia, S. Alegret, New materials for electrochemical sensing IV. Molecular imprinted polymers, Trends Anal. Chem. 21 (2002) 717–725.
[3] T.L. Panasyuk, V.M. Mirsky, S.A. Piletsky, O.S. Wolfbeis, Electropolymerized molecularly imprinted polymers as receptor layers in capacitive chemical sensors, Anal. Chem. 71 (1999) 4609–4613.
[4] S.A. Piletsky, E.V. Piletskaya, A.V. Elgersma, K. Yano, I. Karube, Y.P. Parhometz, A.V. El'skaya, Atrazine sensing by molecularly imprinted membranes, Biosens. Bioelectron. 10 (1995) 959–964.
[5] C. Zhang, K.S. Suslick, A colorimetric sensor array for organics in water, J. Am. Chem. Soc. 127 (2005) 11548–11549.
[6] G. D'Agostino, G. Alberti, R. Biesuz, M. Pesavento, Potentiometric sensor for atrazine based on a molecular imprinted membrane, Biosens. Bioelectron. 22 (2006) 145–152.
[7] M.C. Blanco-López, M.J. Lobo-Castañón, A.J. Miranda-Ordieres, P. Tuñón-Blanco, Voltammetric sensor for vanillylmandelic acid based on molecularly imprinted polymer-modified electrodes, Biosens. Bioelectron. 18 (2003) 353–362.
[8] Y. Fuchiwaki, A. Shimizu, I. Kubo, 6-Chloro-N, N-diethyl-1,3,5-triazine-2,4-diamine(CAT) sensor based on biomimetic recognition utilizing a molecularly imprinted artificial receptor, Anal. Sci. 23 (2007) 49–54.
[9] J.P. Li, J. Zhao, X. P. Wei, A sensitive and selective sensor for dopamine determination based on a molecularly imprinted electropolymer of o-aminopheno, Sens. Actuat. B. Chem 140 (2009) 663–669.
[10] J.P. Li, F.Y. Jiang, A chlortoluron sensor based on molecularly imprinted sensitive membranes, Chem. Lett. 39 (2010) 478–479.
[11] P.Y. Chen, R. Vittal, P.C. Nien, G.S. Liou, K.C. Ho, A novel molecularly imprinted polymer thin film as biosensor for uric acid, Talanta 80 (2010) 1145–1151.
[12] J.P. Li, F.Y. Jiang, X.P. Wei, Molecularly imprinted sensor based on an enzyme amplifier for ultratrace oxytetracycline determination, Anal. Chem. 82 (2010) 6074–6078.
[13] J.P. Li, F.Y. Jiang, Y.P. Li, Z.Q. Chen, Fabrication of the sensitive oxytetracycline molecular imprinted sensor based on competition reaction via GOD-enzymatic amplifier, Biosens. Bioelectron. 26 (2011) 2097–2101.
[14] P. Parmpi, P. Kofinas, Biomimetic glucose recognition using molecularly imprinted polymer hydrogels, Biomaterials 25 (2004) 1969–1973.
[15] C. Yu, K. Mosbach, Influence of mobile phase composition and cross-linking density on the enantiomeric recognition properties of molecularly imprinted polymers, J. Chromatogr. A. 888 (2000) 63–72.

[16] M.M. Wayne, P.C. Edward, Determination of theophylline in serum by molecularly imprinted solid-phase extraction with pulsed elution, Anal. Chem. 15 (1998) 3636–3641.
[17] K. Möller, C. Crescenzi, U. Nilsson, Determination of a flame retardant hydrolysis product in human urine by SPE and LC-MS. Comparison of molecularly imprinted solid-phase extraction with a mixed-mode anion exchanger, Anal. Bioanal. Chem. 378 (2004) 197–204.
[18] B. Sellergren, Direct drug determination by selective sample enrichment on an imprinted polymer, Anal. Chem. 66 (1994) 1578–1582.
[19] R.X. Shi, C.H. Guo, X.H. Zou, C.Y. Zhu, Y.J. Zuo, Y.D. Deng, The development of research in molecular imprinting technique, Prog. Chem. (in Chinese) 14 (2002) 182–191.
[20] G. Wulff, A. Sarhan, Use of polymers with enzyme-analogous structures for the resolution of racemates, Angew. Chem. Int. Ed. Engl. 11 (1972) 341–344.
[21] G. Wulff, A. Akelah, Enzyme-analogue buit polymers synthesis of 5-vinylsalicy-laldehyde and a simplified synthesis of some divinyl derivatives, Die Makromolekulare Chemie 179 (2003) 2647–2651.
[22] T. Mukawa, T. Goto, T. Takeuchi, Post-oxidative conversion of thiol residue to sulfonic acid in the binding sites of molecularly imprinted polymers: Disulfide based covalent molecular imprinting for basic compounds, Analyst 127 (2002) 1407–1409.
[23] B. Sellergren, Noncovalent molecular imprinting: antibody-like molecular recognition in polymeric network materials, Trends Anal. Chem. 16 (1997) 310–320.
[24] M. Kempe, K. Mosbach, Direct resolution of naproxen on a non-covalently molecularly imprinted chiral phase, J. Chromatogr. A. 664 (1994) 276–279.
[25] M.J. Whitcombe, M. E Rodriguez, P. Vilar, E.N. Vulfson, A new method for the introduction of recognition site functionality into polymers prepared by molecular inprinting synthesis and characterization of polymeric receptors for cholesterol, J. Am. Chem. Soc. 117 (1995) 7105–7111.
[26] D.J. O'Shannessy, B. Ekberg, K. Mosbach, Molecular imprinting of amino acid derivatives at low temperature (0°C) using photolytic homolysis of azobisnitriles, Anal. Biochem. 177 (1989) 144–149.
[27] S.S. Milojkovi, D. Kostoski, J.J. Comor, J.M. Nedeljkovi, Radiation induced synthesis of molecularly imprinted polymers, Ploymer 38 (1997) 2853–2855.
[28] T.L. Panasyuk, V.M. Mirsky, S.A. Piletsky, Electropolymerized molecularly imprinted polymers as receptor layers in a capacitive chemical sensors, Anal. Chem. 71 (1999) 4609–4613.
[29] J.F. He, Q.H. Zhu, Q.Y. Deng, Investigation of imprinting parameters and their recognition nature for quinine-molecularly imprinted polymers, Spectrochim. Acta Part A 67 (2007) 1297–1305.
[30] H. Yavuz, V. Karakoç, D. Türkmen, R. Say, A. Denizli, Synthesis of cholesterol imprinted polymeric particles, Intern. J. Biol. Macromol. 41 (2007) 8–15.
[31] Z.Q. Chen, J.P. Li, Y.P. Li, A strategy for constructing sensitive and renewable molecular imprinted carbon paste sensor, Chinese J. Anal. Chem. 39 (7) (2011) 1009–1014.
[32] L. Schweitz, L.I. Andersson, S. Nilsson, Capillary electrochromatography with molecular imprint-based selectivity for enantiomer separation of local anaesthetics, J. Chromatogr. A 792 (1997) 401–409.
[33] L. Ye, P.A.G. Cormack, K. Mosbach, Molecular imprinting on microgel spheres, Anal. Chim. Acta 435 (2001) 187–196.
[34] A.G. Mayes, K. Mosbach, Molecularly imprinted polymer beads: suspension polymerization using a liquid perfluorocarbon as the dispersing phase, Anal. Chem. 68 (1996) 3769–3774.
[35] J. Haginaka, H. Takekira, K. Hosoya, N. Tanaka, Uniform-sized molecularly imprinted polymer for (S)-naproxen selectively modified with hydrophilic external layer, J. Chromatogr. A 849 (1999) 331–339.
[36] C.J. Tan, Y.W. Tong, Preparation of superparamagnetic ribonuclease a surface-imprinted submicrometer particles for protein recognition in aqueous media, Anal. Chem. 79 (2007) 299–306.
[37] M.A. Markowitz, P.R. Kust, J. Klaehn, G. Deng, B.P. Gaber, Surface-imprinted silica particles: the effects of added organosilanes on catalytic activity, Anal. Chim. Acta 435 (2001) 177–185.
[38] A.K. Patel, P.S. Sharma, B.B. Prasad, Voltammetric sensor for barbituric acid based on a sol–gel derivated molecularly imprinted polymer brush grafted to graphite electrode, Int. J. Pharm. 371 (2009) 47–55.
[39] X.L. Xu, G.L. Zhou, H.X. Li, Q. Liu, S. Zhang, J.L. Kong, A novel molecularly imprinted sensor for selectively probing imipramine created on ITO electrodes modified by Au nanoparticles, Talanta 78 (2009) 26–32.

[40] Y.K. Lu, X.P. Yan, An imprinted organic-inorganic hybrid sorbent for selective separation of cadmium from aqueous solution, Anal. Chem. 76 (2004) 453–457.

[41] M.J. Whitcombe, L. Martin, E.N. Vulfson, Predicting the selectivity of imprinted polymers, Chromatographia 47 (1998) 457–464.

[42] M.I. Page, W.P. Jencks, Entropic contributions to rate accelerations in enzymic and intramolecular reactions and the chelate effect, Proc. Natl. Acad. Sci. USA 68 (1971) 1678–1683.

[43] W.P. Jencks, On the attribution and additivity of binding energies, Proc. Natl. Acad. Sci 78 (1981) 4046–4050.

[44] D.H. Williams, J.P.L. Cox, A.J. Doig, M. Gardner, U. Gerhard, P.T. Kaye, A.R. Lal, I.A. Nicholls, C.J. Salter, R.C. Mitchell, Toward the semiquantitative estimation of binding constants. Guides for peptide-peptide binding in aqueous solution, J. Am. Chem. Soc. 113 (1991) 7020–7030.

[45] I.A. Nicholls, Thermodynamic considerations for the design of and ligand recognition by molecularly imprinted polymer, Chem. Lett. 24 (1995) 1035–1036.

[46] R.J. Umpleby II, M. Bode, K. D Shimizu, Measurement of the continuous distribution of binding sites in molecularly imprinted polymers, Analyst 125 (2000) 1261–1265.

[47] B. Sellergren, Imprinted chiral stationary phases in high- performance liquid chromatography, Chromatogr. A 906 (2001) 227–252.

[48] R.J. Umpleby, S.C. Baxter, A.M. Rampey, G.T. Rushton, Y. Chen, K.D. Shimizu, Characterization of the heterogeneous binding site affinity distributions in molecularly imprinted polymers, J. Chromatogr. B 804 (2004) 141–149.

[49] R.J. Umpleby II, S.C. Baxter, M. Bode, J.K. Berch Jr., R.N. Shah, K.D. Shimizu, Application of the Freundlich adsorption isotherm in the characterization of molecularly imprinted polymers, Anal. Chim. Acta 435 (2001) 35–42.

[50] S. Börje, Imprinted chiral stationary phases in high-performance liquid chromatography, J. Chromatogr. A. 906 (2001) 227–252.

[51] C. Baggiani, G. Giraudi, C. Giovannoli, C. Tozzi, L. Anfossi, Adsorption isotherms of a molecular imprinted polymer prepared in the presence of a polymerisable template: Indirect evidence of the formation of template clusters in the binding site, Anal. Chim. Acta 504 (2004) 43–52.

[52] S. Vunnum, V. Natarajan, S. Gallant, S. Cramer, Nonlinear multicomponent gradient chromatography in metal affinity system, Sep. Sci. Technol. 33 (1998) 2465–2489.

[53] S.P. Zhang, Y. Sun, Steric mass-action model for dyeligand affinity adsorption of protein, J. Chromatogr. A. 957 (2002) 89–97.

[54] Q. Lia, H. Su, J. Li, T. Tan, Application of surface molecular imprinting adsorbent in expanded bed for the adsorption of Ni^{2+} and adsorption model, J. Environ. Manage. 85 (2007) 900–907.

[55] G. Wulff, Molecular imprinting in cross-linked materials with the aid of molecular templates—a way towards artificial antibodies, Angew. Chem. Int. Ed. Engl. 34 (1995) 1812–1832.

[56] G. Wulff, S. Schauhoff, Selectivity dependence on the arrangement of functional groups versus spatial requirements, Org. Chem. 56 (1991) 395–400.

[57] B. Sellergren, B. Ekberg, K. Mosbach, Molecular imprinting of amino acid derivatives in macroporous polymers: Demonstration of substrate- and enantio-selectivity by chromatographic resolution of racemic mixtures of amino acid derivatives, J. Chromatogr. A. 347 (1985) 1–10.

[58] S. Cosnier, Affinity biosensors based on electropolymerized films, Electroanalysis 17 (2005) 1701–1715.

[59] B. Sellergren, K.J. Shea, Influence of polymer morphology on the ability of imprinted network polymers to resolve enantiomers, J. Chromatogr. A. 635 (1993) 31–49.

[60] K. Yoshizako, K. Hosoya, Y. Iwakoshi, K. Kimata, N. Tanaka, Porogen imprinting effects, Anal. Chem. 70 (1998) 386–389.

[61] L.I. Andersson, R. Müller, G. Vlatakis, K. Mosbach, Mimics of the binding sites of opioid receptors obtained by molecular imprinting of enkephalin and morphine, Proc. Natl. Acad. Sci. USA 92 (1995) 4788–4792.

[62] S.A. Piletsky, E.V. Piletskaya, A.V. Elgersma, K. Yano, I. Karube, Y.P. Parhometz, A.V. El'skaya, Atrazine sensing by molecularly imprinted membranes, Biosens. Bioelectron. 10 (1995) 959–964.

[63] K. Haupt, A. Dzgoev, K. Mosbach, Assay system for the herbicide 2,4-dichlorophenoxyacetic acid using a molecularly imprinted polymer as an artificial recognition element, Anal. Chem. 70 (1998) 628–631.

[64] J.S. Kenneth, A.S. David, S. Boerje, Polymer complements to nucleotide bases. Selective binding of adenine derivatives to imprinted polymers, J. Am. Chem. Soc. 115 (1993) 3368–3369.
[65] C.E. Dalgliesh, The relation between pyridoxin and tryptophan metabolism, studied in the rat, Biochem. J. 52 (1952) 3–14.
[66] J. Berglund, C. Lindbladh, K. Mosbach, I.A. Nicholls, Selection of phage display combinatorial library peptides with affinity for a yohimbine imprinted methacrylate polymer, Anal. Commun. 35 (1998) 3–7.
[67] G. Wuiff, Selective binding to polymers via covalent bonds. The construction of chiral cavities as specific receptor sites, Pure Appl. Chem. 54 (1982) 2093–2102.
[68] Y. Liu, F. Wang, T.W. Tan, M. Lei, Study of the properties of molecularly imprinted polymers by computational and conformational analysis, Anal. Chim. Acta 581 (2007) 137–146.
[69] O. Vogl, J. Bartus, J.R. Murdoch, Solid-state, optical-rotation measurements on macromolecules using powder-suspensions, Monatshefte Chem. 121 (1990) 311–316.
[70] J. Bartuš, O. Vogl, Optical activity measurements in solids 5. Optical rotation of natural polymers, Poly. International 33 (1994) 25–36.
[71] G. Wulff, H. G Poll, M. Minarik, Enzyme-analogue built polymers. XIX: Racemic resolution on polymers containing chiral cavities, J. Liq. Chromotgr. 9 (1986) 385–408.
[72] G. Wulff, Biorecognition in molecularly imprinted polymers. Concept, chemistry and application, in: T.T. Ngo (Ed.), Molecular interactions in bioseparations, Plenum Press, New York (1993), pp. 363–381.
[73] G. Wulff, W. Vesper, R. Grobe-Einsler, A. Sarhan, Enzyme-analogue built polymers, 4. On the synthesis of polymers containing chiral cavities and their use for the resolution of racemates, Die Makromolekulare Chemie 178 (1977) 2799–2816.
[74] G. Vlatakis, L.I. Andersson, R. Müller, K. Mosbach, Drug assay using antibody mimics made by molecular imprinting, Nature 361 (1993) 645–647.
[75] S.H. Cheong, S. McNiven, A. Rachkov, R. Levi, K. Yano, I. Karube, Testosterone receptor binding mimic constructed using molecular imprinting, Macromolecules 30 (1997) 1317–1322.

CHAPTER

3

The Fabrication and Development of Molecularly Imprinted Polymer-based Sensors for Environmental Application

Shunsheng Cao[1,2], *Juanrong Chen*[3], *Weicheng Sheng*[1], *Weiwei Wu*[1], *Zhiyuan Zhao*[1], *Fang Long*[1]

[1]School of Materials Science and Engineering, Jiangsu University, China

[2]Cranfield Health, Cranfield University, Cranfield, Bedfordshire, UK

[3]School of Environment, Jiangsu University, China

OUTLINE

Molecularly Imprinted Sensors.
DOI: 10.1016/B978-0-444-56331-6.00003-7

1. INTRODUCTION

Nanoscience is exhibiting more and more significance in material science and technology development. Biosensors and chemical sensors are advanced devices resulting from the smart integration of nanotechnology and nanomaterials, and have attracted great attention in the field of modern analytical chemistry, which requires fast ultrasensitive detection methods to respond to the new demands [1–5]. Due to their ultimately controllable size, high surface area, and simulation capacity, they are becoming powerful tools and platforms for making ultra-small, sensitive, and selective devices that offer us new phenomena to tackle key clinical, medical, chemical, pharmaceutical, biological, and environmental issues, and the like. The central part of such sensors is the chemical or biological molecular recognition elements, also named receptors, which can recognize and bind the target analyte specifically [1,2].

A variety of biological recognition elements, including microorganisms, antibodies, enzymes, tissues, DNA probes, and cells are often used as molecular recognition materials in biosensor technology due to their extremely high recognition specificity [2,6,7]. The receptor is in close contact with an interrogative transducer that converts recognizing and binding in an often complex sample [1,2]. One typical example is the immunosensor that involves the use of antibody as a biosensing recognition element, in which reaction takes place between a target analyte and a specific antibody [8]. However, requiring a complex protocol and harsh environmental process are its main challenges for the practical application of such a device because of the poor stability of biomolecule [1,8]. Therefore, it is urgent that the demand for synthesizing artificial recognition receptors should be satisfied in order to overcome these disadvantages of the biological recognition element.

One promising strategy is molecular imprinting technology, which offers a synthetic route to artificial recognition systems by a template polymerization technique, in which they can bind target molecules with similar affinity and specificity to those of their natural counterparts so as to mimic biological receptors [1,8–11]. The technique is simple and cheap, and is more resistant to degradation in harsh environments such as in acids and bases or at high temperatures and pressures than are its biological counterparts [1,8,12]. Furthermore, the introduction of synthetic design into the molecular imprinting method can even make a host element suitable for the analyte, for which the natural receptor does not exist [8]. Since Wulf et al. [13] first described the experiment on the preparation of synthetic polymers with receptor properties for sugar derivatives, many papers have been published on using various kinds of substances, such as sugars, amino acids, peptides, drugs, steroids, metal ions, and dyes as imprinted elements and have achieved great success [1,14–19].

Currently, molecular imprinting polymers are a hot topic in analyte detection and have significant advantages, including superior stability, low cost, and ease of preparation. Most importantly, they can effectively replace natural receptors as the detection component of selective analyte [20–24]. In the last decade, significant efforts have been made in the research and development of molecular imprinting polymers (MIPs). Such sensors now have seen increasing use as recognition elements in analyte detection, as shown in various publications, including Alizadeh et al. [25] Hong et al. [26], Riskin et al. [27], Neto et al. [6], and Michael et al. [28]. Recent extensive research into the design, synthesis, and application of MIPs has led to numerous practical and conceptual developments in that direction, as reflected in several reviews [1,14,29–31]. The review focuses mainly on discussion of some recent advances in a comprehensive way and presents an overview of the synthetic strategies used to fabricate MIPs. In addition, we address works related to their progress in environmental application.

2. GENERAL DESCRIPTION OF MOLECULAR IMPRINTING TECHNOLOGY

As mentioned earlier, biological receptors, such as enzymes, antibodies, and DNA, may suffer from many inherent limitations when they are used as chemical or biochemical recognition elements in sensors and assays due to their poor reproducibility, instability, and problems associated with sterilization. By contrast, synthetic receptors may exhibit a promising alternative to biological receptors for use in biosensors [31,32]. The most typical and promising method for developing synthetic receptors with high affinity and specificity is the molecular imprinting technique [31,33]. In general, this technique can be defined as the formation of specific recognition sites (with binding or catalytic properties) in a polymer through its interaction with a specific template (imprint molecule), where the template directs and reflects the positioning and orientation of the polymer's structural components, namely, as an analyte structure [31].

In detail, this technique involves copolymerization of functional monomer and cross-linking monomers in the presence of target analyte, which acts as an imprint template. The functional monomers initially form a complex with the "imprint" molecule, and following polymerization, their functional groups are held in position by the highly cross-linked structure. After template removal, the polymer can be used as a selective binding medium for the imprint molecule or other structurally related molecules [1,14,31,34]. In that way, a molecular memory is introduced into the polymer, which has a complementary shape and site [35] (i.e., specific recognition site [36] for the template molecule) and can be fabricated in a variety of configurations that are adapted for the structure of target molecule [14,20]. Consequently, MIPs are capable of rebinding and recognizing the analyte with high affinity and specificity and can be used as specific absorbents in separation and recognition techniques [14,34,37,38]. For instance, MIP-based electrochemical sensors can either separate active species to improve the selectivity of the response or accumulate a target analyte on the electrode surface. It has emerged as a good method for on-site selective detection various molecular compounds, as reported by Hrapovic et al. [39], Chrtien et al. [40], Fang et al. [41], Zhang et al. [42], Liang et al. [43], and others. On the other hand, MIP-based biosensors can provide faster analysis with a direct readout without the need to transport samples to a laboratory, in which the recognition element is a molecularly imprinted receptor rather than a biomolecule (i.e., proteins, amino acids, and carbohydrates) [31]. Therefore, it can offer many advantages, such as long-term storage stability, potential reusability, resistance to microbial spoilage, and custom synthesis of selective receptors, and become increasingly more important in medical diagnostic devices, as introduced by Wu et al. [44], Li et al. [20], Zhang et al, [45], and Whitcombe et al. [31]. Although the synthesis strategies of MIPs are simple and cheap, as well as the MIPs obtained offer high selectivity, excellent mechanical strength, and durability to heat, acid, and base conditions [46,47], the synthetic efficiency, high capacity, and specificity of MIPs are still looking for purposes in the imprinting process.

3. SYNTHETIC STRATEGIES FOR MOLECULAR IMPRINTING POLYMERS

Molecular imprinting polymers (MIPs), as promising recognition elements for sensors, have increasingly attracted considerable attention. In MIPs, the specific recognition systems can be spatially organized by positioning functional monomers around the template molecules in a polymeric organic/inorganic matrix through the interactions (i.e., noncovalent and/or covalent interactions) between template and functional monomer [41]. Obviously, their effectiveness depends mainly on the bond nature of the template-monomer complex, the form of imprinted polymer matrix, and the rigidity of the polymeric materials [48–53]. In general, there are three different synthetic strategies for molecular imprinting: noncovalent, covalent, and semi-covalent imprinting, depending on the interaction of the template and functional monomer during both the pre-polymerization and rebinding process [54]. In all these protocols, a template molecule interacts with a functional monomer to create specific recognition sites [14,55]. Since Mosbach et al. [56] first pioneered noncovalent imprinting for the preparation of artificial receptors, MIPs prepared using the above technique have been widely employed for recognizing a variety of molecules, including nucleotides, morphine, hormones, and drugs [57–60]. The noncovalent interaction, such as hydrogen bonding, electrostatic interaction, hydrophobic effect, and van der Waals forces, between substrate and polymer, is used to organize/orient the functional monomers around the template; after polymerization and removal of the template, the functional groups of the MIP matrix can then rebind the target via the same noncovalent interactions [56,61] depicted with many details in Fig. 3.1.

To date, this method is the most widespread in the preparation and application of MIPs, due to the relative simplicity of the synthesis process and the wide range of functionalities [54]. In nature, this technique is also more similar to natural processes in the sense that most biomolecular interactions are noncovalent in nature [14]. Although the noncovalent imprinting of various templates has been extensively reported, it often suffers from several limitations related to heterogeneity in the imprinted sites, especially that the imprinting molecule and target must form a number of noncovalent intermolecular interactions to produce sufficient binding pockets during polymerization [61–63]. Thus, the noncovalent approach cannot be employed for template molecules that lack enough functional groups responsible for strong noncovalent interactions. These difficulties can be effectively overcome by the covalent imprinting method, which utilizes reversible covalent bonding formed between the functional monomer and the imprinted molecule [62], for example, the bulk polymer monoliths were crushed in a mortar and ground in a ball mill and acetone was used to remove the fines. After removal of the template, MIPs were formed; this is more clearly illustrated in Fig. 3.2 [64]. This covalent imprinting technique has various advantages, including complementarities in shape to the imprint molecule, and leads to strong interactions as a result of reformation of the covalent bond between the matrix and the target in comparison with the noncovalent technique. The covalent method is limited, however, by the relatively small number of useful reversible

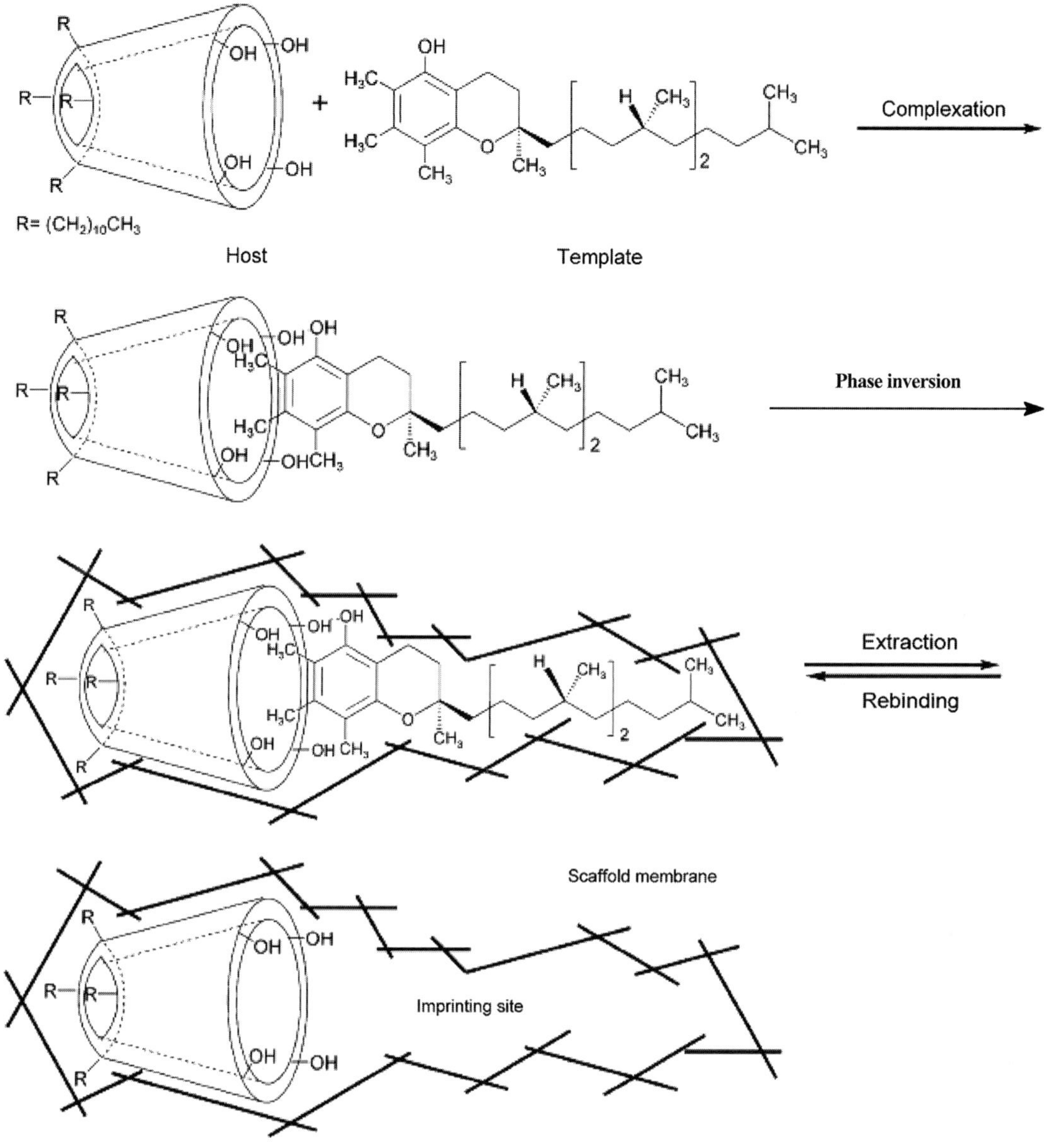

FIGURE 3.1 Illustration of preparation of MIPs via the multiple noncovalent molecular-imprinted technique. *(Reproduced with permission from [61])*

covalent bonding reactions that can be utilized [62,65,66]. This disadvantage makes it impractical for the vast majority of applications in MIPs.

A promising option is referred as semicovalent imprinting. This approach involves all imprinting methods in which a covalent or partly covalent template structure is used in the polymerization step, but the binding and rebinding step is entirely noncovalent in nature

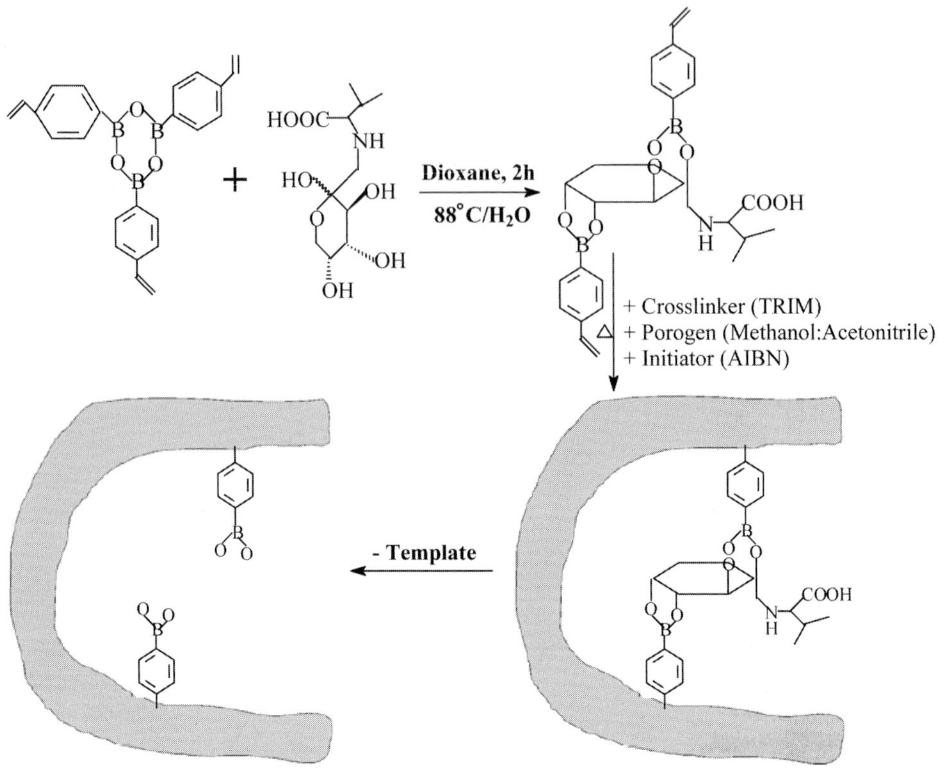

FIGURE 3.2 Generation of fructosyl valine binding polymers by covalent imprinting. *(Reproduced with permission from [64])*

[33,54]. Concretely, this strategy depends mainly on the polymerization of a template that is covalently bound to a functional monomer by a cleavable bond. Following copolymerization, the template is cleaved and leaves an imprint bearing functional groups, which are capable of interacting with the template in a noncovalent sense in the rebinding step [67–69]. The typical preparation is demonstrated in Fig. 3.3. A series of MIPs were synthesized by using semicovalent imprinting with sacrificial spacer [54,67,68]. However, the preparation of highly selective MIP-based receptors with homogeneous binding sites is yet a long standing challenge [68,70]. To overcome these limitations, considerable efforts have been made to develop other polymerization techniques and new imprinting approaches. As a result, several polymerization methods, such as precipitation, emulsion polymerization, sol-gel technique, and electro-assisted deposition, have optimized existing molecular imprinting purposes and have made possible the production of MIPs with better properties [48,71–75].

3.1. Traditional Imprinting Methods

The most common method of preparing MIPs is bulk polymerization, in which all components are polymerized in a mold. Functional monomers and a high concentration

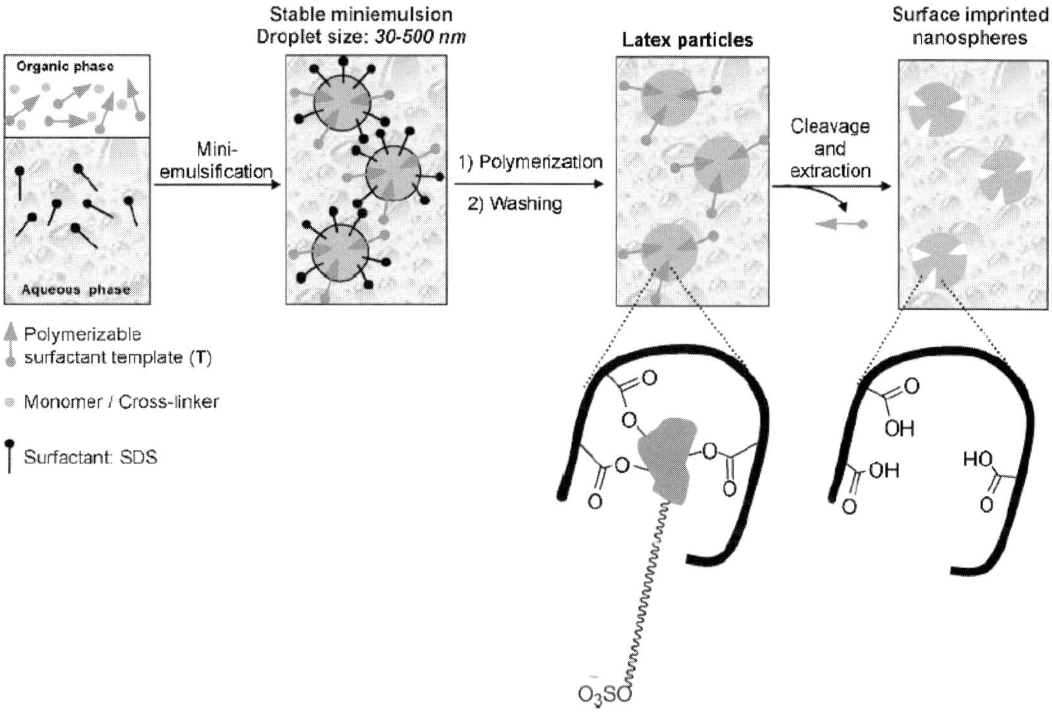

FIGURE 3.3 Schematic representation of the surface semicovalent imprinting approach by mini-emulsion polymerization. *(Reproduced with permission from [68])*

of cross-linker are typically polymerized in the presence of the template molecules. Subsequent removal of templates from the polymeric matrix leaves behind binding sites that are complementary to the size, shape, and functionality of template molecules [76–79]. This traditional imprinting technique is quite simple and easy to carry out in a tailor-made fashion, and all we need are template molecules, functional monomers, solvents, and cross-linking agents. During these processes, a number of functional monomers are assembled in an orderly fashion, with their specific groups oriented toward the desired sites within the cavities of the desired size and shape [26]. In this manner, the MIPs have generally been based on three steps: First, the template molecule interacts with the functional monomer(s) in a predetermined orientation; Subsequently, the resulting polymer is copolymerized in the presence of a large excess of cross-linker, resulting in the formation of a rigid polymeric matrix with the print molecule in a sterically fixed arrangement; Finally, the template molecule is removed from this polymer matrix, generating binding sites with a shape and functional group complementarity to the template molecule, which are able to specifically rebind the same template [29,80]. The production process is outlined in Fig. 3.4. Although conventional methods have been widely used to develop various MIPs with different physical configurations, and have achieved great success [80–83], MIPs prepared through traditional bulk polymerization usually produce deeply embedded binding sites and suffer from some drawbacks, such as slow mass transfer,

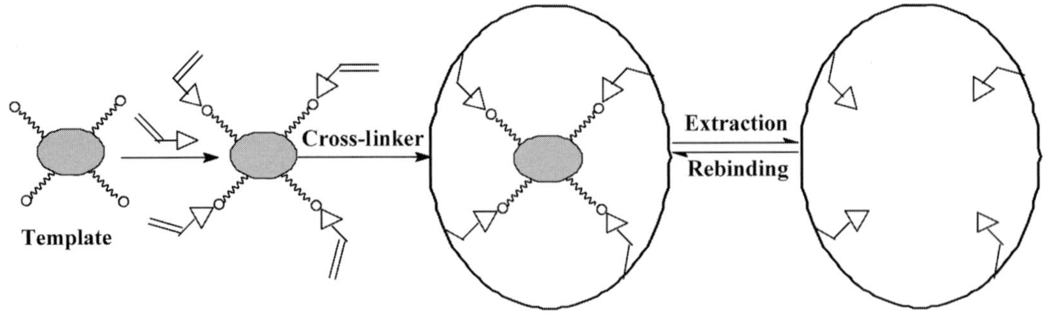

FIGURE 3.4 Schematic representation of MIPs. *(Reproduced with permission from [89])*

moderate sensitivity and selectivity, incomplete template removal, and broad guest affinity [77,78,84–86]. Recently, new polymerization strategies, in which the growing polymer chains exceed the limit of solubility, leading to phase separation during the cross-linking process, have been proposed in dealing with obtaining imprinted materials in order to improve the chemical detection of MIPs. Among these strategies, precipitation polymerization seems to be one of the simplest and most well-suited methods to obtain spherical particles with the desired characteristics, as introduced by Cormack et al. [87], Wang et al. [88], Tamayo et al. [79], and Morelli et al. [80].

3.2. The Sol-gel Technique

In nature, self-assembly is the autonomous organization of components into patterns or structures without human intervention [90,91]. This technique reflects information coded (as shapes, surface properties, charge, polarizability, magnetic dipole, mass, etc.) in individual components; these characteristics determine the interactions among them [92]. As a promising technique, self-assembled monolayers (SAMs) have several attractive features and can play a pivotal role in fabricating the imprinted matrix materials [84,93–95]. Since F.H. Dickey [96] first introduced the demonstration of molecular imprinting in sol-gel matrix, sol-gel imprinting has been developed toward the design of nanostructured materials and solid-state engineering through "bottom-up" procedures [71]. The recognition event may be set up via various reversible attractive intermolecular interactions, such as electrostatic, p–p stacking, van der Waals, hydrogen bonding, and coordination bonding. Sol-gel MIP materials are inorganic (siloxane) based polymer matrix produced by an acid or base-catalyzed hydrolysis and condensation of a series of silane monomers with excellent physical rigidity, chemical inertness, and thermal stability properties, as shown in Fig. 3.5 [48,97,98].

As a convenient and versatile method, sol-gel imprinting has drawn growing interest in molecular imprinting in polymer matrices that may prompt the tailoring of unique MIP materials with controllable pore size, structural rigidity, thermal stability, and enhanced recognition performs [71,99–101]. Therefore, this process is able to entrap a variety of organic, organometallic, and biological molecules (microorganisms, antibodies, DNA, proteins, and enzymes) within a meso/microporous polymeric matrix. The dopant

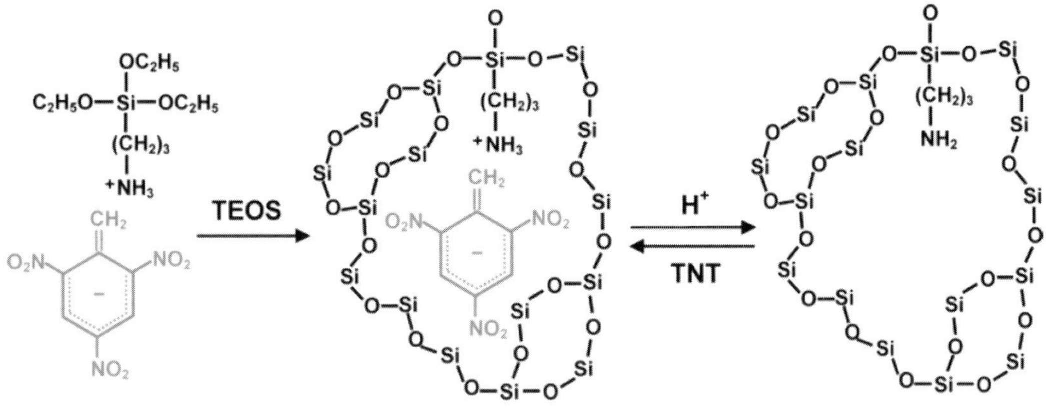

FIGURE 3.5 Schematic illustration for the molecular imprinting mechanism of TNT in a silica matrix through the anion–cation pair and gelation reaction. *(Reproduced with permission from [48])*

encapsulated inside the sol-gel matrix can interact with other molecules or surrounding physical/chemical groups. Importantly, these interactions are easily monitored/detected by appropriate optical methods (absorption and fluorescence). Despite the combination of the molecular imprinting with sol-gel technique, it is ideally propitious to construct MIP-based sensing devices, such a method suffers from several challenges, such as slow diffusion kinetics, long response time, and low sensitivity [97,102–104]. Therefore, to prepare MIP-based sensing materials, the well-design of the signal transducer via sol-gel imprinting remains challenging to enhance the sensitivity of this sensing system [97,105].

3.3. Electro-assisted Deposition

In comparison to other synthetic strategies for the preparation of MIPs, one of the outstanding advantages of electro-assisted deposition is that there is no need to remove the template by the extraction process, because the integration of MIPs with sensors can be easily obtained through *in situ* polymerization, by either grafting with chemical or UV initiation or by photochemical or thermal initiation. The preparation method is briefly demonstrated in Fig. 3.6 [1,53,106,107]. The advantages of electropolymerization lie in the possibility of monitoring and regulating the thickness and density of the MIP layer by modifying polymerization conditions, such as current density and applied voltage, resulting in a uniform coating of the polymer on the electrode surface [1,31,108]. Moreover, this imprinting can offer ease of film-thickness control, fast mass transfer, and rapid response to templates [77]. Evidently, these advantages enable the formation of controlled polymer films on the electrode surface and achieve far more acceptable reproducibilities [108]. Consequently, different cases of an electrosynthesized polymer molecularly imprinted by various neutral templates have been widely explored in biological, medical, and environmental analyses, as reported by Agostino et al. [109], Diltemiz et al. [46], González et al. [110], Tan et al. [77], Alberto et al, [108], and Li et al. [73]. However, this method shows a number of perceived limitations associated with the development

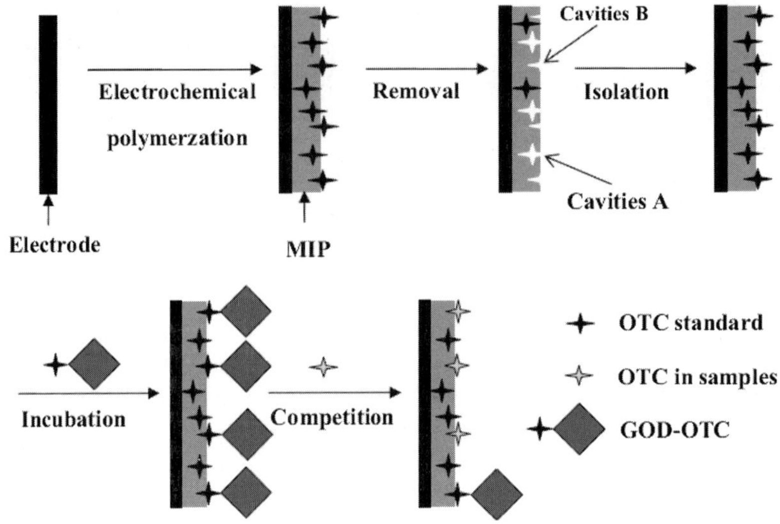

FIGURE 3.6 Scheme of the molecular imprinting technique via electrochemical polymerization. *(Reproduced with permission from [73])*

of MIP sensors: (1) difficulty in integrating them with a transducer and (2) difficulty in transforming the binding event into an electric signal [31]. On the other hand, the function of the transducer is important in the MIP sensing technology because it translates the signal, which was generated upon binding the analyte with the recognition element, into a quantifiable electrical output [1]. To date, electronic nanosensors capable of recognizing proteins or enzyme continue to be a challenge to perform, in part because (1) the MIP film may attenuate signals generated in response to template binding (due to the large thickness); (2) the detection mechanisms do not readily allow for effective signal conversion of template molecule binding; and (3) the sensor platforms do not support highly sensitive detection [72,84,111,112]. Therefore, further research in the field of electrochemical synthesis of MIP-based sensors is focused mainly on following objectives: (a) to achieve high reproducibility in the formation of electrode surface, (b) to obtain stable polymerization, which enables a suitable durability of the sensors, and (c) to enhance the electrode surface in order to achieve greater high mass-transportation efficiency [108,113].

4. ENVIRONMENTAL APPLICATION OF MIPs

The continuous addition of industrial, municipal, and agricultural effluents to the environment has resulted in a considerable increase in toxic contaminants, especially in surface water and groundwater [114,115]. A lot of these substances, such as organic pesticides, heavy metals, phenol compounds, and other pollutants, are hazardous to animals, plants, microorganisms, and other living organisms, and may result in a series of health problems for humans who drink water with such pollutants [103,114]. Therefore, the efficient detection of contaminants in waste streams has created increasing environmental concern due to their acute toxicity. Many analytical methods, such as high-performance liquid chromatography (HPLC), capillary electrophoresis (CE), and gas chromatography (GC), have been widely developed to determine these compounds [116–118]. These techniques, however, do not easily allow continuous on-site monitoring and are both expensive and time-consuming, require highly trained personnel, and sometimes suffer from complicated sample pretreatment steps [3,119,120,121]. Therefore, the development of new, reliable, reproducible, and inexpensive methods of the express-analysis of various toxic compounds in both the environment and foodstuffs is of urgent concern to analytic researchers. MIP-based sensor technology with high selective recognition and affinity for continuous detection in environmental and health care applications has been widely employed because it can provide on-site, real-time detection and quantification of beneficial and toxic compounds. In this chapter, several typical, widely used toxic pollutants have been selected as examples to describe and discuss MIPs for environmental monitoring.

4.1. Phenolic Compounds

Phenols are one of the most important and widely used industrial chemicals. They are present in the products of petroleum thermal decomposition; widely used in drugs, plastics, dyes, disinfectants, synthetic fibers, antioxidants, plasticizers, paper pulp, and pesticides; and directly released into the environment as waste [3,120,122]. Many publications have reported the existence of phenols as pollutants of the air, water, and soil. In nature, they are a class of polluting chemicals and are easily adsorbed by animals, microorganisms, and other living organisms, including humans, through the skin and mucous membranes [3,122]. These pollutants result in disorders of the nervous system, respiration, and blood circulation, as well as irritate the mucous membranes of the eyes and respiratory tract [123,124]. According to widely established occupational safety and health standards, the maximum permissible levels for phenols in water and air vary in a wide range: 0.0005–0.05 mg/l and 0.001–0.05 mg/m^3, respectively [123,125].

Traditionally, the separation and determination processes for low levels of phenol depend mainly on a suitable utilization of available separation/pre-concentration material and techniques. However, the analysis of phenols in complex matrices at a low level of concentration usually requires a sample pretreatment procedure for the accurate detection and analysis of the analytes in trace levels [126,127]. It is universally accepted that solid phase extraction (SPE) is an efficient method for extracting toxic compounds from liquid or solid matrices. The classical SPE sorbents retain analytes through nonselective hydrophobic or polar interactions that result in a partial co-extraction of interfering substances [126]. However, this method requires large and expensive tri-chlorophenol in solid phase extraction and chromatography [128]. To trace the enrichment of analytes and improve the selectivity of the extraction, the development of the express methods of phenol detection is of great importance. MIP is a powerful tool for selective SPE of target compounds from complex matrices due to the selectivity of a MIP for a target compound or a group of compounds [3,126,121]. It has been extensively developed, as introduced by Piletsky et al. [129], An et al. [127], Qi et al. [126], and Sergeyeva et al. [123] Unfortunately, the traditional format of MIPs obtained by the grinding of *in situ* polymerized polymeric blocks or by precipitation/suspension polymerization is not easily compatible with a majority of biotechnological and sensor applications. This explains the growing interest in novel MIP formats [123].

4.2. Nitroaromatic Compounds

Due to the health, ecological, and security needs caused by long- and short-term exposure to explosive compounds, increasing interest has arisen in monitoring environmental samples. Considerable efforts have been devoted to the development of innovative and effective sensors, capable of detecting explosives rapidly in the low-concentration and complex matrices [8,27]. Therefore, sensor researchers and technologists are in a race to develop a robust sensor for detecting nitroaromatic compound-based explosives not only in the vapor phase but also in aqueous solutions [130]. The highly sensitive nature of sensors in the detection of explosives has attracted substantial efforts and attention to various analytical procedures, such as optical, electrochemical, surface acoustic wave, or competitive immunoassay methods [14,28,131]. Among these procedures, MIPs have received increasing interest and been widely used to measure organic compounds, especially explosives such as trinitrotoluene (TNT), dinitrotoluene (DNT), and dinitrodimethylbutane (DMNB) [132–134]. Some chemical sensors, quartz crystal microbalances (QCMs) sensor and electrochemical sensors, designed for TNT determination based on the MIP technology, can be found in the literature [8,135,136]. Although these works are interesting in particular with respect to the reported low-detection limit and high selectivity, their complex instrumentation requirement and complicated sensor preparation protocol, together with the high cost of the methods, are the main disadvantages of these works [8]. In this regard, Alizadeh et al. reported [8] a new, cheap, and simple pathway to design and develop the high selective and sensitive electrochemical sensor for TNT determination at sub-ppb quantities. Therefore, considerable endeavors will be devoted to develop future sensing devices, consolidating good selectivity, fast speed, simplicity, and cost effectiveness, together with a high sensitivity for detecting nitroaromatic compounds.

5. CONCLUSIONS

In the past several decades, MIP-based sensors have attracted considerable attention in the analytical chemistry field and have achieved great success because of their high selectivity, sensitivity, and robustness. This chapter has summarized the mainly synthetic strategies and their environmental applications. In addition, other promising methods, such as suspension, dispersion, precipitation, and controlled/living radical polymerization, have also been used to prepare optimized MIPs. As evidenced by the literature described and discussed in this chapter, a basic understanding of the development of MIPs has enabled the fabrication and preparation of a variety of novel MIP-based sensors with high selectivity, sensitivity, and stability. However, researchers should endeavor to overcome a series of limitations, including small binding capacity, slow binding kinetics, specially designed functional monomers, and the irregular materials shape of MIPs. These researchers should make full use of these existing strategies for building a novel MIP-based sensor platform, which will encourage the continued evolution of these processes and provide more facile synthetic approaches for the development of ideal imprinted materials.

Acknowledgments

The authors would like to thank the Research Directorate-General of European Commission for financial support to carry out this work under the framework of Marie Curie Action FP7-People-IIF-2010 (No. 275336). Clearly, the success of this work has largely depended on the creativity and devotion of many researchers whose names are listed in the References.

References

[1] V. Suryanarayanan, C. Wu, K. Ho, Electroanalysis 22 (2010) 1795.
[2] A.S. Belmont, S. Jaeger, D. Knopp, R. Niessner, G. Gauglitz, K. Haupt, Biosens. Bioelectron 22 (2007) 3267.
[3] S. Cao, J. Chen, X. Jin, W. Wu, Z. Zhao, Enzyme-based biosensor: Synthesis and Applications, in: Li Songjun, Singh Jagdish et al. (Eds.), Biosensor Nanomaterials, Wiley-VCH Publisher, 2011. Chapter 5, 95–115.
[4] D. Shan, M. Zhu, H. Xue, S. Cosnier, Biosens. Bioelectron 22 (2007) 1612.
[5] M.M. Cheng, G. Cuda, Y.L. Bunimovich, M. Gaspari, J.R. Heath, H.D. Hill, C.A. Mirkin, A.J. Nijdam, R. Terracciano, T. Thundat, M. Ferrari, Curr. Opin. Chem. Biol. 10 (2006) 11.
[6] J.R.M. Neto, W. de, J.R. Santos, P.R. Lima, S.M. Tanaka, A.A. Tanaka, L.T. Kubota, Sens. Actuators B: Chem. 152 (2011) 220–225.
[7] S.J. Alcock, A.P.F. Turner (Eds.), Vivo Chemical Sensors: Recent Developments, Cranfield Press, Cranfield, UK, 1993. ISBN 1871315 49 2.
[8] T. Alizadeh, M. Zare, M.R. Ganjali, P. Norouzi, B. Tavana, Biosens. Bioelectron. 25 (2010) 1166.
[9] C.A. Barrios, C. Zhenhe, F. Navarro-Villoslada, D. López-Romero, M.C. Moreno-Bondi, Biosens. Bioelectron 26 (2011) 2801.
[10] Z.Y. Tao, E.C. Tehan, R.M. Bukowski, Y. Yang, E.L. Shughart, W.G. Holthoff, J.L. Urraca, M.D. Marazuela, E.R. Merino, G. Orellana, M.C. Moreno-Bondi, J. Chromatogr, A 1116 (2006) 127.
[11] A.J. Hall, M. Quaglia, P. Manesiotis, E. De Lorenzi, B. Sellergren, Anal. Chem. 78 (2006) 8362.
[12] L. Ye, K. Haupt, Anal. Bioanal. Chem. 378 (2004) 1887.
[13] G. Wulf, A. Sarhan, Angew, Chem. Int. Ed. Eng. 11 (1972) 341.
[14] K. Haupt, K. Mosbach, Chem. Rev. 100 (2000) 2495.
[15] C. Manitesta, I. Losito, P.G. Zambonin, Anal. Chem. 71 (1999) 1366.
[16] A.L. Jenkins, O.M. Uy, G.M. Murray, Anal. Chem. 71 (1999) 373.
[17] S. Kröger, A.P.F. Turner, K. Mosbach, K. Haupt, Anal. Chem. 71 (1999) 3698.
[18] H.S. Ji, S. McNiven, K.H. Lee, T. Saito, K. Ikebukuro, I. Karube, Biosens. Bioelectron 15 (2000) 403.
[19] S.A. Piletsky, S. Subrahmanyam, A.P.F. Turner, Sens. Rev. 21 (2001) 292.
[20] J. Li, J. Zhao, X. Wei, Sens. Actuators B: Chem. 140 (2009) 663.
[21] A. Gomez-Caballero, N. Unceta, M. Aranzazu, J. Goicolea, Ramon-Barrio, Sens. Actuators B: Chem. 130 (2008) 713.
[22] K. Prasad, K.P. Prathish, J. Mary Gladis, G.R.K. Naidu, T. Prasada Rao, Sens. Actuators B: Chem. 123 (2007) 65.
[23] D.Q. Pan, J.H. Chen, L.H. Nie, W.Y. Tao, S.Z. Yao, Anal. Bioanal. Chem. 324 (2004) 115.
[24] G.Z. Hu, Y.C. Liu, J. Zhao, S.Q. Cui, Z.S. Yang, Y.Z. Zhang, Bioelectrochemistry 9 (2006) 254.
[25] T. Alizadeh, M.R. Ganjali, P. Norouzi, M. Zare, A. Zeraatkar, Talanta 79 (2009) 1197.
[26] C. Hong, P. Chang, C. Lin, C. Hong, Biosens. Bioelectron 25 (2010) 2058.
[27] M. Riskin, R. Tel-Vered, I. Willner, Adv. Mater. 22 (2010) 1387.
[28] M. Riskin, Y. Ben-Amram, R. Tel-Vered, V. Chegel, J. Almog, I. Willner, Anal. Chem. 83 (2011) 3082.
[29] C.J. Stephenson, K.D. Shimizu, Polym. Int. 56 (2007) 482.
[30] K.D. Shimizu, C.J. Stephenson, Curr. Opin. Chem. Biol. 14 (2010) 743.
[31] M.J. Whitcombe, I. Chianella, L. Larcombe, S.A. Piletsky, J. Noble, R. Porter, A. Horgan, Chem. Soc. Rev. 40 (2011) 1547.
[32] A.P.F. Turner, B.N. Chen, S.A. Piletsky, Clin. Chem. 45 (1999) 1596.
[33] A.G. Mayes, M.J. Whitcombe, Adv. Drug. Delivery Rev. 57 (2005) 1742.
[34] G. Evtugyn, A. Porfireva, A. Ivanov, O. Konovalova, T. Hianik, Electroanalysis 21 (2009) 1272.
[35] K. Hirayama, Y. Sakai, K. Kameoka, K. Noda, R. Naganawa, Sens. Actuators B: Chem. 86 (2002) 20.
[36] M. Yoshikawa, N. Hotta, J. Kyoumura, Y. Osagawa, T. Aoki, Sens. Actuators B: Chem. 104 (2005) 282.

[37] E.L. Holthoff, F.V. Bright, Anal. Chim. Acta 594 (2007) 147.
[38] V.B. Kandimalla, H. Ju, Anal. Bioanal. Chem. 380 (2004) 587.
[39] S. Hrapovic, E. Majid, Y. Liu, K. Male, J.H.T. Luong, Anal. Chem. 78 (2006) 5504.
[40] J.M. Chrétien, M.A. Ghanem, P.N. Bartlett, J.D. Kilburn, Chem. Eur. J. 15 (2009) 11928.
[41] C. Fang, C. Yi, Y. Wang, Y. Cao, X. Liu, Biosens, Bioelectron 24 (2009) 3164.
[42] Z. Zhang, Y. Hu, H. Zhang, L. Luo, S. Yao, Biosens, Bioelectron 26 (2010) 696.
[43] Y. Liang, L. Gu, X. Liu, Q. Yang, H. Kajiura, Y. Li, T. Zhou, G. Shi, Chem. Eur. J. 17 (2011) 5989.
[44] Z. Wu, X. Hu, C. Tao, Y. Li, J. Liu, C. Yang, D. Shen, G. Li, J. Mater. Chem. 18 (2008) 5452.
[45] W. Zhang, X. He, Y. Chen, W. Li, Y. Zhang, Biosens, Bioelectron 26 (2011) 2553.
[46] S.E. Diltemiz, D. Hür, A. Ersoz, A. Denizli, R. Say, Biosens, Bioelectron 25 (2009) 599.
[47] M. Ávila, M. Zougagh, Á. Ríos, A. Escarpa, Trends Anal. Chem. 27 (2008) 54.
[48] C. Xie, B. Liu, Z. Wang, D. Gao, G. Guan, Z. Zhang, Anal. Chem. 80 (2008) 437.
[49] S.C. Zimmerman, M.S. Wendland, N.A. Rakow, I. Zharov, K.S. Suslick, Nature 418 (2002) 399.
[50] E. Mertz, S.C. Zimmerman, J. Am, Chem. Soc. 125 (2003) 3424.
[51] C.D. Ki, C. Oh, S.G. Oh, J.Y. Chang, J. Am, Chem. Soc. 124 (2002) 14838.
[52] R.H. Schmidt, K. Mosbach, K. Haupt, Adv. Mater. 16 (2004) 719.
[53] M.M. Titirici, B. Sellergren, Chem. Mater. 18 (2006) 1773.
[54] P. Qi, J. Wang, L. Wang, Y. Li, J. Jin, F. Su, Y. Tian, J. Chen. Polymer 51 (2010) 5417.
[55] X. Jiang, C. Zhao, N. Jiang, H. Zhang, M. Liu, Food Chem 108 (2008) 1061.
[56] R. Arshady, K. Mosbach, Macromol. Chem. Phys. 182 (1981) 687.
[57] M. Kempe, K. Mosbach, J. Chromatogr. A. 691 (1995) 317.
[58] L.I. Andersson, R. Muller, G. Vlatakis, K. Mosbach, Proc. Natl. Acad. Sci. USA 92 (1995) 4788.
[59] R. Kublickas, C. Werner, G. Jariene, B. Voit, L. Lasas, Polym. Bull. 58 (2007) 611.
[60] K. Farrington, F. Regan, Biosens. Bioelectron. 22 (2007) 1138.
[61] C.M. Faizal, Y. Kikuchi, T. Kobayashi, J. Memb. Sci. 334 (2009) 110.
[62] S. Farber, B.S. Green, A.J. Domb. J. Polym. Sci. Part A: Polym. Chem. 47 (2009) 5534.
[63] K.D. Shimizu, Post modification of imprinted polymers, in: M. Yan, O. Ramstrom (Eds.), Molecularly Imprinted Materials Science and Technology, Marcel Dekker, New York, NY, 2005, pp. 285–306.
[64] R. Rajkumar, A. Warsinke, H. Möhwald, F.W. Scheller, M. Katterle, Biosens, Bioelectron. 22 (2007) 3318.
[65] G. Wulff Angew, Chem. Int. Ed. Eng. 34 (1995) 1812.
[66] K.J. Shea, Trends. Polym. Sci. 2 (1994) 166.
[67] J.P. Rosengren-Holmberg, J.G. Karlsson, J. Svenson, H.S. Andersson, I.A. Nicholls, Org. Biomol. Chem. 7 (2009) 3148.
[68] P. Curcio, C. Zandanel, A. Wagner, C. Mioskowski, R. Baati, Macromol. Biosci. 9 (2009) 596.
[69] N. Kirsch, M.J. Whitcombe, The Semi-Covalent Approach, in: M. Yan, O. Ramström (Eds.), Molecularly Imprinted Materials, Marcel Dekker, New York (2005) p. 93.
[70] B. Sellergren, Macromolecules 39 (2006) 6306.
[71] G. Arrachart, C. Carcel, P. Trens, J.J.E. Moreau, M.W. Chi Man, Chem. Eur. J. 15 (2009) 6279.
[72] D. Cai, L. Ren, H. Zhao, C. Xu, L. Zhang, Y. Yu, H. Wang, Y. Lan, M.F. Roberts, J.H. Chuang, M.J. Naughton, Z. Ren, T.C. Chiles, Nat. Nanotechnol. 5 (2010) 597.
[73] J. Li, F. Jiang, Y. Li, Z. Chen, Biosens. Bioelectron. 26 (2011) 2097.
[74] D. Vaihinger, K. Landfester, I. Krauter, H. Brunner, E.M. Tovar, Macrom. Chem. Phys. 203 (2002) 1965.
[75] T. de Boer, R.A. de Zeeuw, G.J. de Jong, D.C. Sherrington, P.A.G. Cormack, K. Ensing, Electrophoresis 23 (2002) 1296.
[76] R. Liu, G. Guan, S. Wang, Z. Zhang, Analyst 136 (2011) 184.
[77] X. Tan, B. Li, K.Y. Liew, C. Li, Biosens, Bioelectron 26 (2010) 868.
[78] C.N. Casey, S.E. Campbell, U.J. Gibson, Biosens. Bioelectron 26 (2010) 703.
[79] F.G. Tamayo, J.L. Casillas, A. Martin-Esteban, J. Chromatogr. A 1069 (2005) 173.
[80] I. Morelli, V. Chiono, G. Vozzi, G. Ciardelli, D. Silvestri, P. Giusti, Sens, Actuators B 150 (2010) 394.
[81] V.M. Mirsky, T. Hirsch, S.A. Piletsky, O.S. Wolfbeis, Angew. Chem. Int. Ed. Engl. 38 (1999) 35.
[82] H. Li, Y. Liu, Z. Zhang, H. Liao, L. Nie, S. Yao, J. Chromatogr. A 1098 (2005) 66.
[83] G. Ciardelli, C. Borrelli, D. Silvestri, C. Cristallini, N. Barbani, P. Giusti, Biosens. Bioelectron. 21 (2006) 2329.
[84] Y. Wang, Y. Zhou, J. Sokolov, B. Rigas, K. Levon, M. Rafailovich, Biosens. Bioelectron. 24 (2008) 162.

[85] A. Katz, M. Davis, Nature 403 (2000) 286.
[86] N. Sallacan, M. Zayats, T. Bourenko, A.B. Kharitonov, I. Willner, Anal. Chem. 74 (2002) 702.
[87] L. Ye, P.A.G. Cormack, K. Mosbach, Anal. Commun. 36 (1999) 35.
[88] J. Wang, P.A.G. Cormack, D.C. Sherrington, E. Khoshdel, Angew. Chem. 115 (2003) 5494.
[89] G. Cirillo, F. Puoci, M. Curcio, O.I. Parisi, F. Iemma, U.G. Spizzirri, N. Picci, Colloid Polym. Sci. 288 (2010) 689.
[90] S. Gu, T. Kondo, M. Konno, J. Colloid Interface Sci. 272 (2004) 314.
[91] J. Rodriguez-Hernandez, F. Checot, Y. Gnanou, S. Lecommandoux, Prog. Polym. Sci. 30 (2005) 691.
[92] S. Cao, J. Chen, J. Hu, Aust. J. Chem. 62 (2009) 1561.
[93] S. Flink, F.C.J.M. Veggel, D.N. Reinhoudt, Adv. Mater. 12 (2000) 1315.
[94] K.C. Nirmalya, K. Vijayamohanan, Biosens. Bioelectron. 17 (2002) 1.
[95] M. Lotierzo, O.Y.F. Henry, S. Piletsky, I. Tothill, D. Cullen, M. Kania, B. Hock, A.P.F. Turner, Biosens. Bioelectron. 20 (2004) 145.
[96] F.H. Dickey, Proc. Natl. Acad. Sci. 35 (1949) 227.
[97] Z. Zhang, Y. Hu, H. Zhang, S. Yao, J. Colloid. Interf. Sci. 344 (2010) 158.
[98] A. Olwill, H. Hughes, M. O'Riordain, P. McLoughlin, Biosens. Bioelectron. 20 (2004) 1045.
[99] A.P. Wight, M.E. Davis, Chem. Rev. 102 (2002) 3589.
[100] C. Alexander, H.S. Andersson, L.I. Andersson, R.J. Ansell, N. Kirsch, L.A. Nicholls, J. O Mahony, M.J. Whitcombe, J. Mol. Recognit. 19 (2006) 106.
[101] L. Ye, K. Mosbach, Chem. Mater. 20 (2008) 859.
[102] T. Kunitake, S.W. Lee, Anal. Chim. Acta 504 (2004) 1.
[103] W. Cummins, P. Duggan, P. McLoughlin, Anal. Chim. Acta 542 (2005) 52.
[104] P.J. Skrdla, M. Shnayderman, L. Wright, T.P. O'Brien, J. Non-Cryst, Solids 352 (2006) 3302.
[105] A. Walcarius, D. Mandler, J.A. Cox, M. Collinson, O. Lev, J. Mater, Chem. 15 (2005) 3663.
[106] O.Y.F. Henry, S.A. Piletsky, D.C. Cullen, Biosens. Bioelectron. 23 (2008) 1769.
[107] S.A. Piletsky, E.V. Piletska, B.N. Chen, K. Karim, D. Weston, G. Barrett, P. Lowe, A.P.F. Turner, Anal. Chem. 72 (2000) 4381.
[108] A. Gómez-Caballero, A. Ugarte, A. Sánchez-Ortega, N. Unceta, M.A. Goicolea, R.J. Barrio, J. Electroanal, Chem. 638 (2010) 246.
[109] G.D. Agostino, G. Alberti, R. Biesuz, M. Pesavento, Biosens, Bioelectron. 22 (2006) 145.
[110] G.P. González, P.F. Hernando, J.S.D. Alegría, Biosens. Bioelectron. 23 (2008) 1754.
[111] A. Bossi, F. Bonini, A.P.F. Turner, S.A. Piletsky, Biosens. Bioelectron. 22 (2007) 1131.
[112] J. Rick, T.C. Chou, Biosens. Bioelectron. 22 (2006) 544.
[113] S. Chen, D. Du, J. Huang, A. Zhang, H. Tu, A. Zhang, Talanta 84 (2011) 451.
[114] J. Chen, W. Huang, J. Han, S. Cao, Int. J. Environ. Res. 5 (2011) 515.
[115] A. Khler, S. Hellweg, B.I. Escher, K. Hungerbhler, Environ. Sci. Technol. 40 (2006) 3395.
[116] H. Bagheri, A. Mohammadi, A. Salemi, Anal. Chim. Acta 513 (2004) 445.
[117] M. Suarez, A. Macia, M.P. Romero, M.J. Motilva, J. Chromatogr, A 1214 (2008) 90.
[118] D. Calderon-Preciado, C. Jimenez-Cartagena, G. Penuela, J.M. Bayona, Anal. Bioanal. Chem. 394 (2009) 1319.
[119] B.V. Dorst, J. Mehta, K. Bekaert, E. Rouah-Martin, W.D. Coen, P. Dubruel, R. Blust, J. Robbens, Biosens, Bioelectron. 26 (2010) 1178.
[120] R.K. Kaneto, Curr. Appl. Phys. 5 (2005) 178.
[121] S.V. Dzyadevych, T. Mai Anh, A.P. Soldatkin, N. Duc Chien, N. Jaffrezic-Renault, J.M. Chovelon, Bioelectrochemistry 55 (2002) 79.
[122] E. Han, D. Shan, H. Xue, S. Cosnier, Biomacromolecules 8 (2007) 971.
[123] T.A. Sergeyeva, L.A. Gorbach, O.A. Slinchenko, L.A. Goncharova, O.V. Piletska, O.O. Brovko, L.M. Sergeeva, G.V. Elska, Mat. Sci. Eng. C 30 (2010) 431.
[124] G.C. Jagetia, R. Aruna, Toxicol. Lett. 93 (1997) 205.
[125] G.P. Bespamyatnikov, YuA Krotov, Maximum Permissible Concentrations of Chemical Substances in Environment, Chemistry, Leningrad, 1985.
[126] P. Qi, J. Wang, J. Jin, F. Su, J. Chen, Talanta 81 (2010) 1630.
[127] F. An, B. Gao, X. Feng, J. Hazard, Mater. 157 (2008) 286.
[128] Q. Feng, L. Zhao, J.-M. Lin, Anal. Chim. Acta 650 (2009) 70.

[129] T.A. Sergeyeva, S.A. Piletsky, E.V. Piletska, O.O. Brovko, L.V. Karabanova, L.M. Sergeeva, A.P.F. Turner, A.V. El'skaya, Macromolecules 36 (2003) 7352.
[130] D. Apodaca, R.B. Pernites, F.R. Del Mundo, R.C. Advincula, Langmuir 27 (2011) 6768.
[131] A. Larsson, J. Angbrant, J. Ekeroth, P. Mansson, B. Liedberg, Sens. Actuators, B: Chem. 113 (2006) 730.
[132] G. Bunte, J. Höurttlen, H. Pontius, K. Hartlieb, H.T Krause, Anal. Chim. Acta 591 (2007) 49.
[133] D. Roeseling, T. Tuercke, H. Krause, S. Loebbecke, Organ. Pro. Res. Deve. 13 (2009) 1007.
[134] F.L. Dickert, O. Hayden, Trends Anal. Chem 18 (1999) 192.
[135] N.R. Walker, M.J. Linman, M.M. Timmers, S.L. Dean, C.M. Burkett, A.L. Julie, J.D. Keelor, B.M. Baughman, P.L. Edmiston, Anal. Chim. Acta 593 (2007) 82.
[136] M. Riskin, R. TelVered, T. Bourenko, E. Granot, I. Willner, J. Am. Chem. Soc. 130 (2008) 9726.

CHAPTER

4

Molecularly Imprinted Overoxidized Polypyrrole Films for Sensor Applications from Enantiorecognition to Trace Analysis

Shiho Tokonami[1], *Hiroshi Shiigi*[2], *Tsutomu Nagaoka*[2]

[1]Nanoscience and Nanotechnology Research Center, Research Organization for the 21st Century, Osaka Prefecture University, 1-2 Gakuen-cho, Naka-ku, Sakai, Japan

[2]Department of Applied Chemistry, Graduate School of Engineering, Osaka Prefecture University, 1-2 Gakuen-cho, Naka-ku, Sakai, Japan

OUTLINE

Molecularly Imprinted Sensors.
DOI: 10.1016/B978-0-444-56331-6.00004-9

1. INTRODUCTION

This chapter reviews a novel approach to molecular imprinting technology using conducting polymers and related materials. Conducting polymers have been used in a wide range of applications from batteries to electroanalysis, in which conductivity plays a specific role in their applications. In addition to their electronic properties, the polymers have a unique feature of inserting anionic materials upon polymerization, thus finding another type of utility. Hence, the polymer can retain anionic species as long as they are in the oxidized state, and this property can be conveniently used as matrices and carriers for functionality. Our group started the studies of conducting polymers, focusing mainly on their unique properties as functional carriers more than a decade ago, and successfully introduced molecular recognition properties in the polymers by overoxidation [1]. This finding eventually developed into a new class of molecularly imprinted polymer (MIP), which allowed the highly efficient recognition of template anions through a straightforward synthetic procedure [2].

MIP is a polymeric material that has a built-in functionality for the recognition of a particular chemical substance with its complementary cavity [3–5]. Extensive efforts have already been made to synthesize, characterize, and apply the MIP materials, and the number of related articles has shown a sharp rise during the last two decades. The number of articles related to MIP-based chemical sensors is also increasing, and the ratio of the sensor studies to the total MIP research reports has been significantly increasing during the last decade (Fig. 4.1). The standard synthetic technique of MIP materials consists of the copolymerization of functional and cross-linking monomers in the presence of a template molecule, followed

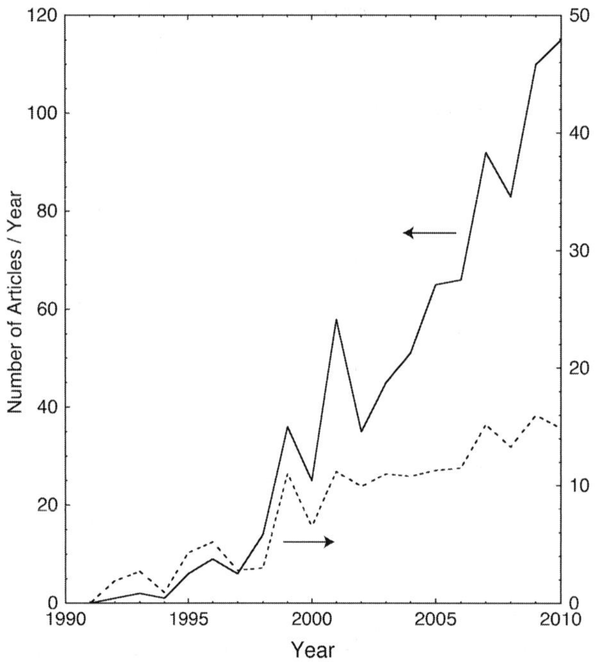

FIGURE 4.1 MIP Research trends for (—) sensors, and for (- - - -) the ratio of the sensor studies to all the MIP studies.

by removal of the template from the polymer matrix to create a receptor cavity that is complementary to the template. This procedure looks quite elegant, but it is often difficult for this protocol to gain sufficient selectivity of a template molecule due to problems in the formation of a complex between the template and functional monomer. To gain selectivity as high as that compared to biomolecules, such as enzymes, it is desirable that the complex have many weak interactions with the template, such as hydrogen bonding, rather than a small number of strong interactions. But it is rather difficult to imagine that such a weakly bound fragile complex is retained intact for a certain amount of time and is successfully polymerized into the matrix without losing much of the initial structure during polymerization. Another problem for the standard protocols is the slow removal of a template molecule based on diffusion processes; the gradual dissolution of the template from the polymer matrix as a result of the incomplete removal could cause a band broadening in chromatographic applications. The leaking of the template also results in the difficult estimation of the physical parameters defining rebinding events.

We have found a new way to circumvent these problems of standard MIP synthetic procedures, using an overoxidized polypyrrole [1,2]. In this review chapter, the unique and attractive features of the overoxidized polypyrrole (PPy) materials for use in chemical sensors are described along with some recent results. Since we presented our first paper in 1999, the number of articles using conducting polymers in the MIP research has been increasing. Among the many conducting polymers, polyaniline [6–15], polypyrrole [16–33], polythiophene [34–36], and their derivatives have been most frequently used as the MIP materials. The conducting polymers are used not only as MIP receptors possessing cavities complementary to a dopant, but also as entrapping matrices for electrically neutral templates [14,15,21,25,26,29,36–38]. Some researchers wash imprinted conducting polymers with solvents in order to remove the dopant template, but this procedure would not properly work as a technique for removing the template as a result of the strong electrostatic interaction between the dopant and the cationic polymer backbone. If an electrolyte solution is used, the dopant may be washed out, but the electrolyte can occupy the cavity instead, which might produce another problem in rebinding. Conducting polymers are sometimes used as a matrix for the standard MIP receptor particle [22,34].

As for the detection platform applied to conducting polymer MIP sensors, it is most convenient to use the electrochemical and quartz microbalance techniques. Spectroscopic techniques are not usually favored, since the conducting polymers except for polythiophene are not optically transparent. However, some researchers use the surface plasmon resonance techniques to evaluate conducting polymer MIP films [21,26,39].

2. OVEROXIDATION OF POLYPYRROLE TO PRODUCE MIP PROPERTIES

As shown in Scheme 1, pyrrole is oxidatively polymerized to polypyrrole (PPy) with a dopant anion, X^-, which acts as a template molecule in our technique, and PPy/X^- (polypyrrole doped with X^-) can be prepared either on an electrode as a film or in solution as a powder or colloidal particle with a chemical oxidizing agent. Elimination of the cationic charge in the polymer backbone by further oxidation in a high pH condition (overoxidation)

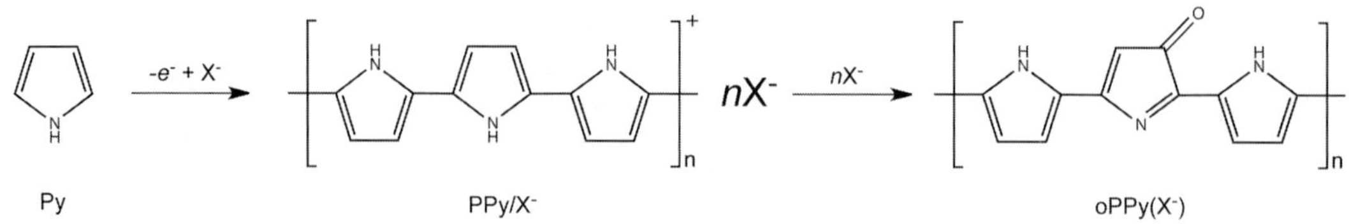

SCHEME 1 Polymerization and overoxidation.

allows the template anion to escape from the overoxidized polypyrrole (oPPy) film to fulfill the electrostatic requirement. It has been pointed out that the nucleophilic attack of hydroxide ion to polypyrrole upon overoxidation destroys its conjugated structure by the insertion of carbonyl groups on the pyrrole rings. Detailed discussions of the overoxidation mechanisms can be found elsewhere [40]. The removal of the template by the electrostatic mechanism uniquely features the oPPy MIP technique in that the quantitative removal of the template almost instantaneously occurs upon application of a potential or addition of an oxidizing agent. This is in comparison to the standard MIP protocols in which washing out a template often takes a long time, ranging from many hours to days, depending on material characteristics such as the product dimensions.

Overoxidation has two effects on selectivity: (1) it can remove the template molecule to create the complementary cavity, and (2) it can cure the polymer texture to make the cavity stable. At the same time, overoxidation is prone to cause exfoliation of the film off the electrode surface. The last effect is not serious when a PPy film is deposited on a carbon electrode but results in a problem for a metallic electrode. This problem can be avoided by roughening the surface of the electrode using a method described in "7. Experimental Remarks."

The overall MIP protocols are illustrated in Scheme 2, in which adenosine triphosphate (ATP) is selected as a template [41]. A solution containing pyrrole and the template (ATP) is oxidized on an electrode to deposit a film of polypyrrole doping ATP (a–b). The film, PPy/ATP, is then oxidized in a NaOH solution typically at 0.5–1.0 V vs. Ag|AgCl (sat. KCl) to convert the PPy/ATP film to oPPy(ATP), where oPPy(ATP) denotes oPPy with a cavity complementary to ATP (c).

Overoxidation of PPy/X^- for different anionic templates, X^-, was confirmed by several analytical techniques, such as voltammetry [1,2,40,42–49], electrochemical quartz

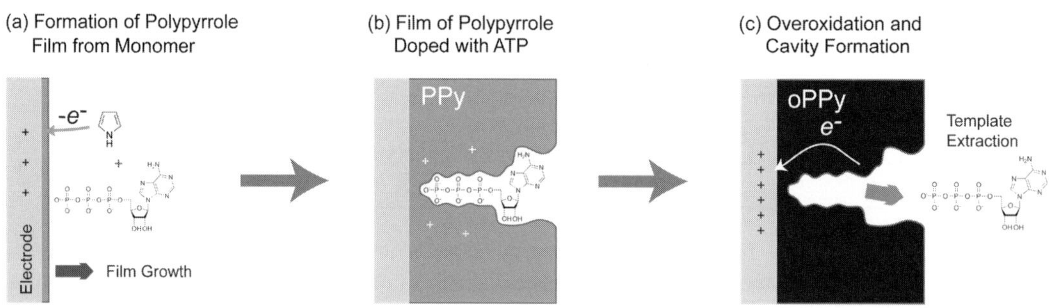

SCHEME 2 MIP protocol by the overoxidation technique: (a) A metallic or carbonaceous electrode contacting with a solution of pyrrole monomer and template anion, (b) a polypyrrole film electrooxidized to dope the template, and (c) overoxidized film with a cavity complementary to the template.

crystal microbalance (EQCM) [2,42,43,50], FTIR spectrometry [42,45], UV/VIS spectrometry [42,45], ESR spectrometry [2], and capillary electrophoresis [46]. Figure 4.2 shows the FTIR spectra of an oPPy colloid imprinted with lactic acid [42,45] for the direct overoxidation from pyrrole to oPPy (Scheme 1) that occurred due to the strong oxidizing ability of the ammonium persulfate. The increasing concentration of persulfate, which overoxidized PPy/lactate, led to two effects in the polymer: (1) removal of the lactate, which was confirmed by the continuous decrease in the 1651 cm^{-1} signal assigned to the template (lactate) and (2) the overoxidation of the polymer, which was indicated by the progressive augmentation of the 1701 cm^{-1} signal assigned to the carbonyl group (Scheme 1). Removal of a template was confirmed by EQCM and X-ray photoelectron spectroscopy (XPS). Figure 4.3 shows an example of the analysis studied by EQCM [50]. Almost complete overoxidation of the film occurring during the very first oxidative scan allowed us to barely observe the overoxidation peak in the second cycle. Interestingly, removal of the template, taurocholate (TC), occurred after completion of the electrochemical overoxidation (a) to cause both the exclusion of TC [resonance frequency (b)] and film curing [resonance resistance (c)], which are some of the exciting features of oPPy as a MIP material in that the shape and size of a template are memorized in the polymer texture concurrently with the exclusion. XPS spectroscopy also confirmed the removal of the template; disappearance of the

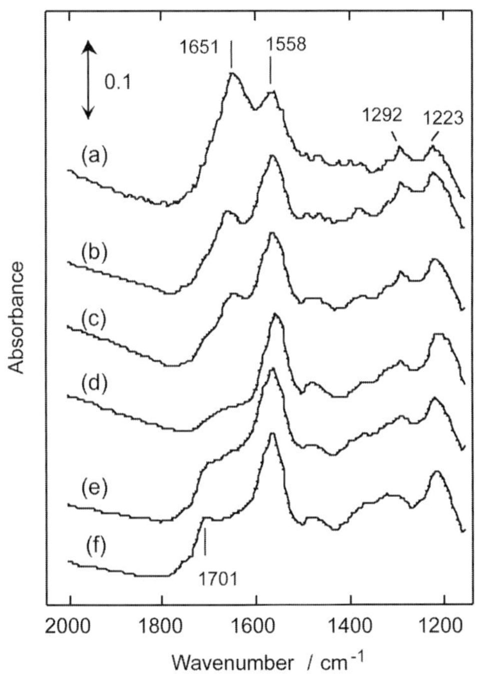

FIGURE 4.2 FTIR spectra of polypyrrole PVP colloids prepared with different concentrations of persulfate at pH 4.0. The concentrations of persulfate in the polymerization solution were 0.010 (a), 0.030 (b), 0.050 (c), 0.10 (d), 0.15 (e), and 0.30 M (f). The composition of the colloid solution was 15 mM of the pyrrole monomer, 0.20 M of L-lactic acid, 0.085 g of PVP (as steric stabilizer), and persulfate (as oxidizing agent) in an aqueous solution (100 mL). *Reprinted with permission from Ref. [45]. © 2002 The Japan Society for Analytical Chemistry.*

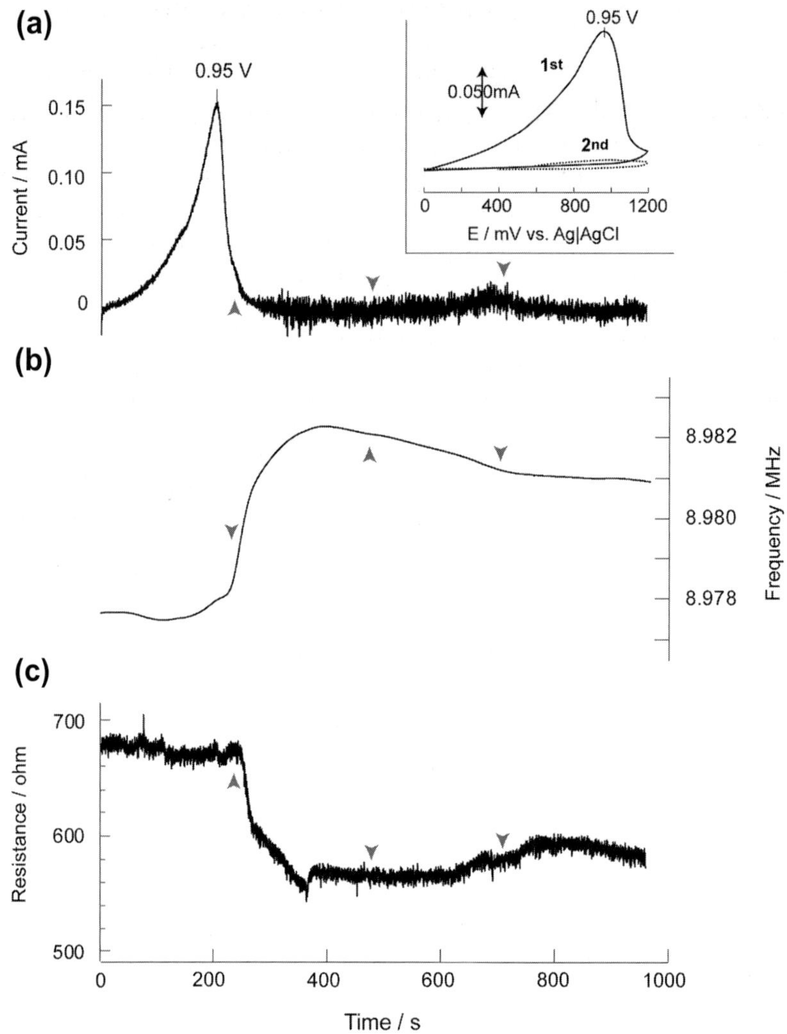

FIGURE 4.3 Time profiles of EQCM characteristics for overoxidation of a PPy/TC film in a 40 mM Li_2CO_3 aqueous solution (pH 9.5) during cyclic voltammetric activation; (a) the current, (b) frequency, and (c) resonance resistance. Potential scan rate: 5.0 mV s^{-1}. The arrow marks given in the figure show the reversal of the potential sweep direction. The inset represents the cyclic voltammogram for a PPy/TC modified gold electrode observed under the same experimental conditions; (——) the first cycle and (- - -) the second cycle. *Reprinted with permission from Ref. [50]. © 2005 The Electrochemical Society, Inc.*

sulfur 2S and 2P signals for the PPy/1-naphthalene sulfonate was observed after the overoxidation [43,44]. The overoxidation MIP protocols are very different from the standard MIP preparation technique in that the removal of a template is made by an electrostatic interaction, which can more thoroughly occur than that for the standard techniques, in which templates are extracted by the diffusional driving force (washing out by a solvent).

3. RESOLVING POWER OF oPPy FILMS FOR MOLECULAR RECOGNITION

In our study, the oPPy materials were prepared as sensor films [1,2,43,49,50], colloids [42,45,46,49], and chromatographic coating [47] to evaluate their analytical performance as a versatile recognition platform. We first applied this technique to the enantiorecognition of amino acids, choosing glutamic acid (glu) as a model template. Glutamic acid exists as a +1-valent cation ($pH < 2.2$), neutral ion ($2.2 < pH < 4.3$), −1-valent anion ($4.3 < pH < 9.7$), and −2-valent anion ($pH > 9.7$), which confines the pH range for the deposition pH of a PPy/glu film above 4.3. On the other hand, pyrrole can be polymerized only in an acidic-neutral pH range, and these two experimental limitations taken together make the preferred polymerization condition weakly acidic. Enantiorecognition of the oPPy(L-glu) and oPPy(D-glu) films deposited on EQCM electrodes are shown in Fig. 4.4. Although the

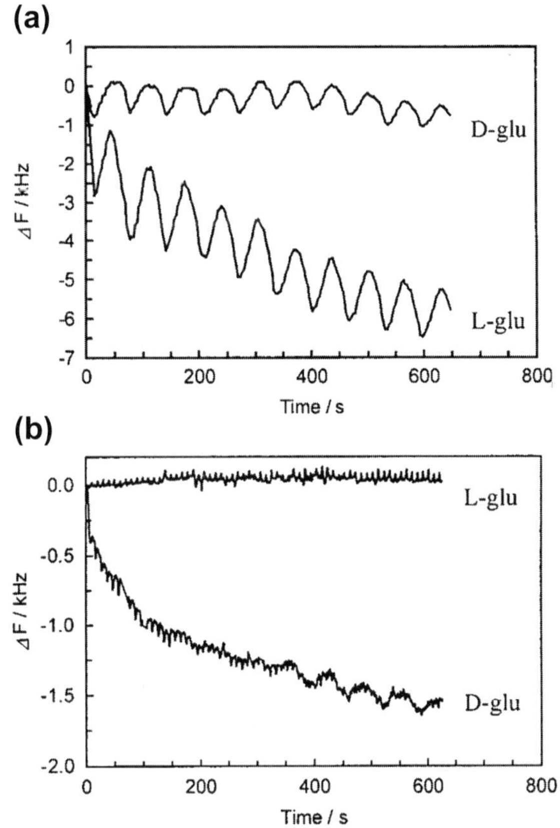

FIGURE 4.4 Time dependence of the EQCM resonance frequency under potentiodynamic polarization of (a) oPPy(L-Glu) and (b) oPPy(D-Glu) films. The films were polarized in the cyclic voltammetric experiment at a scan rate of 40 mV s^{-1} in 10 mM D- and L-Glu solutions (pH 1.7). Each maximum and minimum in the figures corresponds to the frequencies for the potential scanning between +0.6 and 0.0 V. Film preparation condition: 0.050 mA cm^{-2} for 2.00 h. *Reprinted with permission from Ref. [2]. © 2000 American Chemical Society.*

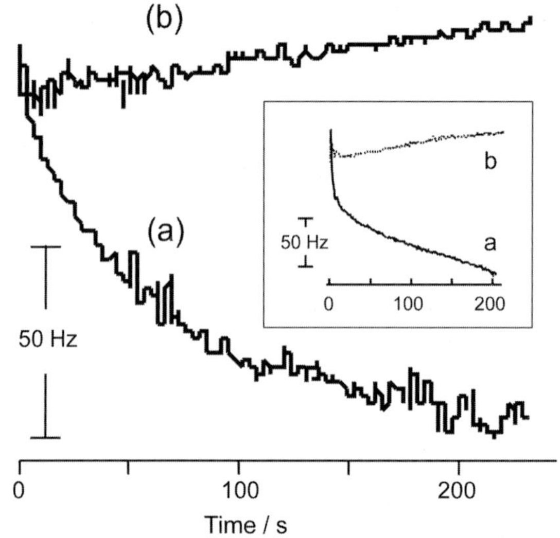

FIGURE 4.5 Time dependence of resonance frequency for an oPPy(1-NS) film in 0.010 M (a) 1-NS and (b) 2-NS solutions at +300 mV. The inset represents the dependence for oPPy(2-NS) in 0.010 M (a) 2-NS and (b) 1-NS solutions under the same conditions. *Reprinted with permission from Ref. [43]. © 2003 The Electrochemical Society, Inc.*

difference in the film thickness caused different sensitivities in the EQCM experiments, the antisymmetrical responses with respect to the L- and D-cavities proved that each cavity is complementary to the template in a highly precise manner.

The oPPy polymer can recognize not only enantiomers, but also structural isomers (Fig. 4.5) [43]. The 1- and 2-naphthalenesulfonates (NS) were then selected as the templates. The oPPy(1-NS) and oPPy(2-NS) films prepared on the EQCM electrodes by the overoxidation technique successfully recognized the template molecules and their counterparts. Recently, other researchers have also studied electrochemical sensors based on the oPPy films to confirm the excellent selectivity [18,19,22].

4. oPPy MIP MATERIALS FOR SEPARATION EVENTS

The oPPy MIP films can be used not only for sensor applications, but also for macroscopic separation events [47]. Figure 4.6 shows an electromodulated separation column packed with glassy carbon fibers modified with the oPPy(L-glu) film. The amino acids were taken up in the column at 0.0 V versus Ag|AgCl (sat. KCl) at pH 1.7 and eluted at 1.2 V. Figure 4.7 summarizes the results of three repetitive enantioseparation experiments, demonstrating that the substantial difference in the enantiouptake was obtained in these medium-scaled uptake experiments, although we have already shown that the enantioselectivity resulted in the finite mass changes in the EQCM framework. Since the oPPy MIP is oxidatively prepared, the film is very stable in harsh chemical environments. This feature is also indicated by no virtual loss in enantioselectivity in the repeated separation experiments. Recently, Chen's group also studied an electromodulated column using the oPPy coating imprinted with L-tryptophan to find an excellent enantioselectivity [17].

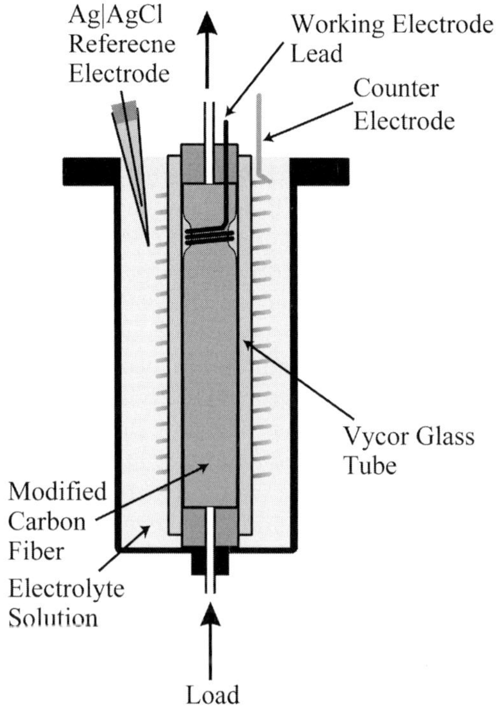

FIGURE 4.6 Cross-sectional diagram of the electromodulated column electrode packed with oPPy(L-glu) carbon fibers. *Reprinted with permission from Ref. [47]. © 2002 Royal Society of Chemistry.*

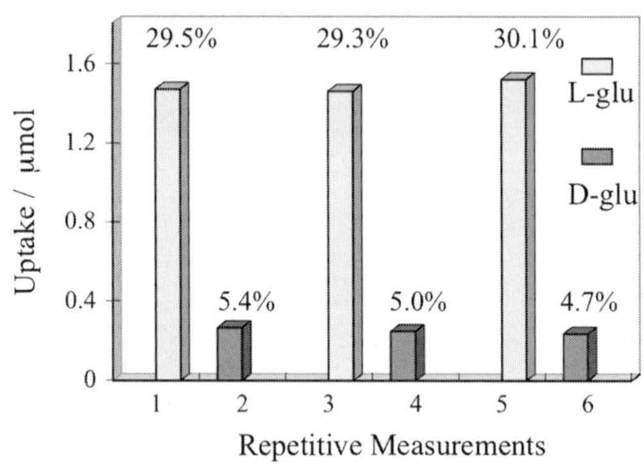

FIGURE 4.7 Repetitive responses of oPPy(L-glu)-modified carbon fiber packed electromodulated column to 5 μmol of L- or D-glutamic acid (1×10^{-3}M, 5 mL). Uptake was carried out at pH 1.7 and 0.0 V. Column washed with 0.1 M KCl between consecutive measurements at 1.2 V. *Reprinted with permission from Ref. [47]. © 2002 Royal Society of Chemistry.*

5. DISPERSED oPPy MIP PARTICLES

The oPPy MIP polymeric materials can be prepared as colloidal particles as well as in insoluble forms such as films [46]. An appropriate surfactant, such as poly(vinylpyrrolidone), PVP, is used to form the polymer into microsized dispersible particles [49]. PPyPVP/L-lactate colloids are prepared from a mixture of pyrrole, L-lactic acid, and PVP, in which persulfate was used as the oxidizing reagent. PPyPVP/L-phenyllactate was synthesized in a similar manner. These colloids were dispersed in aqueous 0.1 M NaOH and overoxidized at 1.2 V versus Ag|AgCl using a flow-through electrolysis cell packed with glassy carbon fibers (cf. Fig. 4.6) to obtain bacilliform oPPyPVP(L-lactate) and oPPyPVP(L-phenyllactate). Different from electrochemical synthesis, anionic species coexisting with a template anion in solution (e.g., sulfate, which is generated from persulfate) is also doped in the PPy texture, which makes the exclusive formation of the complementary cavity by a template difficult, especially when the doping rate of the template is low. The slow doping occurs when the amino acid anion (glutamic or aspartic acid) is employed as a template due to the positive charge carried by the amino group. To avoid this difficulty, we used lactic acid in a colloid preparation due to its fast doping rate, and the enantiorecognition of several amino acids was studied using this pseudocomplementary cavity (Schemes 3 and 4).

Figure 4.8 shows the uptake of amino acids shown in Scheme 3 [49]. Alanine (a) and cysteine (b) were enantiorecognized in the oPPyPVP(L-lactate) cavity due to their structural similarities to lactate; phenylalanine is apparently too large to fit into the cavity, and only a small amount of adsorption nonspecific to the enantiomers was observed. When the uptake of phenylalanine was examined with oPPyPVP(L-phenyllactate), an ample amount of L-phenylalanine uptake was observed in contrast (Fig. 4.9). Figure 4.10 shows quartz crystal

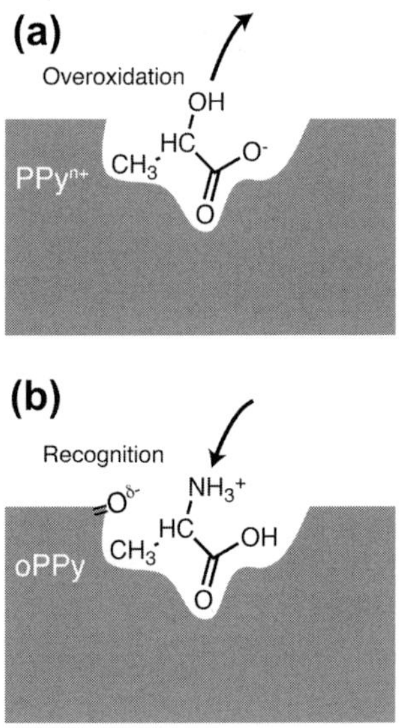

SCHEME 3 Models for (a) overoxidation of polypyrrole doped with L-lactate and (b) for recognition in the cavity. *Reprinted with permission from Ref. [49]. © 2000 Royal Society of Chemistry.*

SCHEME 4 Molecules used in colloid experiments. *Reprinted with permission from Ref. [49]. © 2000 Royal Society of Chemistry.*

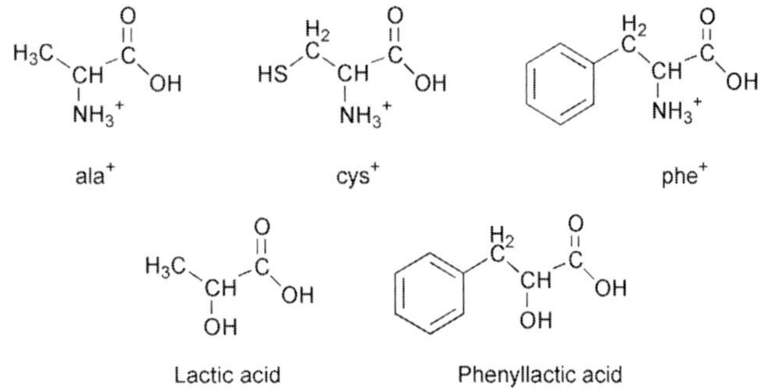

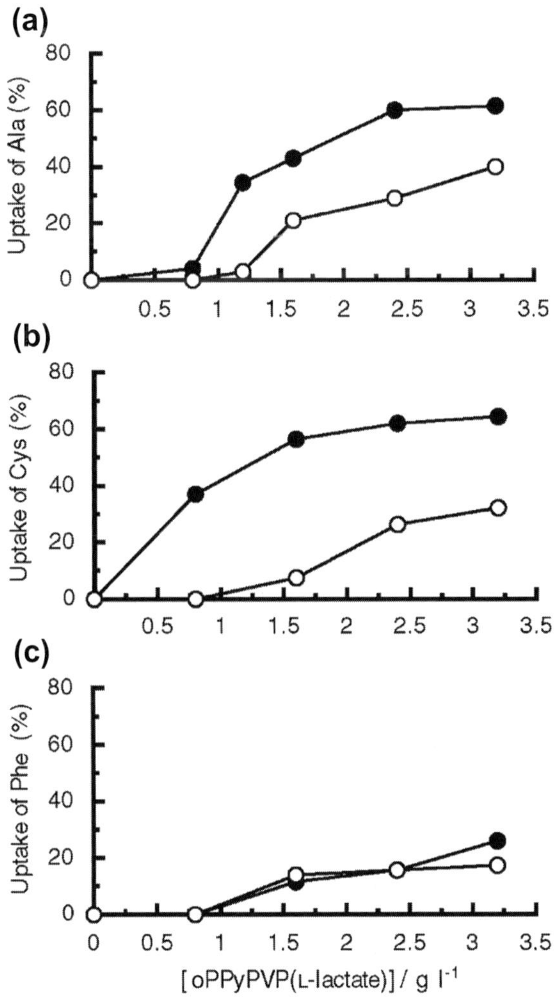

FIGURE 4.8 Uptake of (a) alanine, (b) cysteine, and (c) phenylalanine into oPPyPVP(L-lactate): (●) L- and (○) D-enantiomers. The solution pH was adjusted to 1.0 with HCl. *Reprinted with permission from Ref. [49]. © 2000 Royal Society of Chemistry.*

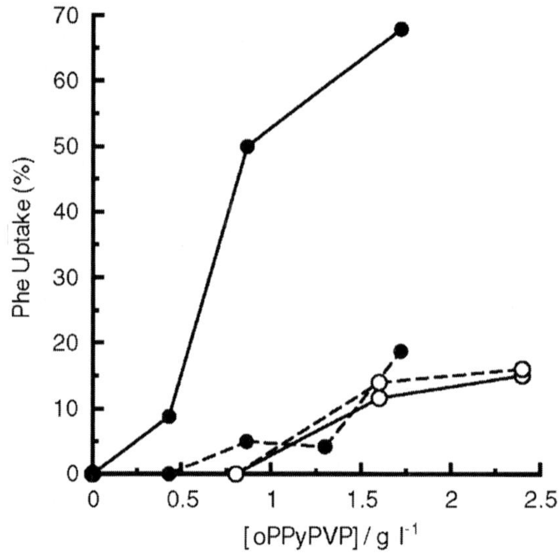

FIGURE 4.9 Uptake of 0.10 mM (—) L- and (---) D-phenylalanine into (●) oPPyPVP(L-phenyllactate) and (○) oPPyPVP(L-lactate) at pH 1.0. *Reprinted with permission from Ref. [49]. © 2000 Royal Society of Chemistry.*

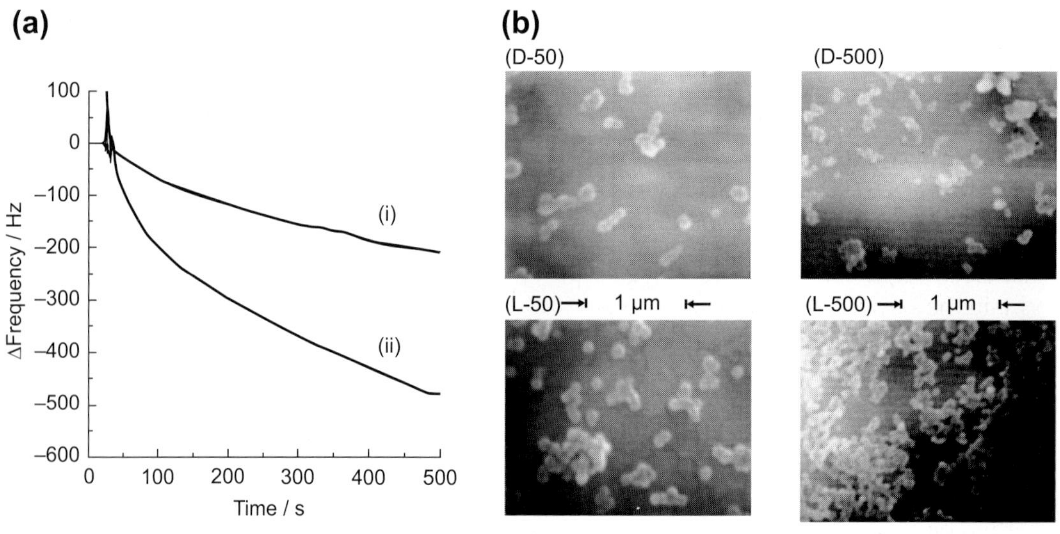

FIGURE 4.10 (a) QCM frequency changes for (i) D- and (ii) L-cysteine modified gold surfaces on the addition of oPPyPVP(L-lactate) at pH 1.0. The spikes in the curves indicate the time of addition. Concentration of colloid after addition, 0.16 g L^{-1}. (b) SEM images of oPPyPVP(L-lactate) adsorbed on Au surfaces modified with L-cysteine (L-50, L-500) and with D-cysteine (D-50, D-500), (D-50), and (L-50); Au surfaces 50 s after immersion in 0.16 g L^{-1} oPPyPVP(L-lactate), (D-500) and (L-500); Au surfaces 500 s after immersion. The pH of the colloidal dispersion was 1.0. *Reprinted with permission from Ref. [49]. © 2000 Royal Society of Chemistry.*

microbalance (QCM) responses and scanning electron microscope (SEM) images for the enantioselective adsorption of oPPyPVP(L-lactate) on gold substrates immobilizing L- and D-cysteine. As expected, enantiorecognition was confirmed for both observations, and the surface modified with L-cysteine adsorbed the oPPyPVP(L-lactate) particle more than that for D-cysteine. The SEM image indicates that the particle number ratio on the surface of L-cysteine to D-cysteine is about 2.7, which agreed with the change in the QCM frequency ratio of 2.8 for the 50-s immersion.

6. APPLICATION IN TRACE ANALYSIS

In this section, we discuss the applicability of an oPPy MIP sensing film for trace analysis [41]. We have developed a MIP-based amperometric sensor, which allows the detection of trace concentrations of adenosine triphosphate (ATP), aimed at food safety applications. The sensor was fabricated on a small carbon electrode covered with an overoxidized polypyrrole film imprinted with an ATP molecule, and operated in the triple pulse amperometric mode to be characterized in both the batchwise and flow injection measurements.

ATP is a signature molecule, which indicates a microbial activity, and its detection in a trace concentration level is useful for the prevention of food poisoning. The commercially available ATP sensors, which rely on chemiluminescence resulting from the enzymatic reaction of luciferase with ATP, have several demanding conditions for on-site use. Those mainly arise from the instability of the expensive enzyme, which should remain inactivated by freezing until used. Accordingly, it is highly desired to develop an easy-to-use and cost-effective real-time sensing system based on a working mechanism different from enzymes. To solve these problems, we have reported an MIP-based sensor, which allows the detection of trace concentrations of ATP.

To achieve a high sensitivity for detecting a trace amount of ATP, we have used triple pulse amperometry (Fig. 4.11) [41]. The sensor film, oPPy(ATP), was synthesized on a coaxial 3-electrode carbon sensor chip according to Scheme 2. The first negative-potential pulse in Fig 4.11 is used for the electrostatic removal of any remaining ATP in the film during the previous pulse cycle, the second pulse for the concentration of ATP into the film, and the third pulse for the detection of ATP. A set of three pulses was continuously applied to the electrode during the analysis, and an ATP solution was applied on the sensor chip to obtain the amperometric responses (Fig. 4.12). The sensor allows the detection of a nanomolar level of ATP thanks to the high affinity of the cavity to ATP as well as the high sensitivity of the triple pulse amperometry, as shown in Fig. 4.12. Figure 4.13 demonstrates the high rejection ratio of the oPPy(ATP) film for molecules other than ATP examined by flow injection analysis. Although ADP and AMP are structurally very similar, except for the number of phosphate residue, the oPPy film recognized the number of residues and another molecular system, L-ascorbic acid, in a highly efficient manner. It should be noted here that the bare carbon electrode scarcely had such selectivity.

7. EXPERIMENTAL REMARKS

The oPPy films adhere well on carbonaceous materials, and we obtained a sufficient stability for long-term use. However, we often observed that the films prepared on metallic

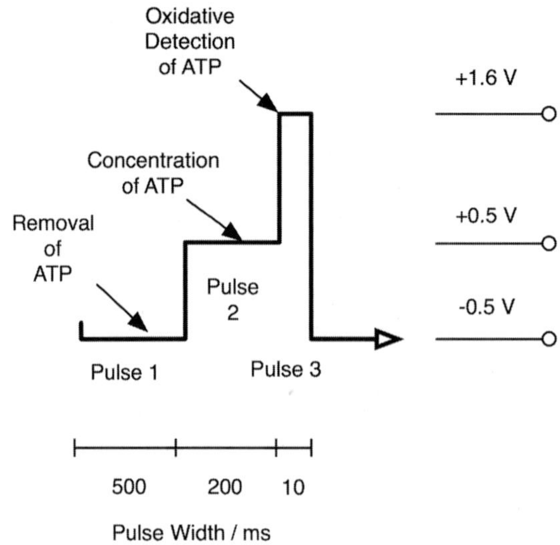

FIGURE 4.11 The triple pulse amperometric sequence for ATP detection. *Reprinted with permission from Ref. [41]. © 2008 The Japanese Association for Flow Injection Analysis.*

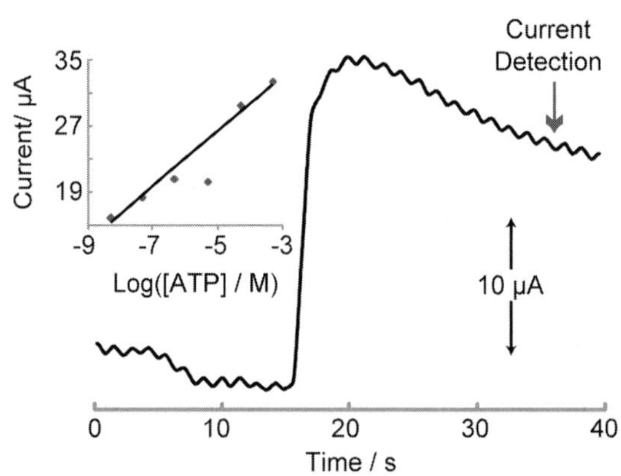

FIGURE 4.12 The sensor response for addition of 1×10^{-7} M ATP: Inset, calibration plots. *Reprinted with permission from Ref. [41]. © 2008 The Japanese Association for Flow Injection Analysis.*

electrodes are prone to peel off the surface during the electrochemical overoxidation. This exfoliation can be avoided by making the metal electrode surface rougher as a general strategy. A recipe for obtaining such a rough surface is as follows:

(1) A gold electrode, for instance, is subjected to plasma etching (air) to make the surface clean and rough. This pretreatment is useful when a vapor-deposited thin metallic layer electrode, such as an EQCM electrode chip, is used.

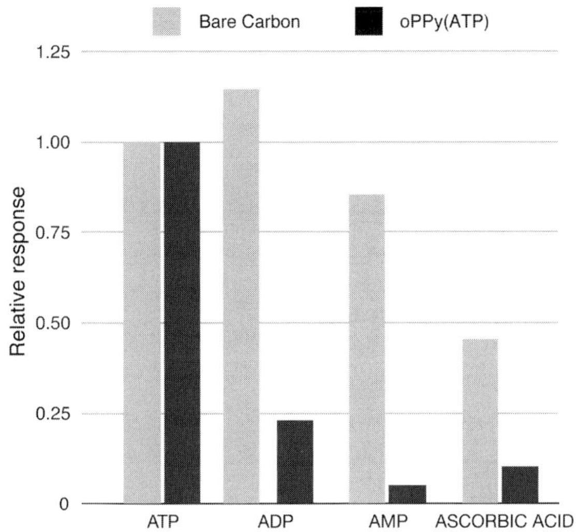

FIGURE 4.13 Relative responses of the unmodified carbon and oPPy(ATP) electrodes for 10 mM adenosine triphosphate (ATP), diphosphate (ADP), monophosphate (AMP) and L-ascorbic acid. The responses of ATP were normalized to unity for both the electrodes.

(2) The etched surface is further roughened by growing Au nanoparticles (approximately 30 to 100 nm in diameter) on the surface by immersing the electrode into a solution containing hexadecyl-trimethylammonium bromide, Au(III) ions, and ascorbic acid. A detailed experimental description is given in Ref. [51].

8. CONCLUSIONS

In this chapter, we have demonstrated the high selectivity of the cavity, which is realized by the simple overoxidation protocols. The important features of the oPPy MIP materials discussed here are as follows:

(1) The cavity complementary to a template is fabricated in a polymer in a straightforward manner by overoxidation.
(2) Overoxidation can concurrently remove the template and cure the polymer for the formation of a stable cavity.
(3) The MIP colloidal particle can be used not only in solution uptake events, but also in the detection of the surface structure. Although the adoption of a pseudocomplementary cavity for the colloidal particles made the selectivity smaller than the genuine cavities formed on the films, the attractive features of the oPPy MIP materials were found for the molecular recognition events.

References

[1] B. Deore, Z. Chen, T. Nagaoka, Anal. Sci. 15 (1999) 827.
[2] B. Deore, Z. Chen, T. Nagaoka, Anal. Chem. 72 (2000) 3989.

[3] S. Tokonami, H. Shiigi, T. Nagaoka, Anal. Chim. Acta 641 (2009) 7.
[4] C. Alexander, H.S. Andersson, L.I. Andersson, R.J. Ansell, N. Kirsch, I.A. Nicholls, J. O'Mahony, M.J. Whitcombe, J. Mol. Recognit. 19 (2006) 106.
[5] M.J. Whitcombe, I. Chianella, L. Larcombe, S.A. Piletsky, J. Noble, R. Porter, A. Horgan, Chem. Soc. Rev. 40 (2011) 1547.
[6] K. Sreenivasan, J. Mater. Sci. 42 (2007) 7575.
[7] M. Najafi, M. Mollazadeh, J. Appl. Polym. Sci. 121 (2011) 292.
[8] Y. Liang, L. Gu, X. Liu, Q. Yang, H. Kajiura, Y. Li, T. Zhou, G. Shi, Chem. Eur. J. 17 (2011) 5989.
[9] Y. Kong, J. Ni, W. Wang, Z. Chen, Electrochim. Acta 56 (2011) 4070.
[10] F. Cao, J. Liao, K. Yang, P. Bai, Q. Wei, C. Zhao, Anal. Lett. 43 (2010) 2790.
[11] F. Berti, S. Todros, D. Lakshmi, M.J. Whitcombe, I. Chianella, M. Ferroni, S.A. Piletsky, A.P.F. Turner, G. Marrazza, Biosens. Bioelectron. 26 (2010) 497.
[12] S.W. Chuang, J. Rick, T.C. Chou, Biosens. Bioelectron. 24 (2009) 3170.
[13] B. Deore, M.S. Freund, Analyst 128 (2003) 803.
[14] K. Sreenivasan, J. Mol. Recognit. 19 (2006) 408.
[15] K. Sreenivasan, React. Funct. Polym. 67 (2007) 859.
[16] E. von Hauff, K. Fuchs, D.C. Hellmann, J. Parisi, R. Weiler, C. Burkhardt, U. Kraushaar, E. Guenther, Biosens. Bioelectron. 26 (2010) 596.
[17] Y. Kong, J. Wei, W. Wang, Z. Chen, Electrochim. Acta 56 (2011) 4770.
[18] C. Xie, S. Gao, Q. Guo, K. Xu, Microchim. Acta 169 (2010) 145.
[19] Y. Kong, W. Zhao, S. Yao, J. Xu, W. Wang, Z. Chen, J. Appl. Polym. Sci. 115 (2010) 1952.
[20] Z.D. Chen, J.X. Wei, W.C. Wang, Y. Kong, Chin. Chem. Lett. 21 (2010) 353.
[21] S.W. Choi, H.J. Chang, N. Lee, J.H. Kim, H.S. Chun, J. Agric, Food. Chem. 57 (2009) 1113.
[22] V. Syritski, J. Reut, A. Menaker, R.E. Gyurcsányi, A. Öpik, Electrochim. Acta 53 (2008) 2729.
[23] J.S. Lenihan, J.C. Ball, V.G. Gavalas, J.K. Lumpp, J. Hines, S. Daunert, L.G. Bachas, Anal. Bioanal. Chem. 387 (2007) 259.
[24] D.R. Albano, F. Sevilla III, Sens. Actuators, B-Chem. 121 (2007) 129.
[25] L. Özcan, Y. Sahin, Sens. Actuators, B-Chem. 127 (2007) 362.
[26] J.C.C. Yu, E.P.C. Lai, React. Funct. Polym. 63 (2005) 171.
[27] H.H. Yang, S.Q. Zhang, F. Tan, Z.X. Zhuang, X.R. Wang, J. Am, Chem. Soc. 127 (2005) 1378.
[28] A. Ramanaviciene, A. Ramanavicius, Biosens. Bioelectron. 20 (2004) 1076.
[29] H.J. Liang, T.R. Ling, J.F. Rick, T.C. Chou, Anal. Chim. Acta 542 (2005) 83.
[30] R. Suedee, W. Intakong, P.A. Lieberzeit, P. Wanichapichart, P. Chooto, F.L. Dickert, J. Appl. Polym. Sci. 106 (2007) 3861.
[31] L. Özcan, M. Sahin, Y. Sahin, Sensors 8 (2008) 5792.
[32] S.P. Özkorucuklu, Y. Sahin, G. Alsancak, Sensors 8 (2008) 8463.
[33] J. Huang, Z. Wei, J. Chen, Sens. Actuators, B-Chem. 134 (2008) 573.
[34] K.C. Ho, W.M. Yeh, T.S. Tung, J.Y. Liao, Anal. Chim. Acta 542 (2005) 90.
[35] C.T. Wu, P.Y. Chen, J.G. Chen, V. Suryanarayanan, K.C. Ho, Anal. Chim. Acta 633 (2009) 119.
[36] E. Pardieu, H. Cheap, C. Vedrine, M. Lazerges, Y. Lattach, F. Garnier, S. Remita, C. Pernelle, Anal. Chim. Acta 649 (2009) 236.
[37] B.S. Ebarvia, S. Cabanilla, F. Sevilla III, Talanta 66 (2005) 145.
[38] C.L. Choong, J.S. Bendall, W.I. Milne, Biosens. Bioelectron. 25 (2009) 652.
[39] R. Pernites, R. Ponnapati, M.J. Felipe, R. Advincula, Biosens. Bioelectron. 26 (2011) 2766.
[40] B. Deore, T. Nagaoka, Curr. Top. Anal. Chem. 2 (2001) 135.
[41] S. Takeda, H. Yagi, S. Mizuguchi, H. Funahashi, H. Shiigi, T. Nagaoka, J. Flow Injection Anal. 25 (2008) 77.
[42] H. Shiigi, H. Yakabe, M. Kishimoto, D. Kijima, Y. Zhang, U. Sree, B.A. Deore, T. Nagaoka, Microchim. Acta 143 (2003) 155.
[43] H. Shiigi, K. Okamura, D. Kijima, A. Hironaka, B. Deore, U. Sree, T. Nagaoka, Electrochem. Solid-State Lett. 6 (2003) H1.
[44] H. Shiigi, K. Okamura, D. Kijima, B. Deore, U. Sree, T. Nagaoka, J. Electrochem. Soc. 150 (2003). H119.
[45] H. Shiigi, M. Kishimoto, H. Yakabe, B. Deore, T. Nagaoka, Anal. Sci. 18 (2002) 41.

[46] H. Okuno, T. Kitano, H. Yakabe, M. Kishimoto, B.A. Deore, H. Siigi, T. Nagaoka, Anal. Chem. 74 (2002) 4184.
[47] B. Deore, H. Yakabe, H. Shiigi, T. Nagaoka, Analyst 127 (2002) 935.
[48] B. Deore, T. Nagaoka, Rev. Polarogr. 47 (2001) 76.
[49] Z. Chen, Y. Takei, B.A. Deore, T. Nagaoka, Analyst 125 (2000) 2249.
[50] H. Shiigi, D. Kijima, Y. Ikenaga, K. Hori, S. Fukazawa, T. Nagaoka, J. Electrochem. Soc. 152 (2005) H129.
[51] M. Oyama, Anal. Sci. 26 (2010) 1.

CHAPTER

5

MIP-based Sensor Platforms for Detection of Analytes in Nano- and Micromolar Range

Frederik Horemans[1], *Hanne Diliën*[1], *Patrick Wagner*[1,2], *Thomas J. Cleij*[1,3]

[1]Hasselt University, Institute for Materials Research (IMO), Agoralaan, Belgium

[2]IMEC, Division IMOMEC, Belgium

[3]Maastricht University, Maastricht Science Programme, The Netherlands

OUTLINE

Molecularly Imprinted Sensors
DOI: 10.1016/B978-0-444-56331-6.00005-0

1. INTRODUCTION

The development of innovative biosensor concepts to detect specific molecules is of considerable interest. Many known biosensor platforms make use of enzymes or antibodies in their recognition layers [1–3]. Such biosensors, which are based on biological recognition elements, can be very sensitive and specific toward their target. Notwithstanding, some significant drawbacks can be identified. First of all, it is time consuming to obtain biological recognition elements in sufficient quantities. In many cases, test animals are required, making the production rather laborious. Furthermore, it is not always possible to obtain suitable biological recognition elements to detect all desired targets [4]. Finally, biomolecules are often unstable in changing chemical and physical environments and have a limited shelf life. These drawbacks mostly limit their applicability to physiological fluids [5] and render them less suitable to, for example, point of care diagnostics.

An answer to these shortcomings can be molecular imprinting [6]. Molecularly imprinted polymers (MIPs) are synthetic polymers containing imprinted nanocavities, which are able to rebind their target in a specific manner. After careful optimization, their synthesis is usually comparatively straightforward and based on conventional polymer chemistry. This is especially true for the noncovalent approach [7,8]. In this approach, a mixture of functional monomers and target molecules will form a pre-polymerization complex due to the noncovalent interactions between target and functional monomer. This complex is stabilized in an appropriate porogen. Subsequently, cross-linkers monomers are added to this mixture to create a stable matrix. After radical polymerization, a bulk polymer is obtained containing the embedded targets. Subsequently, the target is extracted from the polymer matrix resulting in nanocavities, which act as the complementary binding sites. MIPs show similar specific binding characteristics and selectivity toward their target as antibodies and have the advantage of being robust and inert over a wide range of different environments [9]. They are stable in, for example, a wide temperature and pH range. Furthermore, MIPs have the potential to be regenerated, which is not always possible for biological receptors.

In recent years, several research groups have developed MIP-based sensing concepts. Since there is a huge demand for low-cost, fast, and reliable diagnostics, the advantages of molecular imprinting as a method to generate synthetic receptors are tremendous as compared to their biological counterparts. This makes them very suitable for a wide range of sensor applications. In this way, MIPs can be a suitable class of recognition elements for future sensing techniques. In literature, various interesting MIP-based sensors have been described which make use of different readout techniques. Examples include spectroscopic methods, piezoelectric measurements, and electrochemical detection.

The best-described spectroscopic methods include detection based on fluorescence, chemiluminescence, and surface plasmon resonance (SPR). Fluorescence-based [10–19] target detection requires fluorescent functional monomers or the incorporation of some fluorescent comonomers into the MIP matrix in the proximity of the binding sites. In this way depending on the sensor design, target binding will cause either an enhancement or a quenching of the fluorescence signal. An interesting example using fluorescence quenching for target detection is based on a MIP consisting of porphyrin Zn (II) complex monomers and ethylene glycol dimethacrylate (EGDM) as cross-linker in the presence of the target pyrazine [20]. Upon binding of the target, a decrease in fluorescence intensity is detected.

Chemiluminescence [21–32] is also shown to be a valuable technique for characterizing target binding in a MIP matrix. To this end, a suitable chemiluminescent system is selected and combined with the selectivity of MIPs. After target binding, the MIP is exposed to the selected chemiluminescent system, which generates light in function of the amount of target present in the MIP. This technique allows for a flow-through sensor system, which makes it possible to screen a large amount of samples.

SPR [33–37] is also an interesting sensing technique that is compatible with MIPs. The SPR technique is based on the detection of changes in the refractive index of a thin layer. Hence, thin MIP films rather than MIP particles are needed. As a result, the MIP synthesis should be optimized to create thin and uniform films. An example of this approach is a microfluidic system combined with a MIP-based SPR sensor [37]. In this device, MIP films to detect progesterone, cholesterol, and testosterone have been incorporated onto the SPR chip, showing that the outlook for high throughput screening based on MIP sensing is promising.

Furthermore, it is possible to create arrays of MIP microdots (6–12 μm) on surface-enhanced Raman scattering active surfaces [38]. Using a Raman system, S-propanolol binding in single MIP dots can be observed, thus demonstrating the prospect of miniaturized MIP sensors.

For the piezoelectric detection of target binding, microgravimetric methods making use of bulk acoustic wave sensors are of interest [39]. Quartz crystal microbalances (QCMs) are the most widespread sensor examples in the literature [40–49]. For this method of detection, a QCM surface needs to be covered with a thin MIP layer or covered with small-sized MIP particles. Upon target binding, a decrease in resonance frequency is detected corresponding to the mass increase due to target binding. Besides QCM devices, surface acoustic wave sensors have also been combined with MIPs to allow for gas-phase sensing.

The last group of MIP-based sensors is based on electrochemical detection [50] and can be divided into amperometric, voltammetric, and impedimetric transduction methods. The combination of these electrochemical sensing methods and the selectivity of MIPs is very successful. Some important advantages include the possibility of offering good limits of detection, low cost in production of the sensor, and miniaturizing the sensor system. Hence, these techniques are compatible with the development of hand-held devices, which are extremely desirable for applications such as environmental control sensing, quality control in the pharmaceutical industry and food industry, as well as point of care medical diagnostics. Concerning amperometric detection [51–53], an electrode needs to be covered with MIP particles or a MIP film. Target binding is measured by means of current changes at a fixed voltage applied on the electrode. To measure a change in current, the target needs to be electro-active, or upon binding a displacement step coupled with an electrochemical reaction should take place. For voltammetric detection, the current is measured upon the application of a potential sweep [50,54–71]. Different voltammetric techniques are possible, depending on the shape of the applied potential function. The most common techniques used in MIP sensing are cyclic voltammetry and square-wave voltammetry.

It should be noted that both amperometric and voltammetric detection methods sometimes give rise to by-products due to the oxidation or reduction of the target, which can foul the electrode surface. However, this is not the case for impedimetric sensing, since the binding of the target alone is enough to generate the detectable signal [68,72–75]. For impedance spectroscopy an alternating current is applied over a MIP at a certain frequency. Impedance spectroscopy monitors target binding by the detection of changes in the complex resistance. This complex resistance is composed of a combination of capacitance and resistance effects.

This chapter presents a review of our approach to obtain a versatile and straightforward MIP-based sensor setup. In addition, some critical parameters are discussed which are important for the performance of the sensing device. Our approach focuses on developing a universal platform, which is compatible with aqueous measurements, miniaturization, and use in hand-held devices. From the overview outlined above, it becomes evident that microgravimetric and impedimetric detection are especially promising for achieving those aims, thus explaining our focus on these two techniques.

2. MIP SENSOR PLATFORM

2.1. MIP Immobilization onto the Sensor Surface

Conventional MIP synthesis usually results in particles that can bind the target molecule of interest in a specific manner (cf. also Section 3 of this chapter). It is also important to know the magnitude of the aspecific binding to the MIP. To this end, for every MIP a corresponding non-imprinted polymer (NIP) is made. In the sensor both a MIP and an NIP channel can be incorporated, which makes it possible to detect both channels simultaneously. To achieve this, these MIP and NIP particles have to be immobilized onto the sensor surface. For optimal sensor performance, it is important that the MIP and NIP particles are firmly attached to the sensor surface during the measurements. A suitable method to achieve this is matrix entrapment, which ensures a good physical contact between the MIP particles and the sensor surface. This technique is a universal and straightforward method for MIP immobilization onto a surface and allows for different sensing setups. For the entrapment procedure, a polymeric transducer layer is used, which serves as an adhesive to stick the MIP and NIP particles to the surface. This adhesive polymer layer should be sufficiently soluble to allow for spin coating a uniform layer onto the electrodes. Subsequently, the MIP and NIP particles are stamped on top of this polymer layer with the use of a polydimethylsiloxane (PDMS) stamp. When the stamp is removed, a good coverage of MIP and NIP particles on the sensor surface is

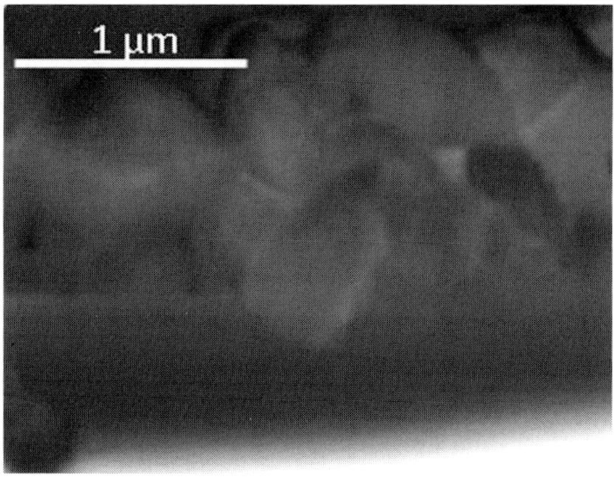

FIGURE 5.1 SEM image of the MIP sensor surface.

created. The next step involves heating the polymer layer above the T_g. This results in a softening of the polymer layer, which in turn will cause the MIP and NIP particles to sink partially into this layer. Afterward the MIP-embedded polymer layer is rinsed and sonicated to remove all non-attached MIP particles. This straightforward method allows for the immobilization of the MIP particles in an efficient manner (Fig. 5.1).

As mentioned earlier, a polymeric transducer is applied between the MIP particles and the electrodes. For the impedance setup this transducer layer serves on one hand as an adhesive and on the other hand as a conductive layer. After all, the transducer layer should be able to conduct electrical charges necessary for the electronic detection. Hence, for this purpose a bifunctional material is needed. A suitable material to fulfill this task is a conjugated polymer, that is, MDMO-PPV (Fig. 5.2) [76]. This polymer not only has appropriate semiconductive properties, but is also stable and straightforward to spincoat. These properties make it a suitable polymer for impedimetric sensing.

For the QCM measurements, there is no need for conducting properties, and only a good contact between MIP and sensor surface is required. A suitable polymer for QCM measurements is the more readily accessible polyvinyl chloride (PVC), which can also be spincoated and allows for a strong immobilization of the MIP particles.

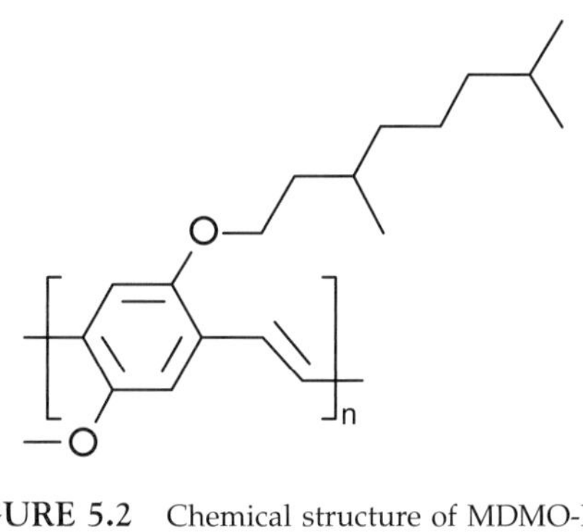

FIGURE 5.2 Chemical structure of MDMO-PPV.

2.2. Impedance Spectroscopy

Impedimetric detection is comparatively straightforward and fast, and the sensor setup has the possibility to be miniaturized. These characteristics make impedance a logical choice as a sensing technique. Impedance (symbol Z) is the frequency-dependent resistance, when an alternating current is applied. Impedance includes the pure resistance and capacitance at an interface layer. Therefore, impedance is often referred to as the complex resistance. The impedance at a given angular frequency consists of a real part and an imaginary part. When plotting the real part of the impedance on the X-axis and the imaginary part on the Y-axis, a Nyquist plot is constructed [77]. For the sensing experiments, such plots typically have the shape of a semicircle, which changes in size depending on a variation in the impedance due to target binding in the MIP binding sites. Each point on the semicircle of the Nyquist plot corresponds to the impedance at a given frequency. Each impedance point is represented by a vector with the length Z. The angle between this vector and the X-axis is the phase Φ or argument of Z. In our impedance based sensor setup, time resolved measurements are applied, in which Z at a given frequency is plotted as a function of time. In this way, it is possible to follow the impedance in time at a chosen frequency, which has the lowest signal-to-noise ratio.

2.3. Quartz Crystal Microbalance (QCM)

QCM is based on a change in the resonance frequency of a quartz crystal due to mass increase. Quartz exhibits piezoelectrical properties. When an oscillating voltage is applied, an oscillating deformation of the crystal will take place. This deformation will oscillate at the natural resonance frequency of the quartz crystal. When mass is applied to the crystal, this resonance frequency will change, in accordance with the amount of mass deposited on the crystal surface. With the Sauerbrey equation it is possible to estimate the mass present on the quartz surface [78]. The quartz crystals with the immobilized MIP particles are connected with a phase-lock oscillator. This is necessary to establish a zero phase between the crystal voltage and current by adjusting the frequency of the oscillator. When target (mass) binds to the MIP, the resonance frequency will change. This change can be measured with a frequency counter and is a measure for the amount of target molecules bound to the MIP.

2.4. MIP Synthesis Optimization for Sensor Applications

The first step toward a working sensor is the MIP synthesis itself. Perhaps most importantly, the MIP should have a large capacity, which makes it possible to measure in a large dynamic range. This is essential since the target can be present in very fluctuating concentration ranges depending on the situation. It is also desirable that the MIP should be porous enough for the target to bind at the binding sites in a timely manner. Insufficient porosity will hamper mass transport and will lead to an increase in the sensor response time. This is undesirable, since a fast sensor response is important for real-time measurements. Furthermore, the MIP should bind selectively to its target molecule and not to other molecules. This property can be referred to as "on/off sensing" and is very important in the sensor domain to exclude false-positive sensing results. Furthermore, for the detection of biomedical relevant molecules, it is important that the MIP binding can take place in aqueous media. Finally, besides the optimization of the binding characteristics, it is important that a suitable MIP morphology is synthesized to allow for an optimal sensor coverage. These MIP properties are required for successful integration into the sensor recognition layer.

A typical MIP synthesis involves many optimization and evaluation steps. Since a considerable variety of monomers can be utilized to form a typical MIP for the same target molecule, potentially a combinatorial process is needed to find the most suitable combination. However, to reduce the amount of work associated with optimization of the MIP synthesis, two alternative approaches are presented in literature. The first approach makes use of computer calculations to predict the optimal composition of the pre-polymerization complex in a MIP. In this way, the MIP synthesis can be optimized with the best suitable selection of monomers. In theory, afterward only little experimental fine tuning is necessary [79–81]. However, a disadvantage of this approach is the complexity of the theoretical implementation of all the different variables. The second approach is semi-empirical and focuses on a "smart" selection of monomers based on the functionalities present in the target molecule. This careful selection of monomers is used to make a series of MIPs in which the ratios of the monomers are changed [82–85]. This technique is more straightforward to work with, but requires more evaluation and optimization steps afterward. Nonetheless, both techniques lead to successful MIP protocols.

For sensor applications, a MIP with a high specificity and selectivity toward its target molecule is needed. Hence, the MIP synthesis for sensor applications will require additional optimization. As was discussed in Section 2.1, in the actual sensor both a MIP and a NIP channel are present. Since the NIP gives a good estimate of the specific adsorption, in this way the specificity of a MIP can be quickly evaluated during every measurement, which will improve MIP optimization.

The MIPs described in this chapter have been prepared using the semi-empirical method to optimize the MIP protocols. Instead of using complex mixtures of monomers, the choice of monomers has been mostly limited to the broadly applicable functional monomer methacrylic acid (MAA) and the cross-linker ethylene glycol dimethacrylate (EGDM) (Fig. 5.3), to reduce the amount of variables during optimization. In literature, these monomers are widely used in the MIP synthesis of a wide range of target molecules [86]. For example, MAA typically will form a stable pre-polymerization complex with target molecules containing nitrogen groups.

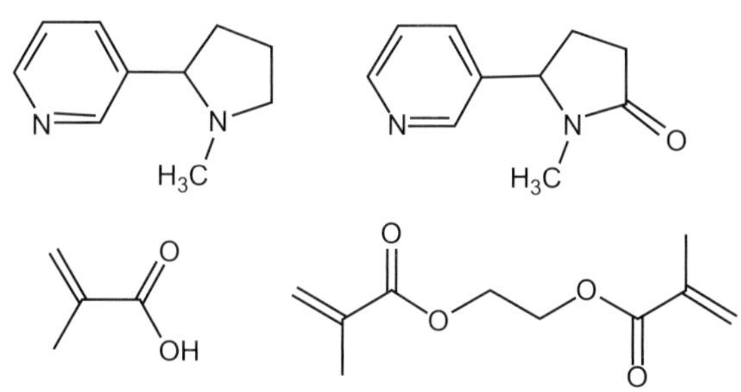

FIGURE 5.3 Chemical structures of L-nicotine (top left), L-cotinine (top right), MAA (bottom left), and EDGM (bottom right).

3. BULK POLYMERIZATION MIPs FOR SENSORS

3.1. Synthesis of Bulk-Imprinted MIPs

The general approach toward the MIP synthesis for sensor purposes can be explained for the target molecule L-nicotine (Fig. 5.3), which contains amine and pyridine functionalities capable of interacting with MAA. The L-nicotine MIPs serve as representative examples of the optimization of MIP systems for sensor applications. The acquired insights can readily be translated to the MIP optimization for other target molecules. To test MIP selectivity, the binding behavior of the MIPs has also been investigated for L-cotinine (Fig. 5.3), which is a structural analogue of L-nicotine. Not only the binding behavior, but also the morphology of the MIP particles is important. For sensor applications, the suitability of several types of MIP particles has been studied, ranging from irregular-shaped crushed monoliths to well-defined spherical particles made with, for example, a suspension polymerization.

Bulk polymerization is the most straightforward method for MIP synthesis and involves fewer optimization steps than other types of polymerizations [87]. First of all, the chemical structure of L-nicotine has been examined to determine the theoretical amount of MAA needed. L-nicotine has two relevant functionalities (nitrogen groups) that are able to form hydrogen bonds with MAA. Therefore it seems reasonable to use two equivalents of MAA for the bulk MIP synthesis. This has been experimentally verified by synthesizing a wide range of MIPs with different compositions, which clearly confirmed the 1:2 ratio of template: MAA. On the other hand, the amount of EGDM has to be experimentally derived from batch rebinding experiments. After thorough optimizations, it was found—perhaps not unexpectedly—that a ratio template:MAA:EGDM of 1:2:4 gave the best results in batch rebinding experiments in acetonitrile.

Furthermore, the amount of porogen needed to create optimal porosity and access for L-nicotine in the imprinted cavities has to be optimized, and in this context the NIPs are of special interest. As mentioned earlier, the performance in a sensor is not only dependent on the binding behavior of a MIP, but also on the binding behavior of the corresponding NIP. A large amount of porogen leads to better porosity, but concomitantly an increased surface area is observed, which leads to more aspecific binding. In the case of L-nicotine MIPs, after optimizations it became clear that 7 mL of porogen (chloroform) for 6.45 mmol of template molecule gave an optimal porosity.

3.2. Sensors with Bulk-Imprinted MIPs

Section 3.1 mentioned that the best performing bulk MIP for L-nicotine has been integrated into our impedimetric sensor setup [73]. To this end, simultaneously, a bare Pt electrode and OC_1C_{10}-PPV coated electrodes without and with MIP and NIP particles were measured in phosphate buffered saline (PBS) solution. It has been found that binding of the target affects the formation of the double layer at the electrode surface, which is reflected in the capacitive changes at the lower frequency range. In addition, it influences the charge transfer from the electrolyte to the electrode. The MIP signal shows an increase of 45% when 10 nM of L-nicotine is introduced in the setup, whereas the signal of the NIP-coated electrode only increased with 30%. Subsequently, the influence of the analogue molecule L-cotinine

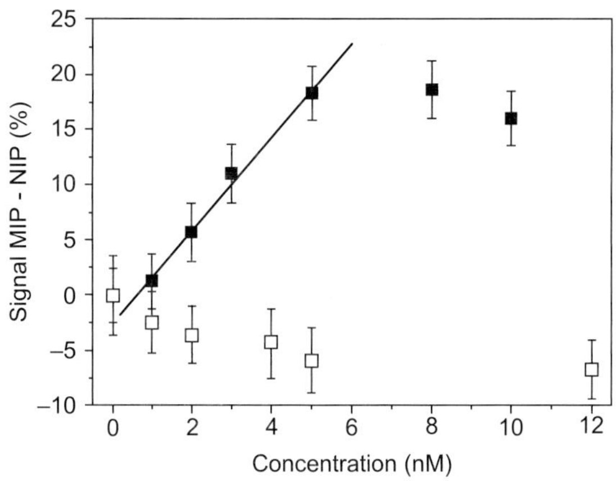

FIGURE 5.4 Difference between bulk MIP and NIP exposed to increasing concentrations of L-nicotine (filled squares) and L-cotinine (open squares).

was investigated on a different sample in the same way as was done for L-nicotine. The response is not specific and is markedly lower than the reaction of L-nicotine of the sensor.

To eliminate the contribution of nonspecific adsorptions to the surface of the MIP, it is better to plot the signal by subtracting the normalized impedance signal of the NIP from the normalized MIP signal (Fig. 5.4). A linear response is found from 0 to 5 nM, with a response of 4.3% increase of the differential signal per nanomole. For concentrations higher than 5 nM, the impedance change is similar for MIP and NIP. This is due to the full occupation of the specific binding places in the MIP, and in this way the MIP reacts in the same way as the NIP.

Although these impedimetric sensor measurements are quite promising, the available concentration range is rather small, the nonspecific binding is large, and the margin of error is quite significant. In addition, several drawbacks can be identified, such as the limited compatibility with aqueous samples, the random morphology and associated irreproducible sensor coverage, and the insufficiently uniform binding characteristics. These drawbacks and possible solutions will be discussed in more detail in the following sections.

4. WATER-COMPATIBLE MIPs FOR SENSORS

4.1. The Intricate Balance between MIP and Porogen Polarity

The applicability of molecularly imprinted polymers (MIPs) is very broad as a result of the fact that they are very robust and can operate under largely different conditions [88]. However, their ability to bind targets selectively in aqueous conditions is an issue that should be addressed [85–95], since many target molecules of interest are only present in aqueous media such as bodily fluids, food products, rivers, and wastewater. As was discussed in Section 3, a bulk MIP made using the porogen chloroform was successfully applied in the recognition layer of our proof of principle MIP-based sensor platform [73]. Detection of L-nicotine was possible in the nanomolar range in PBS buffer using impedimetric detection.

However, a considerable amount of aspecific binding was present. As a result, only a minimal difference in binding was present between the MIP and the corresponding NIP, as was confirmed by the binding isotherms (Fig. 5.5). Hence, it is of interest to further adapt the MIP protocol to enhance the specific binding and reduce the aspecific binding in water.

Our optimizations toward water-compatible MIPs for aqueous sensor applications can serve as a blueprint for the synthesis of other water-compatible MIPs for different target molecules. This is a result of the fact that MIP binding characteristics are significantly different when measured in polar and apolar organic solvents, such as acetonitrile and toluene, or in water. It is well established that in water, target molecules can bind in an aspecific manner to the MIP matrix due to the hydrophobic effect [90]. For sensor purposes this is a considerable problem, since only the specific binding is of interest. Therefore, it is important to develop water-compatible MIPs that can bind their target molecules in a specific and selective manner [74,85,89,91–93].

Although in the literature various MIPs have been presented with sometimes satisfying performance in water, so far no standardized protocol has been presented to obtain good selectivity in aqueous media. This is not straightforward, since one has to overcome the water-induced aspecific hydrophobic binding of targets. Various approaches can be utilized to accomplish this, most of them focusing on making the MIPs either more hydrophilic or hydrophobic, depending on the binding interactions between the target molecule and the functional monomers. After careful analysis of the existing information, we have developed several new and improved procedures for water-compatible MIPs, with an emphasis on tuning the polarity of the pore wall of the MIPs.

As mentioned previously, the most straightforward approach toward MIPs makes use of the monomer methacrylic acid (MAA) and the cross-linker ethylene glycol dimethacrylate (EGDM). The interaction between MAA and the target in a given porogen determines the imprint effect of the MIP binding sites afterward [94–96]. This so-called pre-polymerization complex is strongly influenced by the polarity and hydrogen bonding properties of the

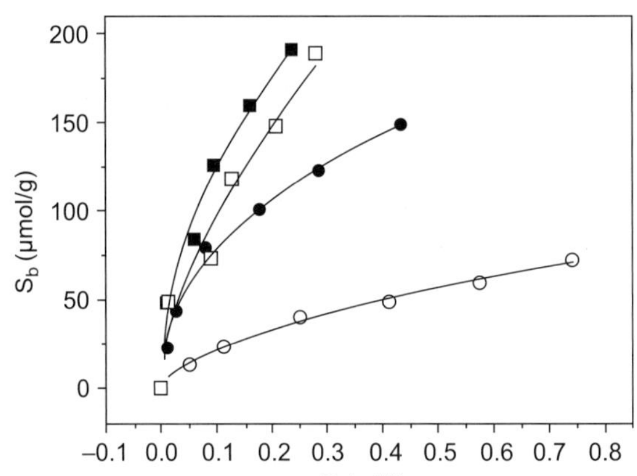

FIGURE 5.5 Binding isotherms of the L-nicotine imprinted bulk MIP (solid symbols) and corresponding NIP (open symbols) prepared with the porogen chloroform. The rebinding solvent was either acetonitrile (circles) or water (squares).

porogen. When a porogen is used, which disrupts the hydrogen bonding between the MAA and the target molecule, no imprinting effect will occur and subsequently no specific binding behavior will be observed. Obviously, this disruption or destabilization of hydrogen bonding by the porogen is dependent on the actual strength of the hydrogen bonding between MAA and the target molecule.

In addition, if the hydrogen bonding between the MAA monomers themselves is stronger as compared to the hydrogen bonding between MAA and the target molecules, no imprinting effect will be possible. This process is referred to as MAA dimerization, which is often present in NIPs, since no target is present during polymerization [97–99]. Dimerization is strongly dependent on the porogen polarity. The more apolar the solvent, the more the process of dimerization of MAA will be favored. Hence, when hydrogen bonding between MAA and the target molecule is not very strong, it is advisable to use an apolar porogen, which will stabilize the weaker pre-polymerization complex. This stabilization allows for a better imprinting effect. On the other hand, when a more polar porogen is used, insufficient stabilization will be present for this complex. This will result in the exposure of free carboxylic acid groups from MAA inside the pore, which changes the polarity of the pore. When an apolar porogen is used, dimerization of MAA will be present along with imprinting. It can be useful for the MIP to have some dimerization between the excess of MAA monomers present, since this will reduce aspecific binding afterward [97]. However, as previously mentioned, this dimerization process is more abundant in the NIP. As a result, when apolar porogens are used in the NIP synthesis, not many free carboxylic groups will be exposed. This will reduce unwanted aspecific binding on the NIP.

As this discussion makes evident, besides the direct stabilization or destabilization of the pre-polymerization complex, the porogen indirectly creates polar or apolar pore walls in the MIP matrix. This microenvironment is very decisive for the further binding behavior of a MIP [85,89,92,100]. The binding behavior in MIPs is usually studied with batch rebinding experiments, which make it possible to differentiate between specific and aspecific binding. A good imprinting effect will create more specific binding sites in a MIP. It is possible to calculate the corresponding binding constant together with the amount of binding sites present for each type of binding site [101,102].

Hence, in order to change the binding behavior of MIPs in water, a strategy has to be developed, which allows for tailoring of the polarity of the pore walls. This section focuses on one method to achieve this, that is, porogen tuning. However, it should be emphasized that a similar effect can also be obtained by carefully switching the polarity of the monomers and/or cross-linkers utilized for the MIP/NIP synthesis. An example of such an approach is the use of the monomer HEMA during the MIP synthesis, using chloroform as the porogen [84]. It is important to keep in mind that increasing the amount of HEMA creates a more polar MIP matrix, which in turn may also lead to more aspecific binding sites. Alternatively, the MIP matrix can also be made apolar with, for example, the cross-linker divinylbenzene (DVB).

4.2. Porogen Tuning

Tuning the MIP matrix polarity, by changing the polarity of the porogen, is expected to have a direct influence on the binding behavior. Therefore, five additional porogens of varying polarity have been utilized to prepare L-nicotine imprinted MIPs: acetonitrile,

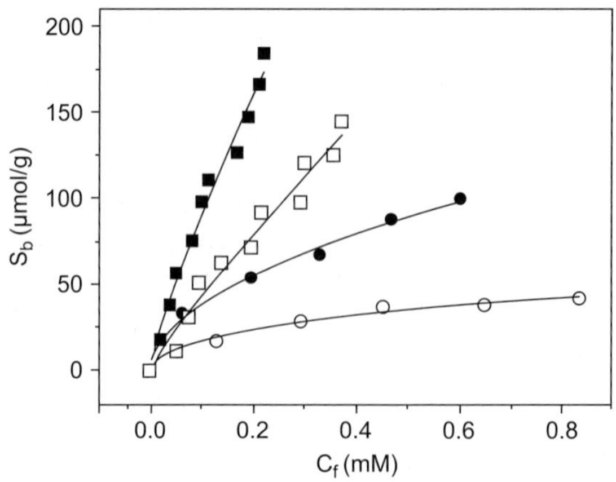

FIGURE 5.6 Binding isotherms of the L-nicotine imprinted bulk MIP (solid symbols) and corresponding NIP (open symbols) prepared with the porogen acetonitrile. The rebinding solvent was either acetonitrile (circles) or water (squares).

dimethylsulfoxide (DMSO), methanol, toluene, and hexane. Batch rebinding experiments have been performed for each MIP/NIP couple in acetonitrile and water. The binding behavior in acetonitrile makes it possible to quantify the specific binding based on hydrogen bonding and other noncovalent interactions, with minimal aspecific binding since no hydrophobic effect is present. The binding behavior in water shows both the specific imprinting effect and the aspecific binding associated with the hydrophobic effect.

As was noted in the previous section, the MIP made with the porogen chloroform performs well in acetonitrile but poorly in water (Fig. 5.5). Hence, it is interesting to see whether a more polar porogens would improve the performance in water. From these experiments it became immediately evident that the MIPs made with the more polar porogens DMSO and methanol exhibit little imprinting. Both in acetonitrile and water only minimal differences in binding behavior are observed between the MIPs and the corresponding NIPs. This suggests that there is insufficient stabilization of the pre-polymerization complex of MAA and L-nicotine during MIP synthesis, resulting in a random distribution of carboxylic acid groups, since no dimerization of MAA has occurred under these conditions. In this way, the MIPs are comparable to the NIPs. The use of the porogen acetonitrile leads to a slightly reduced performance in acetonitrile as compared to the use of chloroform (Fig. 5.6). In contrast, in water the performance of the same MIP is improved, which indicates that the hydrophobically driven aspecific binding is reduced owing to the increased polarity of the pore walls due to the polar porogen. Nonetheless, the amount of binding in the corresponding NIP remains rather high, indicating that a further reduction of the aspecific binding is essential.

In marked contrast, for the MIPs made with apolar porogens, such as toluene and hexane, it is clear that the imprinting effect is very well established and that aspecific binding in acetonitrile is negligible (Fig. 5.7). This can be clearly seen for the NIP, which shows almost no binding of L-nicotine. As discussed earlier, apolar porogens will tend to enhance dimerization of MAA in MIPs. Therefore, the NIPs have almost no free carboxylic groups,

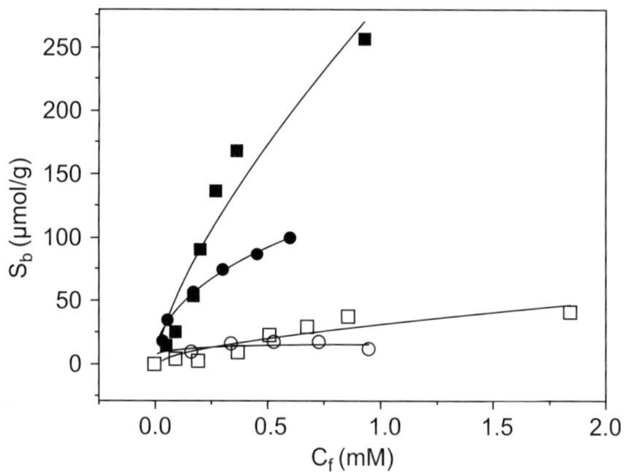

FIGURE 5.7 Binding isotherms of the L-nicotine imprinted bulk MIP (solid symbols) and corresponding NIP (open symbols) prepared with the porogen hexane. The rebinding solvent was either acetonitrile (circles) or water (squares).

creating a totally apolar matrix, to which the polar L-nicotine cannot bind. Besides dimerization, apolar porogens also stabilize the pre-polymerization complex in the MIP. However, it should be noted that these observations are strongly dependent on the target molecule of interest, that is, the stability of the pre-polymerization complex, and that for other targets other porogens may give better results (cf. Section 4.4. for histamine MIPs). In addition, solubility of the template in the porogen is important, since apolar solvents, such as hexane and toluene, are not always optimal for dissolving biologically relevant molecules, which have a tendency to be polar.

4.3. L-Nicotine Sensor with Water-Compatible MIPs

Prior to testing the L-nicotine MIP and NIP prepared with the porogen hexane in the QCM sensor platform, the specificity of the MIP has been confirmed with the structural analogue molecule L-cotinine in water. Subsequently, different L-nicotine and L-cotinine concentrations in water have been added to the sensor, and the frequency changes of the MIP and NIP channels have been measured (Fig. 5.8).

As expected, the MIP channel exhibits a clear response to L-nicotine, whereas the NIP channel responds much less. As a result, it is again possible to subtract the MIP from the NIP channel to compensate for the sensor response associated with the aspecific binding (Fig. 5.9). A linear response is found from 0 to 250 μM before saturation occurs. These measurements confirm that a good sensor performance can be achieved for the detection of L-nicotine in water over a large concentration range, using MIPs made with apolar porogens.

4.4. Histamine Sensor with Water-Compatible MIPs

The advantage of porogen tuning to achieve sensor performance in aqueous media can be further exemplified by our MIP sensor to detect histamine. Histamine is a medically relevant

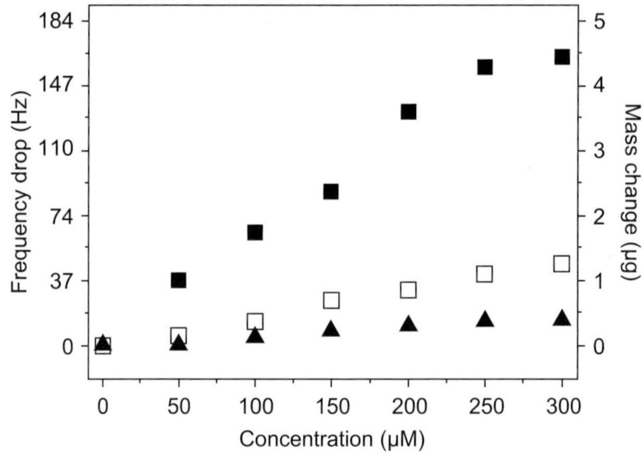

FIGURE 5.8 QCM dose-response curves for the L-nicotine MIP and NIP prepared with the porogen hexane exposed to increasing concentrations of L-nicotine and L-cotinine in water (MIP exposed to L-nicotine: solid squares; MIP exposed to L-cotinine: solid triangles; NIP exposed to L-nicotine: open squares).

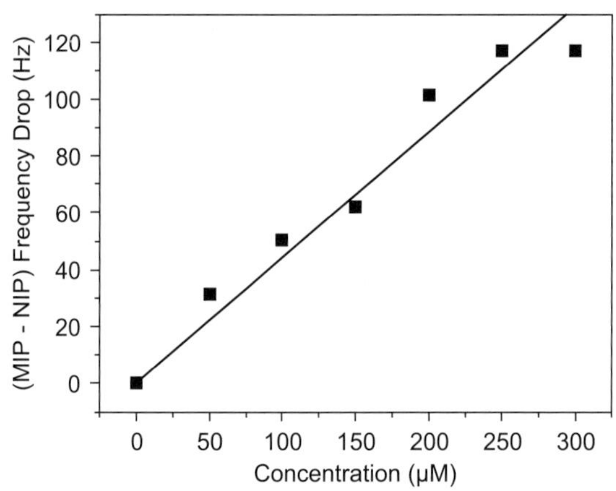

FIGURE 5.9 Difference between the MIP and NIP exposed to increasing concentrations of L-nicotine.

target (Fig. 5.10), which plays a crucial role in allergic reactions. For example, histamine detection is of interest for research into the Irritable Bowel Syndrome (IBS) [103] since a high concentration of histamine in the intestine can be linked to IBS. Since the detection of histamine is important in food science and biomedical testing, the developed sensor platforms may find use in a variety of applications.

Typical histamine concentrations in human plasma range from approximately 1 to 15 nM [104], and the actual histamine concentration can be indicative of certain diseases [105]. Hence, concentration detection in the nanomolar range is of significant importance. Another possible application for a histamine biosensor is the verification of the freshness of food samples. Spoiled foods contain higher histamine levels due to the degradation of proteins

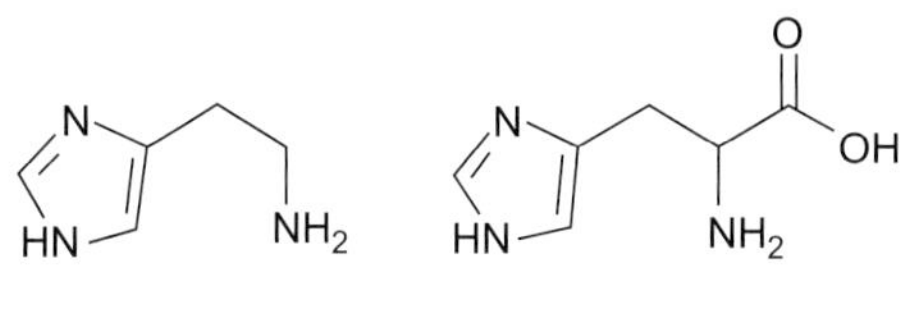

FIGURE 5.10 Structures of histamine and histidine.

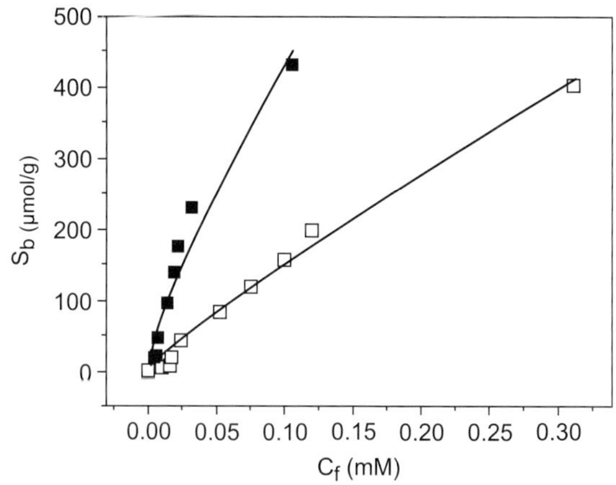

FIGURE 5.11 Binding isotherms for the histamine-imprinted bulk MIP (solid squares) and corresponding NIP (open squares) prepared with the porogen DMSO.

[106]. For example, fish poisoning can occur if spoiled fish contains more than 50 mg of histamine per 100 g of fish [107]. The most common methods used to measure histamine concentrations are high-pressure liquid chromatography (HPLC), gas chromatography (GC), or enzyme-linked immunosorbent assay (ELISA) [108,109]. However, these techniques require extensive sample preparation procedures and are laborious to execute. Since only a limited number of reports exist on MIPs toward histamine as the target molecule [13,110,111], the MIP synthesis was optimized to obtain suitable MIPs for use in sensors to detect histamine in aqueous media.

After careful optimization, it was found that the best performing MIP in this case had the composition ratio Histamine:MAA:EGDM of 1:2:4. In contrast to the L-nicotine MIP (cf. Section 3.1.), it was found that the stability of the pre-polymerization complex is considerably better, probably as a result of the presence of the primary amine, which can interact strongly with the MAA. Hence, the use of polar porogens becomes more favorable. After optimization, also considering the solubility of histamine, it was concluded that the best porogen in this case would be DMSO. The binding isotherms of the resulting MIP and NIP are shown in Fig. 5.11. The results confirm the higher binding affinity of histamine for the MIP as compared to the corresponding NIP.

FIGURE 5.12 Typical SEM image of a small section of the MIP covered surface with the MIP particles (white) visible against the conjugated polymer background (gray).

For comparison, the same MIP and NIP have also been exposed to various concentrations of the structural analogue histidine (Fig. 5.10). Analysis of the binding isotherms shows that the amounts of histidine bound to the MIP and NIP are more or less equal and significantly lower as compared to the histamine binding characteristics. The residual low binding of histidine originates from aspecific interactions with the MIP and NIP. Hence, a satisfactory specificity is achieved, and the sensor can be expected to exhibit no response to even less similar molecules.

Since it is evident that the optimized MIP can bind histamine in a selective manner, the MIP and NIP have been integrated into two different sensor setups. In the first setup, the chemical recognition occurring at the MIP and NIP covered surfaces is directly translated into an electrical signal using impedance spectroscopy. The sensor coverage can be readily seen with optical microscopy and SEM (Fig. 5.12). These techniques also confirm that the coverage of MIP and NIP is very similar, which is important for the sensor measurements to be able to subtract the signal of MIP and NIP from each other to compensate for nonspecific binding.

For the actual measurements, the sensor cell is filled with PBS to simulate a biological environment, and increasing concentrations of histamine have been added in a stepwise manner. Subsequently, the impedance is measured and a clear dose-response curve is obtained (Fig. 5.13, top). Because MIPs typically contain a heterogeneous distribution of different binding sites, with a concomitant large variety in binding constants [102], this curve is not linear, but more closely resembles a typical binding isotherm. The stability of the measured impedance signal is excellent, and the noise level is very low. Due to the low signal-to-noise ratio of the measurements, the limit of detection is well below 1 nM. Since the NIP only exhibits nonspecific binding, the NIP sensor response can be utilized as a reference, which can be subtracted from the MIP sensor response, as was already discussed in the previous sections. The resulting MIP - NIP response therefore corresponds to the specific binding of histamine (Fig. 5.13, bottom).

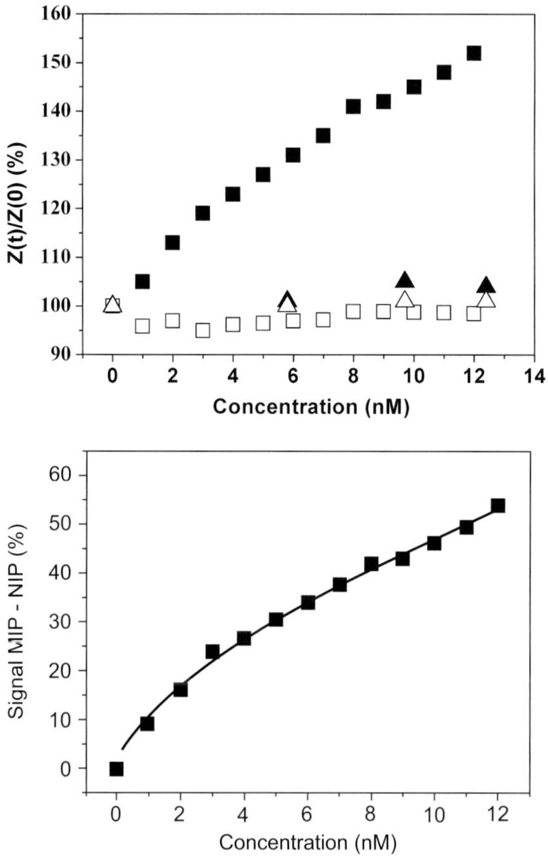

FIGURE 5.13 Impedance spectroscopy dose-response curves for MIP and NIP exposed to histamine and histidine. (MIP exposed to histamine: solid squares; MIP exposed to histidine: solid triangles; NIP exposed to histamine: open squares; NIP exposed to histidine: open triangles) (top) and difference between the MIP and the NIP exposed to histamine (bottom).

Whereas the impedimetric setup exhibits excellent performance in the nanomolar range, interest exists in a wider concentration range. To this end, we have used our second sensor platform based on QCM (cf. Section 2.3). After immobilization of the MIP and NIP particles, the resultant surface coverage has again been assessed using microscopy. The surface coverage obtained from optical microscopy is somewhat lower than observed for the OC_1C_{10}-PPV immobilization. This can be explained by the increased hardness of the PVC as compared to OC_1C_{10}-PPV, which apparently somewhat reduces the matrix entrapment process. Even so, the surface coverage of the MIP and NIP on each crystal is very similar, which is important for the QCM measurements. After integration of the MIP and NIP covered crystals in the sensor setup, the crystals have been exposed to a series of different concentrations (1 μM to 100 μM) of histamine and histidine. This resulted in typical dose-response curves (Fig. 5.14, top).

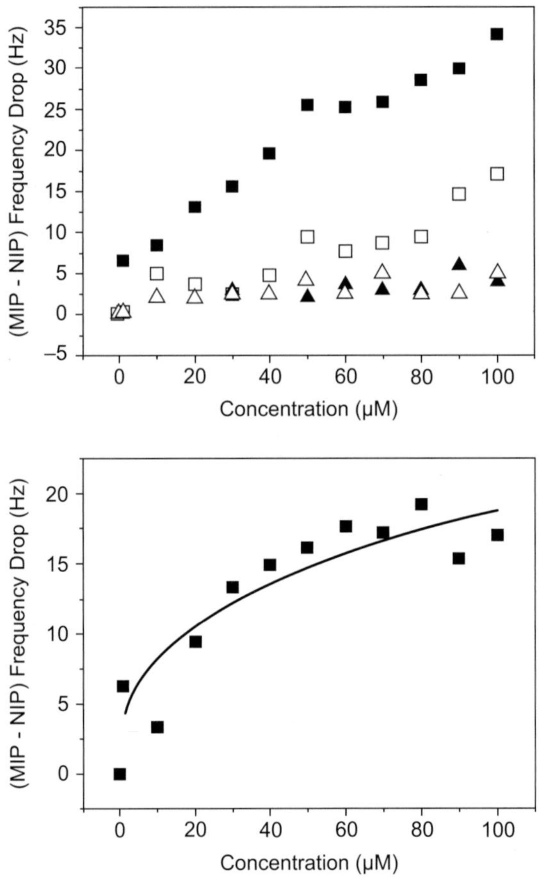

FIGURE 5.14 QCM dose-response curves for MIP and NIP exposed to histamine and histidine (MIP exposed to histamine: solid squares; MIP exposed to histidine: solid triangles; NIP exposed to histamine: open squares; NIP exposed to histidine: open triangles) (top) and difference between the MIP and the NIP exposed to histamine (bottom).

The observed frequency shifts for the MIP exposed to histamine are significantly higher than those for the NIP exposed to histamine. The shape of the dose-response curve is similar as observed for the impedimetric experiments. Due to aspecific binding, the NIP exposed to histamine exhibits a slightly higher frequency shift for the larger concentrations. This rise can be attributed to the significantly higher concentrations used in these experiments, as compared to the impedimetric measurements. This can be explained by the fact that the NIP synthesized from the same optimized mixture of monomers, cross-linkers, and porogen (but without the template being present), can be expected to have also a somewhat higher amount of aspecific binding sites for histamine.

It is again possible to subtract the NIP from the MIP channel to visualize the pure specific binding. The result is a dynamic concentration range from 0 to 60 μM (Fig. 5.14, bottom). In contrast to the results for the exposure to histamine, for both the MIP and the NIP exposed to histidine the frequency remains reasonably constant. As expected, there are no specific binding sites for histidine, and only aspecific binding is observed. This confirms the previous observations on the excellent specificity of MIP-based sensors.

5. SUSPENSION POLYMERIZATION MIPs FOR SENSORS

5.1. Synthesis of Suspension Polymerization MIPs for L-Nicotine

Whereas the compatibility with water can be solved by adapting the synthesis parameters of the bulk polymers, the two other drawbacks identified in Section 3—the random morphology and associated irreproducible sensor coverage and the insufficiently uniform binding characteristics—require additional optimizations. Although often superior in binding properties, the origin of these drawbacks originates from the use of crushed monoliths. Hence, a polymerization strategy with a more uniform distribution of particle size and shape is desirable. We have tested a large variety of synthetic strategies toward a modified MIP morphology and have found that a suspension polymerization is most attractive in the context of MIP sensor applications. The aim of a suspension polymerization is to combine a uniform distribution of round-shaped MIP particles with good enough binding behavior. Therefore the bulk method to synthesize MIPs has been adapted to form small micro reactors inside a dispersing agent. The largest advantage of a suspension over a bulk polymerization is the uniform distribution of MIP beads instead of the ground and sieved irregular-shaped MIP particles. In addition, the MIP polymerization process is more homogeneously taking place, which will result in particles with more homogenous binding characteristics. Therefore, a suspension polymerization combines the MIP synthesis with a very elegant way of producing uniform globular polymer particles.

Two suspension methods are described in the literature, which can be differentiated by the solvents and surfactants used. The first method is water-based [112–114], whereas in the second method a fluorinated solvent has been used [115–117]. Both methods require the use of a suitable surfactant to stabilize the micro containers or micelles. In view of the straightforward methodology, the first method has been further optimized for sensor applications.

Similar to the previously discussed MIPs, also in this case the monomer choice was limited to MAA, with EGDM as the cross-linker. We would like to emphasize once more that this selection is based on the template used and that other templates may require other monomers and cross-linkers to obtain a MIP with satisfactory properties. For the nicotine MIP, the same ratio template:MAA:EGDM of 1:2:4 was again used. To create a suitable dispersion, $CHCl_3$ was used as the porogen and H_2O was added to create a two-phase system. Finally, polyvinylalcohol (PVA) was used as the dispersion agent. It was found in particular that the amount of $CHCl_3$ had a considerable impact on the obtained morphology, as is exemplified in Table 5.1 and Fig. 5.15. This is not surprising, since in order to form stable micelles it is important to find a correct balance between the amount of monomers, porogen, dispersing solvent, and surfactant.

In this case, it has been found that Suspension MIP 1 has the correct morphology, whereas the morphology of MIPs with increased amounts of $CHCl_3$ shows a random distribution of

TABLE 5.1 Relative Molar Composition of Suspension MIPs. The Amount of L-nicotine is Fixed at 6.45 mmol.

	L-nicotine	MAA	EGDM	$CHCl_3$ (mL)	PVA (g)	H_2O (mL)
Suspension MIP 1	1	2	4	7	0.4	60
Suspension MIP 2	1	2	4	14	0.4	60
Suspension MIP 3	1	2	4	50	0.4	60

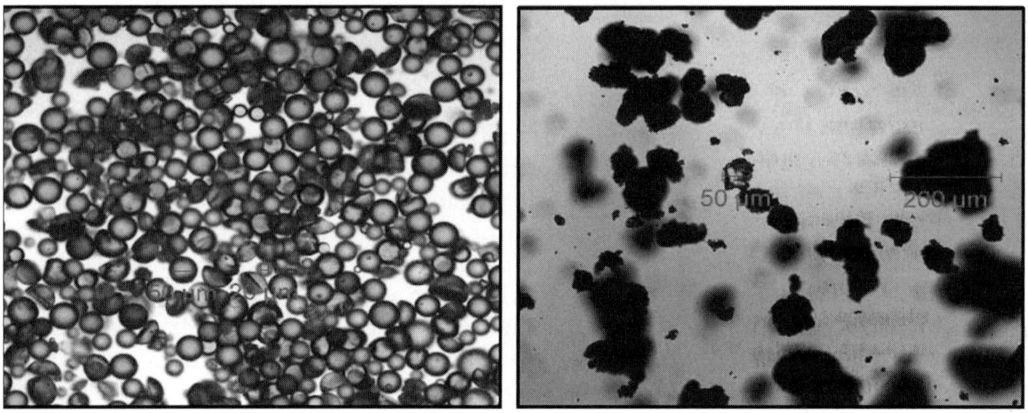

FIGURE 5.15 Light microscopy images of Suspension MIP 1 (left) and Suspension MIP 3 (right). The template for these MIPs is L-nicotine.

particle clusters. However, batch rebinding experiments indicate that the MIP with the most favorable morphology has only a comparatively small amount of specific binding (Fig. 5.16).

This rather small difference in binding site capacity, as compared to bulk MIPs can be explained by the fact that during MIP synthesis the pre-polymerization complex is somewhat disturbed due to the presence of water. Water is known to disturb the hydrogen bonding, and as a result, the outer shell of the MIP beads is expected to have a less well-defined imprinting effect. This suggests that suspension polymerizations do not necessarily lead to improved sensor performance. However, one can anticipate that for templates with stronger pre-polymerization complexes, the increased aspecific binding and associated reduced imprint effect should be considerably less pronounced.

5.2. Synthesis of Suspension Polymerization MIPs for Malachite Green

An example of a target molecule that forms a stronger pre-polymerization complex is Malachite Green (MG), which is of considerable interest for sensing applications, since in many parts of the world it is used as an inexpensive method of controlling fungal attacks and infections in aquaculture. However, this class of dyes is linked to carcinogenesis and respiratory toxicity [118,119]. As a result, it is extremely desirable to monitor the concentration of these

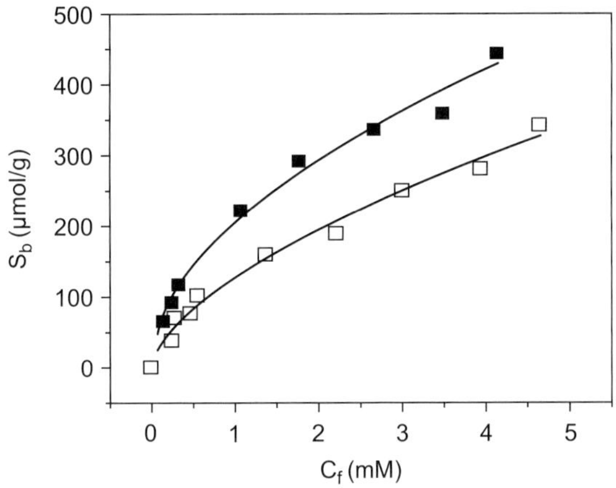

FIGURE 5.16 Binding isotherms of Suspension MIP 1 (filled squares) and corresponding NIP (open squares) for the template L-nicotine.

dyes in water and fish. Currently, state-of-the-art MG detection is performed with HPLC and LC-MS-MS, with a detection limit of 1 µg/kg and 0.2 µg/kg, respectively [120]. However, a sensor alternative with a similar limit of detection would allow for a faster and lower cost detection. Although in the literature many low-cost biomass materials have been tested as adsorbents for MG, such materials do not have sufficient specificity [121–124]. To bind MG in a more specific way, MIPs are a class of artificial receptors with considerable potential [125–129].

In view of the above observations and the fact that for MG a strong pre-polymerization complex can be expected, a MIP suspension polymerization was optimized, similar to the one described in Section 5.1. After all optimizations, an optimal ratio MG:MAA:EGDM of 1:8:16 was used. To obtain optimal morphology and binding characteristics, 1 mL $CHCl_3$, 4 mL H_2O, and 0.02 g of PVA were used (amount of MG is 0.216 mmol). From the batch rebinding experiments it becomes evident that the obtained MIP/NIP combination indeed has considerably improved binding characteristics as compared to the suspension MIP/NIP for L-nicotine from Section 5.1 (Figs. 5.16 and 5.17). The batch rebinding experiments clearly confirm that a suspension polymerization in this case, that is, presence of a strong pre-polymerization complex, provides excellent results.

5.3. Malachite Green Sensors with Suspension Polymerization MIPs

For the QCM measurements, the quartz crystals were coated with the PVC polymer layer, after which the MIPs and NIPs were immobilized by matrix entrapment. Subsequently, the coverage was assessed using scanning electron microscopy (SEM) (Fig. 5.18). Repeated experiments resulted in little spread in the surface coverage. This is in contrast to the ground and sieved bulk MIPs, of which the surface coverage is quite irreproducible. As was mentioned in Section 2.2, such irreproducibility is problematic, since it is important for sensor applications that the surface coverage of the MIP and NIP on each crystal is very predictable.

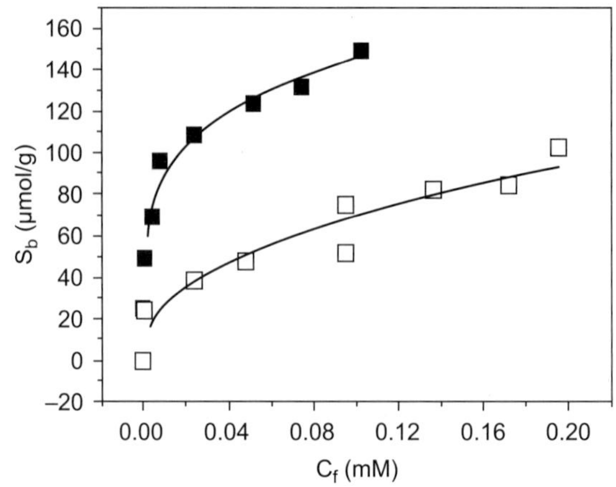

FIGURE 5.17 Binding isotherms of MG suspension MIPs (closed squares) and corresponding NIPs (open squares).

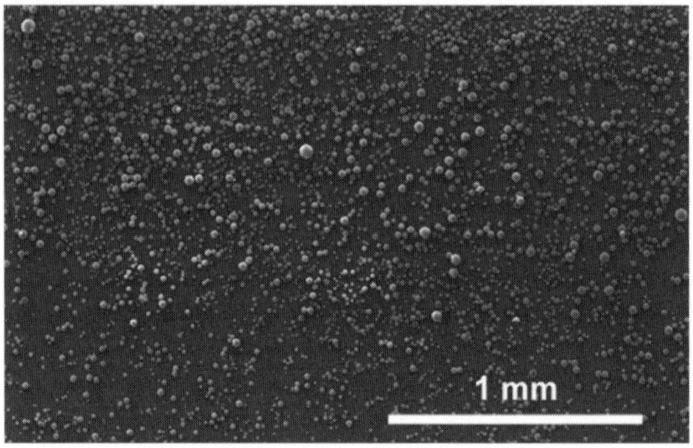

FIGURE 5.18 SEM image of the sensor surface.

Due to the suspension polymerization, each MIP bead has the same binding behavior for MG. This is not the case for a ground and sieved bulk MIP, which can exhibit large variations in the binding properties between different particles. Furthermore, the size distribution of the MIPs in a suspension polymerization is narrower compared to the sieved bulk particles. All these factors contribute to the uniformity and reproducibility of the sensor surface coverage. After matrix entrapment, the MIP and NIP covered crystals were placed in the sensor setup and exposed to different concentrations of MG (Fig. 5.19, top).

From the difference between MIP and NIP, which also here corresponds to the specific binding of MG, it becomes clear that using our current setup, a dynamic range of 200 nM to 1000 nM MG is possible for the QCM sensor (Fig. 5.19, bottom). The QCM measurements were executed in an acidic environment at pH 3 to suppress the aspecific binding of MG to the MIP as a result of hydrogen bonding. A similar strategy is often used in chromatographic

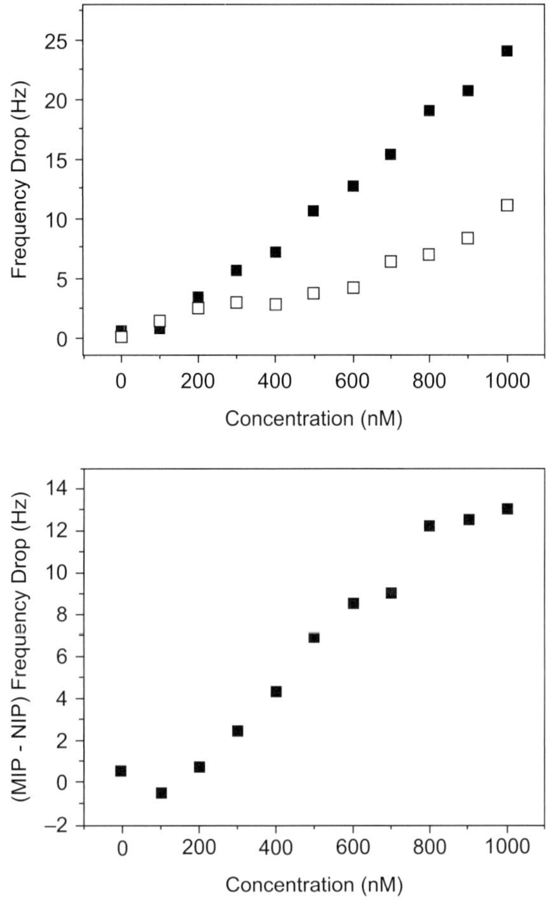

FIGURE 5.19 QCM dose-response curves for both the MIP and NIP exposed to MG (MIP: solid squares; NIP: open squares) (top) and difference between the MIP and the NIP response exposed to MG (bottom).

separation in which MIPs are employed as the solid phase material [130, 131]. Due to the low pH, it can be anticipated that hydrogen bonding between MG and the polymer matrix is disturbed.

To detect MG binding at lower concentrations, impedance measurements have been performed. As has been already discussed in the preceding sections, it has been demonstrated that impedance measurements with MIP-based sensors have a lower limit of detection than QCM measurements [74]. Measurements were performed in PBS at pH 7, since pH 3 is not compatible with the impedance sensor configuration. From the impedance measurements it is found that the MIP-channel exhibits an excellent response to MG (Fig. 5.20, top), whereas the NIP channel, which measures nonspecific binding of MG molecules to the sensor surface, shows a much lower response. The amount of aspecific binding in these experiments is larger than was found for the QCM measurements, which were performed at a lower pH. To visualize the specific binding, also in this case the NIP response was subtracted from the MIP

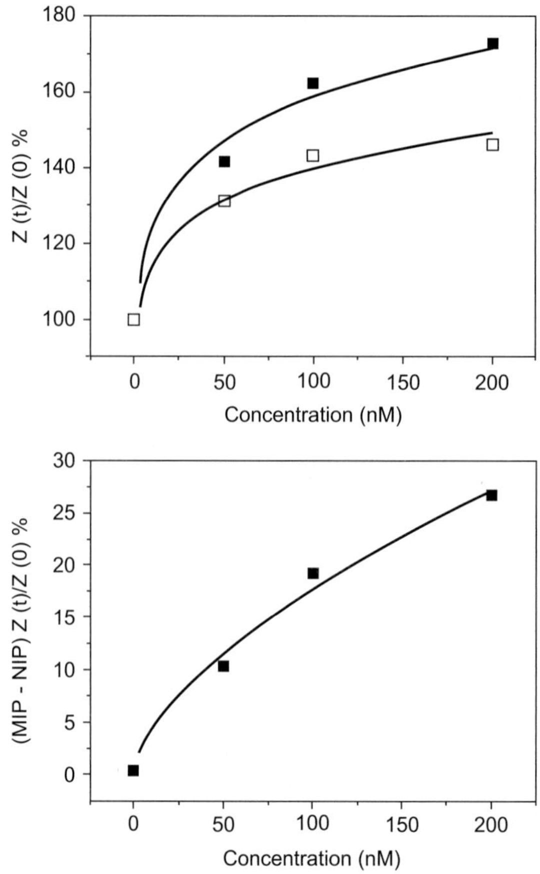

FIGURE 5.20 Dose-response curves for the impedance measurements of both the MIP and NIP exposed to increasing concentrations of MG (MIP: solid squares; NIP: open squares) (top) and difference between the MIP and the NIP exposed to increasing concentrations of MG (bottom).

response (Fig. 5.20, bottom), which makes it clear that use of impedance spectroscopy allows an additional dynamic measuring range from 0 nM to 200 nM.

6. SOME FINAL OBSERVATIONS CONCERNING MIP-BASED SENSING

6.1. MG MIP Binding Kinetics

For sensor measurements, it is important that the response time is as fast as possible. Depending on the measuring principle and the recognition elements used, the response time of a sensor can vary to a considerable degree. This is due to differences in physical and chemical processes occurring upon target binding and the subsequent detection of

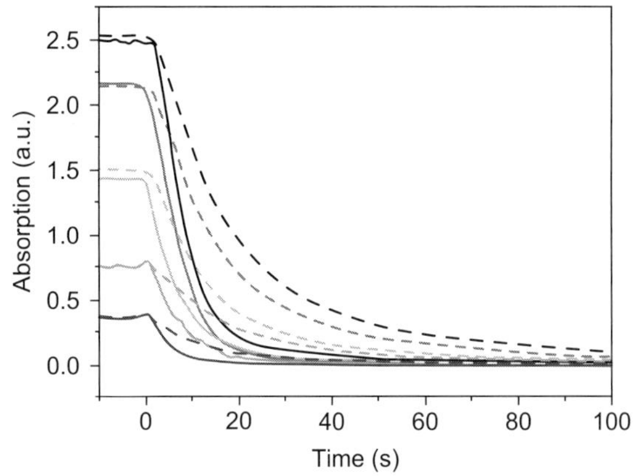

FIGURE 5.21 Kinetic experiments for 0.1 g of MG MIPs (solid) and corresponding NIPs (dashed) exposed to varying concentrations of MG in aquarium water (black 0.04 mM, red 0.03 mM, green 0.02 mM, orange 0.01 mM, pink 0.005 mM). For color version of the figure, the Reader is referred to the Online version of the book.

the binding event. Hence, in addition to the optimization of binding characteristics as described in the previous sections, it is also of interest to optimize the MIP synthesis to allow for good sensor performance in a reasonable time scale [73,74]. One of the most important aspects of a fast sensor response is associated with the binding kinetics between the target molecule and the MIP binding sites [132–134]. Fast binding allows for fast sensing, and, in combination with a strong target/MIP binding, it is possible to detect a given target molecule at very low concentrations. Whereas with batch rebinding experiments qualitative (binding constant) and quantitative (MIP capacity) statements can be made with respect to the binding behavior of a MIP [101,102], binding kinetics cannot be derived from this type of experiments. To study the binding kinetics, the MIPs have been exposed to various concentrations of MG, and the binding has been followed in real time (Fig. 5.21). From these experiments it can be seen that target binding takes longer for the NIPs as compared to the MIPs. However, all curves have about the same shape independent of the concentration.

Besides the addition of different MG concentrations, it is also possible to add different amounts of MIP or NIP to a fixed MG concentration (Fig. 5.22). This makes it possible to further increase the relative MG concentration. The absolute difference between the binding kinetics of MIP and NIP increases at higher ratios of MG to MIP/NIP. This can be represented in a more direct way by plotting the time at which MIP and NIP reach a decrease of 50% in the amount of bound MG (t_{50}) as a function of the amount of MG (Fig. 5.23). This plot confirms the general trend that target binding takes longer for the NIPs. The difference between binding kinetics of MIP and NIP is the largest when the ratio of MG to MIP/NIP is largest, that is, for high concentrations of MG or low amounts of MIP/NIP.

These experiments lead to the general conclusion that at a given point the equilibrium between the MG concentration and the amount of MIP or NIP used is out of balance, resulting in increasing t_{50} values, reflecting the "slower" population of binding sites, which are either less selective or are embedded deeper inside the MIP/NIP. Hence, to allow for a fast and selective response in MIP-based sensor applications, it is important to measure in a suitable

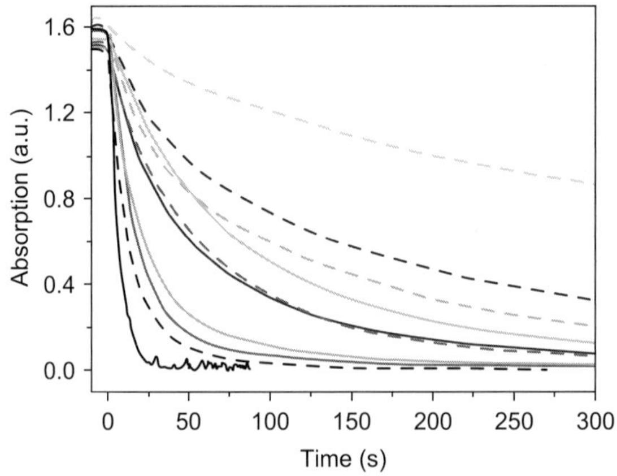

FIGURE 5.22 Kinetic experiments for a concentration of MG (0.02 mM) in aquarium water after addition of varying amounts of MIPs (solid) and NIPs (dashed) (turquoise 0.01 g, blue 0.02 g, green 0.04 g, red 0.06 g and black 0.1 g).

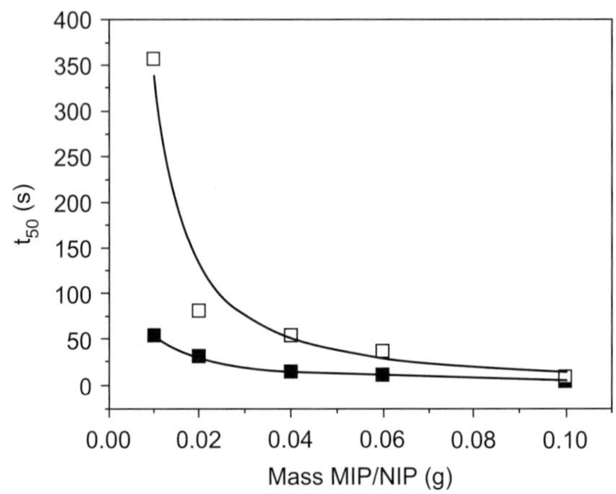

FIGURE 5.23 Graph of the t_{50} of MIP (closed squares) or NIP (open squares) exposed to a fixed MG concentration (0.02 mM) as a function of the amount of MIP or NIP added.

concentration range. It should be noted that all kinetic discussions thus far are associated with measurements in a solution, in which the MIP and NIP particles are freely dispersed. However, on an actual sensor surface these MIP or NIP particles are immobilized, which will further reduce the mass transport and therefore the binding kinetics. To verify this, MIP and NIP beads have been immobilized and exposed to MG (Fig. 5.24). Comparing these data with Figs. 5.21 and 5.22 indeed shows that the binding kinetics are considerably slower.

6.2. L-Nicotine Binding Kinetics of Water-Compatible MIPs

In addition to the studies with MG, kinetic experiments have been carried out with L-nicotine for the new water-compatible MIPs described in Section 4 (Figs. 5.25 and 5.26).

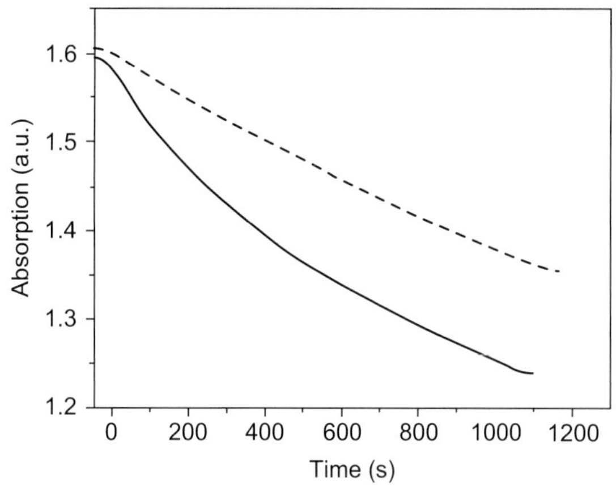

FIGURE 5.24 Kinetic experiment of 0.01g MIP beads (solid line) and 0.01g NIP beads (dashed line) immobilized to a substrate exposed to a MG concentration (0.02 mM).

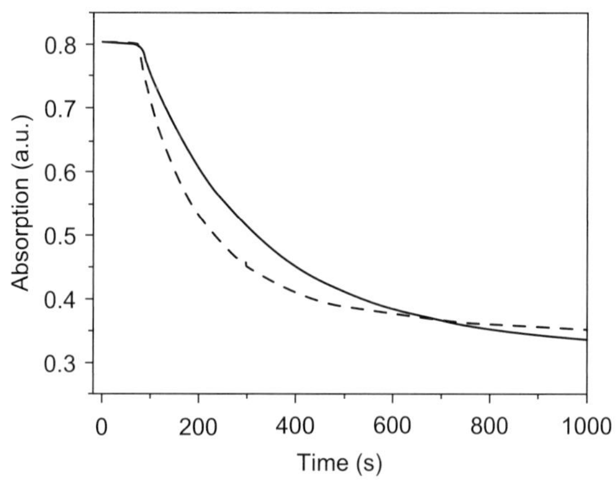

FIGURE 5.25 Kinetic experiments for 0.1 g of L-nicotine MIP and NIP made with the porogen chloroform carried out in water (L-nicotine concentration 0.3 mM).

Whereas, as expected, in aqueous media, the performance of the L-nicotine MIP made with the porogen chloroform exhibits about the same performance as the corresponding NIP, for apolar porogens, such as toluene and hexane, not only a significant difference is found, but these MIPs also reach their binding equilibrium very fast and more or less on the same time scale. This is favorable for the implementation of this MIP/NIP combination for sensor measurements. Besides fast binding kinetics, it is also beneficial for a reusable sensor to be regenerated quickly. Hence, it is also of interest to study the release kinetics of these MIPs by monitoring L-nicotine release as a function of time (Fig. 5.27).

It has been found that the toluene MIP exhibits the fastest release of L-nicotine and that the observed release kinetics are in excellent agreement with the binding kinetics (Figs. 5.25 and 5.26). This experiment shows that binding and release kinetics are very similar and that the

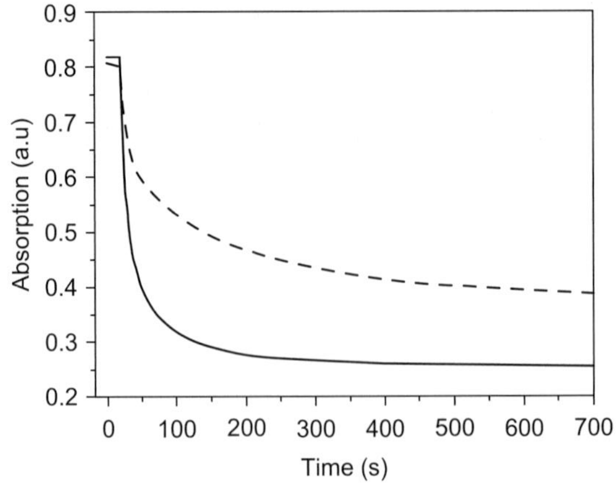

FIGURE 5.26 Kinetic experiments for 0.1 g of L-nicotine MIP and NIP made with the porogen toluene carried out in water (L-nicotine concentration 0.3 mM).

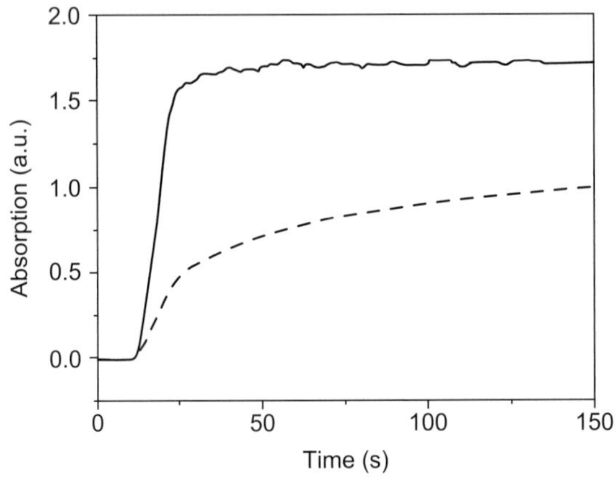

FIGURE 5.27 L-nicotine release from MIPs made with the porogen toluene (solid line) or chloroform (dashed line).

data from a binding experiment can be utilized to predict the release kinetics and vice versa. Finally, the excellent binding and release characteristics of the MIP made with the porogen toluene have been tested with the QCM sensor setup (Fig. 5.28). As expected, the QCM results show a good response time upon the addition of L-nicotine, and regeneration is also possible on the same time-scale. Besides fast binding and regeneration, a reusable sensor needs to be specific toward L-nicotine. This has been confirmed by exposing the QCM sensor to the same concentration of L-cotinine.

6.3. MIP-Based Sensing in Actual Samples

Most of the sensor measurements in this chapter have been performed in pure solvents or buffer solutions. Ultimately, however, these sensors should perform well in actual

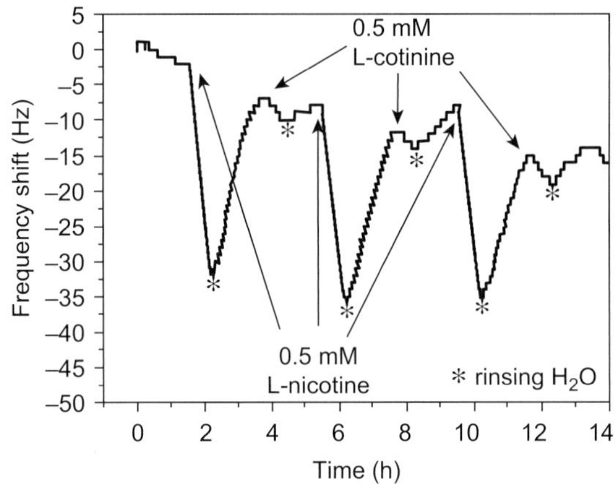

FIGURE 5.28 QCM response of the sensor containing the toluene MIP after the addition of 0.5 mM of L-nicotine and 0.5 mM of L-cotinine. The sensor is rinsed with distilled water to test its reproducibility (rinsing marked with an asterisk).

applications, which usually involve rather complex systems. Fortunately, MIPs have already proven their performance for separation of complex samples [135–137]. They are highly selective and very stable and are not easily contaminated to a degree that would affect their ability to bind relevant target molecules. Over the past years, we have tested the viability of our MIP-based sensors for a variety of applications. Although it is beyond the scope of this chapter to detail these applications, it should be noted that the application range is broad. Examples include histamine in canned fish samples [74], MG in water from a fish tank [138], and serotonin in blood [139]. The sensors not only perform well under these conditions, but interestingly, they also sometimes require dilution of the samples due to the extremely high sensitivity.

A special concern is the pH of the actual samples, since the pH has a large impact on the binding properties of a MIP/NIP combination. Hence, an issue that should be addressed prior to application of the MIP-based sensors is the pH dependence of the sensor response [140]. This is a very important consideration, since actual samples can exhibit a wide range of pH values and it can be anticipated that the sensing results will change when measured at different pH. This is a result of the fact that the MIP binding is based on hydrogen bonding, which is strongly influenced by the pH. Protonation and deprotonation of the target and the polymer matrix at a given pH depend on the pKa of the functional groups present. For every MIP/target couple there is an optimal measuring pH, at which the occurrence of hydrogen bonding between the MIP binding sites and the target molecule is maximal. To understand the pH dependence, it is therefore of interest to investigate the protonation and deprotonation of both the target and the MIP in more detail. Protonation of the MIP binding sites (pKa of the carboxylic acid binding sites in the MIP matrix based on MAA is circa 6.5) is also pH dependent (Fig. 5.29).

It is possible to do a similar calculation for the target molecule. This can be exemplified for histamine, which can be present in three forms with various levels of protonation depending on the pH. This is a result of the fact that it has different pKa values for each

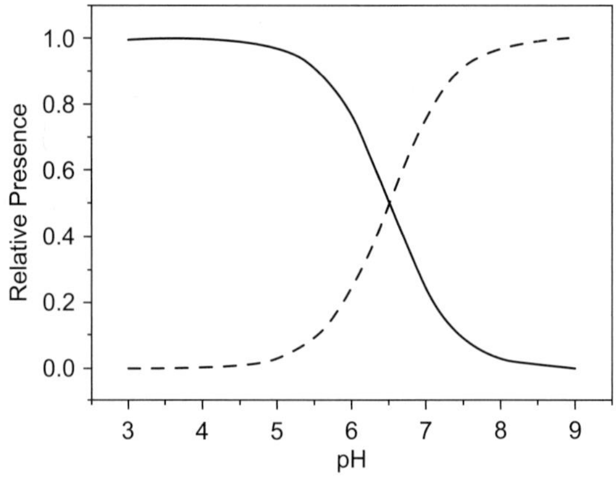

FIGURE 5.29 Sketch of the pH dependent distribution of the protonated (solid) and deprotonated (dashed) form of the carboxylic acid groups (MAA) in the MIP matrix.

of the present nitrogen groups (pKa 6.9 and 10.4) [140–142]. The pH dependence will affect the hydrogen bonding process between histamine and the MIP binding sites, and depending on the pH there will be strong, medium, or weak binding between the target molecule and the MIP matrix. Hence, based on the pKa estimates, it is possible to predict the optimal pH range for sensing. In the case of histamine sensing, measuring at a pH range 7–9 has been found to be optimal since in that range the largest amount of hydrogen bonds is present between the MIP and histamine. These estimates have been confirmed with sensor measurements, and it has also been found that at pH 5 and pH 12 no binding is present between MIP and histamine [140].

By changing the synthesis, this optimal range can be modified. This is important, since for a variety of applications, measurements at somewhat lower pH values can be of considerable interest. Hence, in order to improve the sensor operation at lower pH values, a MIP with a different pKa value, that is, a different combination of functional monomers, is needed. In the case of histamine sensing, the pKa of the functional monomer responsible for the hydrogen bonding should be lower than 6.5 to remain deprotonated at pH 5. This would allow for the formation of strong hydrogen bonds with the protonated histamine at this pH. In this context one can think of, for example, the use of acrylic acid as the functional monomer, since it has a pKa between 4 and 5. The viability of this approach has indeed been verified with batch rebinding studies and sensor measurements.

7. CONCLUDING REMARKS

This chapter focuses on our strategy for a universally applicable MIP-based sensor platform. The intrinsic benefits of MIPs make them very interesting for use as a class of artificial receptors in sensors. Our focus is on MIP particles, and it is shown that immobilization of these particles is straightforward using a stamping method, which results in matrix entrapment of the MIPs. Target binding in the sensors is achieved using two methods, which allow for miniaturization, that is, impedance spectroscopy and QCM. Whereas impedance spectroscopy allows for detection in the nanomolar range, QCM makes it possible to detect targets in the micromolar range. Bulk polymerization with the porogen chloroform is the most straightforward method for MIP synthesis and involves less optimization steps compared to other types of polymerization. After a bulk polymerization, irregular-shaped particles are obtained by grinding and sieving the monolithic MIPs. Although impedimetric sensor measurements with these crushed monoliths are quite promising, the available concentration range is sometimes rather small, the nonspecific binding is large, and the margin of error quite significant. In addition, several drawbacks can be identified, such as the limited compatibility with aqueous samples, the random morphology and associated irreproducible sensor coverage, and the insufficiently uniform binding characteristics.

Further improvements of the MIP synthesis towards sensor applications provide a variety of solutions for these drawbacks. Two of the most promising approaches are outlined in this chapter. First of all, it is very important to design a MIP capable of detecting the target of interest in aqueous environments without losing specificity. This is possible by carefully adjusting the polarity of the MIP matrix by porogen tuning. In this way, it is possible to detect low concentrations of target molecules in, for example, food samples and plasma. MIP morphology is also an important issue, since for sensor purposes it is better to integrate spherical MIP beads. Not only is the imprinting effect in these MIP particles more uniform due to the more homogeneous character of the polymerization, but also the reproducibility of the MIP immobilization process in the recognition layer is enhanced. The latter is especially important, since for the fabrication of reliable and reproducible sensors a more uniform covered sensor surface is desirable. In conclusion, we have demonstrated that MIP-based sensors can be readily constructed and that these sensors have considerable potential as low-cost, mass-produced, hand-held devices for a wide range of applications.

References

[1] T. Vo-Dinh, B. Cullum, Anal. Chem. 366 (2000) 540.
[2] F. Ricci, G. Volpe, L. Micheli, G. Palleschi, Anal. Chim. Acta 605 (2007) 111.
[3] P.D. Skottrup, M. Nicolaisen, A.F. Justesen, Biosens. Bioelectron. 24 (2008) 339.
[4] K. Yano, I. Karube, TrAC Trend, Anal. Chem. 18 (1999) 199.
[5] V.B. Kandimalla, H. X Ju, Anal. Bioanal. Chem. 380 (2004) 587.
[6] C. Alexander, H.S. Andersson, L.I. Andersson, R.J. Ansell, N. Kirsch, I.A. Nicholls, M.J. J.O'Mahony, Whitcombe, J. Mol. Recognit. 19 (2006) 106.
[7] B. Sellergren, C.J. Allender, Adv. Drug Deliver. Rev. 57 (2005) 1733.
[8] H. Yan, K. Row, Int. J. Mol. Sci. 7 (2006) 155.
[9] P.K. Owens, L. Karlsson, E. Lutz, L.I. Andersson, Trac-Trend, Anal. Chem. 18 (1999) 146.
[10] A. Valero-Navarro, P.C. Damiani, J.F. Fernández-Sánchez, A. Segura-Carretero, A. Fernández-Gutiérrez, Talanta 78 (2009) 57.
[11] A. Rachkov, S. McNiven, A. El'skaya, K. Yano, I. Karube, Anal. Chim. Acta 405 (2000) 23.
[12] J.L. Suárez-Rodríguez, M.E. Díaz-García, Anal. Chim. Acta 405 (2000) 67.
[13] A. Tong, H. Dong, L. Li, Anal. Chim. Acta 466 (2002) 31.
[14] C Chow M.H., W. Lam, M.K.P. Leung, Anal. Chim. Acta Rev. 466 (2002) 17.
[15] J. Tan, H. Wang, X. Yan, Anal. Chem. 81 (2009) 5273.
[16] G. González, P. Hernando, J. Durand Alegría, Anal. Bioanal. Chem. 394 (2009) 963.
[17] M.C. Moreno-Bondi, J.L. Urraca, E. Benito-Pea, F. Navarro-Villoslada, S.A. Martins, G. Orellana, B. Sellergren, Proc. SPIE 6619 (2007) 66190C.
[18] C.J. Stephenson, K.D. Shimizu, Polym. Int 56 (2007) 482.
[19] E.L. Holthoff, F.V. Bright, Anal. Chim. Acta 594 (2007) 147.

[20] J. Matsui, T. Sodeyama, Y. Saiki, T. Miyazawa, T. Yamada, K. Tamaki, T. Murashima, Biosens. Bioelectron. 25 (2009) 635.
[21] W. Thongchai, B. Liawruangath, S. Liawruangrath, G.M. Greenway, Talanta 82 (2010) 560.
[22] P. Chang, Z. Zhang, C. Yang, Anal. Chim. Acta 666 (2010) 70.
[23] J. Yu, C. Zhang, L. Ge, P. Dai, S. Ge, Anal. Sci. 25 (2009) 1351.
[24] J. Yu, F. Wan, D. Ping, S. Ge, L. Bo, J. Huang, Anal. lett. 42 (2009) 746.
[25] Y. Fang, S. Yan, B. Ning, N. Liu, Z. Gao, F. Chao, Biosens. Bioelectron. 24 (2009) 2323.
[26] L. Wang, Z. Zhang, Sens. Actuators B: Chem. 133 (2008) 40.
[27] Q. Feng, L. Zhao, W. Yan, F. Ji, Y. Wei, J. Lin, Anal. Bioanal. Chem. 391 (2008) 1073.
[28] L. Wang, Z. Zhang, L. Huang, Anal. Bioanal. Chem. 390 (2008) 1431.
[29] L. Wang, Z. Zhang, Anal. Chim. Acta 592 (2007) 115.
[30] O.Y.F. Henry, D.C. Cullen, S.A. Piletsky, Anal. Bioanal. Chem. 382 (2005) 947.
[31] G. Guan, B. Liu, Z. Wang, Z. Zhang, Sensors 8 (2008) 8291.
[32] G.E. Southard, K.A. Van Houten, E.W. Ott Jr., G.M. Murray, Anal. Chim. Acta 581 (2007) 202.
[33] T. Matsunaga, T. Hishiya, T. Takeuchi, Anal. Chim. Acta 591 (2007) 63.
[34] A. Kugimiya, T. Takeuchi, Biosens. Bioelectron. 16 (2001) 1059.
[35] M. Lotierzo, O.Y.F. Henry, S. Piletsky, I. Tothill, D. Cullen, M. Kania, B. Hock, A.P.F. Turner, Biosens. Bioelectron. 20 (2004) 145.
[36] R.B. Pernites, R.R. Ponnapati, R.C. Advincula, Macromolecules 43 (2010) 9724.
[37] S.C. Huang, G.B. Lee, F.C. Chien, S.J. Chen, W.J. Chen, M.C. Yang, J. Micromech. Microeng. 16 (2006) 1251.
[38] K. Kantarovich, I. Tsarfati, L.A. Gheber, K. Haupt, I. Bar, Biosens. Bioelectron. 26 (2010) 809.
[39] W. Wen, H. Shitang, L. Shunzhou, L. Minghua, P. Yong, Sens. Actuators B: Chem. 125 (2007) 422.
[40] M. Romero Guerra, I. Chianella, E.V. Piletska, K. Karim, A.P.F. Turner, S.A. Piletsky, Analyst. 134 (2009) 1565.
[41] R. Say, A. Gültekin, A.A Özcan, A. Denizli, A. Ersöz, Anal. Chim. Acta 640 (2009) 82.
[42] C. Sontimuang, R. Suedee, B. Canyuk, N. Phadoongsombut, F.L. Dickert, Anal. Chim. Acta 687 (2011) 184.
[43] H. Ji, S. McNiven, K. Lee, T. Saito, K. Ikebukuro, I. Karube, Biosens. Bioelectron. 15 (2000) 403.
[44] S. Yan, Y. Fang, Z. Gao, Biosens. Bioelectron. 22 (2007) 1087.
[45] K. Reimhult, K. Yoshimatsu, K. Risveden, S. Chen, L. Ye, A. Krozer, Biosens. Bioelectron. 23 (2008) 1908.
[46] A. Pietrzyk, S. Suriyanarayanan, W. Kutner, R. Chitta, M.E. Zandler, F. D'Souza, Biosens. Bioelectron. 25 (2010) 2522.
[47] N. Iqbal, G. Mustafa, A. Rehman, A. Biedermann, B. Najafi, P.A. Lieberzeit, F.L. Dickert, Sensors 10 (2010) 6361.
[48] T. Kobayashi, Y. Murawaki, P.S. Reddy, M. Abe, N. Fujii, Anal. Chim. Acta 435 (2001) 141.
[49] N. Tsuru, M. Kikuchi, H. Kawaguchi, S. Shiratori, Thin Solid Films 499 (2006) 380.
[50] M.C. Blanco-López, M.J. Lobo-Castañón, A.J. Miranda-Ordieres, P. Tuñón-Blanco, TrAC Trend, Anal. Chem. 23 (2004) 36.
[51] C. Choong, J. Bendall, W. Milne, Biosens. Bioelectron. 25 (2009) 652.
[52] P. Chen, P. Nien, C. Wu, T. Wu, C. Lin, K. Ho, Anal. Chim. Acta 643 (2009) 38.
[53] M. Pan, G. Fang, B. Liu, K. Qian, S. Wang, Anal. Chim. Acta 690 (2011) 175.
[54] D. Lakshmi, A. Bossi, M.J. Whitcombe, I. Chianella, S.A. Fowler, S. Subrahmanyam, E.V. Piletska, S. A.Piletsky, Anal. Chem. 81 (2009) 3576.
[55] Y. Fuchiwaki, R. Shoji, I. Kubo, H. Suzuki, Anal. lett. 41 (2008) 1398.
[56] T. Alizadeh, M. Zare, M.R. Ganjali, P. Norouzi, B. Tavana, Biosens. Bioelectron. 25 (2010) 1166.
[57] C. Fang, C. Yi, Y. Wang, Y. Cao, X. Liu, Biosens. Bioelectron. 24 (2009) 3164.
[58] E. Mazzotta, R.A. Picca, C. Malitesta, S.A. Piletsky, E.V. Piletska, Biosens. Bioelectron. 23 (2008) 1152.
[59] W. Santos, M. Santhiago, I.V. Yoshida, L.T. Kubota, Anal. Chim. Acta 695 (2011) 44.
[60] Y. Wang, Y. Cao, C. Fang, Q. Gong, Anal. Chim. Acta 673 (2010) 145.
[61] E. Pardieu, H. Cheap, C. Vedrine, M. Lazerges, Y. Lattach, F. Garnier, S. Remita, C. Pernelle, Anal. Chim. Acta 649 (2009) 236.
[62] I. Kubo, N. Yokota, Y. Nakane, Y. Fuchiwaki, Int. J. Electrochem. (2011).
[63] T. Alizadeh, Thin Solid Films 518 (2010) 6099.
[64] E. Mazzotta, C. Malitesta, Sens. Actuators B: Chem. 148 (2010) 186.

[65] Y. Qu, Y. Liu, T. Zhou, G. Shi, L. Jin, Chin. J. Chem. 27 (2009) 2043.
[66] A.K. Patel, P.S. Sharma, B.B. Prasad, Electroanalysis 20 (2008) 2102.
[67] C. Weng, W. Yeh, K. Ho, G. Lee, Sens. Actuators B: Chem. 121 (2007) 576.
[68] X. Liu, C. Li, C. Wang, T. Li, S. Hu, J. Appl. Polym. Sci. 101 (2006) 2222.
[69] T. Alizadeh, M.R. Ganjali, M. Zare, P. Norouzi, Electrochim. Acta 55 (2010) 1568.
[70] T. Alizadeh, M.R. Ganjali, P. Norouzi, M. Zare, A. Zeraatkar, Talanta 79 (2009) 1197.
[71] T. Alizadeh, Electroanalysis 21 (2009) 1490.
[72] T. Panasyuk-Delaney, V.M. Mirsky, M. Ulbricht, O.S. Wolfbeis, Anal. Chim. Acta 435 (2001) 157.
[73] R. Thoelen, R. Vansweevelt, J. Duchateau, F. Horemans, J. D'Haen, L. Lutsen, D. Vanderzande, M. Ameloot, M. vandeVen, T.J. Cleij, P. Wagner, Biosens. Bioelectron. 23 (2008) 913.
[74] F. Horemans, J. Alenus, E. Bongaers, A. Weustenraed, R. Thoelen, J. Duchateau, L. Lutsen, D. Vanderzande, P. Wagner, T.J. Cleij, Sens. Actuators B 148 (2010) 392.
[75] J. Broeders, S. Duchateau, B. Van Grinsven, W. Vanaken, M. Peeters, T. Cleij, R. Thoelen, P. Wagner, W. De Ceuninck, Phys. Status. Solidi-R 208 (2011) 1357.
[76] F. Louwet, D. Vanderzande, J. Gelan. Synth. Met. 69 (1995) 509.
[77] S. Grimmes, O.G. Martinsen, Bioimpedance and bioelectricity basics, London Academic Press, 2000.
[78] G. Sauerbrey, G., Verwendung von Schwingquarzen zur Wagung dünner Schichten und sur Mikrowagung, Z. Phys. 155 (1959) 206–222.
[79] D. Batra, K.J. Shea, Curr Opin, Chem. Biol. 7 (2003) 434.
[80] S. Subrahmanyam, S.A. Piletsky, E.V. Piletska, B. Chen, K. Karim, A.P. Turner, Biosens. Bioelectron. 16 (2001) 631.
[81] S.A. Piletsky, K. Karim, E.V. Piletska, C.J. Day, K.W. Freebairn, C. Legge, A.P. Turner, Analyst 126 (2001) 1826.
[82] T. Takeuchi, D. Fukuma, J. Matsui, Anal. Chem. 71 (1998) 285.
[83] F. Lanza, B. Sellergren, Macromol. Rapid. Commun. 35 (2004) 59.
[84] C. Chassaing, J. Stokes, R.F. Venn, F. Lanza, B. Sellergren, A. Holmberg, C. Berggren, J. Chromatogr. B, 804 (2004) 71.
[85] B. Dirion, Z. Cobb, E. Schillinger, L.I. Andersson, B. Sellergren, J. Am. Chem. Soc. 125 (2003) 15101.
[86] L.I. Andersson, J. Chromatogr. B 745 (2000) 3.
[87] N. Perez-Moral, A.G. Mayes, Anal. Chim. Acta 504 (2004) 15.
[88] P.K. Owens, L. Karlsson, E.S.M. Lutz, L.I. Andersson, TrAC Trend, Anal. Chem. 18 (1999) 146.
[89] O. Ramström, L. Ye, P. Gustavsson, Chromatographia 48 (1998) 197.
[90] D.S. Janiak, P. Kofinas, Anal. Bioanal. Chem. 389 (2007) 399.
[91] E.V. Piletska, M. Romero-Guerra, A.R. Guerreiro, K. Karim, A.P.F. Turner, S. A Piletsky, Anal. Chim. Acta 542 (2005) 47.
[92] E.V. Piletska, A.R. Guerreiro, M. Romero-Guerra, I. Chianella, A.P.F. Turner, S.A. Piletsky, Anal. Chim. Acta 607 (2008) 54.
[93] F.A. Villamena, A.A. De La Cruz, J. Appl. Polym. Sci. 82 (2001) 195.
[94] K. Haupt, Anal. Chem. 75 (2003) 376A.
[95] J.O. Mahony, K. Nolan, M.R. Smyth, B. Mizaikoff, Anal. Chim. Acta 334 (2005) 31.
[96] M. Sibrian-Vazquez, D.A. Spivak, J. Org. Chem. 68 (2003) 9604.
[97] Y. Zhang, D. Song, L.M. Lanni, K.D. Shimizu, Macromolecules 43 (2010) 6284.
[98] R.J. Ansell, K.L. Kuah, Analyst 130 (2005) 179.
[99] J. Svenson, J.G. Karlsson, I.A. Nicholls, J. Chromatogr. A 1024 (2004) 39.
[100] S. Piletsky, E. Piletska, K. Karim, G. Foster, C. Legge, A. Turner, Anal. Chim. Acta 504 (2004) 123.
[101] R.J. Umpleby, M. Bode, K.D. Shimizu, Analyst 125 (2000) 1261.
[102] D.A. Spivak, Adv. Drug Deliver. Rev. 57 (2005) 1779.
[103] G. Barbara, V. Stanghellini, R. De Giorgio, C. Cremon, G.S. Cottrell, D. Santini, G. Pasquinelli, A.M. Morselli-Labate, E.F. Grady, N.W. Bunnett, S.M. Collins, R. Corinaldesi, Gastroenterology 126 (2004) 693.
[104] L.G. Harsing, H. Nagashima, D. Duncalf, E.S. Vizi, P.L. Goldiner, Electrochem. Clin. Chem. 32 (1986) 1823.
[105] J. von Mach-Szczypinski, S. Stanosz, K. Sieja, M. Stanosz, Metabolism 58 (2009) 867.
[106] L. Auerswald, C. Morren, A.L. Lopata, Food Chem. 98 (2006) 231.
[107] L. Lehane, J. Olley, Int. J. Food Microbiol. 58 (2000) 1.
[108] F. Gosetti, E. Mazzucco, V. Gianotti, S. Polati, M.C. Gennaro, J. Chromatogr. A, 1149 (2007) 151.

[109] M. Claeys-Bruno, O. Vandenabeele-Trambouze, M. Sergent, M. Geffard, D. Bodet, M. Dobrijevic, A. Commeyras, Chemometr. Intell. Lab 80 (2006) 186.
[110] C.J. Allender, C. Richardson, B. Woodhouse, C.M. Heard, K.R. Brain, Int. J. Pharm. 195 (2000) 39.
[111] A. Pietrzyk, S. Suriyanarayanan, W. Kutner, R. Chitta, F. D'Souza, Anal. Chem. 81 (2009) 2633.
[112] N. Pérez-Moral, A.G. Mayes, Bioseparation 10 (2002) 287.
[113] C. Lin, W. Chu, J. Med, Bio. Eng. 23 (2003) 53.
[114] M. Hui, W. Jinfang, Z. Liangmo, W. Qinghai, Z. Daoqian, Chromatography 21 (2000) 217.
[115] R.J. Ansell, K. Mosbach, J. Chromatogr. A 787 (1997) 55.
[116] A.G. Mayes, K. Mosbach, Anal. Chem. 68 (1996) 3769.
[117] N. Perez-Moral, A.G. Mayes, Biosens. Bioelectron. 21 (2006) 1798.
[118] S. Srivastava, R. Sinha, D. Roy, Aquat. Toxicol. 66 (2004) 319.
[119] A. Stammati, C. Nebbia, I.D. Angelis, A.G. Albo, M. Carletti, C. Rebecchi, F. Zampaglioni, M. Dacasto, Toxicol. in Vitro 19 (2005) 853.
[120] A.A. Bergwerff, P. Scherpenisse, J. Chromatogr. B 788 (2003) 351.
[121] W.T. Tsai, H.R. Chen, J. Hazard, Mater. 175 (2010) 844.
[122] B.H. Hameed, M.I. El-Khaiary, J. Hazard, Mater. 154 (2008) 237.
[123] B.H. Hameed, M.I. El-Khaiary, J. Hazard, Mater. 153 (2008) 701.
[124] Z. Bekçi, C. Özveri, Y. Seki, K. Yurdakoç, J. Hazard, Mater. 154 (2008) 254.
[125] S. Yan, Z. Gao, Y. Fang, Y. Cheng, H. Zhou, H. Wang, Dyes. Pigments 74 (2007) 572.
[126] L.Q. Su, S. Qiao, W.B. Zhang, Chin. Chem. Lett. 18 (2007) 229.
[127] Y.H. Li, T. Yang, X.l. Qi, Y.W. Qiao, A. Deng, Anal. Chim. Acta 624 (2008) 317.
[128] M.J. Martínez Bueno, S. Herrera, A. Uclés, A. Agüera, M.D. Hernando, O. Shimelis, M. Rudolfsson, A.R. Fernández-Alba, Anal. Chim. Acta 665 (2010) 47.
[129] C. Long, Z. Mai, Y. Yang, B. Zhu, X. Xu, L. Lu, X. Zou, J. Chromatogr. A, 1216 (2009) 2275.
[130] C. Cacho, L. Schweitz, E. Turiel, C. Pérez-Conde, J. Chromatogr. A, 1179 (2008) 216.
[131] H. Yan, K.H. Row, J. Ind, Eng. Chem. 13 (2007) 552.
[132] W. Li, S. Li G., Luo, K. Ding, B. Korean. Chem. Soc. 28 (2007) 1346.
[133] S. Li, X. Huang, M. Zheng, W. Li, K. Tong, Sensors 8 (2008) 2854.
[134] Y. Wang, J. Zhang, X.X. Zhu, A. Yu, Polymer 48 (2007) 5565.
[135] M. Pan, J. Wang, G. Fang, W. Tang, S. Wang, J. Chromatogr. B, 878 (2010) 1531.
[136] R.A. Anderson, M.M. Ariffin, P.A.G. Cormack, E.I. Miller, Forensic Sci. Int. 174 (2008) 40.
[137] H.H. Yang, W.H. Zhou, X.C. Guo, F.R. Chen, H.Q. Zhao, L.M. Lin, X.R. Wang, Talanta 80 (2009) 821.
[138] F. Horemans, J. Alenus, A. Weustenraed, K. Eersels, L. Lutsen, D. Vanderzande, P. Wagner, T.J. Cleij, Anal. Chem. Acta (2011). submitted.
[139] M. Peeters, F.J. Troost, B. van Grinsven, F. Horemans, J. Alenus, D. Keszthelyi, A. Ethirajan, R. Thoelen, T.J. Cleij, P. Wagner, Sens. Actuators B (2011). submitted.
[140] E. Bongaers, J. Alenus, F. Horemans. A. Weustenraed, L. Lutsen, D. Vanderzande, T.J. Cleij, F.J. Troost, R.J. Brummer, P. Wagner, Phys. Status Solidi-R 207 (2010) 837.
[141] H.A. De Abreu, W.B. De Almeida, H.A. Duarte, Chem. Phys. Lett. 383 (2004) 47.
[142] L. Alderighi, P. Gans, A. Ienco, D. Peters, A. Sabatini, A. Vacca, Coordin. Chem. Rev. 184 (1999) 311.

CHAPTER

6

Molecularly Imprinted Polymers for Sensors: Comparison of Optical and Mass-Sensitive Detection

Adnan Mujahid[a,b], *Franz L. Dickert*[a]

[a]Department of Analytical Chemistry, University of Vienna, Austria

[b]Institute of Chemistry, University of the Punjab, Pakistan

OUTLINE

Molecularly Imprinted Sensors.
DOI: 10.1016/B978-0-444-56331-6.00006-2

1. INTRODUCTION

Modern human development greatly depends on the industrial revolution, but at the same time industrial advances have also brought some serious environmental problems. Along with industrial waste pollution, incomplete combustion from automobiles is also producing various toxic hydrocarbons and other aromatic compounds. Most of them are carcinogenic, and their presence in air and water is extremely poisonous to living beings. This concern has led environmental protection authorities to adopt efficient detection mechanism against these threats. Already established analytical techniques such as gas chromatography mass spectrometry (GCMS) and high-performance liquid chromatography (HPLC), provide suitable detection pathways. However, chemical sensor technology [1] offers many more advantages over classical techniques in environmental analysis. Their high precision, small size, low cost, and working ability in harsh environments such as at high temperatures, low pH, makes them ideally fit for the sensing of various environmental pollutants. Moreover, they are highly useful for remote sensing applications. Generally, chemical sensors are small devices comprised of three major parts: a chemical-sensitive layer, a transducer, and a data processing unit. Whenever they are exposed to an analyte environment, the chemical-sensitive interface of the sensor tends to undergo some changes. Transducer devices convert this change into an appropriate electrical signal, and ultimately the third unit processes this information and displays results. Figure 6.1 demonstrates schematic representation of optical and mass-sensitive sensor setup using the same recognition layer.

In sensor development, the selection of a chemical-sensitive layer is the most crucial aspect that needs to be considered thoroughly. For this purpose various natural and synthetic materials are available that can be integrated on a sensor interface. This includes natural antibodies [2], self-assembled monolayers (SAMs) [3], affinity layers of polymerized phthalocyanines [4], labeled biomolecules such as DNA [5], and some others. Besides all these methods, molecular imprinting [6] is a relatively innovative approach to synthesize highly sensitive and selective artificial receptor surfaces. The analyte interaction sites are generated by a self-organization [7] process of polymer chains around template molecules, thus producing adapted cavities that are capable of reversible re-inclusion of analyte species. The technique is versatile and can be applied to a variety of different bio-analytes ranging from bacteria [8], viruses [9], yeast cells [10], blood cells [11], enzymes [12], and many others. In the case of analytes having a stiff structure such as polycyclic aromatic hydrocarbons (PAHs), herbicides, and mid-chain carbonic acids, molecular imprinting yields substantial sensitivity and selectivity.

Apart from coating interface, the design of transducer is also important as the chemical changes in layer material can be converted into an appropriate measurable signal. A broad range of transducers are available for sensing purposes, that is, metal oxide semiconductor (MOS) devices [13], field-effect transistors [14], potentiometric [15], optical [16], and mass-sensitive devices [17]. However, our interest is in optical and mass-sensitive systems to which molecular imprinted polymers can be integrated, thus designing powerful sensor setup. Both optical and mass-sensitive sensor devices possess some significant advantages over each other; for example, optical sensors provide much lower detection limits and high sensitivity during trace analysis, while mass-sensitive or acoustic systems offer more suitable dynamic range.

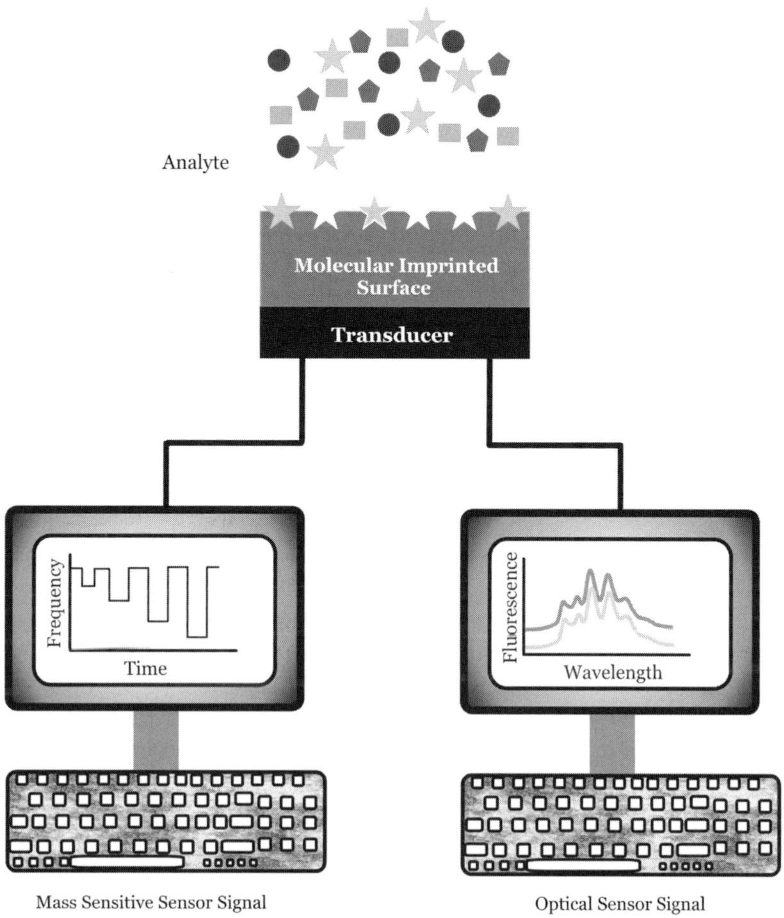

FIGURE 6.1 Schematic representation of optical and mass-sensitive sensor setup.

Although molecularly imprinted polymers (MIPs) have been successfully used in designing optical sensors, they have been more extensively studied for mass-sensitive devices. The combined features of optical and acoustic transducers accomplished complementary information about analyte interaction with the MIP surface and enhanced sensor signal; therefore, they are very valuable in modern sensor development. This chapter seeks to address all the important issues related to the structuring of MIP surfaces, working principles involved for optical and mass-sensitive sensor setups, and discusses their individual and combined applications in the sensor field.

2. MIPs AS SENSOR INTERFACE

In designing modern sensors, the chemical-sensitive layer material should be capable of molecular recognition, resistive toward organic solvents, stable over time, and applicable to virtually every type of analyte. MIPs meet all these prerequisites and, therefore, are preferred over their natural competitors, that is, natural antibodies and are considered ideally suited for sensor interface. Molecular imprinting [18] is a straightforward technique for producing sensitive and robust materials in which template molecules, along with the monomers and cross-linkers, are polymerized under specific conditions. Polymer chains undergo a self-organization process and assemble themselves around the template. After polymerization, template molecules are washed away from the matrix, thus producing tailor-made cavities suitable for reversible analyte re-inclusion. The process of molecular imprinting is shown in Fig. 6.2.

Molecular hollows can also be generated via supramolecular [19] compounds such as calixarenes [20] and cyclodextrins [21] for sensitive coating material. These schemes are frequently assisted by computer modeling for getting optimized cavity size and dimension. However, the synthetic route is relatively complex and requires extensive labour. In contrast, molecular imprinting provides a fairly simple and uncomplicated synthesis course, creating analyte analogous cavities. In the first step, template molecules along with functional monomers and cross-linkers are mixed together. The reaction conditions are adjusted in order to acquire the desired degree of polymerization. Finally, after reaching the gel point, template molecules are washed away carefully from the matrix. This process is known as *bulk-imprinting* [22] as the cavities are produced in the whole matrix. Another approach to generate molecular cavities on the surface of imprinted polymer is called *surface imprinting* [23]. This scheme is especially designed for imprinting larger bio-analytes such as yeast cells, red blood cells, bacteria, and others on MIP surface. In surface imprinting, a closely packed stamp of template molecules is designed and pressed over the pre-polymerized surface. In this way, patterned polymer surfaces are generated having memory of stamped template specie. This technique has two significant advantages over bulk imprinting. First, the surface imprinting is useful for achieving complete reversibility as in bulk-imprinted polymer the release of large bio-analytes from the bulk is difficult and consequently the sensor signal

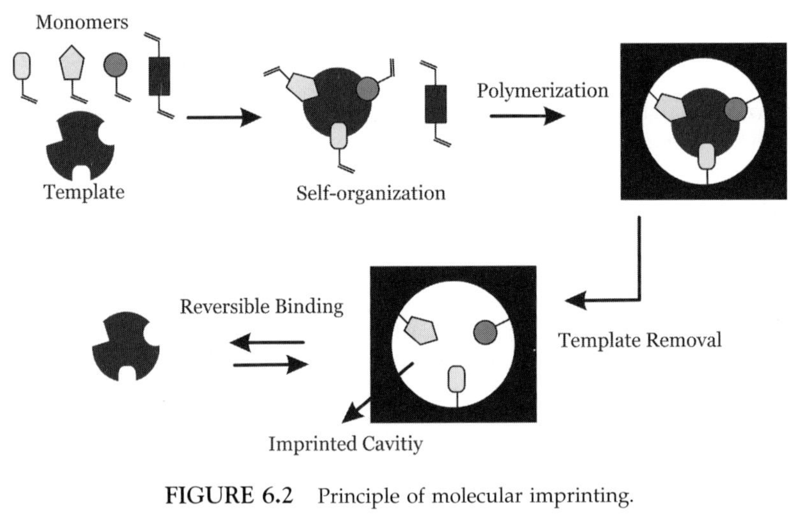

FIGURE 6.2 Principle of molecular imprinting.

never returns to the baseline value. Second, the diffusion pathways are too long for these analytes to reach the inner core of the polymer matrix that causes a longer response time during measurement. Because of these facts, the surface imprinting technique is considered complementary for all large analytes, particularly for bio-species.

3. IMPRINTING APPROACHES

In general, molecular imprinting can be classified into two major types depending on the nature of interaction: covalent and noncovalent binding with analytes. *Covalent imprinting* [24,25] is useful for those molecules that do not possess any functional groups. The template is covalently bonded to monomers prior to the imprinting step and thus is not allowed to escape during the polymerization reaction. The covalent bond has to break every time in releasing the template from the polymer matrix. This practice is quite useful, however, literature reports limited applications for sensing purpose. The needle eye in this procedure is the removing of the template from the imprinted polymer.

A relatively modified method has been adopted for preparing imprinted polymers in which the template with a covalent or partially covalent structure is used in MIP synthesis. However, in this case binding with analyte is completely reversible and noncovalent. This strategy has been named *semicovalent imprinting* [26]. At first, a template of methacrylate ester is used for preparing semicovalent imprinted polymer. The template is removed by the simple hydrolysis method following which the MIP can reversibly bind unesterified template molecules. Nevertheless, the removal of template from polymer is not feasible by straightforward hydrolysis as it is hindered by some sterical effects due to hydrogen bonding interactions. These factors strongly influence the imprinting process. This limitation can be overcome by introducing a linker group in the polymer system. This linker group possesses a double character; first, it helps in the binding of the template with the polymer during the imprinting process and then afterward behaves like a spacer between the polymer and the template to avoid steric problems, thereby allowing a smooth, reversible binding of analyte with MIP. Whitecomb et al. [27] used the carbonyl group of carbonate ester as a sacrificial spacer group for cholesterol imprinting. Besides carbonyls, some salicylates had also been reported as spacers for imprinted polymers. The semicovalent imprinting approach is practical for synthesizing imprinted emulsions and dendrimers [28].

In covalent imprinting, the nature of covalent linkage between functional monomer and template prohibits nonspecific interactions. However, in covalent and semicovalent imprinting methods, extensive synthetic protocols have to be followed for both the binding and releasing of the template by the polymer.

Conversely, *noncovalent imprinting* [29,30] is a more versatile and flexible technique for designing recognition materials. The removal of template molecules is done in a more elegant way without breaking any bond. Template molecules should have a functional group so that affinity interactions can be created with monomers. Noncovalent forces, including hydrogen bonding, dipole–dipole interactions, van der Waals forces, and $\pi-\pi$ interactions are responsible for achieving selectivity. Solvent polarity and porogenic properties, hydrogen bonding interactions, and polymerization temperature have a strong impact on the imprinting process. Low-polarity solvents are complementary for achieving hydrogen bonding in the matrix, while highly polar solvents are best suited for developing strong polar interactions.

A relatively modern method of imprinting is introducing two different template species instead of a single one; this concept is known as *double imprinting* [31]. The chemical interactions are the same as in noncovalent imprinting. It has been reported that the combination of two different templates leads to producing a more suitable size cavity structure and better diffusion pathways for analyte interaction compared to the single imprinted polymer system. The *epitope approach* [32] is another imprinting strategy designed for selective recognition of proteins and peptides.

The template monomer ratio [33] in MIP synthesis is an important aspect inasmuch as a high percentage of functional monomer leads to more nonspecific binding, whereas, on the other hand, a low percentage of monomer produces poor imprinting sites due to insufficient complexation with the template. Takeuchi et al. [34] proposed combinatorial methods to achieve optimized polymer composition for crafting synthetic receptors. Temperature is another important parameter that controls the gel point of the polymer. It has been observed that at higher temperatures the polymer matrix is more closely packed around template molecules, thus producing more rigid cavities. In general, carefully controlling polymerization conditions such as the nature of functional monomer, amount of cross linking, template percentage, solvent polarity, and temperature can be exploited to get desired MIP properties. There are also some drawbacks regarding noncovalent imprinting, including nonselective single-point binding with analytes and formation of heterogeneous receptor surfaces. These factors contribute more to noise and reduce sensitivity in MIP sensor applications. However, on the whole noncovalent imprinting is a much more versatile and straightforward method that is best suited for generating artificial sensor surfaces.

4. OPTICAL TRANSDUCERS

Optical sensor technology [35] has found significant applications in various fields such as biotechnology, environmental monitoring, pharmaceuticals, and clinical analysis. Due to the extraordinary features of MIPs, they can easily be coupled with optical devices. There are several possible designs for optical sensors, depending on the optical properties of analyte molecules. If an analyte possesses some optical features, their recognition can be executed by MIPs. In optical transducers, the information of binding events on MIP surface is not directly converted into measurable electronic signals. In fact, the change in optical properties, including absorption, reflection, or fluorescence of layer material is recorded upon analyte interactions. Various spectroscopic approaches have been followed for designing MIP optical sensors.

4.1. UV/Vis Absorption Methods

UV/Vis spectroscopy is one of the most simplified and economical methods for examining analyte interactions with MIPs where only the change in absorbance is measured as a function of wavelength. The technique is versatile and gives rapid response regarding quantitative information on template binding. Besides pure sensing application, this method is quite suitable for screening [36] MIPs and choosing the finest polymer composition. With the help of the UV/Vis spectrum, the thorough mechanism of complexation between templates, monomer, and cross-linker during polymerization can also be better understood. It has been observed that after complexation, an absorbance shift toward shorter wavelengths takes place. The procedure makes it easy to compare the spectrum of free template and functional monomer with that of the complex formed. This strategy is equally suitable for monitoring metal polymer complexation in visible regions [37]. Although UV/Vis spectroscopy is not as selective as the fluorescence method, it is nevertheless quite suitable for designing low-cost MIP sensors with moderate sensitivity.

4.2. Surface Plasmon Resonance (SPR) Transducers

The SPR technique was initially used for measuring the change in thickness and refractive index of thin films upon analyte interaction. Conventionally, it was adopted for exploring recognition sites in natural antibodies [38,39] for the adsorption of larger molecules such as proteins, thus designing biosensors. It can also be combined with imprinted polymers; however, small-size template molecules do not significantly contribute to a change in the refractive index, and therefore, the sensitivity is often limited. In contrast to natural receptors, the primary advantages in using MIPs are their robustness and their excellent storage over long periods of time. Lotierzo et al. [40] designed homogenized thin MIP films for the detection of domoic acid and compared the sensor robustness and sensitivity against natural receptors, that is, monoclonal antibodies. It has been observed that the recognition properties are not disturbed upon regeneration in MIPs compared to natural counterparts. Moreover, MIPs can continuously perform measurements for about two months without losing performance, whereas antibody can be denatured after some regeneratiuons.

4.3. Fluorescence-based Transducers

The most frequent method for combining optical devices with MIPs is done by means of fluorescence spectroscopy [41,42]. Enhanced sensitivity, excellent selectivity, and low-detection limits make them ideally suited to combine with imprinted layer materials for sensor development. Optical waveguides can be used as fluorescence transducers as this setup is highly significant for fluorescent analytes. The working principle is illustrated in Fig. 6.3. The thickness of the layer material and the large surface size of waveguides drastically contribute to fluorescence emission.

Fluorescence microscopy [43,44] is also a useful way to access the MIP structures, calculating the number of binding sites and their efficiency for analyte recognition. There are several ways to design fluorescent sensors using MIPs. One very prominent method is using fluorescence templates or labeled analytes. This scheme is quite useful not only for sensor development but also for evaluating MIP structure, homogeneity, and binding properties. However, this sensing scheme is no more practical when dealing with nonfluorescence analyte molecules.

The problem can be solved through the formation of fluorescent complexes between monomers [45,46] possessing inherent fluorescent properties. This actually allows the use of fluorescence spectroscopy in optical sensing for those analytes that are not intrinsically

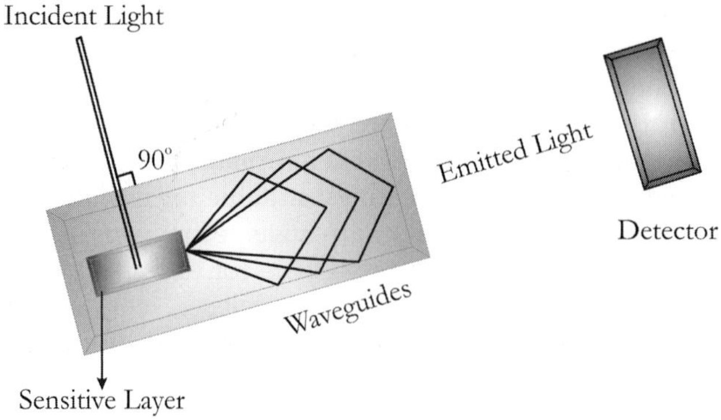

FIGURE 6.3 Principle of a planar waveguide sensor.

fluorescence active. The relative shifts in emission wavelength can be recorded in the presence of the initial amount of target analyte molecules; therefore, the information is quantified. This approach has been extensively studied for synthesizing new functional monomers that are capable of separating closely related compounds.

Rathbone et al. [47] successfully developed hydrogen bonding monomers and their use for fluorescence assay of drugs. Another very elegant way of designing MIP optical sensors is the integration of fluorescent dyes into the polymer matrix. This method is suitable for those analytes without having any specific optical property. When the analyte molecules interact with the polymer system, the fluorescence signal is quenched and therefore the sensing becomes possible. The scheme has been proven effective in designing chemosensors for essential biological molecules such as cyclic adenosine monophosphate (cAMP). Chow et al. [48] focused on the derivatization of nonfluorescent analytes such as homocysteine by tagging it with pyrenyl maleimide which is fluorescent active. In general, these approaches are very promising for detection of no-fluorescent analytes.

Some other optical phenomena are also studied for introducing them with MIPs in sensor designs, including chemiluminescence [49], reflectometric interference spectroscopy [50] analogues to SPR, and surface-enhanced Raman scattering techniques. Besides all presented methods, fluorescence is the most widely adopted technique and the best choice available for integrating MIPs in optical sensor devices.

5. MASS-SENSITIVE TRANSDUCERS

The inherent properties of analytes are the focal point in almost every sensing system; therefore, one should select that property which is more fundamental and easier to measure without any labeling protocols. Mass is the inborn property of every analyte and can directly be measured, so that MIPs can better be used as sensor coatings with mass-sensitive transducers. The sensing mechanism of these devices depends solely on piezoelectric phenomena. The history of piezoelectricity [51] goes back to 1880 when Jacques and Pierre Curie found that certain crystalline substances are subjected to mechanical stress, the material is polarized and some voltage is generated. Conversely, by applying electrical voltage to these materials, mechanical deformation takes place. Materials showing such behavior are known as piezoelectric materials as they transform mechanical signal into electrical one. Naturally occurring materials include quartz and Rochelle salt, while lithium niobate and lithium tantalate are synthetically designed ceramics. The selection of the material is concerned with price, availability, robustness, and the nature of its ultimate use. For example, lithium tantalate is preferred when designing shear wave devices, gallium phosphate is the most suitable material for working in high-temperature environments, and polyvinylidene fluoride exhibits higher piezoelectricity. Among all the crystalline substances, quartz crystal microbalance (QCM) is the most stable and widely used piezoelectric material for designing sensors, transducers, and other important applications. The mass-sensing concept of piezoelectric materials was first discovered by a German physicist, G. Sauerbrey [52] in 1959. He observed that an oscillating piezocrystal undergoes resonance frequency shifts when a certain mass is loaded onto its surface, provided that the deposition is of a rigid nature. This quantitative relationship of loaded mass and frequency shift has been expressed as follows.

$$\Delta f \propto -f_o^2 \Delta m$$

It is obvious that change in frequency is directly proportional to change in deposited mass and square of the fundamental resonance frequency. The second parameter, that is, fundamental frequency, is noticeable as, by doubling the fundamental frequency, the change in frequency would be four times. This indeed is a very attractive feature for designing highly sensitive transducers. The fundamental frequency depends on the plate thickness of the piezoelectric substance; that is why acoustic devices are often referred to as mass sensitive because thickness decides the resonance frequency. The classification of acoustic devices is made on the basis of the wave propagation mode through the substrate, that is, the bulk or surface of the material. Brief descriptions of these devices are as follows.

5.1. Bulk Acoustic Wave (BAW) Devices

Bulk acoustic wave devices are those in which waves propagate through the bulk of the substrate. QCM is the best example of these devices. They are available in different fundamental resonating frequencies ranging from 5 MHz to nearly 50 MHz. However, 10 MHz QCM is the most widely used mass-sensitive device with sensitivity up to 1Hz/ng. Going above this frequency value, the thickness of the quartz sheet decreases considerably as mechanically it becomes fragile. Generally, AT-cut quartz is preferred to compensate for temperature dependence over sensor signal. With regard to the Sauerbrey equation, it was basically designed when working with piezoelectric materials and rigid coatings in gas phase where the viscoelastic effects of medium are negligible. While in liquid phase the situation is different: frequency shifts are not entirely due to the mass loads but also depending on the viscosity of the medium. So in liquid phase, exact frequency shifts due to analyte adsorption have to be separate from the medium's interference, for example, viscosity. Temperature is the other important factor that influences frequency shifts. In the gaseous phase, the per degree rise in temperature leads to negligible frequency change, while in water this change is much higher—nearly 40 Hz. Apart from medium viscosity and temperature, several other

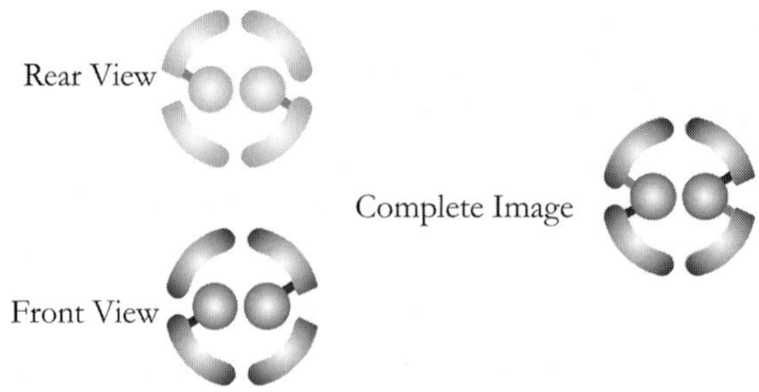

FIGURE 6.4 AT-cut quartz crystal having dual electrodes patterned on both sides.

parameters contribute to frequency shifts during analyte measurement. The pure frequency shifts due to analyte interaction can be determined by performing differential measurements on dual-electrode QCMs. In this manner, one electrode acts as a working electrode coated with the sensitive layer, while the other works as reference so as to eliminate all the unspecific frequency shifts in sensor design. A dual-electrode QCM is shown in Fig. 6.4.

5.2. Surface Acoustic Wave (SAW) Devices

As its name makes obvious, in SAW devices the wave propagation is associated with the surface of the piezoelectric material. A comblike interdigital electrode structure is fabricated on the surface of SAW devices [53] as shown in Fig. 6.5. In this type of sensor, particles are displaced from their positions and surface waves are produced much the same as Rayleigh waves [54]. These waves travel along the surface from one electrode finger to another where the sensitive layer is coated and in this way an alternating voltage is produced that can be measured quantitatively.

Dimension and the distance between electrode centers determine the operating frequency and wavelength of the device. SAW devices are produced by ST-cut (temperature

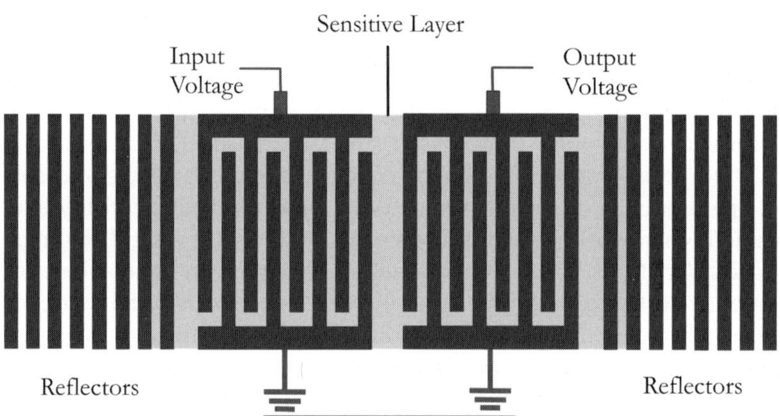

FIGURE 6.5 Surface acoustic wave (SAW) resonator.

compensated cut for SAWs) from a quartz sheet; however, they operate at relatively high resonance frequency—up to 2.5 GHz compared to Bulk acoustic wave (BAW) devices. This ultimately results in high sensitivity and extends the detection limit down to a picogram. The major concern in using these devices is their substantial damping in high-viscosity mediums such as in liquid phases. This limits their sensor applications in liquid mediums.

5.3. Shear Transverse Wave (STW) Resonators

The drawback of excessive damping in SAW devices can be overcome by shear transverse wave (STW) resonators. Two critical points should be made in this respect. First, in STW [55] devices the cutting angle of piezoelectric material is changed so as to shift the shear vertical waves into shear horizontal waves. The shear waves travel completely within the medium and produce less damping as compared to Rayleigh waves. In this way high-viscosity solvents do not influence STW oscillations. The second important point is related to the selection of piezoelectric substance. For example, quartz has a relatively low dielectric constant value (i.e., 3.8), so it is very much influenced by polar solvents especially when working in water of a dielectric constant of nearly 80. Lithium tantalate is a suitable piezomaterial for manufacturing STW resonators as it has a dielectric constant of nearly 40, which is somewhat comparable to water. Therefore, in lithium tantalate STW devices, mass-sensitive measurements are not strongly damped by solvent polarity.

6. EXEMPLARY APPLICATIONS OF MIP OPTICAL SENSORS

Molecular imprinted materials, including thin films and microparticles, have been extensively investigated for optical sensor development. In some cases where the template has some optical properties, there is no need for an additional labeling procedure. However, where the analyte is lacking optical features, it needs to be tagged such as with fluorescent indicators. Some potential applications of MIP optical sensors are highlighted next.

6.1. MIP Optical Sensors for Biological Species

Conventional methods for the detection of important hormones such as cyclic adenosine monophosphate (cAMP) rely on natural recognition elements that specifically bind target analyte. Nonetheless, these natural sensing materials lack long-term stability and cannot be stored for longer times. Moreover, the analyte interaction is not fully reversible. MIPs provide an alternative way to synthesize bio-mimetic sensing material for fluorescent detection of cAMP. Turkewitsch et al. [56] have crafted template selective binding sites using *trans*-4-[*p*-(*N*,*N*-Dimethylamino)styryl]-*N*-vinylbenzylpyridinium chloride as fluorescent monomer and 2-hydroxyethyl methacrylate as functional monomer. This leads to developing both recognition and measuring elements in the imprinted polymer matrix. The selectivity was investigated by exposing the sensing material to structurally analogous hormone cGMP, and the differences from nonimprinted polymer were evaluated. It was found that both imprinted and nonimprinted polymer exhibit a slight change in fluorescence intensity for cGMP compared to cAMP. This same research group [45] has also studied time-resolved fluorescence spectroscopy decay analysis for exploring MIPs' specificity against cAMP. Furthermore, they embedded cAMP-imprinted polymer in a thin film [57] to avoid mechanical grinding and compared it with results for bulk polymer. They concluded that thin-layer film offers a high affinity for adsorption of cAMP. Consequently, this causes fluorescence quenching in MIPs, indicating highly specific functionalized cavities. This system was examined by steady-state and time-resolved fluorescence spectroscopy as previously both techniques were adopted individually for MIP characterization.

A. Rachkov et al. [58] had designed a MIP-based fluorescence sensing system for the detection of steroidal hormone β-estradiol. Imprinted particles were examined in two ways: first, by making direct fluorescence measurements and second, by using them as stationary phase in HPLC columns. Experimental results indicate that the direct fluorescence detection method is more successful and enables detection of μM concentrations of β-estradiol.

MIPs have also been explored for the selective binding of cinchona as their attachment to the imprinted polymer shifts the fluorescence spectra [59]. This indeed is very attractive for using such a system as an antibody receptor for assaying cinchonidine. Piletsky et al. [60] reported a displacement chromatographic method for ligand receptor binding by displacement of nonspecific fluorescent dyes from MIPs. The technique is quite interesting, yielding highly sensitive and reproducible measurements for optical isomers. Comparative studies between imprinted and blank polymer were carried out to understand the affinity interactions. These findings reveal that the imprinted matrix contains a large number of binding sites with high affinity toward analyte in contrast to blank. Furthermore, titration studies also suggest that there is no substantial difference between imprinted and blank polymer. It was the patterned spatial arrangement in MIPs that was not present in the nonimprinted matrix. The same group [61] also reported MIP-based fluorescent assays of epinephrine. The difference in dissociation constants of imprinted and control polymer shows the highly specific nature of synthetic materials, thus making them a real alternative to the natural antibodies used in enzyme linked immunosorbent assay (ELISA). The relative affinity of different analytes used in assay is shown in Table 6.1. Although substances such as norepinephrine and phenylephrine had a very closely related structure to epinephrine, they had much weaker binding interactions with MIPs.

Lin et al. [62] have designed MIP chemiluminescent sensors for selective recognition of amino acid derivatives. Dansyl-L phenylalanine (dns-L-Phe) was imprinted in acrylic-based polymer; on rebinding with MIP, it decomposes and yields chemiluminescent emission. This technique is suitable not only for achieving molecular recognition but also for providing a chemiluminescent reaction medium.

Homocysteine is a nonessential amino acid; its elevated level in human blood serum is an indication of serious cardiovascular diseases. Apart from already established methods, Chow et al. [48] designed a MIP fluorescence sensing system for its determination in blood samples. They introduced a thiol-specific fluoro-tagging agent, that is, *N*-(1-pyrenyl)maleimide, for the derivatization of DL-homocysteine and fabricated into MIPs. This bio-mimetic material has been found to be highly specific for homocysteine sensing in human blood serum and also very effective to catalyze the derivatization of homocysteine with *N*-(1-pyrenyl)maleimide. The luminescent response of exactly the same amount of imprinted and nonimprinted materials for different concentrations of DL-homocysteine is shown in Fig. 6.6. The imprinted

TABLE 6.1 Relative Affinity of Structurally Related Analytes to Epinephrine-imprinted Polymer as Determined by Their Dissociation Constants. *Adapted from [61]*

Analytes	Dissociation Constant K_D (μM)
Epinephrine	9.2 ± 2
Norepinephrine	46 ± 4
Isoproterenol	21 ± 2
Phenylephrine	>1000
Catechol	>1000
Clenbuterol	300 ± 15
Pindolol	464 ± 33
Propranolol	157 ± 4

material shows a significant linear sensor response, whereas the nonimprinted material is insensitive to increasing DL-homocysteine concentrations.

Determination of antibiotics in biological samples is another exciting field of MIP optical sensors that is highly desirable in clinical analysis and the pharmaceutical industry as well. For example, antibiotic analyses are usually performed by immunological or enzymatic assay and are often confirmed by liquid chromatography mass spectroscopic (LCMS) technique. Benito-Pena et al. [63] had designed a series of methacrylic acid polymers for imprinting

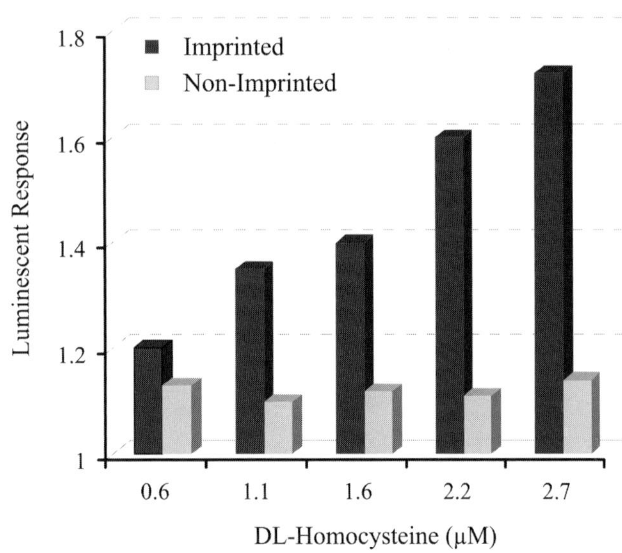

FIGURE 6.6 Luminescent responses of imprinted and no-imprinted materials toward increasing homocysteine concentration. *Adapted from [48].*

β-lactams antibiotics such as penicillin G following fluorescent labeling for their detection. MIP binding sites were found to be highly specific for penicillin G recognition, in comparison to other related antibiotics; the general assay method is also adoptable for analyzing commercial pharmaceutical formulations. Kubo et al. had used 2,6-bis(acrylamido)pyridine as a functional monomer for the imprinting of cyclobarbital. The affinity interactions were developed via hydrogen bonding. Previously, this functional monomer had been used for forming hydrogen bonds with barbiturates and fluorouracil.

Digoxin is an important cardiac drug most commonly used in treatment of heart failure and atrial fibrillation. Already established methods for determination of digoxin in blood and urine samples are based on radioimmunoassays (RIAs), HPLC assay in combination with RIA, or fluorescence LCMS methods. Recently, Gonzalez et al. [64] designed MIP membranes for a digoxin-sensitive fluorescence sensor. Scanning electron microscopic (SEM) images suggest that MIPs had much higher porosity in contrast to nonimprinted. The selectivity of the membrane was tested by exposing codeine, morphine, and heroin concentrations that were nearly 100 times higher than digoxin in the calibration curve. There was no significant interference observed, demonstrating highly selective affinity of MIP membranes. This analysis technique for determining digoxin in serum samples was compared to another reference method, microparticle enzyme immunoassay (MEIA). There was no substantial difference between these two approaches as the relative standard deviation was $<10\%$ and the confidence interval was 95%. The data for both these methods are shown in Fig. 6.7.

The short analysis time of 13 minutes is very attractive for getting prompt response. The authors have reported that the lifetime of MIP membrane is 50 assays, drawing attention to designing accurate and cost-effective kits for routine assaying of digoxin in serum specimens.

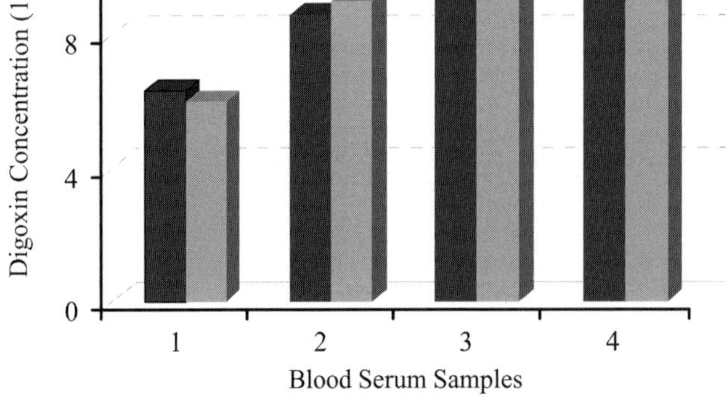

FIGURE 6.7 Comparison of human serum sample analysis results by MIP membranes and reference method (MEIA). *Adapted from [64].*

6.2. MIP Optosensors in Plants

Molecular imprinted materials possess outstanding flexibility in crafting sensor surfaces for various plant viruses. This could be very beneficial for diagnosis of plant diseases and thus reaching toward their treatment. Tobacco mosaic virus (TMV) is the first discovered virus with the ability to damage the tobacco plant leaves causing serious infections. TMV is quite rigid and resistive to extreme pH and temperature. Bolisay et al. [65] synthesized TMV bulk-imprinted hydrogels and optically monitored its binding capacity for target virus. The dry MIP hydrogels were mixed with TMV suspension for four hours in a micro centrifuge. After filtration, the concentration of virus was measured by UV/VIS spectrophotometer. Imprinted gels have double the binding capacity of nonimprinted hydrogels. Selectivity was evaluated by mixing another competing virus, tobacco necrosis virus (TNV), with imprinted and nonimprinted hydrogels. It is clear from Fig. 6.8 that the binding capacity of TNV is nearly the same for both imprinted and nonimprinted gel materials. This proves that imprinted hydrogel specifically binds the TMV virus.

MIPs have been extensively studied for the casting image of plant-based food products such as flavonols, an important ingredient in daily diet. Suarez-Rodriguez et al. [66] designed fluorescence flow through a sensing scheme for flavonols using the methacrylate polymer system. The selection of flavonol as a model template has two major advantages. First, it has aromatic π electrons, making it fluorescent. Second, it possesses carbonyl and hydroxyl functional groups that are suitable for developing intermolecular hydrogen bonding interactions that yield selective cavities. Polymer composition and optical sensing parameters were optimized to obtain the best fluorescence signal with relatively shorter recovery time to the base line point. Cross-selectivity studies of flavonol-imprinted polymer were conducted by examining the binding of morine and quercetin. Flavonol-imprinted polymers exhibit a high specificity for target analyte against the other competing flavonoids. Sensor performance was evaluated by applying analytical models that include linearity over a wide concentration range, detection limit, and quantification, and also by calculating relative standard deviation. The authors reported the astonishingly high reproducibility of the system up

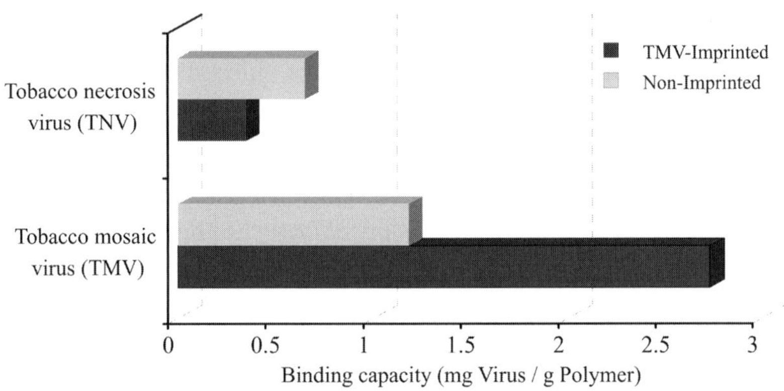

FIGURE 6.8 Binding capacity of TMV-imprinted hydrogels for 0.1 mg/mL solution of TMV and TNV. Response of nonimprinted hydrogel has also been shown. *Adapted from [65].*

to 93%, keeping in view all the steps starting from synthesis until the final measurements. Furthermore, MIPs exhibit excellent storage stability as they had been used continuously for more than two months with almost the same efficiency. Comprehensive studies regarding MIP optical flow through sensing suggest that this system can be used for direct sensor applications without performing any separation or extraction steps. Lulka et al. [67] utilized the optical MIP sensing approach to detect toxic plant proteins such as the castor bean lectin, Ricin. Silanes imprints of *Ricin, Ricin A,* and *Ricin B* chains were prepared and characterized by the steady-state fluorescence method. The detection limit for this plant's toxins reaches up to the micromolar concentration level, which is encouraging. These species are fluorescence active; in addition, their binding to MIPs protects them from solvent effects and also does not change their optical properties.

6.3. MIP Optical Sensors for Metal Ions

MIP metal ion sensors can be designed in two ways: first, by taking polymerizable metal ion chelates which is a covalent approach, and second, by being based on a noncovalent approach following fluorescent metal chelates as templates. Metal ion sensing applications are largely related to clinical toxicology, wastewater treatment, and other environmental concerns. Guney et al. [68] followed the first approach of covalent imprinting and designed a lead (Pb) templated fluorescent sensor capable of determining a very low concentration of metal ion. Fish et al. [69] designed very sophisticated MIP coatings for selectively extracting Cu^{+2} in the presence of other ions, that is, Co^{+2}, Ni^{+2}, Zn^{+2}as the charge density of these ions is quite similar to Cu^{+2}. Intriguing this approach by using same binding ligand, the authors synthesized a highly selective matrix for Hg^{+2} ion sensing in comparison to the related ions Cd^{+2}, Ag^{+1}, Pb^{+2}, Fe^{+3}, and Cu^{+2}. This system can be combined with optical fibers to develop real-time sensors for potentially hazardous metals. 8-Hydroxyquinoline sulfonic acid has an inherent tendency for Al^{+3} to form a stable complex with suitable optical properties. Narayanaswamy et al. [70] selected this ligand and polymerized it with acrylamide, hydroxyethyl methacrylate in the presence of Al^{+3} as the template. In order to estimate the rebinding efficiency of imprinted polymer, the template was completely removed, and the other reaction conditions were also optimized including pH, temperature, and ionic strength. The sensor efficiency was investigated by a series of tests including linearity, repeatability, recovery response, and selectivity over other ions. The fluorescence response for Al^{+3} ions is linear up to 1.0×10^{-4} M under optimal experimental conditions as shown in Fig. 6.9.

The limit of detection was found to be 3.62 μM. In general, the Al^{+3} imprinted polymer was found to be exceptionally sensitive and selective and to give a linear response that can be regenerated.

Al-Kindy et al. [71] also designed an MIP optical flow-through sensor for Al^{+3} ions using aluminum-morin chelate as the template. This polymer system is selective over the other di- and trivalent metal ions such as Be^{+2}, Ca^{+2}, Mg^{+2}, Eu^{+2}, Zn^{+2}, and Fe^{+3}, signifying the highly specific nature of a recognition matrix. The observed detection limit was 0.01μg/mL for Al^{+3} ions.

James et al. [72] reported uranyl ion-imprinted polymer particles embedded in a polymethyl methacrylate matrix for uranium ion sensing in natural water samples. The operation of the sensing system was optimized by controlling the pH, response time, and composition

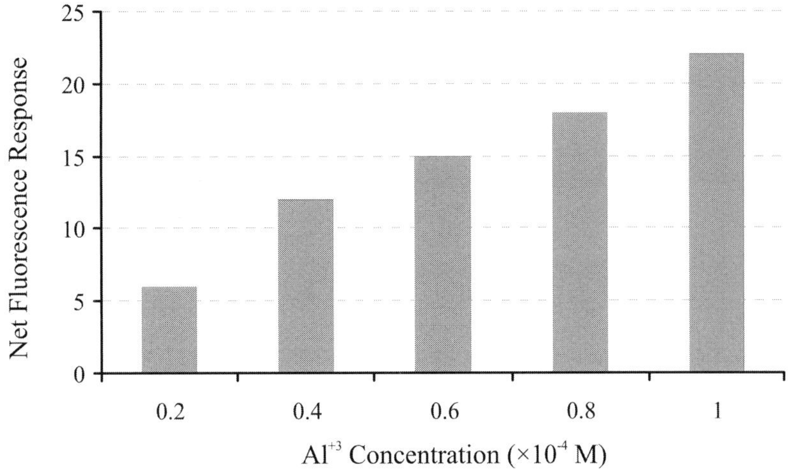

FIGURE 6.9 Fluorescence sensor response for increasing concentrations of Al^{+3} ions, showing a linear behavior in dynamic range. *Adapted from [70].*

of the sensing element. By improving these parameters, the sensor performance for uranyl ions, reproducibility and stability was evaluated. James et al. concluded that this ion-imprinted polymer is effective for pre-concentration and detection of uranium ions from complex water samples, thus ultimately using it for routine water treatment analysis.

6.4. MIP Optical Sensors for Environmental Toxins and Explosives

One of the key assignments of MIPs is to design highly sensitive and selective sensors for recognizing various toxic, carcinogenic, and explosive compounds in environment due to public health and safety. A number of MIP optical sensors have been proposed for this task. Polycyclic aromatic hydrocarbons (PAHs) are produced from industrial plants' exhaust and incomplete combustion. The ash particles containing PAHs can reach the lungs through breathing, causing cancer; in drinking water, their presence is also very dangerous. The Environmental Protection Agency (EPA) has recommended a threshold level of 0.2 ppb [73] of all PAHs to minimize health risks. Because PAHs have fused ring structure, therefore, their imprinting is better suited in a polymer that has an aromatic monomer unit such as polyurethane, which has the desired aromatic composition. The affinity interactions in polymer matrix are generated by noncovalent $\pi-\pi$ interactions and van der Waals forces [41]. The polymerization reaction can be completed on a transducer surface to produce a fine thin film. Template molecules are washed away from the matrix, leaving behind well-adapted cavities in polymer and distinct diffusion pathways. As PAHs possess a rigid structure, molecular recognition depends on the size and shape of the imprinted cavity structure. The imprinted cavities are selective toward the template molecules and show maximum fluorescence only to target analyte. Because the cavities are size oriented, larger analytes do not fit the cavities and smaller ones are washed away. Apart from the eminent selectivity of MIP polyurethanes, they demonstrate a pronounced optical sensitivity, lowering the detection

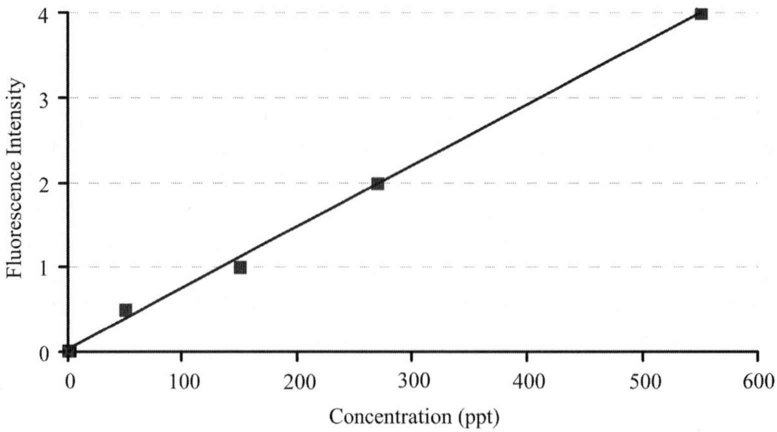

FIGURE 6.10 Fluorescence signal of imprinted polyurethane layer for different concentrations of pyrene in ppt range. *Adapted from [41].*

limit to parts per trillion (ppt) ranges. The fluorescence sensor signal for increasing pyrene concentrations is demonstrated in Fig. 6.10.

Furthermore, these synthetic coatings [74] are optimized to incorporate only the target molecule in the presence of fluorescence quenchers such as humic acid, which is often found in groundwater samples. It has been reported that a 14mg/L concentration of humic acid causes more than a 50% decrease in the intensity during direct measurements in cuvette, while decline is less than 10% in the optical sensor signal. The fluorescence quenching against the increasing concentration of humic acid for MIP sensitive layer and for direct measurements is shown in Fig. 6.11. This indicates that optical sensor measurements are not disturbed by quenching agents and perform effectively in the presence of other interfering agents.

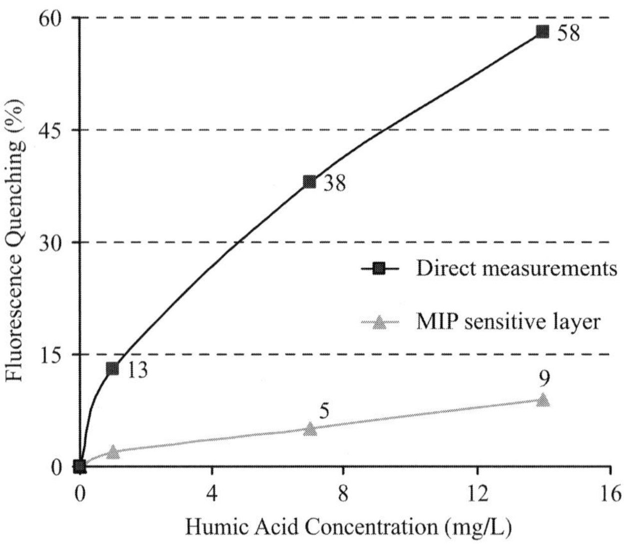

FIGURE 6.11 Fluorescence quenching by humic acid using the MIP sensor layer and comparing to the direct measurement in aqueous solution. *Adapted from [74].*

Haupt et al. [75] combined noncovalent-imprinted polymers with Fourier transform infrared (FT-IR) spectroscopy for the selective recognition of the herbicide 2,4-dichlorophenoxyaceticacid (2,4-D). MIP sensor layers were prepared directly on zinc selenide attenuated total reflection (ATR) interface. The sensor coatings were found to be sensitive enough to respond to lower ppb concentrations of (2,4-D). MIP selectivity was investigated by performing control experiments with nonimprinted and other structurally analogous compounds. As a transducing element, FT-IR spectroscopy enhances the discrimination between different analyte species, thus avoiding cross-sensitivity problems. In general, MIPs combined with FT-IR have the ability to work efficiently in harsh and complex mixtures for environment pollutant screening. Leung et al. [76] exploited the induced electron transfer photo induced electron transfer (PET) mechanism for luminescent sensing of (2,4-D) by MIP-based sol-gel material. This analyte is inherently nonfluorescent, so making use of the PET approach, optical properties are induced, being very valuable in signal transduction. Haupt et al. [77] performed fluorescent polarization sorbent assay by adopting imprinted polymers as a recognition element. This scheme provides suitable selectivity for screening hazardous (2,4-D) and other related compounds.

Apart from environmental toxins, one of the major concerns is to design a smart sensor for detecting different explosive compounds. Classical instruments such as GC, HPLC, and mass spectrometry provide adequate sensitivity, though they cannot be used for field measurements. Moreover, the sensing material should be miniaturized, capable of working in different environments and stable enough to function continuously. Stringer et al. [78] synthesized molecularly imprinted microparticles and cross linked them with fluorescent quantum dots for the detection of 2,4,6-trinitrotoluene (TNT) and 2,4-dinitrotoluene (DNT). The imprinted particles were characterized by transmission electron microscope (TEM) before and after the quantum dot labeling. Imprinted and nonimprinted microparticles were exposed to different concentrations of TNT, and subsequent quenching in fluorescent intensity was observed as a function of time. It has been observed that the fluorescence signal remains constant for nonimprinted microparticles, while in MIPs the signal drops exponentially as shown in Fig. 6.12. A similar pattern was observed for DNT. This indicates

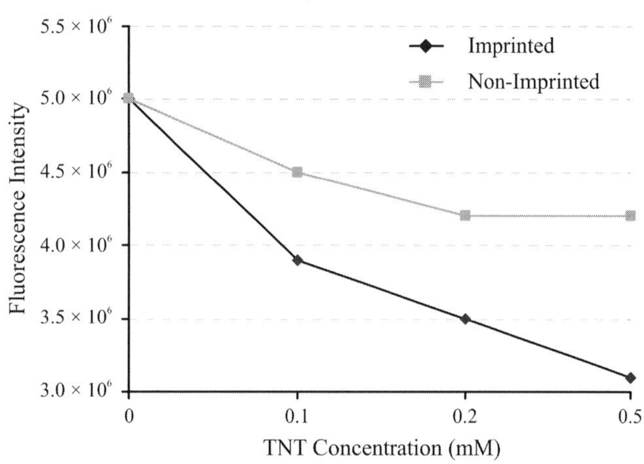

FIGURE 6.12 Sensor response of fluorescent-labeled molecularly imprinted material against nonimprinted control on exposure to various concentrations of TNT template. *Adapted from [78].*

template-specific sites present in MIPs that lead to fluorescence quenching upon analyte binding. The authors [78] had reported a detection limit of 40.7 μM and 30.1 μM for TNT and DNT, respectively, that corresponds to approximately 0.5 to 1 ppm. However, commercially available devices exhibit a further lower detection limit from 1 to 5 ppb.

This finding suggests that additional optimization is required in order to improve sensitivity and achieve lower detection limits.

7. EXEMPLARY APPLICATIONS OF MIP MASS-SENSITIVE SENSORS

Up to now, we have discussed imperative applications of MIP optical sensors in various fields, including biological, pharmaceutical, and environmental monitoring. In the next sections, we will focus on major applications of MIP mass-sensitive sensor devices in the same domains as those of optical sensors and then eventually analyze these approaches.

7.1. MIP Mass-Sensitive Biological Sensors

Mass-sensitive devices along with the imprinted technique has a very broad market for designing a variety of biosensors. They are equally efficient as optical sensors for sensing different body hormones. For example, MIP films fabricated on a QCM surface have been used for size-based selective recognition of catecholamines [79]—that is, an important body hormone acting as neurotransmitters in the brain. Catecholamine was imprinted in a sol-gel matrix of tetraethyl orthosilicate (TEOS) along with aluminum chloride as a functional monomer to enhance the rebinding efficiency. The recognition abilities of the imprinting system were optimized by controlling the pH on rebinding, composition of Al/Si ratio, capping of surface groups, and the pH of Al/Si solutions. Three different templates—dopamine (DA), norepinephrine (NEP), and epinephrine (EP)—were imprinted, and their relative rebinding efficiency was monitored as illustrated in Fig. 6.13.

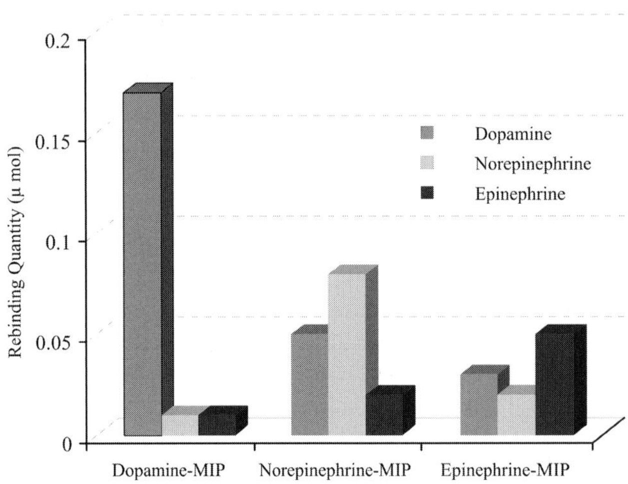

FIGURE 6.13 Rebinding of dopamine (DA), norepinephrine (NEP), and epinephrine (EP) imprinted polymers with different analytes. *Adapted from [79].*

This graph indicates that each imprinted material selectively adsorbs its own template in comparison to other analytes. The selectivity is size oriented as DA is smallest in size and can reincorporate more suitably to DA-imprinted polymer cavities, in contrast to other larger size analytes. MIP films with both an uncapped and capped surface were subjected to mass-sensitive measurements to determine relative rebinding selectivity.

Although the MIP uncapped surface exhibits sufficient binding selectivity, it has been observed that after surface capping, specificity improves astonishingly. The frequency shift of surface-capped DA-imprinted film for DA is 13.1 Hz, while the change for NEP and EP is only 0.4 and 0.2 Hz, respectively. The results of both DA-MIP and EP-MIP surface-capped and uncapped films are displayed in Fig. 6.14.

In general, both the imprinted powder material and thin films coated on QCM are equally competent for selective recognition of catecholamines.

Mass-sensitive devices with MIPs as a sensor interface have gained considerable significance in various bio-recognition systems and a variety of bio-analyses. Surface-imprinted polymers, as explained earlier, have been thoroughly studied for selective sensing of different bacteria and viruses. Conventional methods for detection of viruses and bacteria are based on culturing and counting bacterial colonies that are reliable but time consuming. In contrast, modern techniques such as ELISA or polymerized chain reaction (PCR) are highly sensitive, yield lower detection limits, and give prompt response. These features are desirable in routine clinical analysis; however, there is still space for more sensitive, robust, and low-cost sensor chips for rapid analysis of various bacterial and viral species.

Surface-imprinted polyurethanes [80] have been identified as a powerful tool for selective detection of *E. coli*; that is, gram negative bacteria caused illness through food contamination. Atomic force microscopy (AFM) images of *E. coli*-imprinted polyurethane revealed fine surface patterning at micrometer scale. Furthermore, the imprinted polymer surface was

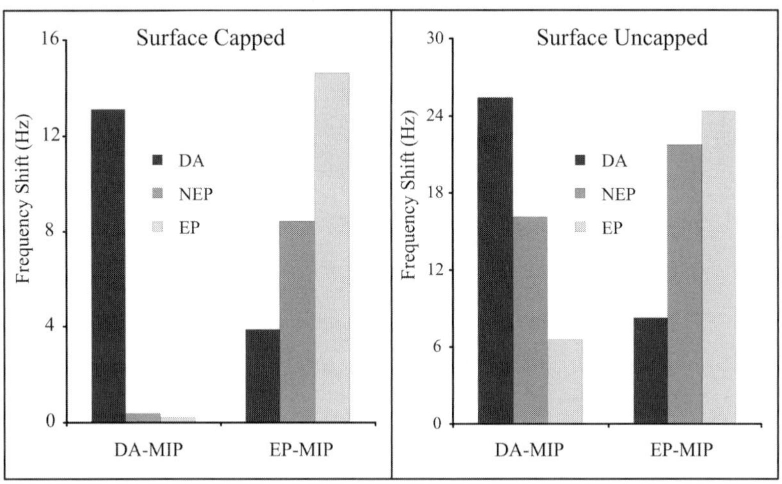

FIGURE 6.14 Relative rebinding selectivity of DA-MIP and EP-MIP surface capped and uncapped by QCM measurements. *Adapted from [79].*

exposed to 1mg/mL *E. coli* cell solution at 25 °C for mass-sensitive measurements. The frequency shift was nearly 70 Hz on imprinted channels and about 5 Hz for nonimprinted.

Viruses are much smaller in size than bacteria, and as a result their sensing is even more challenging. Besides ELISA and PCR techniques, some inexpensive kits are also available for screening viral species such as human immunodeficiency virus (HIV) and hepatitis C virus (HCV). However, the screening tools lack quantitative information and are virtually not applicable to a broader range of viruses. Surface imprinting provides a more general platform for molecular recognition of a variety of viral species.

For example, pico viruses [81] such as human rhino virus (HRV) and foot and mouth disease virus (FMDV) have been imprinted by the stamping method on a polyurethane surface. These layers demonstrate highly selective and reversible interactions specifically toward the target virus. The difference between imprinted and nonimprinted layers is also noticeable for two different concentrations of HRV2 as shown in Fig. 6.15.

It has been noticed that the MIP polyurethane layer shows nearly a half sensor response to HRV2 samples heated at 54 °C for 15 minutes when compared to untreated virus suspension. Furthermore, interestingly both imprinted and nonimprinted layers yield nearly the same sensor signal for HRV2 suspension when heated at 60 °C. This reveals that the heating leads to denatured HRV2 and that there is a conformational change in virus surface that makes it unfit to show any specific interaction with the MIP surface. AFM results also confirm that there is no integral virus particles on a polyurethane surface treated with heated suspension. Combining this information, it is evident that MIPs can easily distinguish between native and denatured species due to highly specific cavities. Intragroup selectivity studies were also carried out by imprinting polyurethane with HRV1, HRV2, and HRV14 virus, respectively.

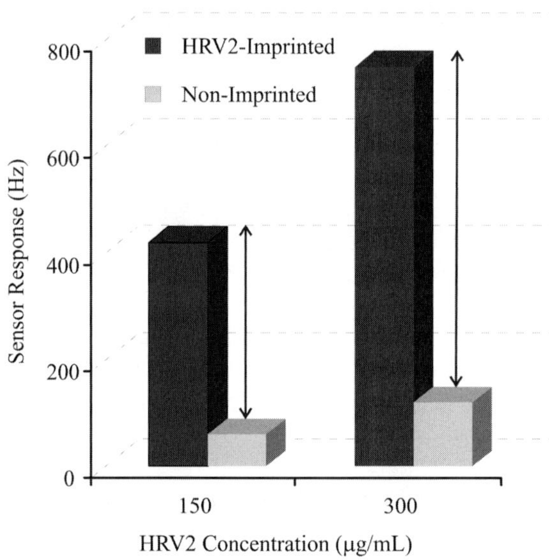

FIGURE 6.15 QCM sensor response of imprinted and nonimprinted polyurethane layers for different concentrations of HRV2. *Adapted from [81].*

After imprinting, polyurethane surfaces were exposed to these templates' suspensions of concentration. It has been observed that each imprinted layer respond highest sensor signal for that particular virus which was used in imprinting procedure. Apart from intragroup selectivity, interspecies selectivity studies were also performed by exposing HRV and FMDV suspensions of the same concentration to the HRV-imprinted layer.

Figure 6.15 shows that HRV-imprinted layers display a frequency shift of 270 Hz for HRV, while for FMDV the response is 35 Hz only. Summarizing the intergroup and intragroup selectivity results, we can conclude that MIPs are capable of recognizing extremely small species—for example, rhinoviruses at a nanoscale.

MIPs can be tuned for the sensing of larger bio-analytes such as red blood cells [11,82] and other microorganisms [10]. Mass-sensitive MIP sensors provide a straightforward method for determining blood group type. Cross-selectivity studies conducted with polyurethane imprinted with blood groups A, B, AB, and O suggest that, in any case the highest sensor signal was observed when the analyte and imprinted template belonged to the same blood group. Furthermore, MIPs can be tuned to distinguish subblood group types such as A1 and A2.

7.2. The Mass-Sensitive MIP Sensor in Plants

MIP-based mass-sensitive systems have been exclusively studied for plant viral species and important hormones. Takeuchi et al. [83] used N,N-dimethylaminoethyl methacrylate as a functional monomer for bulk imprinting of indoleacetic acid (IAA)—a phytohormone important for plant growth and development. The imprinted polymer layer was fabricated on platinum-coated 9 MHz QCM and treated with different concentrations of IAA ranging from 10 nmol to 200 nmol. This measurement was performed on three different QCMs to investigate sensor reproducibility. Results indicate that the coefficient of variation is 5.3%, 6.1%, and 8.8%, respectively, for the highest concentration of IAA, indicating almost the same performance by each device. It proves not only that all three layers are uniformly coated on QCM electrodes but also that the binding sites for IAA are also distributed evenly. This is a very valuable finding in terms of designing a sensor that yields a reproducible sensor response.

Selectivity studies were carried out by treating IAA-imprinted QCMs to other analogous plant hormones—that is, indole-3-butyric acid (IBA), indole, and indole-3-ethanol (IET) of the same concentrations. The highest sensor response was monitored for IAA hormones in contrast to other competing analytes. The selectivity data of IAA-imprinted QCM is shown in Fig. 6.16. Selectivity at the molecular level was achieved through electrostatic interactions between the carboxylic group of IAA and the functional monomer N,N-dimethylamino groups. In addition, the sensor layer was used repeatedly more than 50 times in organic solvents, keeping in view that biosensors having natural biomolecules can only be used in aqueous media. In general, the designed synthetic sensor material possesses high sensitivity, selectivity, and robustness that distinguish them from their natural counterparts.

Bulk-imprinted hydrogels have already been discussed for optical detection of TMV. Nevertheless, surface imprinting is a more versatile technique and therefore, along with mass-sensitive devices, it can be developed for TMV sensing [84]. At first attempt, acrylic-based polymers were applied for TMV recognition using dual-electrode 10 MHz

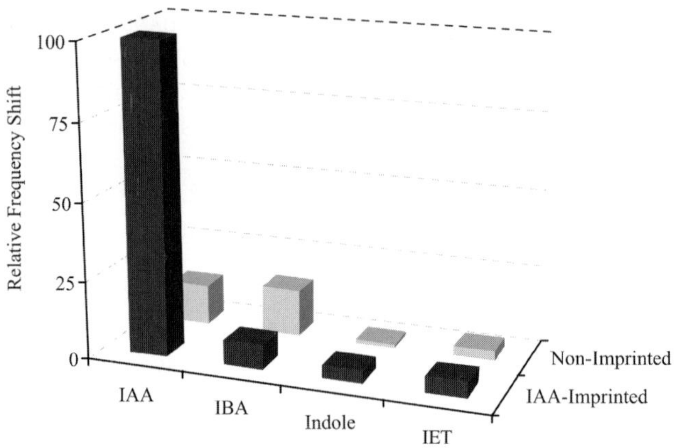

FIGURE 6.16 Relative frequency shift of IAA-imprinted and nonimprinted polymer for different plant hormones to evaluate selectivity. *Adapted from [83].*

QCM as the sensor device. The imprinted electrode showed suitable sensitivity for TMV as compared to the nonimprinted electrode. This finding was attributed to surface imprinting which enables nano patterning of polymers for mass-sensitive sensor applications. Imprinted polyurethane sensor layers exhibited much more improved sensor responses than polyacrylate. Moreover, they were found to be highly specific while examining cross-selectivity. The TMV-imprinted polyurethane [9] layer yields a 70 Hz sensor signal upon exposing 8 ng/mL TMV solution, while on the same surface other nontarget viruses such as HRV2 solution of 10 μg/mL give only a 10 Hz response. It is evident that HRV2 adsorbs nonspecifically and therefore produces a small effect even at higher concentration than TMV.

7.3. MIP Mass-Sensitive Sensors for Metal Ions

Ion imprinted polymers (IIPs) are analogous to MIPs since they specifically focus on the recognition of metal ions. Various approaches have been reported for synthesizing IIPs and exploring their recognition properties. T.P. Rao et al. [85] comprehensively reviewed these strategies for IIP synthesis, characterization, and their applications. These include pre-concentration of analytes in trace amounts, membrane separations, solid phase extraction (SPE), and sensor development. Although IIPs have been widely utilized for sensing purposes, their contribution to the development of mass-sensitive systems is limited. Yang et al. [86] designed a MIP-based QCM setup for Cu ion sensing in aqueous solution. They synthesized Cu-IIP using methacrylic acid as a functional monomer and directly coating on a QCM gold electrode. Then nonimprinted polymer was prepared in a similar fashion excluding the template. The Cu-IIP surface was characterized by SEM, and the image was compared to an uncoated QCM electrode surface. It was observed that the Cu^{+2}-IIP surface is relatively rough and contains a large number of holes that may be related to the diffusion paths for Cu^{+2} ions to reach the bulk material. The performance of a Cu^{+2}-IIP sensor was investigated by adjusting the amount of cross-linker, pH effect on frequency shifts, the dynamic range of sensors and, most important, parameter selectivity. The optimal molar ratio of cross-linker to monomer was found to be 1.43, which is best suited for getting a broad

TABLE 6.2 Selectivity Comparison of Different Ion-imprinted Polymers by Measuring Relative Frequency Shifts. *Adapted from [86]*

	Frequency Shift [Hz]		
Ion-imprinted Polymers	Cu^{+2}	Co^{+2}	Ni^{+2}
Cu^{+2}- IIP	95	35	22
Co^{+2}- IIP	12	78	15
Ni^{+2}- IIP	10	20	72
Nonimprinted	18	16	17

linear range and a lower detection limit. The highest frequency shift was observed at pH 6.0; therefore, all sensor measurements were conducted at this pH. Different Cu^{+2} ion concentrations were exposed to the IIP layer, and it was observed that the sensor response is linear in the range of 0.001 to 50 μM, while the lower detection limit was found to be 0.0008 μM, which is appreciable (Table 6.2).

Sensor selectivity was evaluated by performing differential measurements. As Table 6.2 makes clear, each IIP film that is coated responds with the highest sensor signal to its own template metal ions. For example, Cu^{+2}-IIP, Co^{+2}-IIP, and Ni^{+2}-IIP exhibit higher frequency shifts for Cu^{+2}, Co^{+2}, and Ni^{+2} ion solutions, respectively. It is evident that high selectivity and pronounced sensitivity are hallmarks of IIPs. There was no substantial change in sensor characteristics after several analysis cycles, so that it can be used for many times. Furthermore, wastewater analysis by a Cu^{+2}-IIP QCM sensor provides a highly selective and low-cost sensing platform for Cu^{+2} ion detection.

Recently, a MIP-based tetra electrode QCM sensor [87] system has been designed for sensing ionic and neutral analytes. In the case of ionic species, a polyurethane matrix was modified by potassium picrates for K^+ ion sensing, while aminopropyl silanes were tuned for selective recognition of Cu^{+2} and Ni^{+2} ions. The templating of picrate in polyurethane imparts high sensitivity for K^+ sensing. In the case of Cu^{+2} and Ni^{+2} ions, sensitivity was attributed to the ligand field stabilization effect rather depending on metal ion size or imprinted cavity dimensions. This means that weak coordinative forces between metal ions and monomers are decisive in pattern recognition. The Ni-imprinted polymer layer shows the highest sensor signal for Ni^{+2} ions and following for Zn^{+2} and Ca^{+2} ions, respectively. Similar results were obtained in the case of Cu-imprinted layers. The order of sensitivity is attributed to the Pearson concept as Ni^{+2} is harder than Zn^{+2} and Ca^{+2} ions; therefore, higher frequency shifts were observed for Ni^{+2} ions, whereas the Ca^{+2} ions are much softer than both Ni^{+2} and Zn^{+2} and so exhibit the least sensor effect. Therefore, the general sensor effect in order of hardness is $Ni^{+2} > Zn^{+2} > Ca^{+2}$.

7.4. MIP Mass-Sensitive Sensors for Environmental Monitoring

Environmental monitoring is one of the key areas of modern research since the waste material from different industries is directly polluting the air and water. Exhaust from power plants and automobiles produces numerous hydrocarbons and other toxic aromatic compounds.

Moreover, chloroflourohydrocarbons are destroying the ozone layer and pose a global hazard. Therefore, the ultimate goal of researchers is to design sensitive, miniaturized, stable, and cost-effective chemical sensors for continuous online monitoring of environmental toxins [88] and other poisonous gases. Acoustic devices with molecular-imprinted coating material are suitable candidates for various applications such as detection of volatile organic compounds (VOCs) [7,20], sensing of compost gases [89], and PAH determination in water [31].

For instance, during composting a number of malodorous gases and VOCs are emitted, and so their continuous surveillance is essential for process control. The degradation of bio-waste takes place in several phases and yields a very complex mixture of gases. Also, there is a certain pattern of emission of various gases from compost material that needs to be investigated in order to understand the maturity of the compositing process. E-noses [90] or a multisensor array deliver the concept of multi-analyte sensing. Classical instruments such as GCMS are highly sensitive and efficient for detection of compost gases; however, they cannot be adopted for continuous analysis. MIPs coated on QCMs provides a relatively different way to deal with composting because their high sensitivity, small size, and reversible sensor signal make them ideally suited for this task. Aiming at a complex mixture, imprinted sensor layers can be tuned to enhance sensitivity and selectivity. MIP-based QCM e-noses [91] have shown considerable potential for online compost monitoring. These sensor arrays provide the general response profile of a different analyte class rather than giving information about individual analyte molecules. The sensor array can be calibrated by exposure to known concentrations of target analytes and at a defined relative humidity level.

A six-electrode QCM array coated with MIP layers was positioned in a compost bin to monitor the degradation of grass and pine in the presence of compost accelerating material. The sensor array measurements were continued 14 days for grass and 40 days for pine. The sensor responses of six channels were processed by principal component analysis (PCA) and by removal of the outliers. The artificial neuronal network (ANN) was trained by calibration, thus ultimately evaluating real-time QCM array signals. Besides the statistical treatment of sensor results, the compost samples were analyzed by GCMS two to three times a day in order to validate the performance of the array. After processing of sensor array data sheets and comparing them with GCMS, it was observed that emission of alcohols, ethyl acetate, and terpenes follows almost the same compost pattern, which is astonishing. Although there are some minor deviations, the profile of analytes is very similar for both grass and pine composting. Furthermore, for elucidating quantitative information, the sensor array data was compared with GCMS results. For this purpose during composting, two points were selected—after 100 hours and 700 hours, respectively. The results of sensor array and GCMS are presented in Fig. 6.17.

Figure 6.17 illustrates that at the initial stage—100 hours—the sensor response for alcohols and terpenes was exactly the same as determined by GCMS, but after 700 hours, the combined sensor effect for alcohols and terpenes was slightly different from GCMS results. This was explained by the steric effect for alcohols leading to a reduced sensor signal and the emission of some other terpenes, careen, for example, that produce somewhat higher signals. However, the QCM sensor array proves to be a suitable alternative to GCMS for real-time online degradation studies with high precision and relatively much lower cost.

VOCs have serious health effects as they are continuously released in the environment by different industries, including petrochemicals, cosmetics, and many others. Consequently,

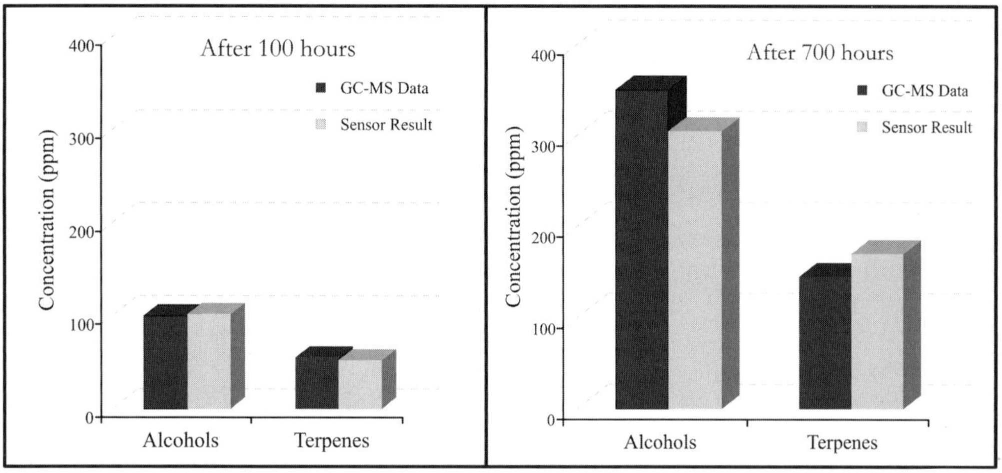

FIGURE 6.17 Comparison of GC-MS data with QCM sensor array results for compost monitoring. *Adapted from [91].*

their detection in air samples is essential and more important for those VOCs that do not have any odor. Various spectroscopic techniques such as FT-IR, UV/Vis, and Raman spectroscopy have been reported for their detection. Imprinted materials with mass-sensitive transducers can be designed for the sensing of various VOCs. Polystyrene imprinted [92,93] with toluene exhibits excellent sensitivity for toluene vapors, while the sensor response for other related compounds such as benzene and o-xylene is much lower. The imprinted hollows are adaptive for toluene, whereas both of the nontarget analytes have different sterical arrangements and affinity that do not match with MIP cavities. In addition, imprinted material can differentiate between different xylene isomers, as seen in Table 6.3. It is clear from the table that the o-xylene-imprinted layer gives the highest sensor signal for o-xylene vapors, and the same is the case with other polymers. This selectivity pattern is indeed attributed to molecular imprinting.

Recently [94], molecular imprinted titanate sol-gel layers coated on dual-electrode QCM have been developed for sensing different VOCs, including alcohols and alkanes. The sensitivity and selectivity are attributed to the precursors, such as tetrapropoxy titanium (TPT) and tetrabutoxy titanium (TBT), used for generating MIP layers. Although the end product is the same in both cases, it was observed that careful tuning of these functionalized materials

TABLE 6.3 QCM Selectivity Pattern of Xylene-imprinted Polymers for their Respective Isomers. *Adapted from [92]*

	Relative Sensor Effect [%]		
Imprinted Polymers	**o-Xylene**	**m-Xylene**	**p-Xylene**
Ortho-Xylene	100	35	15
Meta-Xylene	5	100	45
Para-Xylene	20	30	100

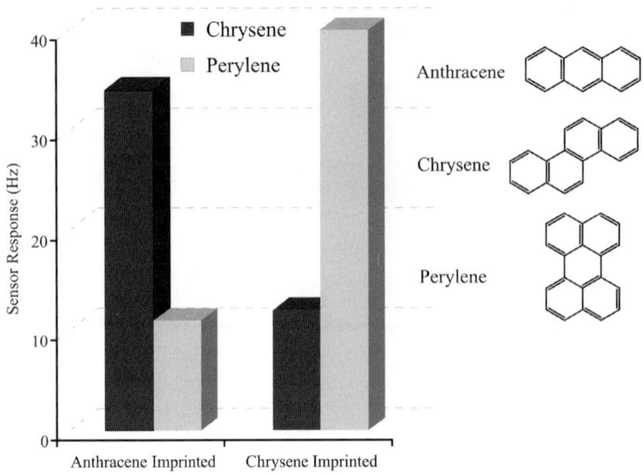

FIGURE 6.18 QCM sensor response of anthracene- and chrysene-imprinted polyurethane layer to 2 μg/L concentration of chrysene and perylene, respectively. *Adapted from [74].*

allows distinguishing between octane isomers—n-octane and iso-octane. Furthermore, the mass-sensitive effects of different vapors were correlated to their Wiener index, and it was attributed to highly selective binding cavities produced by imprinting. This sensing approach is attractive in designing highly selective surfaces that can recognize isomeric analytes.

MIPs along with mass-sensitive transducers have been explicitly studied for the sensing of PAHs, as polyurethane possesses aromatic chains that are favorable for imprinting PAHs of different sizes and shapes. After extensive study [74], it has been concluded that the best adaptive cavities are not generated by the analyte itself in the case of polyurethanes; imprinting sites can be optimized to yield enhanced sensitivity by choosing suitably sized single template molecules or a combination of two templates. It has been observed that the template for molecular imprinting should be of a relatively smaller size than analyte molecules. For example, if we want to design a MIP sensor for chrysene, it should be imprinted with smaller size anthracene, while in the case of perylene the imprint molecule is chrysene. The result of mass-sensitive measurements performed with anthracene- and chrysene-imprinted polyurethane layers, respectively, are shown in Fig. 6.18.

It is obvious that the anthracene-imprinted polyurethane layer exhibits a larger frequency shift for chrysene than perylene. Whereas chrysene imprinted layer shows higher sensor signal for perylene than chrysene. Therefore, it can be concluded that template size plays important role for getting pronounced sensor effect.

These findings confirm that in the case of PAHs, the geometry of the imprinted molecule signifies the sensitivity and selectivity. Concerning template imprinting in polyurethane, PAHs possess a special character. It has been reported that using a combination of two templates—a larger and a smaller one—creates highly specific molecular cavities. The concept was named double imprinting [31]; it sets up a new dimension in molecular imprinting technique. Generally, the two different templates are mixed in an appropriate ratio where the larger template acts as porogen, thereby creating more suitable diffusion pathways for the reversible re-inclusion of analyte.

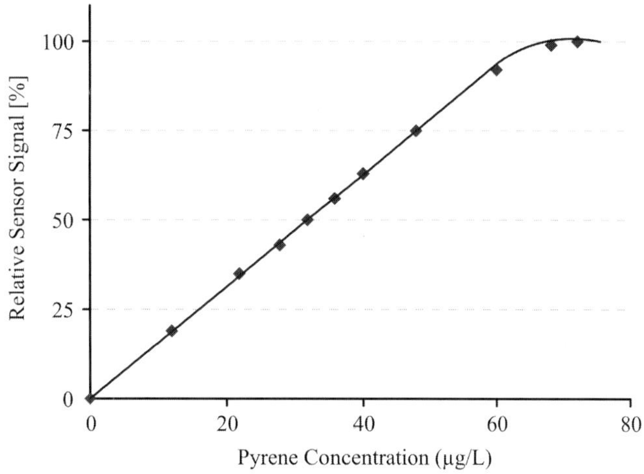

FIGURE 6.19 Relative sensor effect of a 10 MHz QCM coated with double imprinted, that is, acenaphthene-pyrene (20:80%) polyurethane layer, when exposed to different concentrations of pyrene. *Adapted from [95].*

Mass-sensitive devices using a template mixture (20% naphthalene and 80% pyrene) can detect a concentration of 100 ng/L of pyrene, which is appreciable. In contrast, a fluorescence sensor provides a much lower detection limit using the same double imprinted polyurethane as coating material.

It has been observed that double imprinted polymers [95] can enhance the sensor signal by a factor of five as compared to single imprinted sensor materials. Furthermore, these sensor materials can perform linearity in a broad concentration range indicating a more favorable dynamic range. The sensor response of the double imprinted polyurethane layer coated on 10 MHz QCM for different pyrene concentrations is shown in Fig. 6.19. This graph shows that the sensor layer becomes saturated after 60 μg/L pyrene concentration. Compared to single-template MIPs, in double imprinted materials the imprinted cavities are more easily accessible due to the rapid evaporation of smaller size templates; as a result, the overall number of interaction centers increases.

In addition, the temperature and solvent for polymerization also play very important roles in crafting a suitable cavity structure. At higher temperatures, that is, 70 °C, polyurethane chains are more closely arranged around the template, thus yielding highly structured cavities for selective analyte incorporation. Tetrahydrofuran (THF) was used exclusively as a solvent for the synthesis of polyurethane. It was observed that polymerization under solvent-saturated vapors inhibits rapid vaporization, thus ultimately producing lower viscosity coating material. The lower viscosity enhances the mobility of polymer chains, causing them to be organized around analyte molecules.

8. CRITICAL COMPARISON BETWEEN OPTICAL AND MASS-SENSITIVE SENSORS

MIP-based optical and mass-sensitive sensors have been comprehensively discussed according to their potential applications in important fields. Both of these sensor systems have intrinsic features that make one preferable to the other, although there are some other examples where optical and mass-sensitive sensors are directly compared using cholesteric liquid crystals (CLC) [96], polysiloxanes [97], and other sensitive materials. However, the combination of MIPs with optical and mass-sensitive sensors for the very same analyte is not fully investigated.

The general characteristics of a sensor system include sensitivity, selectivity, linear range, and lower detection limit. In addition to these characteristics, some other very important aspects need to be considered. For example, in the case of optical sensors, if the target analyte does not have any optical features, then it becomes difficult to execute MIPs as sensor coatings. In order to solve this problem, labeling indicators are introduced in the system attached either to analyte or to MIPs, so that the binding event can be monitored by optical sensors. Nevertheless, this labeling step adds more cost and time to the overall optical analysis. In some cases, the labeling reagent interferes with the target analyte that leads to inaccurate results, thus making the sensing mechanism more complicated and less reliable. That is why MIP optical sensors are somewhat limited to a certain class of analyte.

On the other hand, acoustic or mass-sensitive sensors are virtually applicable to all types of analytes. Since mass is a fundamental property of all compounds, these devices are considered universal transducers. Molecular imprinting allows design of adaptive cavities for reversible binding of target molecules, and whenever the analyte interacts with MIPs, a direct change in frequency is observed. For analytes that do not possess optical properties, usually mass-sensitive devices can be preferred to optical transducers. Conversely, if the target compound has some optical features, then it can be detected by both schemes by designing a suitable imprinted polymer. In this case, they can be compared with regard to sensitivity and selectivity, lower detection limits, and sensor linearity. The reversibility and reusability factors also have a very strong impact on the ultimate cost of the sensor.

Although it is not easy to fully compare both of these sensing techniques for a particular analyte, there are a few examples where both approaches overlap in sensor design. In this regard, PAHs have been studied extensively by both optical and mass-sensitive transducers using MIPs as sensor interface. Polyurethanes have already proven the right choice for PAH imprinting since the monomer building blocks are very similar to PAHs in size that yield patterned interaction sites. Other factors influencing the imprinting process of PAHs, such as temperature of polymerization, nature of solvent, size, and combination of the template molecules, have all been discussed earlier. As previously mentioned, that selectivity is governed by the geometry of imprint molecule; best results are obtained when the imprint molecule is smaller in size than the target molecule. Fluorescence and QCM measurements of polyurethanes imprinted with anthracene and chrysene are presented in Fig. 6.20.

It is clear from the figure that both the optical enrichment factor and mass-sensitive measurements support this idea and exhibit a higher sensor signal for the larger size analyte. Mass-sensitive measurements were performed for two larger size analytes, while fluorescence data give a much broader view [74]. In this way, the results of one sensor scheme can be validated by the other. The two different sensor approaches provide complementary information about the binding event on MIP interface.

Our focus is on evaluating and comparing these two sensor technologies having same coating interface. In the case of optical systems, fluorescence measurements of PAHs in water samples reveal that the technique is highly sensitive and yields lower detection limits.

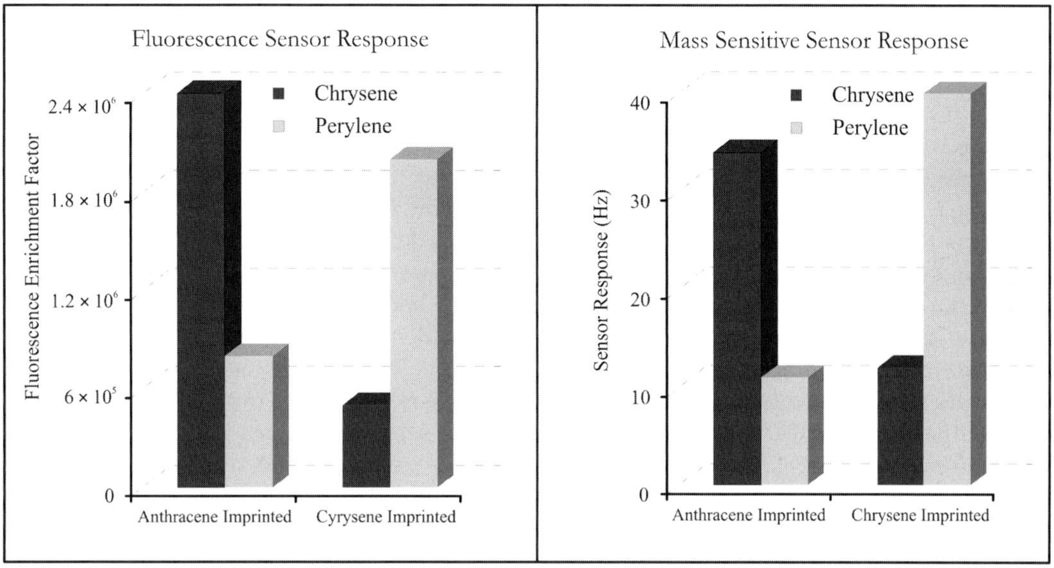

FIGURE 6.20 Comparison of fluorescence and mass-sensitive sensor measurements using the same MIP coatings. The selectivity is indeed attributed to the size of the template. *Adapted from [74].*

However, fluorescence sensor reaches saturation much earlier than mass sensitive setup. This does not correspond that the polyurethane sensor layer is saturated and there are no more cavities left for analyte molecules since mass-sensitive transducers still show linear behavior at this point. The results of optical and mass-sensitive measurements for pyrene determination are shown in Fig. 6.21.

The figure shows that at the beginning sensor signals of both systems increase linearly with the increasing analyte concentration and then reach their respective saturation points.

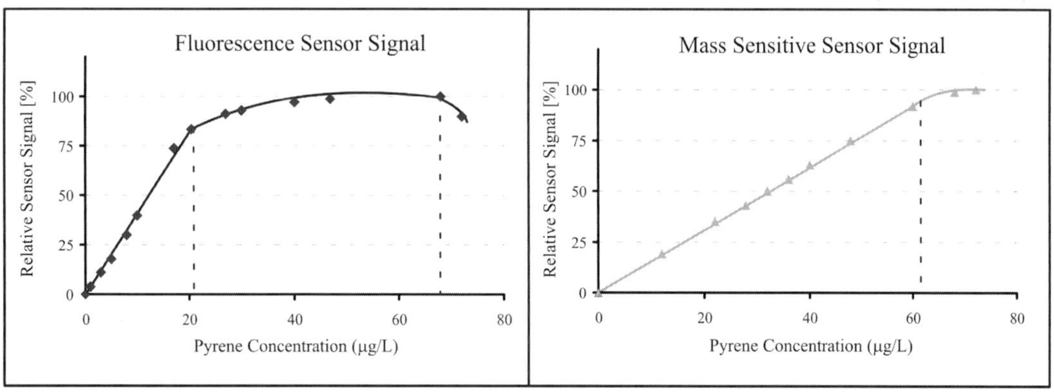

FIGURE 6.21 Comparison of optical and mass sensor signals for different concentrations of pyrene by the same double imprinted polyurethane coatings. The optical sensor is more sensitive, whereas the QCM sensor has a broader dynamic range. *Adapted from [95].*

The dynamic range of the fluorescence sensor is relatively limited; thus plateau is attained at nearly 15 μg/L concentration, whereas QCM sensor is linear at the same value. The mass-sensitive sensor reaches the saturation point at 60 μg/L, and interestingly at this position the fluorescence signal starts dropping. Such a different behavior of two different sensing schemes using the same polymeric layers can be explained in the following way. As mass is the universal property of every analyte, so as soon as the cavities are available the sensor response increases with increasing analyte concentration. In addition, there are no quenching problems for mass-sensitive sensors that can interfere on the sensor signal.

In the case of fluorescence measurements, two emission spectra of the MIP layer were recorded, one at the beginning and the second at the end of plateau. It has been observed that after reaching a certain concentration level, further increase in pyrene concentration produces an excimer. Although these excimers have similar dimensions, they do not contribute to pyrene emissions. As a result, the plateau is attained at much lower concentration compared to the QCM sensor signal. When mass-sensitive sensors become saturated at higher concentrations, a large number of excimers are produced that significantly decrease fluorescence intensity. PAHs integrate in imprinted cavities following the $\pi-\pi$ interactions, whereas excimers are produced by introducing a second PAH molecule into the already occupied cavity. So when all the imprinted sites are occupied, further increase in pyrene concentration eventually leads to substantial quenching, thus decreasing the fluorescence sensor signal. Comparing the limit of detection, fluorescence measurements can go down to 10 ng/L, while for QCM the detection limit is 100 ng/L, which is ten times higher. In general, both sensing approaches possess intrinsic features; the QCM sensor is suitable for working in a broad dynamic range, whereas fluorescence is best suited for achieving lower detection limits. It is important to note that in this comparison of optical and mass-sensitive sensors, no labeling indicator is used for fluorescence detection. If the labeling step is added to optical detection, then the overall comparison becomes more complex.

9. CONCLUDING REMARKS

We have comprehensively discussed the design and working of both optical and mass-sensitive devices and their integration with MIPs for sensor development. Both of these setups have conferred numerous applications in various fields ranging from biological species to environmental toxins. In optical sensing, detection is made mostly through fluorescence. It is important to note that 85% of the known compounds are not fluorescence active, and therefore their optical sensing is not as straightforward as in case of mass sensors. In order to make these compounds fluorescence active, additional labeling is necessary. The labeling approach is good for bringing optical character, but it also adds more cost and time, thereby making analysis somewhat complicated. Conversely, mass is the most fundamental property of every analyte, and so realizing this fact, one can design a universal sensor setup no matter what the structure of the analyte is. Both sensor strategies have characteristic advantages, including sensitivity, lower detection limits, and broader dynamic range that can be used according to one's interest. The ultimate objective of this study is to investigate the potential of MIPs for optical and mass sensing and then compare these approaches in different fields for selecting the right device for the desired application.

References

[1] P.A. Lieberzeit, F.L. Dickert, Anal. Bioanal. Chem. 387 (2007) 237.
[2] F. Caruso, E. Rodda, D.N. Furlong, J. Colloid Interface Sci. 178 (1996) 104.
[3] X.L. Su, Y. Li, Biosens. Bioelectron. 19 (2004) 563.
[4] S. Kurosawa, E. Tawara-Kondo, N. Minoura, N. Kamo, Sens. Actuators, B 43 (1997) 175.
[5] R. Doong, H. Shih, S. Lee, Sens. Actuators, B 111/112 (2005) 323.
[6] G. Wulff, A. Sarhan, Angew. Chem., Int. Ed. 11 (1972) 341.
[7] F.L. Dickert, M. Keppler, Adv. Mater. 7 (1995) 1020.
[8] E. Howe, G. Harding, Biosens. Bioelectron. 15 (2000) 641.
[9] O. Hayden, P.A. Lieberzeit, F.L. Dickert, Adv. Funct. Mater. 16 (2006) 1269.
[10] J.A. Barnett, Microbiology 149 (2003) 557.
[11] O. Hayden, K-J. Mann, S. Krassnig, F.L. Dickert, Angew. Chem. Int. Ed. 45 (2006) 2626.
[12] O. Hayden, R. Bindeus, C. Haderspock, K.-J. Mann, B. Wirl, F.L. Dickert, Sens. Actuators, B 91 (2003) 316.
[13] T. Inoue, K. Ohtsuka, Y. Yoshida, Y. Matsuura, Y. Kajiyama, Sens. Actuators, B 25 (1995) 388.
[14] X. Wang, X. Wang, J. Zhou, J. Song, J. Liu, N. Xu, Z.L. Wang, Nano Lett. 6 (2006) 2768.
[15] G.D. Agostino, G. Alberti, R. Biesuz, M. Pesavento, Biosens. Bioelectron. 22 (2006) 145.
[16] O.Y.F. Henry, D.C. Cullen, S.A. Piletsky, Anal. Bioanal. Chem. 382 (2005) 947.
[17] F.L. Dickert, P.A. Lieberzeit, Springer Ser Chem. Sens. Biosens. 5 (2007) 173.
[18] G. Wulff, Angew. Chem. Int. Ed. 34 (1995) 1812.
[19] F.L. Dickert, R. Sikorski, Mater. Sci. Eng. C 10 (1999) 39.
[20] F.L. Dickert, O. Schuster, Microchim. Acta 119 (1995) 55.
[21] F.L. Dickert, A. Haunschild, Adv. Mater. 5 (1993) 887.
[22] J.L. Defreese, A. Katz, Chem. Mater. 17 (2005) 6503.
[23] O. Hayden, C. Haderspock, S. Krassnig, X. Chen, F.L. Dickert, Analyst 131 (2006) 1044.
[24] G. Wulff, R. Vesper, R. Grobe-Einsler, A. Sarhan, Makromol. Chem. 178 (1977) 2799.
[25] G. Wulff, S. Schauhoff, J. Org. Chem. 56 (1991) 395.
[26] S.H. Cheong, S. Mcniven, A.E. Rachkov, R. Levi, K. Yano, I. Karube, Macromolecules 30 (1997) 1317.
[27] M.J. Whitcombe, M.E. Rodriguez, P. Villar, E.N. Vulfson, J. Am. Chem. Soc. 117 (1995) 7105.
[28] L.G. Schultz, Y. Zhao, S.C. Zimmerman, Angew. Chem. Int. Ed. 40 (2001) 1962.
[29] O. Norrlow, M. Glad, K. Mosbach, J. Chromatogr. 299 (1984) 29.
[30] K. Mosbach, Trends Biochem. Sci. 19 (1994) 9.
[31] F.L. Dickert, P. Achatz, K. Halikias, Fresenius J. Anal. Chem. 371 (2001) 11.
[32] D.F. Tai, C.Y. Lin, T.Z. Wu, L.K. Chen, Anal. Chem. 77 (2005) 5140.
[33] H.S. Andersson, J.G. Karlsson, S.A. Piletsky, A.C. Koch-Schmidt, K. Mosbach, I.A. Nicholls, J. Chromatogr., A 848 (1999) 39.
[34] T. Takeuchi, D. Fukuma, J. Matsui, Anal. Chem. 71 (1999) 285.
[35] P. Guenther, Optical Sensors: Industrial, Environmental and Diagnostic Applications, in: R. Narayanaswamy, O.S. Wolfbeis (Eds.), Anal. Bioanal. Chem. 381, 2005, p. 1.
[36] F. Lanza, B. Sellergren, Anal. Chem. 71 (1999) 2092.
[37] M.C. Burleigh, S. Dai, E.W. Hagaman, J.S. Lin, Chem. Mater. 13 (2001) 2537.
[38] C.K.O. Sullivan, Anal. Bioanal. Chem. 372 (2002) 44.
[39] P. Leonard, S. Hearty, J. Brennan, L. Dunne, J. Quinn, T. Chakraborty, R. O'Kennedy, Enzyme Microb. Technol. 32 (2003) 3.
[40] M. Lotierzo, O.Y.F. Henry, S. Piletsky, I. Tothill, D. Cullen, M. Kania, B. Hock, A.P.F. Turner, Biosens. Bioelectron. 20 (2004) 145.
[41] F.L. Dickert, H. Besenbock, M. Tortschanoff, Adv. Mater. 10 (1998) 149.
[42] F.L. Dickert, P.A. Lieberzeit, M. Tortschanoff, Sens. Actuators. B 65 (2000) 186.
[43] M.E. Byrne, E. Oral, J.Z. Hilt, N.A. Peppas, Polym. Adv. Technol. 13 (2002) 798.
[44] C. Sulitzky, B. Ruckbert, A.J. Hall, F. Lanza, K. Unger, B. Sellergren, Macromolecules 35 (2002) 79.
[45] B. Wandelt, P. Turkewitsch, S. Wysocki, G.D. Darling, Polymer 43 (2002) 2777.
[46] B. Wandelt, A. Mielniczak, P. Turkewitsch, S. Wysocki, J. Lumin. 102-103 (2003) 774.
[47] D.L. Rathbone, D.Q. Su, Y.F. Wang, D.C. Billington, Tetrahedron Lett. 43 (2000) 123.

[48] C.-F. Chow, M.H.W. Lam, M.K.P. Leung, Anal. Chim. Acta 466 (2002) 17.
[49] K. Haupt, React. Funct. Polym 41 (1999) 125.
[50] D. Nopper, O. Lammershop, G. Wulff, G. Gauglitz, Anal. Bioanal. Chem. 377 (2003) 608.
[51] J. Curie, P. Curie, C R Acad. Bulg. Sci. 91 (1880) 294.
[52] G. Sauerbrey, Z. Phys. Chem. 155 (1959) 206.
[53] F.L. Dickert, P. Forth, W.E. Bulst, G. Fischerauer, U. Knauer, Sens. Actuators, B 46 (1998) 120.
[54] L. Rayleigh, Proc. Lond. Math. Soc. s1-17 (1885) 4.
[55] A. Mujahid, A. Afzal, G. Glanzing, A. Leidl, P.A. Lieberzeit, F.L. Dickert, Anal. Chim. Acta 675 (2010) 53.
[56] P. Turkewitsch, B. Wandelt, G.D. Darling, W.S. Powell, Anal. Chem. 70 (1998) 2025.
[57] B. Wandelt, A. Mielniczak, P. Cywinski, Biosens. Bioelectron. 20 (2004) 1031.
[58] A. Rachkova, S. McNiven, A. Elskaya, K. Yano, I. Karube, Anal. Chim. Acta 405 (2000) 23.
[59] J. Matsui, H. Kubo, T. Takeuchi, Anal. Chem. 72 (2000) 3286.
[60] S.A. Piletsky, E. Terpetschnig, H.S. Andersson, I.A. Nicholls, O.S. Wolfbeis, Fresenius J. Anal. Chem. 364 (1999) 512.
[61] S.A. Piletsky, E.V. Piletska, A. Bossi, K. Karim, P. Lowe, A.P.F. Turner, Biosens. Bioelectron. 16 (2001) 701.
[62] J.M. Lin, M. Yamada, Anal. Chem. 72 (2000) 1148.
[63] E. Benito-Pena, M.C. Moreno-Bondi, S. Aparicio, G. Orellana, J. Cederfur, M. Kempe, Anal. Chem. 78 (2006) 2019–2027.
[64] G.P. Gonzalez, P.F. Hernando, J.S.D. Alegria, Anal. Chim. Acta 638 (2009) 209.
[65] L.D. Bolisay, J.N. Culver, P. Kofinas, Biomacromolecules 8 (2007) 3893.
[66] J.L. Suarez-Rodriguez, M.E. Diaz-Garcia, Anal. Chim. Acta 405 (2000) 67.
[67] M.F. Lulka, S.S. Iqbal, J.P. Chambers, E.R. Valdes, R.G. Thompson, M.T. Goode, J.J. Valdes, Mater. Sci. Eng. C 11 (2000) 101.
[68] O. Guney, Y. Yilmaz, O. Pekcan, Sens. Actuators, B 85 (2002) 86.
[69] H. Chen, M.M. Olmstead, R.L. Albright, J. Devenyi, R.H. Fish, Angew. Chem. Int. Ed. 36 (1997) 642.
[70] M.N. Sing, R. Narayanaswamy, Anal. Bioanal. Chem. 386 (2006) 1235.
[71] S. Al-Kindy, R. Badia, Diaz-Garcia Me, Anal. Lett. 35 (2002) 1763.
[72] D. James, J.M. Gladis, A.K. Pandey, G.R.K. Naidu, T.P. Rao, Talanta 74 (2008) 1420.
[73] S. Stanley, C.J. Percival, Anal. Chem. 75 (2003) 1573.
[74] F.L. Dickert, M. Tortschanoff, Anal. Chem. 71 (1999) 4559.
[75] M. Jakusch, M. Janotta, B. Mizaikoff, K. Mosbach, K. Haupt, Anal. Chem. 71 (1999) 4786.
[76] M.K.P. Leung, C.F. Chow, M.H.W. Lam, J. Mater. Chem. 11 (2001) 2985.
[77] C.E. Hunt, P. Pasetto, R.J. Ansell, K. Haupt, Chem. Commun. (2006) 1754.
[78] R.C. Stringer, S. Gangopadhyay, S.A. Grant, Anal. Chem. 82 (2010) 4015.
[79] T.R. Ling, Y.Z. Syu, Y.C. Tasi, T.C. Chou, C.C. Liu, Biosens. Bioelectron 21 (2005) 901.
[80] A. Afzal, X. Chen, M. Jenik, S. Krassnig, F.L. Dickert, Commercial and Pre-Commercial Cell Detection Technologies for Defence against Bioterror, in: L.M. Lechuga, et al. (Eds.), IOS Press, 2008, p. 60.
[81] M. Jenik, R. Schirhagl, C. Schirk, O. Hayden, P. Lieberzeit, D. Blaas, G. Paul, F.L. Dickert, Anal. Chem. 81 (2009) 5320.
[82] A. Seifner, P. Lieberzeit, C. Jungbauer, F.L. Dickert, Anal. Chim. Acta 651 (2009) 215.
[83] A. Kugimiya, T. Takeuchi, Electroanalysis 11 (1999) 1158.
[84] F.L. Dickert, O. Hayden, P.A. Lieberzeit, C. Haderspoeck, R. Bindeus, C. Palfinger, B. Wirl, Synth. Met. 138 (2003) 65.
[85] T.P. Rao, R. Kala, S. Daniel, Anal. Chim. Acta 578 (2006) 105.
[86] Z.P. Yang, C.J. Zhang, Sens. Actuators, B 142 (2009) 210.
[87] U. Latif, A. Mujahid, A. Afzal, R. Sikorski, P.A. Lieberzeit, F.L. Dickert, Anal. Bioanal. Chem. 400 (2011) 2507.
[88] F.L. Dickert, P.A. Lieberzeit, Encyclopedia of Analytical Chemistry, in: R.A. Meyers (Ed.), John Wiley & Sons Ltd, Chichester, 2000, p. 3831.
[89] P.A. Lieberzeit, A. Rehman, N. Iqbal, B. Najafi, F.L. Dickert, Monatsh. Chem. 140 (2009) 947.
[90] K. Persaud, G.H. Dodd, Nature 299 (1982) 352.
[91] P.A. Lieberzeit, A. Rehman, B. Najafi, F.L. Dickert, Anal. Bioanal. Chem. 391 (2008) 2897.

[92] P.A. Lieberzeit, S. Gazda-Miarecka, K. Halikias, C. Schirk, J. Kauling, F.L. Dickert, Sens. Actuator, B 111–112 (2005) 259.
[93] F.L. Dickert, P.A. Lieberzeit, S. Gazda-Miarecka, K.J. Mann, O. Hayden, C. Palfinger, Biosens. Bioelectron 20 (2004) 1040.
[94] U. Latif, A. Rohrer, P.A. Lieberzeit, F.L. Dickert, Anal. Bioanal. Chem. 400 (2011) 2457.
[95] P.A. Lieberzeit, K. Halikias, A. Afzal, F.L. Dickert, Anal. Bioanal. Chem. 392 (2008) 1405.
[96] A. Mujahid, H. Stathopulos, P.A. Lieberzeit, F.L. Dickert, Sensors 10 (2010) 4887.
[97] A. Hierlemann, K. Bodenhofer, M. Fluck, V. Schurig, W. Gopel, Anal. Chim. Acta 346 (1997) 327.

CHAPTER

7

Discrimination of Analytes with Fluorescent Molecular Imprinting Sensor Arrays

Jin Tan[1], *Xiu-Ping Yan*[2]

[1]Tianjin Key Laboratory of Food Biotechnology, College of Biotechnology and Food Science, Tianjin University of Commerce, China

[2]Research Center for Analytical Sciences, College of Chemistry, and State Key Laboratory of Medicinal Chemical Biology, Nankai University, Tianjin, China

OUTLINE

1. INTRODUCTION

Chemical sensor arrays have been demonstrated to be highly effective tools for the discrimination of analytes [1–4]. The inspiration for chemical sensor arrays arises from the superb performance of the cross-responsive mammalian olfactory system. Compared with

Molecularly Imprinted Sensors.
DOI: 10.1016/B978-0-444-56331-6.00007-4

selective biosensors and chemical receptors designed for particular analytes, which tend to suffer from cross-reactivity, chemical sensor arrays conversely take advantage of such cross-reactivity to achieve differential sensing for multi-analytes. The elements in a sensor array do not need individual selectivity toward any given analyte, but instead the collected sensors should present as much sensing diversity as possible so that the array responds to different analytes with the largest possible cross-reactivity. A distinct pattern of responses produced by the sensors in the array can provide a fingerprint that allows classification of the analytes.

Molecularly imprinted materials [5–8] possess high chemical and thermal stability, and can be generated rapidly and inexpensively, making them amenable to an array setting. The major drawbacks associated with imprinted materials—relatively low overall affinities and high levels of cross-reactivity toward structurally highly similar species—can actually be advantageous for differential sensor arrays [4,9]. The relationship of templates and imprinted materials synthesized with the templates can be compared to that of antigens and antibodies. This biomimetic feature of molecular imprinting just fits the inspiration of chemical sensor arrays. The combination of molecular imprinting and sensor arrays can be expected to simulate the mammalian tongue and nose for the senses of taste and smell. As long as signals are generated for an analyte from one or more imprinted material sensor, a characteristic fingerprint for this analyte can be obtained [10–16]. Pattern recognition can be completed by mathematical multivariate analyses such as principal component analysis (PCA) and linear discriminant analysis (LDA) [2,4].

2. MOLECULAR IMPRINTING IN ARRAY APPLICATION

2.1. Synthesis of Molecularly Imprinted Materials

Choice of functional monomer and cross-linker is an important step in the preparation of molecularly imprinted materials. Silica-based inorganic molecularly imprinted materials have been shown to be effective matrices for sensor application and can be prepared in a variety of physical forms such as thin films, nanoparticles, and bulk materials to suit the final application desired [17,18]. Due to their high porosity, concomitant large surface area, facile synthesis, and robustness as well as their optical transparency in the visible to UV range, mesoporous silicas are in principle ideally suited as support materials for sensitive probes and have been widely used as matrices for optical sensors [19–27].

For fluorescent sensor application, organic functional groups are usually introduced into mesoporous silicas and the so-called organic-inorganic hybrid materials are obtained. There are two basic strategies for the fabrication of functional mesoporous silica: (1) one-pot strategy via the co-condensation of organic modified functional monomer and cross-linker, and (2) postsynthetic covalent grafting of organic compounds [28]. Compared with postsynthetic functionalization, one-pot co-condensation offers more homogeneously distributed organic units and avoids the pore blocking problem that is encountered in the postsynthetic method.

In our first study of fluorescent imprinting sensor array for metal ions [15], a direct synthetic route involving the co-condensation of 8-hydroxyquinoline (8-HQ) functionalized monomer with tetraethyl orthosilicate (TEOS) in the presence of the surfactant cetyltrimethylammonium bromide (CTAB) was utilized to fabricate the nonimprinted material (NIM) and imprinted materials (Fig. 7.1). NIM was prepared without the addition of any metal ion template, while for imprinted materials, a 100-fold excess of template metal ion with respect to the fluorescent monomer was used to guarantee that a substantial proportion of the monomers was engaged in the interaction with the template ion during cross-linking. In our further work on imprinting sensor array for saccharides [16], a phenylboronic acid-appended triethoxysilane was synthesized through the reaction of 3-isocyanatopropyltriethoxysilane and 3-aminophenylboronic acid. The co-condensation of the obtained functionalized triethoxysilane with TEOS in the presence of the surfactant CTAB then produced the organic-inorganic hybrid saccharide-imprinted mesoporous silica (Fig. 7.2).

2.2. Selection of Templates

In traditional molecular imprinting, target analytes are always used as the templates. As the aim of the preparation of imprinted materials is to selectively adsorb or bind the analyte of interest, it is easy to understand that employing the analyte as the template can leave a binding site that is exactly tailor-made for the analyte. However, as the end is multi-analyte sensing in the design of the molecular imprinting sensor array, the selection of templates plays an important role in the discrimination performance of the imprinting sensor array, but is not that simple and easy. It is true that the most reliable solution is to synthesize different imprinted materials using all the analytes to be differentiated as templates, respectively. Actually, the synthetic work in the fabrication of molecular imprinting sensor arrays need not be that tedious. Rational selection of templates can save plenty of time and effort by using less imprinted materials to discriminate more analytes.

For instance, to fabricate an imprinting sensor array for Zn^{2+}, Cd^{2+}, Mg^{2+}, Ca^{2+}, and Al^{3+}, we chose Zn^{2+} and Cd^{2+} as the templates. Zn^{2+} imprinted material (ZnIM) and Cd^{2+} imprinted material (CdIM) as well as NIM were synthesized as the sensor elements [15].

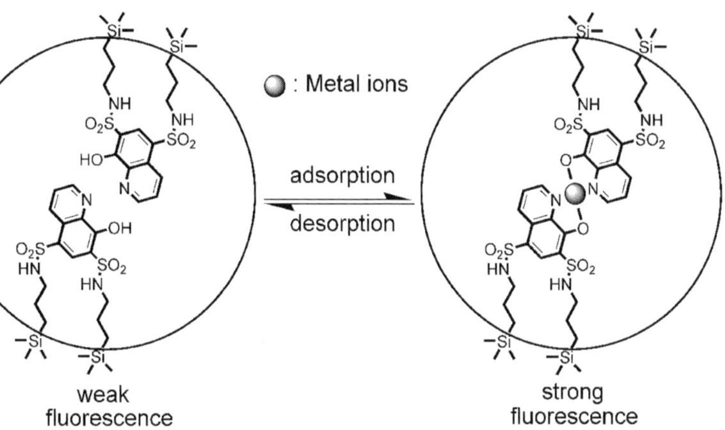

FIGURE 7.1 Schematic representation of fluorescent ion-imprinted mesoporous silica. *Adapted from Tan, J., Wang, H.-F., Yan, X.-P., 2009. Biosens. Bioelectron. 24, 3316-3321. Copyright (2009), with permission from Elsevier.*

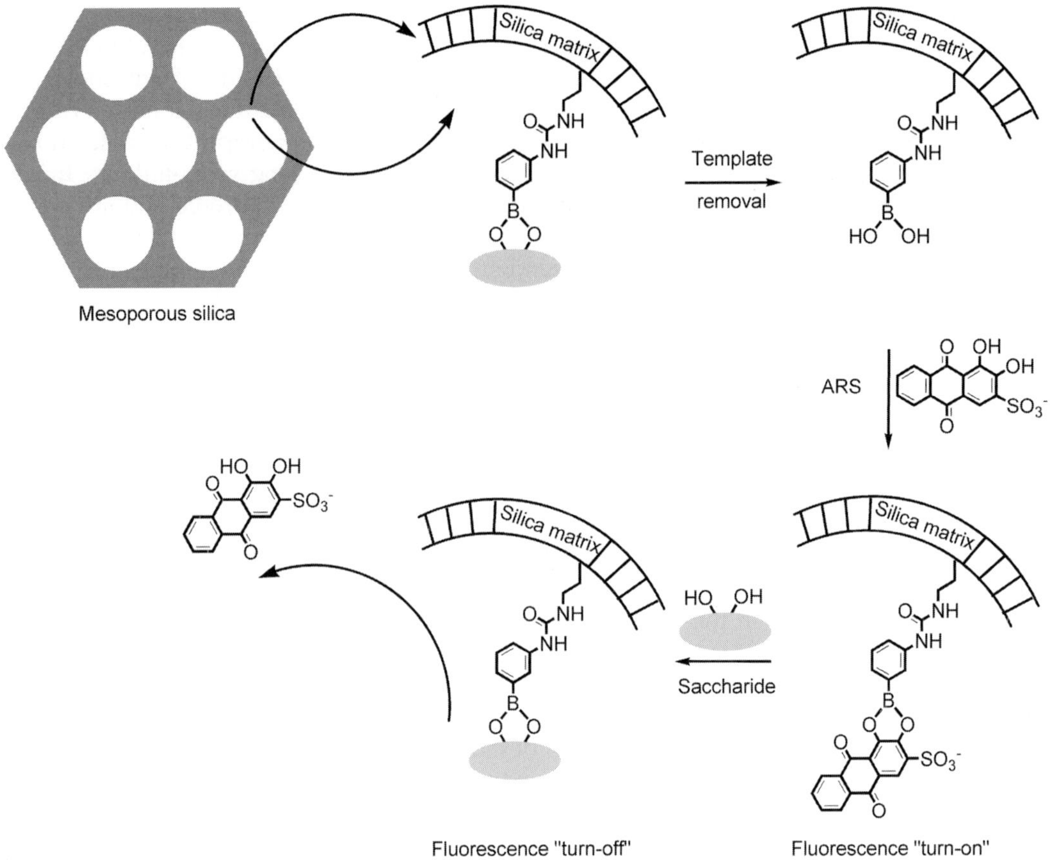

FIGURE 7.2 Schematic representation of molecularly imprinted mesoporous silica as indicator-displacement sensor for saccharides. *Reprinted with permission from Tan, J., Wang, H.-F., Yan, X.-P., 2009. Anal. Chem. 81, 5273-5280. Copyright 2009 American Chemical Society.*

Zn and Cd belong to the same main group in the Periodic Table of Elements. Compared with Mg^{2+}, Ca^{2+}, and Al^{3+}, both Zn^{2+} and Cd^{2+} possess higher affinity for the functional moiety 8-HQ [29]. Furthermore, Cd^{2+} has a larger ionic radius (0.095 nm) and covalent radius (0.148 nm) than Zn^{2+} (ionic radius of 0.075 nm and covalent radius of 0.131 nm), as Cd is in the next period to Zn. The high affinity of the templates for the functional monomers guarantees the high binding performance of imprinted materials, while the difference between the two templates in size renders the diverse response of ZnIM and CdIM to different metal ions with different sizes. Additionally, the imprinting effect makes the two imprinted materials possess different responses to metal ions compared with NIM. Thus, the sensor array composed of the two imprinted materials and NIM is anticipated to be able to discriminate several metal ions of interest. Our results show that the imprinting sensor array composed of ZnIM, CdIM, and NIM is able to discriminate the five metal ions, including both template ions of Zn^{2+} and Cd^{2+} and nontemplate ions of Mg^{2+}, Ca^{2+}, and Al^{3+} [15].

In our next work for saccharides [16], D-fructose and D-xylose were chosen as the templates as they are quite different in both molecular size and binding affinity toward boronic acid and are also relatively common and inexpensive. D-Fructose is a six-carbon ketose with relatively strong affinity toward boronic acid, while D-xylose, a five-carbon monosaccharide, forms moderate ester with boronic acid. Because the two saccharide templates leave distinct binding sites in their own templated materials, the two imprinted materials are expected to respond differently to several saccharides with different molecular size, stereochemistry, and affinity to boronic acid. Similar to our first work, a three-element sensor array was fabricated utilizing NIM and two imprinted materials. The use of only three materials enabling differentiation of ten saccharides, which demonstrates that reasonable selection of templates allows the maximal acquisition of discrimination power with less synthetic and analytical effort.

3. SIGNALING TYPES

A major concern in the development of molecular imprinting sensors is signal transduction. To date, various kinds of molecular imprinting sensors have been reported, such as optical sensors, electrochemical devices, and sensors based on mass changes [6,18]. In principle, all the signaling types in the published works on molecular imprinting sensors can be employed in the design of molecular imprinting sensor arrays. In our researches, fluorescence was chosen from the several mechanisms owing to its well-documented advantages such as simplicity, high selectivity, and low detection limits [9]. Depending on the property of the analytes of interest, either the indicator-spacer-receptor method or the indicator-displacement strategy can be employed.

3.1. Indicator-Spacer-Receptor Approach

To directly visualize the binding events of metal ions, 8-HQ was incorporated into the imprinting cavity as a fluorescent receptor. 8-HQ and its derivatives are famous for their fluorescent chelates with a number of metal ions, including Al^{3+}, Cd^{2+}, Mg^{2+}, and Zn^{2+} [29], and have shown great potential in the design of cross-reactive arrays for metal ions because they present a turn-on signal and are highly cross-reactive. In the excellent works from Palacios et al. [30,31], 8-HQ-based ligands with extended conjugated fluorophores synthesized by the Suzuki-Miyaura cross-coupling reaction were rationally designed for fluorescent sensor arrays for metal ions. In our investigation [15], sulfonylchloride moieties were introduced into the 8-HQ skeleton at the 5- and 7-positions, and the obtained intermediate product reacted with 3-aminopropyltriethoxysilane (APTES) to form the fluorescent functional monomer.

Both the fluorescence intensities of the as-synthesized imprinted materials were significantly stronger than those of NIM, demonstrating that the 8-HQ moieties planted in the imprinted materials were in chelation with the metal ion templates. After rinsing with 0.1 M EDTA solution, the imprinted materials showed weaker fluorescence similarly to NIM, indicating that the metal ion templates were extracted out of the materials. Similarly, the metal ion-loaded materials can be easily recovered to the blank materials by treatment with 0.1 M EDTA solution, making the materials recyclable [15].

3.2. Indicator-Displacement Strategy

With the well-known property of reversible covalent cyclic ester formation with diols, boronic acids are intensively used as receptors in saccharide sensors [32–36] and saccharide-imprinted materials [37-39]. Unlike the chelates of 8-HQ and metal ions, the examination of the binding of boronic acids to saccharide is often difficult as the cyclic ester of boronic acid and saccharides does not show a strong fluorescent signal. Therefore, a fluorophore is usually appended to the boronic acid to convert the binding event into fluorescence signal. Compared with the indicator-spacer-receptor approach, indicator-displacement assay (IDA) [40,41] does not require covalent incorporation of a fluorophore into the receptors and hence circumvents the difficult synthesis.

In an IDA, an indicator first binds reversibly to a receptor, and a competitive analyte is then introduced into the system to displace the indicator from the host, in turn to modulate an optical signal. Two kinds of approaches are employed in indicator-displacement differential sensors for saccharides. The first approach involves the ensemble of organic indicator dyes and phenylboronic acid/boric acid receptors as probe set [42,43]. Although the compounds used in this approach are commercially available, a lot of test work is required to collect the effective indicator-receptor probe pairs. Compared with the first approach, the second approach puts more emphasis on the organic synthesis of various boronic acid receptors with different binding behavior to a group of analyte saccharides [44–47]. Both approaches possess high discrimination performance toward saccharides, but each has its own intrinsic shortcoming. Compared with conventional methods, molecular imprinting provides an alternative approach for saccharide sensor arrays, with easy and facile procedures both analytically and synthetically.

In our investigation [16], a commonly used fluorescent dye Alizarin Red S. (ARS) was employed as the indicator because Springsteen and Wang [48,49] have fully examined the system of phenylboronic acid and ARS to investigate the interaction between boronic acid and diols. The catechol-type dye ARS shows weak fluorescence at neutral pH, while the formation of boronate ester of ARS drastically enhances the fluorescence. As shown in Fig. 7.2, the binding of ARS into the synthesized materials increases the fluorescence due to the ester formation. The addition of analyte saccharides changes the equilibrium and drives some of the ARS molecules out of the binding site, which turns off the fluorescence intensity to some extent.

4. EVALUATION OF IMPRINTING EFFECT

The binding responses of sensor elements to different analytes can be examined by directly measuring the fluorescence of the solid imprinted materials suspended in solution, which can provide valuable data to investigate fluorescent molecular imprinting sensor arrays. By comparing the kinetics and thermodynamics of the binding of imprinted materials and NIM, the imprinting effect can be easily evaluated and the different imprinting effects of imprinted materials on their templates and other highly related analytes can provide the discrimination power for the imprinting sensor array.

Prior to the fluorescence test, the mixture solutions, including the solid materials and the analytes, were ultrasonicated for 3 minutes to disperse the particles of the materials and to make the suspension homogeneous. The suspensions were then directly introduced into the cuvette, and the fluorescence was measured immediately without stirring. The emission of the suspensions remained constant within 2–3 minutes, and the fluorescence then decreased gradually as a consequence of the particle settling.

4.1. Binding Kinetics

The time-dependent response of the NIM and ZnIM to Zn^{2+} and Cd^{2+} is shown in Fig. 7.3 [15]. The fluorescence intensity of ZnIM reached 85% of the equilibrium value within 0.5 min after the addition of Zn^{2+}, while for NIM, both of the metal ions showed a relatively slow binding kinetics. The fluorescence intensity increased gradually to no more than 40% of the equilibrium value within 5 minutes after the addition of Zn^{2+} or Cd^{2+}, and it took more than 3 hours to reach the full equilibrium. The faster binding kinetics of imprinted materials can be ascribed to the templating process. In NIM, the functional groups are randomly distributed in the matrix, while in imprinted materials the functional monomers are "frozen" in the position where they initially form the chelates with the template metal ions. Thereby, the memory effect of imprinted materials allows them to own a shorter response time to metal ions than NIM.

The binding kinetics of the IDA sensors for saccharides [16] is not as fast as those of the metal ion-imprinted sensors [15]. The time-dependent fluorescence responses of NIM, D-fructose-imprinted material (FruIM), and D-xylose-imprinted material (XylIM) toward D-fructose show that the process is relatively slow and that it takes about 10 hours to reach full equilibrium. Considering that the relatively large dye molecule ARS needs a period of time to penetrate into the gel matrix, the slow binding kinetics in this work is understandable, similar to some other saccharide sensors using boronic acid functionalized gels and ARS, which require several hours for the gels to become dye saturated [36,50]. As the binding kinetics is decided by how fast ARS can get into the binding site, it is not surprising that the saccharide-imprinted materials do not respond to the template saccharide faster than NIM.

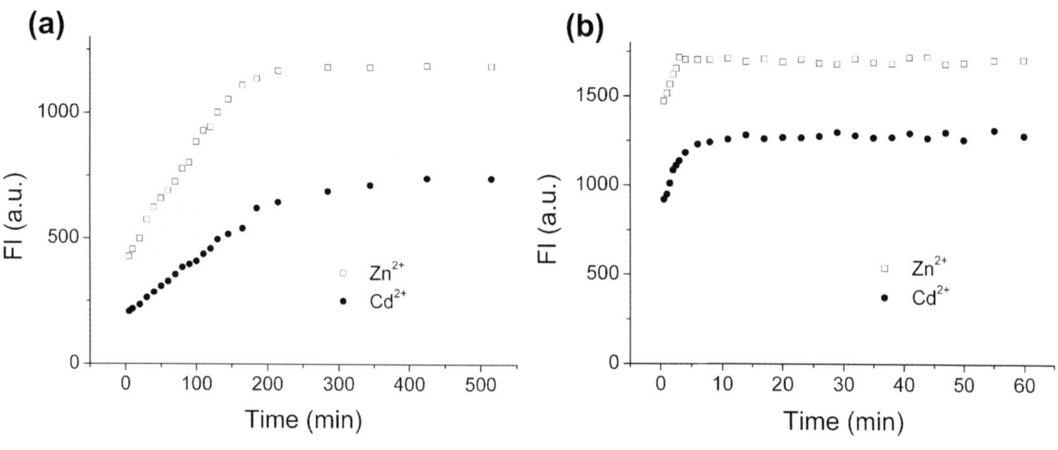

FIGURE 7.3 Time-dependent fluorescence response of NIM (a) and ZnIM (b) toward Zn^{2+} and Cd^{2+}. *Reprinted from Tan, J., Wang, H.-F., Yan, X.-P., 2009. Biosens. Bioelectron. 24, 3316–3321. Copyright (2009), with permission from Elsevier.*

4.2. Binding Thermodynamics

Fluorescence titration can be carried out to investigate the binding thermodynamics of the imprinted materials and NIM. By performing a Langmuir-type analysis of the fluorescence titration data using equation (7.1) [21,22], apparent binding constants (adsorption constants) are obtained.

$$F = F_0 + (F_1 - F_0)\frac{Kc}{1 + Kc} \tag{7.1}$$

$$c = \frac{-\left(\frac{1}{K} + \frac{n_M}{V} - c_0\right) + \sqrt{\left(\frac{1}{K} + \frac{n_M}{V} - c_0\right)^2 + \frac{4c_0}{K}}}{2}$$

where K is the adsorption constant, F is the measured fluorescence intensity of the material, F_0 and F_1 are the observed intensities of the material in the absence of analyte and in the presence of excessive analyte, respectively, c_0 is the added analyte (mol L^{-1}), V is the volume of the solution (L) and n_M is the maximum amount of adsorbed analyte in the monolayer (mol).

The spectrofluorimetric titrations of ZnIM with Zn^{2+} and Cd^{2+}, CdIM with Zn^{2+} and Cd^{2+}, and NIM with Zn^{2+} and Cd^{2+} were performed. Typical spectrofluorimetric titration spectra of ZnIM with Zn^{2+} are shown in Fig. 7.4. Addition of Zn^{2+} or Cd^{2+} led to slight red-shifts in the fluorescence spectra of ZnIM and CdIM due to excimer formation between adjacent fluorophores in the solid matrix [51]. In imprinted materials, the templating process fixed the 8-HQ moieties in the position where they initially formed the complex with the template metal ion and hence made them in close proximity. No obvious red-shift was found in the case of NIM because 8-HQ moieties were randomly distributed in NIM without such close proximity.

Both of the two imprinted materials show stronger binding affinity toward their corresponding template ion than NIM. On the other hand, the two imprinted materials are to some extent cross-responsive to the nontemplate metal ions. ZnIM shows some imprinting effect toward Cd^{2+}, and CdIM exhibits some imprinting effect toward Zn^{2+}. The cross-reactivity of the imprinted materials provides the feasibility to differentiate the metal ions by sensor array composed of the materials.

In the imprinting IDA array for saccharides [16], the interaction between the materials and the indicator ARS was studied. The binding constants of NIM, FruIM, and XylIM with ARS are 2.1×10^5, 3.7×10^5, and 3.2×10^5 M^{-1}, respectively. The response of an imprinted material to an analyte can be assigned to the cooperation of the thermodynamics and imprinting effect. Theoretically, molecular imprinting generates pockets in a polymer matrix that could only be filled by the template molecule or some structurally highly similar species. However, as the functional groups are fixed at the position where they form a complex with the template, removal of the template molecule leaves a route to the imprinting cavity. Some small molecules that have strong binding affinity with the functional monomer can also penetrate into the imprinting cavity and hence react with the functional monomer. Thus, the response of the imprinted materials to ARS can be attributed to the cooperation of the superiority of ARS in thermodynamics and the steric hindrance effect.

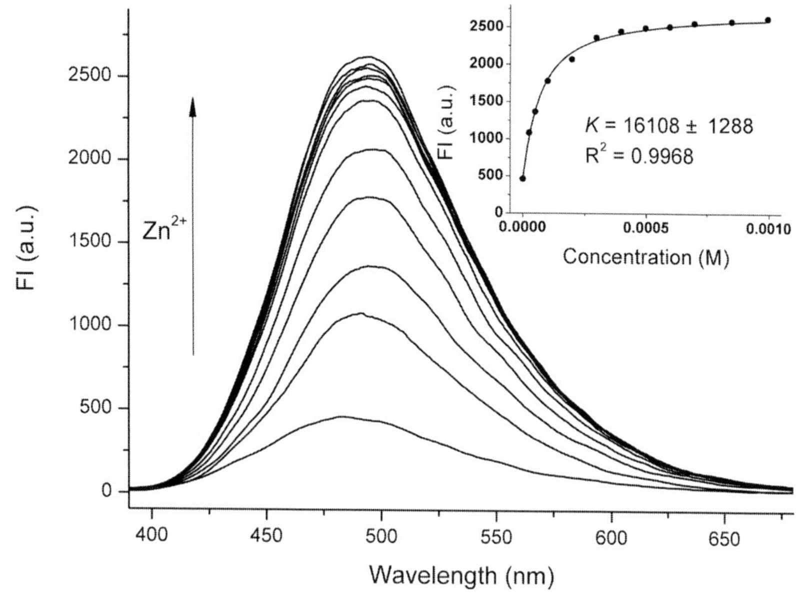

FIGURE 7.4 Changes in the spectrofluorimetric titration spectra of ZnIM (5 mg/10 mL) with Zn^{2+} at pH 5.0 (λ_{ex} = 340 nm). Inset: Plot and nonlinear fit of fluorescent intensity at λ_{em} as a function of the zinc concentration. *Reprinted from Tan, J., Wang, H.-F., Yan, X.-P., 2009. Biosens. Bioelectron. 24, 3316–3321. Copyright (2009), with permission from Elsevier.*

Studies on the IDA performance toward several saccharides reveal that the extent of fluorescence quenching in the case of FruIM is larger than that in the case of NIM, indicating that FruIM shows a more positive imprinting effect for D-fructose than for ARS. FruIM, XylIM, and NIM respond differently to eight selected saccharides in spectrofluorimetric quenching titration. Stern–Volmer quenching constants (K_{SV}) of NIM, FruIM, and XylIM with eight selected saccharides are listed in Table 7.1. All the materials show relatively small, but distinct, quenching constants for the eight saccharides.

TABLE 7.1 Stern–Volmer Quenching Constants (K_{SV}) of NIM, FruIM, and XylIM with Eight Selected Saccharides. Saccharides: D-Fructose (Fru), D-Xylose (Xyl), D-Arabinose (Ara), D-Glucose (Glu), D-Galactose (Gal), D-Mannose (Man), L-Sorbose (Sor), and D-Ribose (Rib). *Adapted with permission from Tan, J., Wang, H.-F., Yan, X.-P., 2009. Anal. Chem. 81, 5273–5280. Copyright 2009 American Chemical Society.*

	K_{SV} (M^{-1})[a]							
Material	**Fru**	**Xyl**	**Ara**	**Glu**	**Gal**	**Man**	**Sor**	**Rib**
NIM	2.3	2.1	1.5	1.0	1.1	2.3	6.5	6.0
FruIM	3.9	2.5	1.7	0.81	0.58	1.4	5.4	5.5
XylIM	1.5	3.0	1.9	0.93	0.52	1.4	3.0	2.2

[a] *All errors are < 10%.*

5. DISCRIMINATION OF ANALYTES

5.1. Metal Ions

A sensor array composed of NIM, ZnIM, and CdIM was designed to discriminate Zn^{2+}, Cd^{2+}, Mg^{2+}, Ca^{2+}, and Al^{3+} [15]. Each sensor of this array responded to a different degree to a particular metal ion at 10^{-4} M or 5×10^{-5} M; hence, a characteristic fingerprint for each metal ion was obtained [Fig. 7.5(a)]. In order to reduce the dimensionality of the data set, to find underlying variables, and to expose the fingerprints more clearly, PCA was applied [Fig. 7.5(b)]. The PCA plot shows a clear discriminative and reproducible pattern for the five tested metal ions at two different concentrations, and the first two components contained 93.6% of the data variation. The responses for each ion are clustered into tight distinct groupings, indicating the reproducibility of the response for each metal ion. Owing to the cross-reactivity of molecular imprinting, use of a common fluorescent receptor 8-HQ allows not only the identification of the two template metal ions, but also the discrimination of three nontemplate metal ions.

5.2. Saccharides

Encouraged by the discrimination performance of the metal ion-imprinted sensor array, we employed our saccharide-imprinted sensor array composed of FruIM, XylIM, and NIM to discriminate ten saccharides: D-fructose, D-xylose, D-arabinose, D-glucose, D-galactose, D-mannose, L-sorbose, D-ribose, L-rhamnose, and sucrose [16]. The PCA plot shows a clear discrimination of the ten tested saccharides (Fig. 7.6). These specific binding-based unique

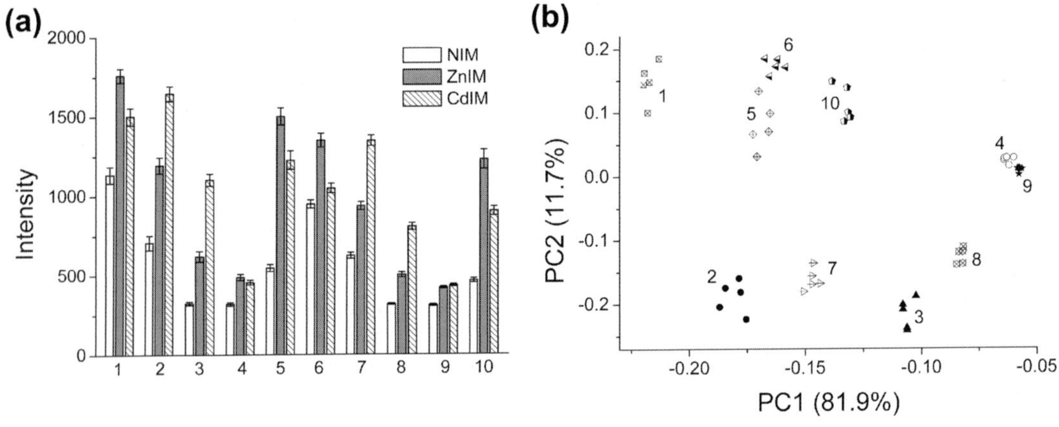

FIGURE 7.5 (a) Plot of the response of an array composed of NIM, ZnIM, and CdIM to five metal ions at two different concentrations at λ_{em} (λ_{ex} = 340 nm) tested in replicate (5 times). (b) Two-dimensional PCA plot for five metal ions at two different concentrations tested in replicate (5 times) to an array composed of NIM, ZnIM, and CdIM. Metal ions: 10^{-4} M Zn^{2+} (1), 10^{-4} M Cd^{2+} (2), 10^{-4} M Mg^{2+} (3), 10^{-4} M Ca^{2+} (4), 10^{-4} M Al^{3+} (5), 5×10^{-5} M Zn^{2+} (6), 5×10^{-5} M Cd^{2+} (7), 5×10^{-5} M Mg^{2+} (8), 5×10^{-5} M Ca^{2+} (9), and 5×10^{-5} M Al^{3+} (10). *Reprinted from Tan, J., Wang, H.-F., Yan, X.-P., 2009. Biosens. Bioelectron. 24, 3316–3321. Copyright (2009), with permission from Elsevier.*

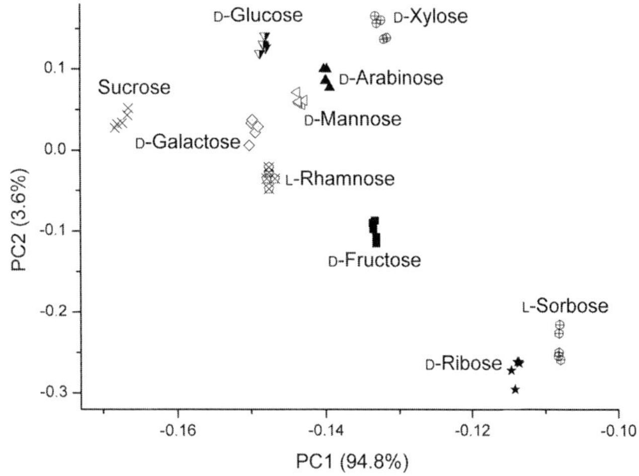

FIGURE 7.6 Two-dimensional PCA plot for the identification of ten saccharides (100 mM) based on an array composed of NIM, FruIM, and XylIM. *Adapted with permission from Tan, J., Wang, H.-F., Yan, X.-P., 2009. Anal. Chem. 81, 5273–5280. Copyright 2009 American Chemical Society.*

fingerprints generated by PCA allow the identification of these saccharides. It is of great significance to achieve the classification of large numbers of analytes using imprinting arrays with small numbers of sensors.

5.3. Orange Juice Samples

To test the applicability of the molecular imprinting sensor array to real-world samples, we employed the saccharide sensor array to three brands of sugar containing orange juice beverage, including Coca-Cola Minute Maid, President Orangeate, and Huiyuan orange juice beverage [16]. Orange juice beverages are complex samples with several major ingredients, including saccharides (mainly fructose, glucose, and sucrose), citric acid, sodium citrate, β-carotene, and ascorbic acid. Citric acid, sodium citrate, and β-carotene without diols do not significantly change the fluorescence intensity of the system of ARS-boronic acid. Ascorbic acid containing a diol moiety was examined to show little quenching effect on the fluorescence intensity of the IDA system, even at the maximum concentration indicated by the ingredient description on the label of the beverage. The two major colorants in orange juice, Sunset Yellow FCF and Tartrazine, did not cause significant changes in either the excitation or emission of the test solution at an average level of concentration added to the beverage. Thus, the inner filter effect exerted by the colored media on both the exciting light and emitted light was negligible. As the other main ingredients do not significantly affect the fluorescence intensity of the IDA system, fluorescence quenching can be mainly assigned to the saccharides in the beverage.

In the tests, the three brands of orange juice beverage were tested against the array (NIM, FruIM, and XylIM) five times, and the raw data matrix of fluorescence intensity was introduced to PCA. As a result, the three samples are clearly separated from each other (Fig. 7.7). Thus, the three brands of orange juice beverage can be differentiated by the response patterns obtained from the imprinting sensor array. The diverse binding affinities of NIM, FruIM, and XylIM toward fructose, glucose, and sucrose, as well as differences in

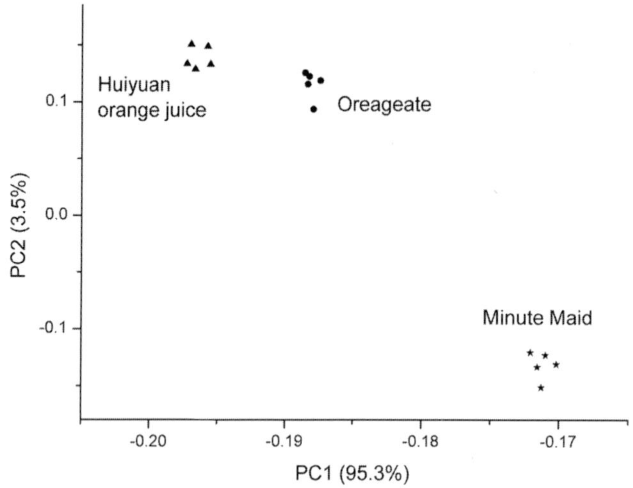

FIGURE 7.7 Two-dimensional PCA plot for the discrimination of three brands of orange juice beverage based on an array composed of NIM, FruIM, and XylIM. *Adapted with permission from Tan, J., Wang, H.-F., Yan, X.-P., 2009. Anal. Chem. 81, 5273–5280. Copyright 2009 American Chemical Society.*

the composition ratio of fructose, glucose, and sucrose in the three brands of orange juice, enable us to distinguish them.

6. CONCLUSIONS

Cross-reactivity to structurally related species that are highly similar to the template in size, shape, and functionality is always encountered and hard to avoid in molecular imprinting, but imprinting sensor arrays are particularly suitable for discrimination of such highly similar analytes such as metal ions and saccharides. As can be expected, amino acids, peptides, nuclear acids, proteins, and other small and medium-sized analytes, and large bioanalytes may be added to the analyte pool. The extension of fluorescent molecular imprinting sensor arrays to various kinds of analytes may lead to exciting developments in differential sensing with synthetic arrays. Another noteworthy advantage of molecular imprinting sensor arrays is the use of less imprinted materials for discriminating more analytes. We can take full advantage of molecular imprinting to design sensor arrays for various kinds of analytes with less analytical and synthetic efforts, making molecular imprinting sensor arrays effective electronic noses and tongues.

References

[1] K.J. Albert, N.S. Lewis, C.L. Schauer, G.A. Sotzing, S.E. Stitzel, T.P. Vaid, D.R. Walt, Chem. Rev. 100 (2000) 2595.
[2] P.C. Jurs, G.A. Bakken, H.E. McClelland, Chem. Rev. 100 (2000) 2649.
[3] J.J. Lavigne, E.V. Anslyn, Angew. Chem. Int. Ed. 40 (2001) 3118.
[4] A.T. Wright, E.V. Anslyn, Chem. Soc. Rev. 35 (2006) 14.
[5] P.A.G. Cormack, K. Mosbach, React. Funct. Polym 41 (1999) 115.
[6] K. Haupt, K. Mosbach, Chem. Rev. 100 (2000) 2495.
[7] M.J. Whitcombe, E.N. Vulfson, Adv. Mater 13 (2001) 467.

[8] C. Alexander, H.S. Andersson, L.I. Andersson, R.J. Ansell, N. Kirsch, I.A. Nicholls, J. O'Mahony, M.J. Whitcombe, J. Mol. Recognit. 19 (2006) 106.
[9] L. Basabe-Desmonts, D.N. Reinhoudt, M. Crego-Calama, Chem. Soc. Rev. 36 (2007) 993.
[10] T. Hirsch, H. Kettenberger, O.S. Wolfbeis, V.M. Mirsky, Chem. Commun. (2003) 432.
[11] F.L. Dickert, P.A. Lieberzeit, P. Achatz, C. Palfinger, M. Fassnauer, E. Schmid, W. Werther, G. Horner, Analyst 129 (2004) 432.
[12] N.T. Greene, S.L. Morgan, K.D. Shimizu, Chem. Commun. (2004) 1172.
[13] N.T. Greene, K.D. Shimizu, J. Am. Chem. Soc. 127 (2005) 5695.
[14] T. Takeuchi, D. Goto, H. Shinmori, Analyst 132 (2007) 101.
[15] J. Tan, H.-F. Wang, X.-P. Yan, Biosens. Bioelectron 24 (2009) 3316.
[16] J. Tan, H.-F. Wang, X.-P. Yan, Anal. Chem. 81 (2009) 5273.
[17] M.E. Díaz-García, R.B. Laíño, Microchim. Acta 149 (2005) 19.
[18] E.L. Holthoff, F.V. Bright, Anal. Chim. Acta 594 (2007) 147.
[19] M. Ganschow, M. Wark, D. Wöhrle, G. Schulz-Ekloff, Angew. Chem. Int. Ed. 39 (2000) 161.
[20] M. Comes, M.D. Marcos, R. Martínez-Máñez, F. Sancenón, J. Soto, L.A. Villaescusa, P. Amorós, D. Beltrán, Adv. Mater 16 (2004) 1783.
[21] A.B. Descalzo, K. Rurack, H. Weisshoff, R. Martínez-Máñez, M.D. Marcos, P. Amorós, K. Hoffmann, J. Soto, J. Am. Chem. Soc. 127 (2005) 184.
[22] A.B. Descalzo, M.D. Marcos, R. Martínez-Máñez, J. Soto, D. Beltrán, P. Amorós, J. Mater, Chem. 15 (2005) 2721.
[23] S. Tao, G. Li, H. Zhu, J. Mater. Chem. 16 (2006) 4521.
[24] L.-L. Li, H. Sun, C.-J. Fang, J. Xu, J.-Y. Jin, C.-H. Yan, J. Mater. Chem. 17 (2007) 4492.
[25] A.B. Descalzo, M.D. Marcos, C. Monte, R. Martínez-Máñez, K. Rurack, J. Mater. Chem. 17 (2007) 4716.
[26] D. Chandra, T. Yokoi, T. Tatsumi, A. Bhaumik, Chem. Mater 19 (2007) 5347.
[27] M.H. Lee, S.J. Lee, J.H. Jung, H. Lim, J.S. Kim, Tetrahedron 63 (2007) 12087.
[28] F. Hoffmann, M. Cornelius, J. Morell, M. Fröba, Angew. Chem. Int. Ed. 45 (2006) 3216.
[29] K. Soroka, R.S. Vithanage, D.A. Phillips, B. Walker, P.K. Dasgupta, Anal. Chem. 59 (1987) 629.
[30] M.A. Palacios, Z. Wang, V.A. Montes, G.V. Zyryanov, B.J. Hausch, K. Jurstíková, P. Anzenbacher Jr., Chem. Commun. (2007) 3708.
[31] M.A. Palacios, Z. Wang, V.A. Montes, G.V. Zyryanov, P. Anzenbacher Jr., J. Am. Chem. Soc. 130 (2008) 10307.
[32] A.-J. Tong, A. Yamauchi, T. Hayashita, Z.-Y. Zhang, B.D. Smith, N. Teramae, Anal. Chem. 73 (2001) 1530.
[33] H. Fang, G. Kaur, B. Wang, J. Fluoresc. 14 (2004) 481.
[34] X.-Y. Sun, B. Liu, Y.-B. Jiang, Anal. Chim. Acta 515 (2004) 285.
[35] G.A. Baker, R. Desikan, T. Thundat, Anal. Chem. 80 (2008) 4860.
[36] W.M.J. Ma, M.P. Pereira Morais, F. D'Hooge, J.M.H. van den Elsen, J.P.L. Cox, T.D. James, J.S. Fossey, Chem. Commun. (2009) 532.
[37] N. Fujita, S. Shinkai, T.D. James, Chem. Asian J 3 (2008) 1076.
[38] E. Granot, R. Tel-Vered, O. Lioubashevski, I. Willner, Adv. Funct. Mater 18 (2008) 478.
[39] R. Rajkumar, A. Warsinke, H. Möhwald, F.W. Scheller, M. Katterle, Talanta 76 (2008) 1119.
[40] S.L. Wiskur, H. Ait-Haddou, J.J. Lavigne, E.V. Anslyn, Acc. Chem. Res. 34 (2001) 963.
[41] B.T. Nguyen, E.V. Anslyn, Coord. Chem. Rev. 250 (2006) 3118.
[42] J.W. Lee, J.-S. Lee, Y.-T. Chang, Angew. Chem., Int. Ed. 45 (2006) 6485.
[43] S.H. Lim, C.J. Musto, E. Park, W. Zhong, K.S. Suslick, Org. Lett. 10 (2008) 4405.
[44] Y. Koshi, E. Nakata, H. Yamane, I. Hamachi, J. Am. Chem. Soc. 128 (2006) 10413.
[45] F. Zaubitzer, A. Buryak, K. Severin, Chem.-Eur. J 12 (2006) 3928.
[46] N.Y. Edwards, T.W. Sager, J.T. McDevitt, E.V. Anslyn, J. Am. Chem. Soc. 129 (2007) 13575.
[47] R.A. Schiller, B. Wessling, Singaram, Angew. Chem. Int. Ed. 46 (2007) 6457.
[48] G. Springsteen, B. Wang, Chem. Commun. (2001) 1608.
[49] G. Springsteen, B. Wang, Tetrahedron 58 (2002) 5291.
[50] M.V. Kuzimenkova, A.E. Ivanov, C. Thammakhet, L.I. Mikhalovska, I.Y. Galaev, P. Thavarungkul, P. Kanatharana, B. Mattiasson, Polymer 49 (2008) 1444.
[51] T.A. Hopkins, K. Meerholz, S. Shaheen, M.L. Anderson, A. Schmidt, B. Kippelen, A.B. Padias, H.K. Hall, N. Peyghambarian, N.R. Armstrong, Chem. Mater 8 (1996) 344.

CHAPTER

8

Chiral Sensors Based on Molecularly Imprinted Polymers

Marzena Kaniewska, Marek Trojanowicz

Department of Chemistry, University of Warsaw, Pasteura 1, 02-093 Warsaw, Poland

OUTLINE

1. INTRODUCTION

Many naturally occurring and synthetic chemicals exist in an optically active form. They therefore have pairs of enantiomers. Enantiomers of the same compound, indistinguishable on the basis of physical and chemical properties, differ, sometimes significantly so, in physiological effects. The differences may be subtle, but sometimes they have a tremendous importance. They can vary in taste, odor, or activity [1–3]. Research on the individual enantiomers of the chosen compounds can be a valuable indicator of the quality of different food products. The presence of compounds with a specific configuration may indicate the source of food contamination [4–6]. Among pharmaceuticals, in some cases even minor enantiomeric impurities can cause severe toxic side effects [7–9]. The trend has been toward ensuring that medicines invented, approved, and commercialized as a racemate, or a mixture of diastereomers, are remarketed as single enantiomers called chiral switches [10,11]. The presence of some compounds naturally occurring in living organisms in a different configuration can also be an indication of disease [12].

Also in the group of pesticides one can find many examples of the different reactivity of enantiomers to target or nontarget species. Several cases in which enantiomers of pesticides

Molecularly Imprinted Sensors.
DOI: 10.1016/B978-0-444-56331-6.00008-6

varied in activity, rate of reaction, or degradation time are described [13–16]. Chiral analysis can also be used to identify the sources of water pollution [17,18] or to determine the age of archaeological finds [19]. So, the development of enantioselective analytical methods and determination of the enantiomeric purity or the quantification of one enantiomer in the mixture are extremely important in the pharmaceutical, food, and pesticide industries, as well as in clinical analysis.

The enantiomers of the same compound exhibit almost identical physical properties, for example, melting point, boiling point, or solubility. They cannot be separated by commonly used methods such as fractional distillation or fractional crystallization, except when the solvent used is optically active. Currently used methods for analysis of optically active compounds are mainly high-performance separation methods such as gas chromatography (GC) and high-performance liquid chromatography (HPLC) using chiral stationary phases, chiral selectors in the mobile phase, or flow reactors for derivatization. Also, efficient electromigration techniques such as capillary electrophoresis (CE) employing chiral selectors are widely used for determination of enantiomers. Other methods are mass spectroscopy, NMR (Nuclear Magnetic Resonance) to the study of molecular recognition, as well as some spectroscopic techniques such as circular dichroism. Those techniques require expensive equipment, and the analysis is often time consuming.

In addition to separation methods in the literature, one can find reports describing other methods of detection of enantiomers such as piezoelectric sensors based on self-assembled monolayers with chiral functional groups or polymer with chiral substituents [20,21], or the largest group of potentiometric membrane electrodes containing a chiral selector in the membrane [22–24]. There are also examples of voltammetric sensors that use the difference in Gibbs energy of ion transfer chiral from the aqueous phase to chiral organic phase [25–26] and optical sensors [27]. They can be based on different mechanism effects such as adsorption or chiral enantiomers or different membrane permeability under the influence of the applied pressure. A large group of sensors are also chiral biosensors containing biological recognition elements such as enzymes [28–30], antibodies [31–33], bioreceptors [34,35], and aptamers [36]. These compounds are naturally enantioselective and are often specific to a target molecule. The sensors and biosensors that use different detection methods can be the alternative instrumentation for chiral analysis [37]. However, sensors and biosensors also present some difficulties and limitations, such as limited long-term stability and reusability of these devices. Strictly defined measurement conditions such as pH and temperature must be kept. Biological compounds employed as molecular recognition elements are often quite expensive and sometimes difficult to obtain. For these reasons, the chemical literature describes numerous attempts to develop new methods of molecular recognition alternative to bioreceptors.

One of the well-established technologies to produce materials exhibiting molecular recognition is molecular imprinting. Molecularly imprinted polymers (MIPs) are highly targeted, specific polymeric materials. The first attempts at preparation of specific adsorbents can be dated back to 1956 [38]. One year later Beckett and Anderson presented "stero-selective adsorbents" for determining the configuration of organic molecules [39]. The first patent for a MIP is dated 1978 and belongs to Wulff and co-workers. Generally, conventional MIPs are fabricated by conducting the polymerization process in the presence of target analyte or compounds similar in size and shape. Two forms of monomers are also necessary. The first is functional and allows binding the analyte, and the other is called a cross-linker.

Functional monomers create complex with the target analyte prior to polymerization by different interactions such as the reversible covalent bonds, covalently attached polymerizable binding groups activated for noncovalent interactions by template cleavage, coordination with a metal center or electrostatic, hydrophobic, or van der Waals interactions. That causes stable emplacement of their functional groups in the structure of polymer. The next step after polymerization with the cross-linking agent is the extraction of template and liberation of the binding site that is well-fitting considering size and shape to analyte. When the template is an optically active compound, the obtained polymer often possesses the chiral properties. MIPs can be obtained by different methods. The most frequently used methods are bulk polymerization and surface grafting, electropolymerization, self-assembling molecularly imprinted nanofiber, imprinting sol-gel layers but also coating on nanoparticles, entrapping MIP nanoparticles in carbon paste or plasticized membranes, imprinting in ionic liquid-modified porous polymer, or casting on membranes (MIPs). There were also untypical attempts such as imprinting inside dendrimer with cross-linkable double bonds at the outer shell and covalently attached porphyrin template in the core [40].

MIPs have a lot of advantages such as low-cost production, ease of preparation, chemical and physical resistance, and stability against a wide range of environments. They can be shaped in various formats, and they most of all present a binding affinity comparable to biological recognition elements.

MIPs have found a wide range of applications. They are used in chromatographic methods, in capillary electrochromatography, and as a sorbent in solid phase extraction. They have also been used in binding assays, catalysis, and drug delivery [41,42]. MIPs are used to prepare sensitive and selective sensors. Over 50% of published work concerns optical sensors. Two other main types of detection are electrochemical and piezoelectric sensors [43].

2. ELECTROCHEMICAL SENSORS

Electrochemical sensors involve various types of signal transduction such as voltammetric, conductometric, capacitive, amperometric, ion-sensitive field effect transistors (ISFETs), and potentiometric [44]. They have found many applications owing to their high sensitivity, low operating costs, and possibility of miniaturization. Data on chiral electrochemical sensors employing MIPs reported in the literature are shown in Table 8.1.

MIPs were employed in the construction of chiral potentiometric sensors for amino acids. An octadecylsiloxane (ODS) layer was covalently bound onto an indium tin oxide (ITO) surface in the presence of chiral N-carbobenzoxy-aspartic acid (N-CBZ-Asp) molecules [45]. Interaction between amino acid and ITO involved two factors: (1) hydrophobic interaction with the ODS layer that proves chiral selectivity; and (2) electrostatic binding with surface oxides. The sensor exhibited high sensitivity and enantiomeric selectivity in the range of 5×10^{-6}–1.2×10^{-2} M of N-CBZ-Asp. Similar studies also carried out for non–N-protected amino acids proved that such a strategy works not only for N-protected amino acids. Calibration curves presented in Fig. 8.1 show the difference of potential response for enantiomers of aspartate and glutamate.

The sensor based on the use of field-effect transistors with the imprinting of chiral molecular recognition sites in TiO_2 films was developed for detection of anions of carboxylic

TABLE 8.1 Electrochemical Enantioselective Sensors

Method of detection	Analyte/template	Polymer	Enantioselectivity ratio	Linear response and [LOD]	Reference
Potentiometric	L-NCBZ-Asp L-Glutamic acid L-aspartic acid	Octadecyltrichlorosilane (OTS)	0.009	5×10^{-6}–1.2×10^{-1} μmol L^{-1}	45
Potentiometric	R-Methylferrocene carboxylic acid R- 2-Phenylbutanoic acid R- 2-Propanoic acid	$Ti(O\text{-}^nBu)_4$	n.d.	0.05–6.25 mmol L^{-1} 0.0625–1.25 mmol L^{-1} 0.05–1.25 mmol L^{-1}	46
Amperometric	Tyrosine	Polypyrrole	25.8	n.d.	47
Amperometric	D-Glucose, D-mannose	Phenol, 3-hydroxyphenyl boronic acid	n.d.	0.2-5 mmol L^{-1}	48
DPASV	D or L Thyroxine	DAU, EGDMA Cu II/bpy complex, TEA, chloroform	n.d.*	0.06–15.00 ng mL^{-1}	49
DPASV	D or L Thyroxine	DAU, EGDMA, Cu II/bpy complex, TEA, chloroform, carbon powder	n.d.*	0.010–17.2 ng mL^{-1}	50
DPASV	L or D Tryptophan	4-Nitrophenyl acrylate, EGDMA, Cu II/bpy complex, TEA, chloroform, (carbon powder)	n.d.*	0.15–30.00 ng mL^{-1}	51
DPASV	L or D Tryptophan	4-Nitrophenyl acrylate, EGDMA, Cu II/bpy complex, TEA, chloroform, carbon powder	n.d.*	0.90–18.60 ng mL^{-1} 24.23–840.22 ng mL^{-1}	52
Cyclic voltammetry	D- and L-Dopa R-Fc and S-Fc	TMOS, PTMOS ethoxyethanol	0.14	[0.2 μg L^{-1}] [2 mg L^{-1}]	53
Capacitance	D or L Glutamic acid	*o*-PD, DA	24	16.7–250 μmol L^{-1}	54
Cyclic voltammetry	L or D Phenylalanine anilide	MAA, EDMA, AIBN	4.83	0.5–1 μmol L^{-1} >1 μmol L^{-1}	55

** No interaction with the opposite enantiomer was observed.*

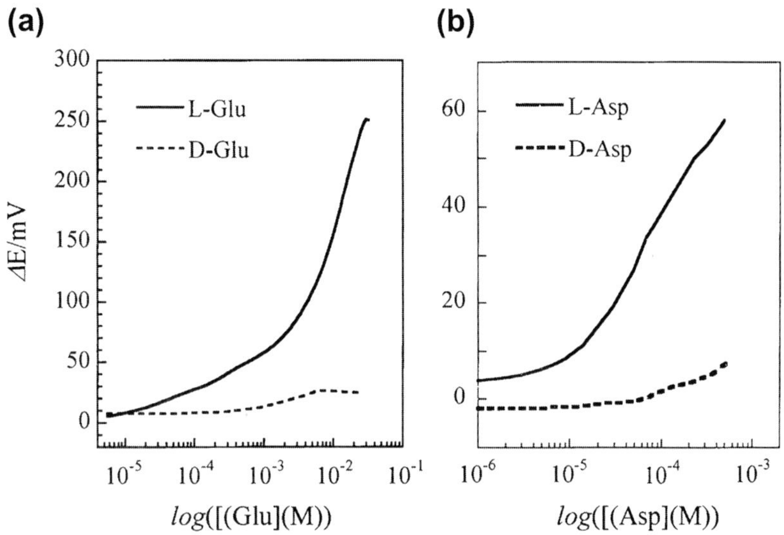

FIGURE 8.1 Potentiometric response of MIP sensors: (a) L-Glu ODS/ITO sensor for L-Glu (solid line) and D-Glu (dashed line), and (b) L-Asp ODS/ITO sensor for L-Asp (solid line) and D-Asp (dashed line) [45].

acids [46]. Three pairs of enantiomers of methylferrocene carboxylic acid, 2-phenylbutanoic acid, and 2-propanoic acid were chosen for investigation. The imprinting was carried out by hydrolyzing the respective carboxylate titanium (IV) butoxide complex on the gate surface of ISFETs, followed by washing with an ammonia solution to exclude the carboxylate units. The structure of chemically assembled dielectric membrane films on the gate surface was characterized by impedance measurements. The results obtained show that the response of the sensor is not only selective toward the other enantiomer (Fig. 8.2), but also specific for the chosen analyte, even in the case of compounds of similar structure.

The enantioselectivity has also been reported for amperometric and voltammetric sensors. The amperometric sensor was built with a nickel electrode by coating the active surface with polypyrrole in the presence of tyrosine enantiomers [47]. After extraction of the template from the polymer by a water/methanol solution, the complementary cavity was formed. The rebinding of tyrosine anions to the polymer film was induced by application of positive potential and was controlled electrochemically. The template enantiomer of tyrosine favorably entered into the sensor active layer when the opposite enantiomer rebinding was significantly slower.

A monosaccharide sensor was assembled by co-electropolymerization of phenol with 3-hydroxyphenyl boronic acid complex on a gold electrode in the presence of D-glucose or D-mannose [48]. The dynamics of association of different monosaccharides was examined using ferrocene-labeled molecules as redox indicators. The recorded response shown in Fig. 8.3 indicates a higher affinity of D-glucose-imprinted film to ferrocene-functionalized D-glucose compared to the L-glucose imprinted sensor. The polymer imprinted by D-glucose showed enantioselective interaction with enantiomers of glucose favoring the enantiomer D.

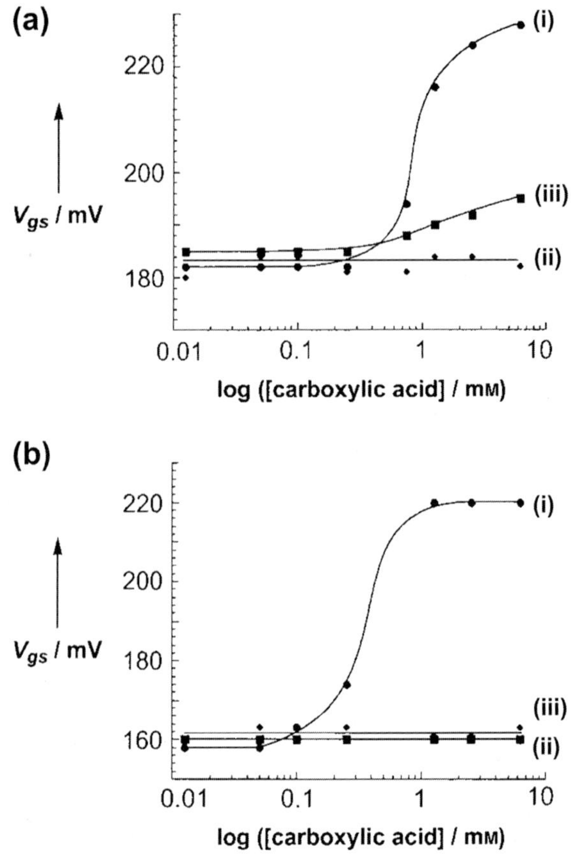

FIGURE 8.2 The gate source voltage as a function of carboxylic acid concentration obtained for MIP based field-effect transistor sensor [46]. (a) Polymer imprinted with (R)-2-propanoic acid, response for (R)-2-propanoic acid (i), response for (S)-2-propanoic acid (ii), response for (R) 2-phenylbutanoic acid (iii). (b) Polymer imprinted with (S)-2-propanoic acid, response for (S)-2-propanoic acid (i), response for (R)-2-propanoic acid (ii), response for (S) 2-phenyl-butanoic acid (iii).

Differential pulse anodic stripping voltammetric measurements were used for detection with a MIP-based sensor sensitive for D- or L-thyroxine [49]. For MIP preparation, 1,3-diacroloylurea (DAU) as monomer and ethylene glycol dimethacrylate (EGDMA) as cross-linker were mixed and polymerized by electron transfer−atom transfer radical polymerization in the presence of D- or L-thyroxine. As reducing agent, the TEA (triethylamine) was used, which converted Cu(II)-bpy complex to Cu(I) complex to catalyze polymerization in the presence of chloroform as an initiator. The polymer was coated on a silver wire electrode modified by a vinyl functionalized self-assembled monolayer. The same polymer was also used as sensor material for extraction fiber in the micro-solid phase extraction. The voltammetric sensor exhibited excellent limit of detection 8 ng L^{-1} for thyroxine. The enantioselectivity of the polymer was determined by measuring the extraction yields of D- and L-thyroxine and several other compounds. All examined interferences were found to be nonresponsive with

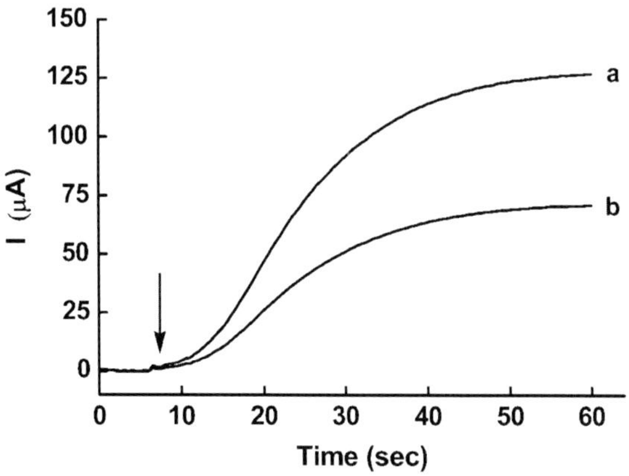

FIGURE 8.3 Time dependent current responses obtained for MIP based amperometric sensor upon the incorporation of ferrocene-labeled-D-glucose into D-glucose (a) and L-glucose (b) imprinted polymer film [48].

appropriate L- or D-tyroxine-imprinted polymer. The same template was also used in the construction of layer-by-layer assembled molecularly imprinted silver electrode. The polymer was obtained by thermal cross-linking from the mixture containing besides the template DAU, EGDMA, Cu(II) complex as a catalyst, reducing agent TEA, chloroform as an initiator, and carbon powder [50]. The electrode exhibited an even lower detection limit 6 ng L^{-1}. According to the suggested enantioselective binding mechanism, in the case of L-tyroxine-imprinted polymer, the rebinding of L-tyroxine occurred rapidly. D enantiomer could not enter into the L-imprinted cavity due mainly to its shape difference and functional group mediated constraints.

The same research group published two reports describing MIP-fiber sensors for enantioselective determination of D- and L-tryptophan [51,52]. The 4-nitrophenyl acrylate/EGDMA mixture was polymerized in dimethyl sulfoxide (DMSO) via activator generated by electron transfer—atom-transfer radical polymerization with the tryptophan enantiomer as a template [51]. A monolithic MIP fiber was produced by injecting the pre-polymer solution in a home-made glass capillary, plugged with Teflon tape and incubation at 60 °C for 4 hours. The enantioselectivity of sensor was demonstrated by differential pulse anodic stripping voltammetric response of electrode molecularly imprinted by particular enantiomer of tryptophan to L- or D-tryptophan or the racemic mixture. The L-Trp sensor showed an analytical response only in the presence of L-enantiomer or the mixture, and in latter case the signal was not significantly higher than in the presence of pure enantiomer. The D-Trp-imprinted sensor interacts only with D-tryptophan. Apart from the enantioselectivity, structural analogues of Trp such as tyrosine and 3-indole acetic acid show no binding affinity with the sensor. In another work [52] a polymer-carbon composite fiber sensor for enantiomers of tryptophan was reported with sensing material, which was prepared in a similar way as in the previous cited work, except that a mixture of carbon powder was added before polymerization. Figure 8.4 presents an enantioselective response for sensors with MIP-carbon imprinted for D-Trp recorded by differential pulse anodic stripping voltammetry. Both sensors have enabled trace analysis of L-tryptophan in biological fluids at concentration levels 0.9 to 18.6 mg L^{-1}.

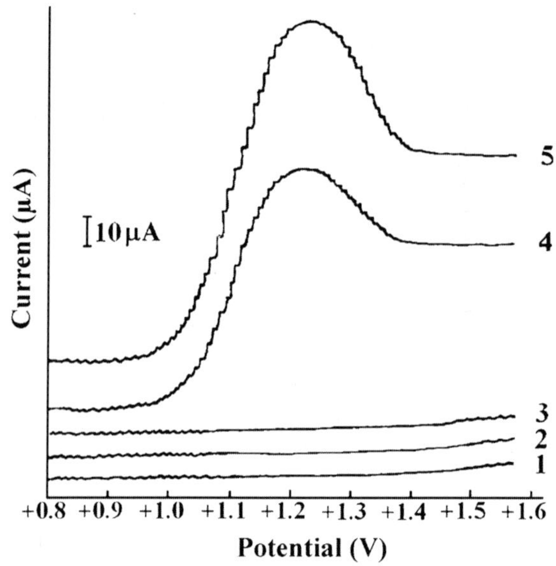

FIGURE 8.4 Differential pulse anodic stripping voltammograms recorded for MIP-carbon composite fiber sensor imprinted with D-Trp: (1,2) blank solution, (3) L-Trp (16.5 ng mL^{-1}), (4) D-Trp (7.5 ng mL^{-1}), (5) L-Trp and D-Trp (both 11 ng mL^{-1}) [52].

ITO electrodes spin-coated with sol-gel molecularly imprinted film were examined as an enantioselective voltammetric sensor for D- and L-3,4-dihydroxyphenylalanine (D- and L-dopa) or R- and S-N,N′-dimethylferrocenylethylamine [(R)-Fc and (S)-Fc] [53]. For the film preparation, different amounts of tetramethoxysilane-TMOS, phenyl trimethoxysilane-PTMOS, and ethoxyethanol were mixed with HCl and water. The authors suggested that in the case of sol-gel polymers, choosing suitable functional organosilane monomers is a crucial factor, as well as film thickness. Very thin films (70 nm) used to construct sol-gel sensors in comparison with films of 700 nm thickness presented higher sensitivity superior to larger electrochemical signals and faster removal of the template by solvent extraction. Figure 8.5 shows high enantioselectivity of sol-gel films and also excellent selectivity against compounds that have similar structure. Both sensors exhibit high sensitivity and low nonspecific adsorption.

Molecularly imprinted polymer was also employed as a recognition element for a capacitance sensor for enantioselective recognition of glutamic acid [54]. Polymer was synthetized on the surface of a gold electrode by electrochemical polymerization from the mixture of *o*-phenylenediamine (*o*-PD) and dopamine (DA) in the presence of D- or L-glutamic acid. Surface morphology of MIP was characterized with atomic force microscopy and X-ray photoelectron spectroscopy. MIP capacitive sensors showed good stability, satisfactory repeatability (RSD-Relative Standard Deviation of 3.5%), and excellent enantioselectivity (shown in Fig. 8.6). The slope of linear response for L-glutamic acid was 24 times larger than that for the D-Glu response. The sensor preparation method seems to be simple, convenient, and low cost.

A chiral discriminative gate effect of MIP in organic solvents was also reported [55]. Polymer imprinted with L- or D- phenylalanine anilide (PPA) was polymerized on an indiumtin

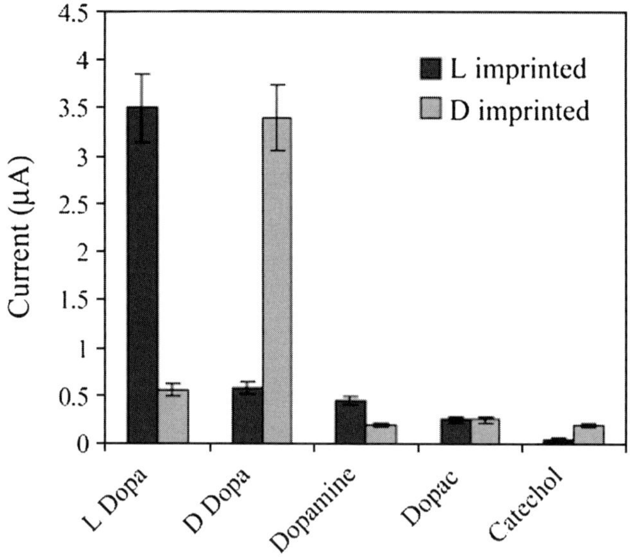

FIGURE 8.5 Selectivity of uptake of D- and L-dopa, dopamine, dopac and catechol by sol-gel molecularly imprinted by L- and D-dopa [53]. All data corresponding to cyclic voltammograms recored after incubation of electrodes in a solution of appropriate compound.

oxide electrode. Cyclic voltammograms of ferrocene in a chosen organic solvent were registered in the presence of a template molecule or its opposite enantiomer. The greater change in the anodic current in the presence of the template was observed. The gate effect was explained as a result of the change in a diffusion rate of ferrocene in the MIP layer induced by the specific interaction of polymer with the template. The sensor gave a linear response

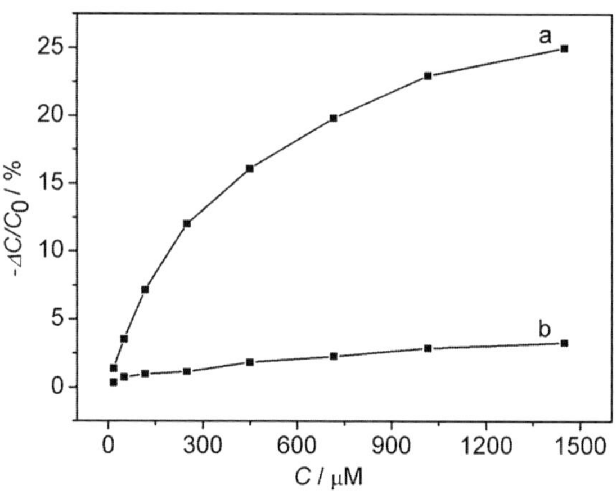

FIGURE 8.6 Response of capacitance senor with MIP imprinted with D-glutamic acid to changes of concentration of D-glutamic acid (a) and L-glutamic acid (b) [54].

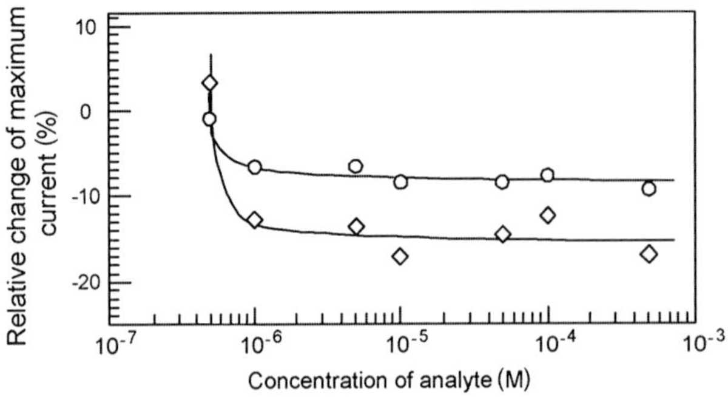

FIGURE 8.7 Relative changes of maximum anodic current observed for the electrode imprinted with L-phenylalanine anilide in dichloromethane in the presence of L-PAA (diamonds) and D-PAA (circles) [55].

only in the small range of concentration (0.5–1 μmol L^{-1}). Above 1 μmol L^{-1} analyte concentration, the sensor presented a stable difference in the relative change of maximum current for both enantiomers (Fig. 8.7); hence, the presented sensor was considered suitable for highly sensitive qualitative analysis. The enantioselectivity was expressed by the ratio of the change in current of the template to the change of the current of its enantiomer (α). The highest enantioselectivity was observed for the polymer imprinted with L-PPA in toluene where α was 4.83. The authors of the study also suggest that the presented gate effect could have limited applications with the enantioselective amperometric sensors working in nonpolar solvents where biosensors cannot work.

3. PIEZOELECTRIC SENSORS

Molecular imprinting technology is widely applied in the development of piezoelectric sensors based on quartz crystal microbalance. As in all other sensors, a crucial factor determining the polymer enantioselectivity is the selection of appropriate monomers and cross-linkers, as well as type of polymerization procedure, to form an MIP layer on the surface of QCM electrodes. The developed enantioselective piezoelectric sensors reported so far in the literature are listed in Table 8.2.

In the case of chiral QCM sensors for dansylphenylalanine, a photopolymerization was used [56]. The polymeric thin film was formed on a gold electrode surface of QCM modified with a photoactive precursor-vinyl group terminated monolayer via the self-assembly process. The influence of template concentration as well as the effect of pH value on the sensor response were investigated and optimized. The obtained film was insoluble and chemically resistive, and exhibited high selectivity and stability. The sensor exhibited satisfactory enantioselectivity, and the enantiomeric composition of mixtures of dansylphenylalanine could be determined quantitatively from the frequency changes.

A highly specific noncovalently imprinted polymer was used as a recognition element in a piezoelectric sensor for determining L-menthol [57]. Prior to polymerization by UV light radiation, the QCM electrodes were coated by the mixture of L-menthol, methacrylic acid (MA), ethylene glycol dimethacrylate (EDMA), and chloroform with 2,2′-azobis(2-methylpropionitrile) (AIBN) using the sandwich casting method. The enantiomeric selectivity coefficient for obtained sensor was evaluated as 3.6. The authors indicated that the lower selectivity compared to previously reported sensors [58] can be

TABLE 8.2 Piezoelectric Enantioselective Sensors

Analyte/template	Polymer	Type of polymerization	Enantioselectivity ratio	Linear response [LOD]	Reference
Dansylphenylalanine	4-Vpy, MAA, EDMA, AIBN	Photopolymerization	n.d.	5–500 μg mL^{-1}	56
L-Menthol	MAA, EDMA, AIBN	Photopolymerization	3.6	0.2–1 mg L^{-1}	57
L-Serine	MAA, EDMA, AIBN	Photopolymerization	4.8	2–400 μg L^{-1}	59
Propranolol	MAA, TRIM	Photopolymerization	n.d	0.19–3.8 mmol L^{-1}	60
L-Tryptophan	AM, TRIM	Photopolymerization	6.4	[8,8 μmol L^{-1}]	61
D-Methamphetamine	MAA, EDMA, AIBN	Polymerization[b]	15.9	10^{-5}–10^{-1}μg mL^{-1}	62
L-Glutamic acid	Overoxidized polypyrrole	Electropolymerization	30[a]	n.d.	63
L-Aspartic acid	Overoxidized polypyrrole	Electropolymerization	20[a]	n.d.	64
L-Tryptophan	Overoxidized polypyrrole	Electropolymerization	n.d.	n.d.	65
L-Histidine	TEOS, PTMOS, MTMOS	Polymerziation[b]	0.24	5×10^{-8}–1×10^{-4} mol L^{-1}	66

[a] *Uptake ratio*
[b] *Bulk polymerization in solution left for evaporation*
n.d.—not determined

explained by the reduced number of binding sites in the polymer. The same research group presented molecularly imprinted acoustic wave sensor sensitive for L-serine [59]. The optimization of cross-linker concentration suggested that its increase can significantly reduce adherence to the electrode surface. The L-serine QCM sensor exhibited the enantiomeric selectivity coefficient 4.8.

Photopolymerization was also used for preparation of piezoelectric sensor imprinted with propranolol. The possibility of controlling the thickness of the MIP film to be below 50 nm was obtained by use of a photo-initiator covalently coupled to a self-assembled monolayer of carboxyl terminated alkanethiol on a gold surface [60]. The carboxyl groups on the self-assembled monolayer (SAM) were activated by 2-ethyl-5-phenylisoxazolium-3′-sulfonate and covered with 2,2′-azobis(2-amidinopropane) hydrochloride (ABAH). Polymer was prepared from the mixture of methacrylic acid (MA) and −trimethylolpropane trimethacrylate (TRIM). The influence of the monomer concentration and the film thickness for the chiral selectivity was determined. The ultrathin films used in a flow injection analysis system under a high flow rate could detect chiral differences in a concentration range 0.38−3.8 mM.

A high selective and sensitive sensor for L-tryptophan was also built using noncovalent molecular imprinting [61]. The acrylamide (AM) was used as functional monomer to promote hydrogen bonding with the L-tryptophan molecule and allow obtaining good selectivity and reversibility. The high sensitivity was correlated with the concentration of the cross-linking agent TRIM. The sensor presented the detection limit of L-tryptophan 8,8 μmol L^{-1}. The enantioselectivity coefficient of the polymer was 6.4.

Bulk polymerization was used to produce a piezoelectric quartz crystal sensor for a quantitative analysis of D-methamphetamine [62]. The MIP was polymerized from the mixture of methacrylic acid, ethylene glycol dimethacrylate, template molecule, and 2,2′-azobis (isobutyronitrile) used as initiator. The polymer was mixed with cyanoacrylate ester and THF, applied to the gold electrode by spin-coating, and was allowed to dry. The sensor presented rapid, stable response and highly reversible response with the compatibility to the GC-MS method. The enantioselectivity of this sensor is demonstrated by the calibration graphs shown in Fig. 8.8.

A different type of MIP can be produced using overoxidized polypyrrole doped with a chiral compound [63]. The enantioselective uptake of glutamic acid into an overoxidized

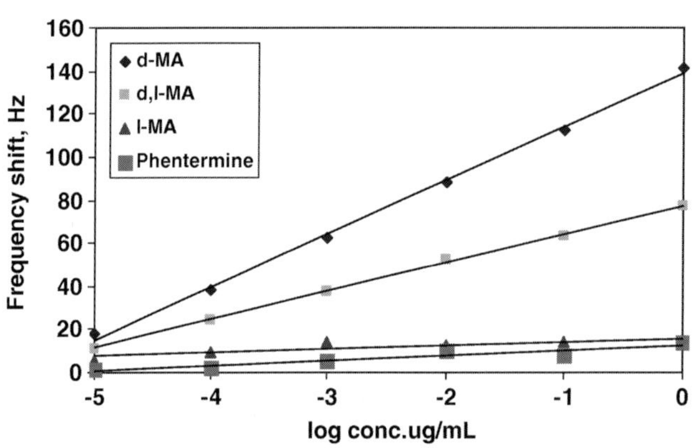

FIGURE 8.8 Frequency change observed for piezoelectric sensor modified by polymer imprinted with D-methamphetamine for enantiomers of methamphetamine and its homolog (phentermine) [62].

polypyrrole was investigated. The film of polypyrrole doped with glutamic acid was deposited galvanostatically on Pt-coated quartz crystal. During the overoxidation performed by scanning the potential until the proper value of current was obtained, the polypyrrole lost its electroactivity due to ejection of dopant. The oxygen-containing groups were introduced to the pyrrole units. Selective uptake or release of the template molecule was induced by the film charging or discharging. The chiral selectivity dependent of the surface morphology was greatly influenced by the overoxidation potential. The polymer exhibited a high selectivity over tested amino acids. The enantioselectivity determined in the chosen conditions and presented as the uptake ratio was about 30 for the polymer doped with L-glutamic acid. The overoxidized polypyrrole templated with enantiomers of aspartic acid was also developed to create a piezoelectric sensor for aspartic acid [64]. Conditions of synthesis such as electrolyte composition and pH strongly influence the enantioselectivity of the polymer. The polymer imprinted with L-Asp showed uptake favorable to L-Asp, with the uptake ratio of L/D enantiomers about 20 (Fig. 8.9).

Properties of the polypyrrole film imprinted with L-tryptophan were investigated using cyclic voltammetry and the quartz crystal microbalance technique [65]. In order to obtain the highest enantioselectivity of the polymer, parameters such as electropolymerization potential, overoxidation potential, and time and thickness of the film were studied. The effect of the scanning rate in the case of voltammetry experiments was also examined. For the L-Trp doped polymer the peak current obtained in CV -Cyclic Voltammetry measurements was over 10 times higher for L enantiomer compared to D for the same chosen concentration. The QCM measurements proved the enantioselective character of the imprinted polymer.

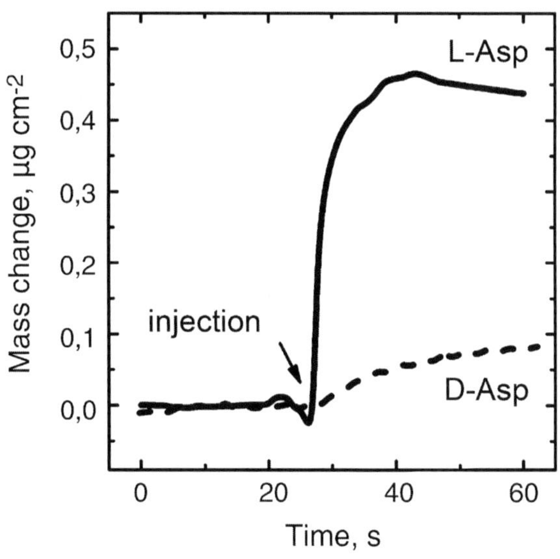

FIGURE 8.9 Mass changes of the piezoelectric MIP sensor with the overoxidized polypyrrole layer imprinted with L-Asp observed for injection of 10 mM L-Asp and D-Asp into KCl–HCl solution of pH 1.6 at a constant potential of −0.4 V [64].

A piezoelectric chiral sensor based on molecularly imprinted sol-gel was also developed [66]. L-histidine was mixed with a hybrid mixture of functionalized organosilicon precursors (tetra ethylorthosilicate—TEOS, phenyltrimethoxysilane—PTMOS, methyltrimethoxysilane—MTMOS) and spread on the surface of a quartz crystal electrode. Selectivity of the L-imprinted sensor was determined over several compounds. The enantioselectivity factor, defined as the ratio of the sensor's frequency shift toward the D-histidine to that toward the template molecule, was calculated as 0.24. For comparison, the same factor for the nonimprinted polymer was 0.96.

There are also examples of molecularly imprinted chiral sensors based on different sensing materials than polymers. For instance, the electrochemical sensor based on L-cysteine self-assembled monolayer for the enantioselective determination L-serine was reported [67]. The sensing layer was prepared by immersing a gold microelectrode in the solution of L-cysteine and L-serine in HCl. The analytical signal was registered as a decrease of a peak current intensity of potassium ferricyanide in the presence of analyte at different concentrations by differential pulse voltammetry (DPV). The sensor presented high selectivity toward other amino acids and an enantioselectivity coefficient of about 3.0.

4. OPTICAL SENSORS

The third group of sensors where molecular imprinting is used to create chiral recognition are optical sensors, and data on their design are listed in Table 8.3. One of the first examples of such sensors is a fluorescence sensor based on the use of polymer produced by thermal polymerization from the mixture of fluorescent monomer (dansyl derivative of 3,3-dime-thylacrycic acid), cross-linker (ethyleneglycol dimethacrylate—EGDMA) and free radical initiator 2,2′-azobis-(2-methylpropionitrile) in the presence of L-tryptophan as template [68]. The fluorimetric measurements were carried out with polymer suspension in chloroform placed into a cuvette. The fluorescence of polymer was quenched by 4-nitrobenzaldehyde. The template molecule was associated more strongly with binding sides of the polymer than the quencher. After addition of the template, the quencher was replaced by L-Trp, which resulted in an increase of the fluorescence intensity in the concentration-dependent way. In the case of enantiomer D, the same effect was reduced to 70% compared to L-Trp.

Reflectometric interference spectroscopy was used to examine optical sensors imprinted with (R,R)- or (S,S)-2,3-di-O-benzoyltartaric acid [69]. The modified glass plates were spin-coated by the mixture of ethylene-dimethacrylate, diethyl-4-vinylbenzamidine, di-O-benzoyl-tartaric, and 2,2′-azobis(isobutyronitrile) (AIBN). The enantioselective polymer was obtained by photopolymerization. After removing the template, the polymer was exposed to different concentrations of both enantiomers. It presented linear response with a difference in sensitivity toward the template and its antipode. Enantioselectivity was described by the separation factor α, which is the ratio of the sensitivity for the template and the antipode. For the imprinted polymers α was 1.2.

Inverse opal hydrogel film imprinted with chiral dopa was used to create a label-free sensitive optical sensor [70]. A highly ordered 3D structure made from colloid silica crystals was generated on a glass substrate. The polymer prepared as a mixture of the complex of

TABLE 8.3 Optical Enantioselective Sensors

Method of detection	Analyte/template	Polymer	Enantioselectivity ratio	Linear response [LOD]	Reference
Fluorescence	L-tryptophan	dansyl-3,3-dimethylacrycic acid, EGDMA, 2,2′-azobis-(2-methylpropionitrile)	0.7 (70% of signal)	1–10 mmol L^{-1}	68
Reflectometric interference spectroscopy	(R,R)-2,3-di-O-benzoyltartaric acid (S,S)-2,3-di-O-benzoyltartaric acid	ethylene-dimethacrylate, diethyl-4-vinylbenzamidine, di-*O*-benzoyl-tartaric, AIBN	1.23 1.19	0.75–3 mmol L^{-1} 3–5.5 mmol L^{-1}	69
Fluorescence	L-Dopa	MA, EDMA, AIBN	n.d.	n.d., [0.01 μmol L^{-1} – 10 mmol L^{-1}]	70
Chemiluminescence	Dansyl-L-phenylalanine	MAA, 4-VP, EGDMA, AIBN	2.12	0.025–125 μmol L^{-1}	71
Fluorescence	Dansyl-L-phenylalanine	2-VP, MMA, EGDMA, AIBN	1.6	10–100 μg mL^{-1}	72
UV-Vis absorption spectra	R-Propranolol S-Propranolol	$Ti(O\text{-}^nBu)_4$	4.46 2.38	10–40 μmol L^{-1}	74
Surface plasmon resonance	L-Glutamic acid	Au nanoparticles, 4-mercapto-aniline, mercaptoethane sulfonic acid, cysteine	0.05*	2–200 nmol L^{-1}	75
Surface plasmon resonance	D-Glucose L-Glucose	Au nanoparticles, 4-mercapto-aniline, mercaptoethane sulfonic acid, mercaptophenyl boronic acid	5.3×10^{-4}* 1.1×10^{-4}*	0.04–100 nmol L^{-1}	76

* *Ratio between the association constants.*

monomer mathacrylic acid (MMA) with a template (L-dopa), cross-linker (ethylene glycol dimethylacrylate EGDMA), and initiator (2,2′-azobis(isobutyronitrile) (AIBN), was infiltrated into colloid crystals and photopolymerized. Then the silica particles and template were removed, leaving a well-defined macroporous structure with the number of cavities complementary to L-dopa. Because the colloidal hydrogel film has a characteristic spacing on the order of hundreds of nanometers, it interacts strongly with visible and infrared light, leading to spectral shifts in the Bragg diffraction. Interactions of the film with dopa enantiomers were detected by fluorescence and UV-Vis spectroscopy. The results indicated that molecularly imprinted film presented chiral selectivity with the preference to L-dopa.

In other designs, polymer microspheres templated with dansyl-L-phenylalanine were immobilized in 96-well microliter plates. The polymer was prepared from the mixture of methacrylic acid with template molecule, 4-vinylpyridine (4-VP), EDGMA, and AIBN by thermal polymerization [71]. The chemiluminescence signal was achieved by the reaction of bis (2,4,6-trichloro-phenyl) oxalate (TCPO) with hydrogen peroxide in the presence of analyte adsorbed on the microspheres. The sensor allowed for 96 independent measurements in 30 minutes. The control experiment was also carried out using capillary electrophoresis with no significant difference between analytical results. Polymer imprinted with dansyl-L-Phe presented high selectivity toward few dansylated amino acids and enantioselectivity. The highest adsorption was obtained for the template molecule (Fig. 8.10).

An innovative sensor chip with the polymer imprinted with dansyl-L-phenylalanine was also reported [72]. Polymer dots of about 300 μm diameter were fabricated in situ on a glass cover slip by CO_2 laser pulse initiated polymerization. The 14 monomers (as 1-vinyl-imidazole, 2-vinylpyridine, 2-acryloamino-2-methyl-1-propanesulfonic acid, 2-hydroxyethyl methacrylate, acrylamide, acrylic acid, acrylonitrile, allylamine, ethyleneglycol dimethacrylate, itaconic acid, m/p-divinylbenzene, methacrylic acid, 2-(trifluoromethyl) acrylic acid, and combination of 2-vinylpyridine and methacrylic acid) were tested for the preparation of the sensor chip from which a combination of methacrylic acid and 2-vinylpyridine

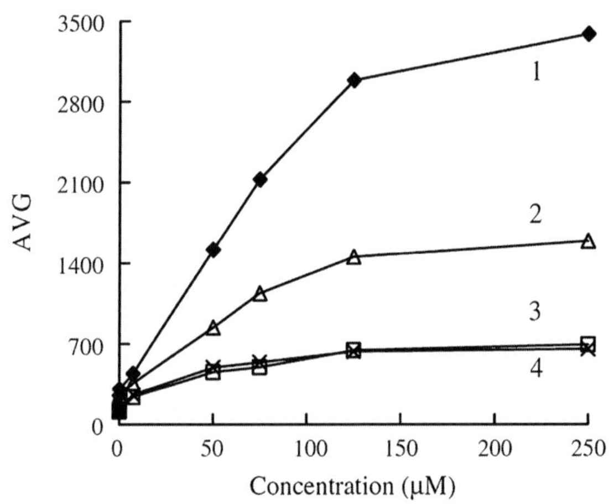

FIGURE 8.10 Adsorption curves of dansyl-L-Phe (1), (3) and dansyl-D-Phe (2), (4) on a L-imprinted polymer (1),(2) and non-imprinted control polymer (3), (4) [71].

showed the largest affinity to dansyl-L-phenylalanine. The selectivity of the sensor was examined with fluorescence. At a chosen sample concentration of 0.5 mg mL^{-1} the fluorescence response measured for the L enantiomer was 1.6 times higher than for the opposite enantiomer of danlyl phenylalanine.

The polymer molecularly imprinted with L-phenylalanine anilide was employed for measurement of enantiomeric excess [73]. The assay was based on differences in concentration of the solution after the interaction with MIP. L-isomer-imprinted polymer strongly interacts with L-isomer, resulting in low concentration in solution. D-isomer of phenylalanine anilide resulted in high concentration in the tested solution. Measurements were made by UV spectroscopy. The presented method was rapid and accurate.

One recently presented design is the optical sensor based on the molecularly imprinted TiO_2 nanofilm [74]. The TiO_2 gel film was prepared by spin-coating the mixture of titanium(IV)-n-butoxide, and one of the enantiomers of propranolol was used as a template on a quartz plate. UV-Vis absorption spectra were observed for the specific binding of the template molecule in the imprinted sites of the polymer gel and nonspecific binding onto the TiO_2 gel surface. As a definition of enantioselectivity, a ratio of absorbance increases due to the binding of a given chiral template molecule on the template, and its enantiomer-imprinted TiO_2 films, $(\Delta Abs_{t\text{-film}}/\Delta abs_{e\text{-film}})_{template}$ were used. T-film and e-film mean template- and enantiomer-imprinted films, respectively. The polymer gel film imprinted with R-propranolol- and S-propranolol presented an enantioselectivity of 4.46 and 2.38, respectively.

Functionalized Au nanoparticles (NPs) were used to assemble Au-NP composites molecularly imprinted with L-glutamic acid SPR for construction of the SPR (surface plasmon spectroscopy) sensors [75]. In order to prepare the sensing film, Au nanoparticles were modified with 4-mercaptoaniline acting as electropolymerizable component, mercapto-ethane sulfonic acid—the stabilizer and cysteine that can interact with analyte. The modified Au NPs were electropolymerized on a Au-coated glass surface in the presence of L-glutamic acid sites. The association constants of L- and D-glutamic acid for the L-glutamic-imprinted film were obtained as ${K_{ass}}^{L/L} = 5 \times 10^8\ mol^{-1}$ and ${K_{ass}}^{D/L} = 2.5 \times 10^7\ mol^{-1}$, respectively. Similar constructions of sensors were developed for the enantioselective discrimination of saccharides [76]. Au nanoparticles functionalized with 4-mercaptoaniline, mercaptoethane sulfonic acid, and mercaptophenyl boronic acid were imprinted with D- or L-glucose. Sensors with film imprinted with L-glucose presented association constants of L- and D-glucose as ${K_{ass}}^{L/L} = 1.2 \times 10^9\ mol^{-1}$ and ${K_{ass}}^{D/L} = 1.3 \times 10^5\ mol^{-1}$, respectively.

5. CONCLUSONS

Chemical sensors and biosensors in recent decades are important areas of instrumentation in modern analytical chemistry. Based on different mechanisms of molecular recognition and various transducers [77], they are widely employed first of all in clinical and environmental analysis, where of great importance is miniaturization and remote sensing is needed. One of the recently increasing trends in this field is the development of enantioselective chemical sensors and biosensors [78], which for some analytical purposes may eventually replace more complex instrumentally and costly high-performance chromatographic or electrophoretic methods. There is a widespread need for such devices, especially in pharmaceutical analysis. Designing such devices is much more challenging than developing separation methods, where a single operation unit of enantiomeric differentiation is multiplied thousands of times during chromatographic or electrophoretic runs.

The most efficient chiral discrimination in nature is considered in immunochemical interactions, hence increasing interest in applications for the same purpose synthetic materials mimicking antibodies, namely, MIPs. In the present development of these materials, besides conventional and well-developed bulk polymerization or surface grafting methods, several other methods have been reported for their preparation, including, for example, electropolymerization [79], self-assembling on nanofibers [80], imprinting of colloids [81], or surface imprinting on quantum dots [82]. MIPs can be tethered to the sol-gel layers [83], formed in ionic liquid-modified porous polymer [84], or casted on membranes [85]. Based on attempts reported in the literature and reviewed in this chapter, one can expect that, in using different transducers, those different methods of MIP fabrication may allow design of enantioselective sensors satisfactory for practical applications—for example, in the sensing of unwanted enantiomer in monitoring the chiral purity of target compound in pharmaceutical or environmental applications.

References

[1] M. Laska, A. Liesen, P. Teubner, Am. J. Physiol. Reg. Interactive Comp. Physiol. 277 (1999) 1098.
[2] R. Zawirska-Wojtasiak, Acta Sci. Pol. Technol. Aliment. 5 (2006) 21.
[3] R. Herráez-Hernández, P. Campíns-Falco, Anal. Chim. Acta 434 (2001) 315.
[4] R. Marchelli, A. Dossena, G. Palla, Trend Food Sci. Technol. 7 (1996) 113.
[5] J. Csapo, Z. Csapo-Kiss, J. Stefler, T.G. Martin, S. Nemethy, J. Dairy Sci. 78 (1995) 2375.
[6] G. Gandolifi, R. Palla, A. Marchelli, S. Dossens, C. Puelli, J. Salvadori, Food Sci. 59 (1994) 152.
[7] T. Nakanishi, N. Yamakawa, T. Asahi, N. Shibata, B. Ohtani, T. Osaka, Chirality 16 (2004) S36.
[8] M. Eichelbaum, S.A. Gross, Adv. Drug Res. 28 (1996) 1.
[9] R.M. Lane, G.B. Baker, Cell. Mol. Neurobiol. 19 (1999) 355.
[10] H. Caner, E. Groner, L. Levy, Drug Discovery Today 9 (2004) 105.
[11] A.J. Hutt, J. Valentová, Acta Facult. Pharm. Univ. Comenianae 50 (2003) 7.
[12] M. Friedman, J. Agric. Food Chem. 47 (1999) 3457.
[13] I.J. Buerge, T. Poiger, M.D. Müller, H.-R. Buser, Env. Sci. Technol. 37 (2003) 2668.
[14] A.W. Garrison, P. Schmitt, D. Martens, A. Kettrup, Env. Sci. Technol. 30 (1996) 2449.
[15] A. Miyazaki, T. Nakamura, M. Kawaradani, S. Marumo, J. Agric, Food. Chem. 36 (1988) 835.
[16] Q. Zhou, C. Xu, Y. Zhang, W. Liu, J. Agric. Food Chem. 57 (2009) 1624.
[17] L.J. Fono, D.L. Sedlak, Env. Sci. Technol. 39 (2005) 9244.
[18] T. Bucheli, S.R. Müller, A. Voegelin, R.P. Schwarzenbach, Env. Sci. Technol. 32 (1998) 3465.
[19] J. Robins, M. Jones, E. Matisoo-Smith, Auckland University, Private Bag 92019.
[20] C. S-Ng, T. Sun, H.S.O. Chan, Tetrahedron Lett. 43 (2002) 2863.
[21] M.C. Tanese, L. Torsi, N. Cioffi, L.A. Zotti, D. Colangiuli, G.M. Farinola, F. Babudri, F. Naso, M.M. Giangregorio, L. Sabbatini, P.G. Zambonin, Sens. Actuat. B 100 (2004) 17.
[22] H. Tsukube, H. Sohmiya, Tetrahedron Lett. 31 (1990) 7027.
[23] P. Bates, R. Kataky, D. Parker, J. Chem. Soc. Perkin Trans. 2 (1994) 669.
[24] H. Chibvongodze, K. Hayashi, K. Toko, Sens. Materials 13 (2001) 99.
[25] F. Scholz, R. Gulaboski, Faraday Discuss. 129 (2005) 169.
[26] F. Scholz, R. Gulaboski, V. Mirčeski, P. Langer, Electrochem. Commun. 4 (2002) 659.
[27] H. He, G. Uray, O.S. Wolfbeis, Anal. Chim. Acta 246 (1991) 251.
[28] T. Kullick, R. Ulber, H.H. Meyer, T. Scheper, K. Schlügerl, Anal. Chim. Acta 293 (1994) 271.
[29] Y. Inaba, K. Mizukami, N. Hamada-Sato, T. Kobayashi, C. Imada, E. Watanabe, Biosens. Bioelectron. 19 (2003) 423.
[30] M. Wcisło, D. Compagnone, M. Trojanowicz, Bioelectrochemistry 71 (2007) 91.
[31] A. Tsourkas, O. Hofstetter, H. Hofstetter, R. Weissleder, L. Josephson, Ang. Chem. Int. Ed. 43 (2004) 2395.

[32] O. Hofstetter, J.K. Hertweck, H. Hofstetter, J. Biochem. Biophys. Meth. 63 (2005) 91.
[33] M. Chen, Y. Fu, X. Ciu, L. Wang, M. Li, Electroanalysis 21 (2009) 2339.
[34] W.C. Su, W.G. Zhang, S. Zhang, J. Fan, X. Yin, M.L. Luo, S.C. Ng, Biosens. Bioelectron. 25 (2009) 488.
[35] P. Sandblad, R. Arnell, J. Samuelsson, T. Fornstadt, Anal. Chem. 81 (2009) 3551.
[36] K. Ohsawa, T. Kasamatsu, J. Nagashima, K. Hanawa, M. Kuwahara, H. Ozaki, H. Sawai, Anal. Sci. 24 (2008) 167.
[37] M. Trojanowicz, M. Kaniewska, Electroanalysis 21 (2009) 229.
[38] Z. Vysotskii, M.V. Polyakov, Zh. Fiz. Khim. 30 (1956) 1901.
[39] A.H. Beckett, P. Anderson, Nature 179 (1957) 1074.
[40] K. Haupt, Anal. Chem. 75 (2003) 377A.
[41] N.M. Maier, W. Lindner, Anal. Bioanal. Chem. 389 (2007) 377.
[42] C. Alexander, H.S. Andersson, L.I. Andersson, R.J. Ansell, N. Kirsch, I.A. Nicholls, J. O'Mahony, M.J. Whitcombe, J. Mol. Recognit. 19 (2006) 106.
[43] M. Avila, M. Zougagh, A. Escarpa, A. Rios, Trends Anal. Chem. 27 (2008) 54.
[44] M.C. Blanco-Lopez, M.J. Lobo-Castanon, A.J. Miranda-Ordieres, P. Tunon-Blanco, Trends Anal. Chem. 23 (2004) 36.
[45] Y. Zhou, B. Yu, K. Levon, Chem. Mater. 15 (2003) 2774.
[46] M. Lahav, A.B. Kharitonov, I. Willner, Chem. Eur. J. 7 (2001) 3992.
[47] H.-J. Liang, T.-R. Ling, J.F. Rick, T.-C. Chou, Anal. Chim. Acta 542 (2005) 83.
[48] E. Granot, R. Tel-Vered, O. Lioubashevski, I. Willner, Adv. Funct. Mater. 18 (2008) 478.
[49] B.B. Prasad, M.P. Tiwari, R. Madhuri, P.S. Sharma, J. Chromatogr. A 1217 (2010) 4255.
[50] B.B. Prasad, R. Madhuri, M.P. Tiwari, P.S. Sharma, Anal. Chim. Acta 681 (2010) 16.
[51] B.B. Prasad, M.P. Tiwari, R. Madhuri, P.S. Sharma, J. Chromatogr. B 879 (2011) 364.
[52] B.B. Prasad, R. Madhuri, M.P. Tiwari, P.S. Sharma, Talanta 81 (2010) 187.
[53] S. Fireman-Shoresh, I. Turyan, D. Mandler, D. Avnir, S. Marx, Langmuir 21 (2005) 7842.
[54] R. Ouyang, J. Lei, H. Ju, Y. Xue, Adv. Funct. Mater. 17 (2007) 3223.
[55] S. Sekine, Y. Watanabe, Y. Yoshimi, K. Hattori, K. Sakai, Sens. Actuat. B 127 (2007) 512.
[56] L. Cao, X. Chun Zhou, S. Fong, Y. Li, Analyst 126 (2001) 184.
[57] C.J. Percival, S. Stanley, M. Galle, A. Braithwaite, M.I. Newton, G. McHale, W. Hayes, Anal. Chem. 73 (2001) 4225.
[58] K. Haupt, K. Noworyta, W. Kutner, Anal. Commun. 36 (1999) 391.
[59] S. Stanley, C.J. Percival, T. Morel, A. Braithwaite, M.I. Newton, G. McHale, W. Hayes, Sens. Actuat. B 89 (2003) 103.
[60] T. Piachama, A. Josell, H. Arwin, V. Prachayasittikul, L. Ye, Anal. Chim. Acta 536 (2005) 191.
[61] F. Liu, X. Liu, S.-C. Ng, H.S.-O. Chan, Sens. Actuat. B 113 (2006) 234.
[62] L.F. Arenas, B.S. Ebarvia, F.B. Sevilla III, Anal. Bioanal. Chem. 397 (2010) 3155.
[63] B. Deore, Z. Chen, T. Nagaoka, Anal. Chem. 72 (2000) 3989.
[64] V. Syritski, J. Reut, A. Menaker, R.E. Gyurcsányi, A. Öpik, Electrochim. Acta 53 (2008) 2729.
[65] Y. Kong, W. Zhao, S. Yao, J. Xu, W. Wang, Z. Chen, J. Appl. Polymer Sci. 115 (2010) 1952.
[66] Z. Zhang, H. Liao, H. Li, L. Nie, S. Yao, Anal. Biochem. 336 (2005) 108.
[67] S. Huan, G. Shen, R. Yu, Electroanalysis 16 (2004) 1019.
[68] Y. Liao, W. Wang, B. Wang, Bioorg. Chem. 27 (1999) 463.
[69] D.O. Nopper, G. Lammershop, G. Wulff, Gauglitz, Anal. Bioanal. Chem. 377 (2003) 608.
[70] X. Hu, Q. An, G. Li, S. Tao, J. Liu, Angew. Chem. Int. Ed. 45 (2006) 8145.
[71] L. Wang, Z. Zhang, L. Huang, Anal. Bioanal. Chem. 390 (2008) 1431.
[72] O.Y.F. Henry, S.A. Piletsky, D.C. Cullen, Biosens. Bioelectro. 23 (2008) 1769.
[73] Y. Chen, K.D. Shimizu, Org. Lett. 4 (2002) 2937.
[74] N. Mizutani, D.-H. Yang, R. Selyanchyn, S. Korposh, S.-W. Lee, T. Kunitake, Anal. Chim. Acta 694 (2011) 142.
[75] M. Riskin, R. Tel-Vered, M. Frasconi, N. Yavo, I. Willner, Chem. Eur. J. 16 (2010) 7114.
[76] Y. Ben-Amram, M. Riskin, T. Willner, Analyst 135 (2010) 2952.
[77] M. Trojanowicz, Main concepts of chemical sensing. Chapter, in: R. Potyrailo, V. Mirsky (Eds.), Combinatorial methods for chemical and biological sensors, Springer (2009), pp. 25–60.
[78] M. Kaniewska, M. Trojanowicz, Chiral biosensors and immunosensors. Chapter, in: P.A. Serra (Ed.), Biosensors – Emerging Materials and Applications. In Tech, (2011), pp. 99–122.
[79] F. Berti, S. Todros, D. Lakshmi, M.J. Whitcombe, I. Chianella, M. Ferroni, S.A. Piletsky, A.P.F. Turner, G. Marrazza, Biosens. Bioelectron. 26 (2010) 497.

[80] F. Cao, J. Liao, K. Yang, P. Bai, Q. Wei, C. Zhao, Anal. Lett. 43 (2010) 2790.
[81] H. Okuno, T. Kitano, H. Yakabe, M. Kishimoto, B.A. Deore, H. Siigi, T. Nagaoka, Anal. Chem. 74 (2002) 4184.
[82] H. Wang, Y. He, T. Ji, X. Yan, Anal. Chem. 81 (2009) 1615.
[83] A.K. Patel, S. Sharma, B.B. Prasad, Thin Solid Films 518 (2010) 2847.
[84] M. Tian, W. Bi, K.H. Row, Anal. Bioanal. Chem. 399 (2011) 2495.
[85] M. Ulbricht, J. Chromatogr. B 804 (2004) 113.

CHAPTER

9

Artificial Receptors for Mass-Sensitive Sensors: Targeting Analytes from Surfaces, Nanoparticles, and Bioanalytes by Molecular Imprinting

Naseer Iqbal, Peter A Lieberzeit

University of Vienna, Department of Analytical Chemistry, Vienna, Austria

OUTLINE

1. INTRODUCTION

In living systems, membrane proteins are usually responsible for perception and show a wide range of multispecific receptors [1]. For instance, hundreds of different olfactory

Molecularly Imprinted Sensors.
DOI: 10.1016/B978-0-444-56331-6.00009-8

receptors are encoded in the human genome that can recognize and distinguish different chemical substances and their constituents even at the molecular level [2]. Hence, the interaction between chemical species and recognition membranes generates multireceptor responses that are converted to adequate signals within the living organism. In technological systems, their analogues are transducers and receptors of chemical sensors [3]. In recent years, materials science and chemistry have become increasingly interested in mimicking such natural receptor strategies. The main idea is to provide them with properties provided by technological materials, such as thermal rigidness, processabilty, and robustness, thereby taking the best from two worlds.

For reaching this goal, molecular imprinting provides a synthetically rather straightforward path [4]. Over the last few decades, molecularly imprinted polymers (MIPs) have fascinated scientists because of their considerable advantages over natural receptors, biomacromolecules, enzymes, and antibodies. These advantages include high stability, low cost, comparable ease of preparation, and flexibility regarding the target compound (MIPs in practice can be prepared for virtually any kind of template) [5]. Molecular imprinting involves the formation of adduct between the template species and functional monomers in a suitable solvent. This is based either on noncovalent interactions, that is, hydrogen bonding, ionic interactions, van der Waals forces, metal coordination interactions, hydrophobic interactions, or covalent strategies, that is, reversible chemical bond formation [4,6]. During polymerization (Fig. 9.1) this interaction network is stabilized and finally fixed in its geometrical position by the hardening polymer matrix. After template removal, binding locations within the polymer possessing both the shape and the correct orientation of functional groups are created which allow selective recognition of the template species. MIP technology has been successfully used in separation sciences, generating artificial antibodies, as alternatives to the enzymes, artificial catalysis, and polymer-assisted synthesis and in biomimetic sensors [4–9].

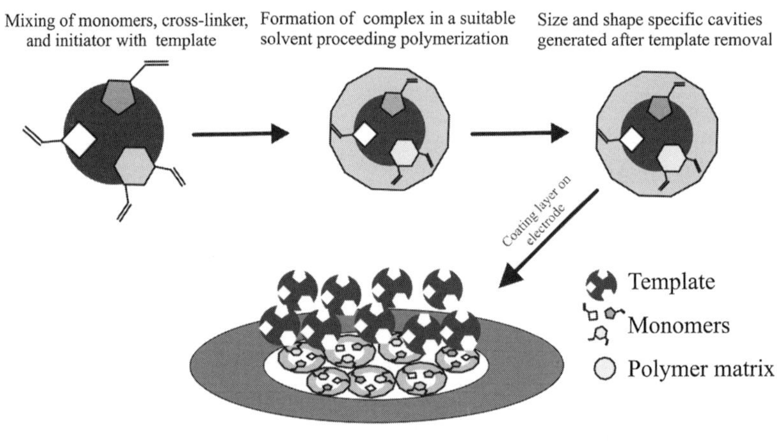

FIGURE 9.1 A schematic of molecular imprinting and MIP layer on QCM transducer.

1.1. Molecular Imprinting Strategies for Chemical Sensing

A variety of parameters, such as the nature of the analyte, the solvent used for polymerization, functional monomers, cross-linking agents, and initiators have an influence on the performance of the final imprinted polymer. The following strategies have been developed for optimization:

Covalent Imprinting is an imprinting strategy based on reversible covalent bond formation in monomeric solution before proceeding to polymerization, as depicted in Fig. 9.2. It was introduced by G. Wulff [10] and used mainly for compounds lacking functionalities [11,12]. Initially, covalent bonds are formed between template molecules and functional monomers. After polymerization, template removal indeed requires the breaking of that bond to achieve recognition cavities [4,13].

Noncovalent Imprinting: Klaus Mosbach and co-workers promoted the noncovalent imprinting [14,15] strategy as presented in Fig. 9.3, which is still the most frequently used MIP technique in developing artificial receptors. The monomer-template complex is formed

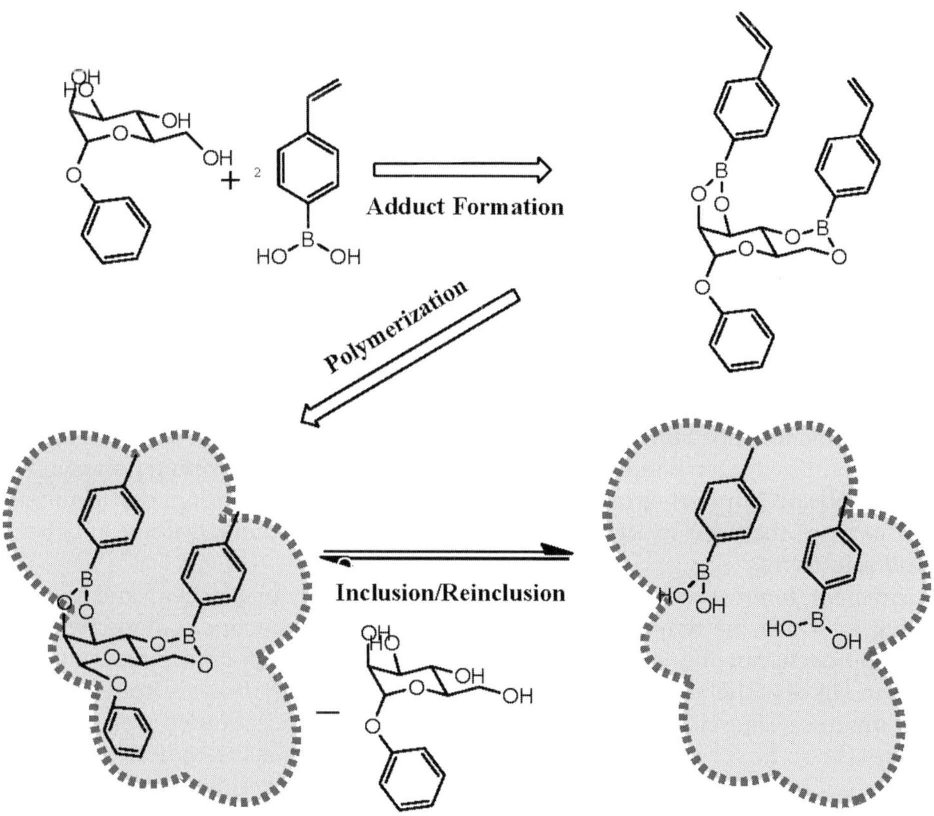

FIGURE 9.2 Covalent Imprinting of phenyl-D-mannoside using vinylphenylboronic acid [4]. © Elsevier 1997.

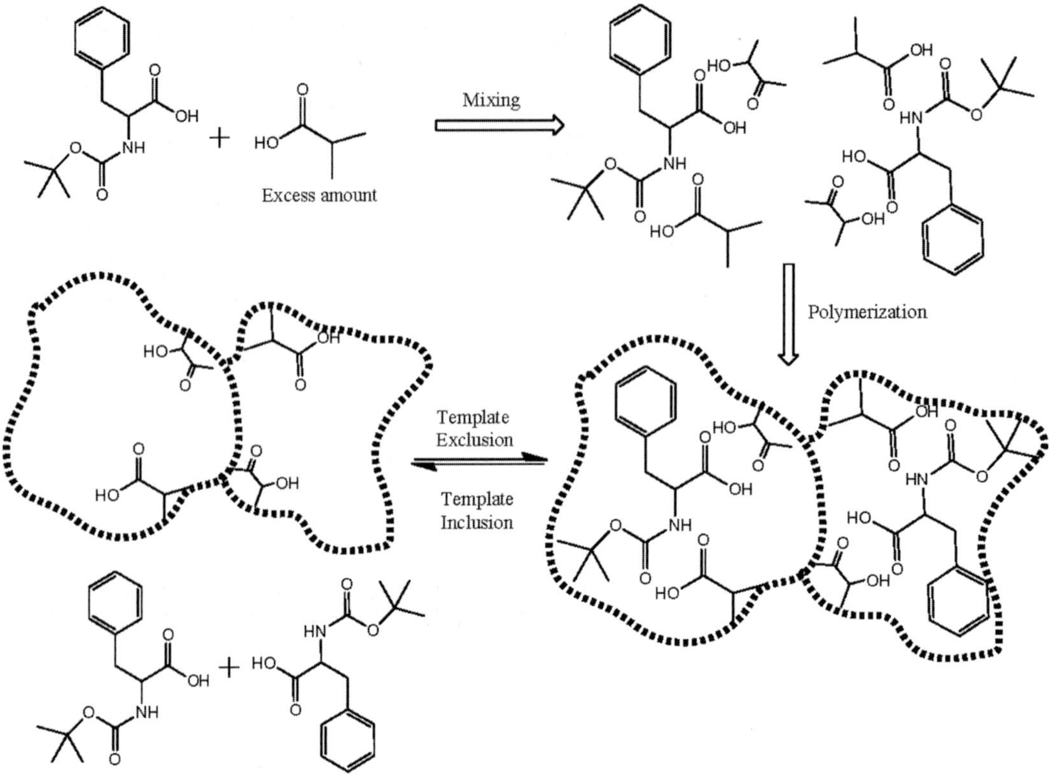

FIGURE 9.3 Non-covalent imprinting of Boc-phenylalanine using methacrylic acid [4]. © Elsevier 1997.

by noncovalent interactions [13]. As removing the template molecules does not involve any bond breaking, it is usually straightforward and can be achieved by washing or evaporation. Generally speaking, imprinting yields optimal results for template molecules with pronounced functionality. However, the influence of the solvent cannot be negated: less polar solvents support hydrogen bonding between the template and the forming polymer, whereas more polar solvents support stronger nonpolar interactions. In addition, porogenic solvents can help increase the porosity of the resulting imprinted polymers, leading to better sensitivity and selectivity.

Semicovalent Imprinting: This strategy is a modification of covalent and noncovalent imprinting in which the template is bound covalently to a functional monomer, as in the covalent approach, but the template rebinding is based only on noncovalent interactions, as shown in Fig. 9.4. The template is typically washed out via hydrolysis from a cross-linked polymer matrix [16]. Generally, two synthetic routes are adopted for semicovalent imprinting: (1) by direct connection of template and monomer through an ester or amide linkage or (2) by introduction of a spacer between the template and polymerizable recognition component in the imprinting step [13]. This approach has numerous advantages: association of functional monomers to template molecules, random distribution

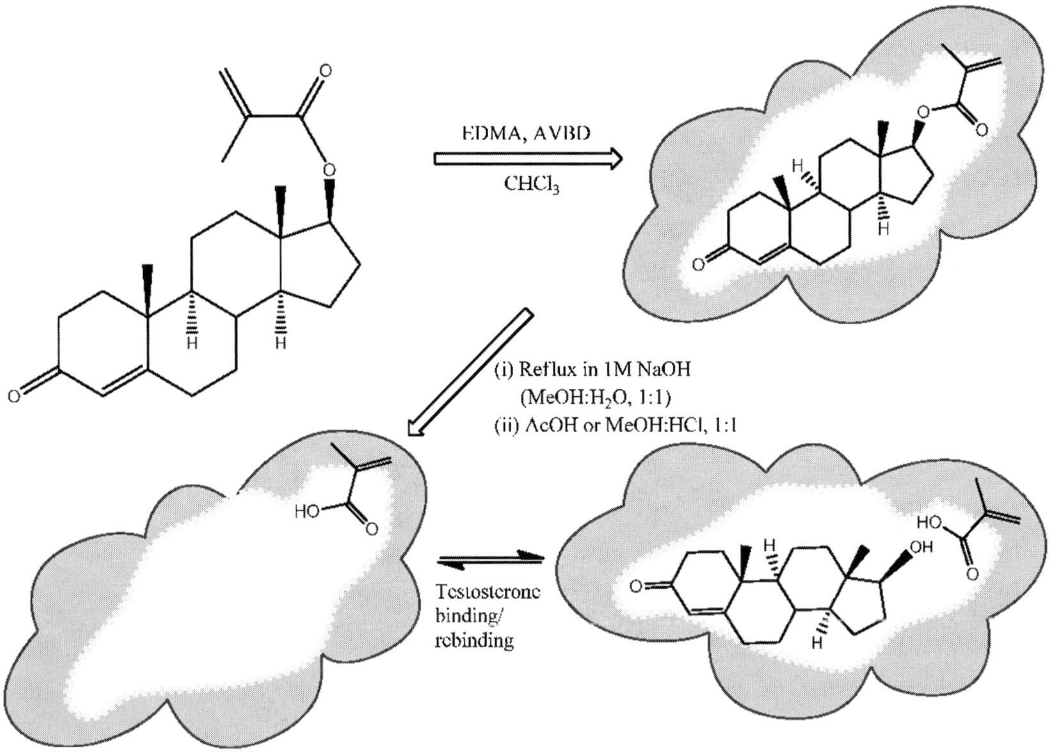

FIGURE 9.4 Schematic of semi- covalent imprinting of testosterone methacrylate [13].

of functional monomers; production of uniform recognition sites; and finally, reinclusion of template molecule, which is free from any kinetic hindrance except diffusion. Despite these attributes of the semicovalent imprinting approach, "template bleeding" can be observed for those imprinted polymers where hydrolysis cannot remove template, as it could be covalently bound or possibly by some sterical hindrance due to hydrogen bonding. Nonetheless, this behavior typically can be observed for noncovalently imprinted polymers as well [22]. Incorporating a linking group into a polymer mixture can solve this issue, which has been addressed in detail by Whitcombe et al. [17]: they introduced the carbonate ester group, for example, which initially helped in template monomer binding and then latterly functions as spacer in the growing polymer chain.

Imprinting via Metal Coordination: In this strategy, coordinative bonds between metal ions and the respective ligands are highly directed and define a clear 3D structure [13] (see Fig. 9.5). Such an approach is highly suited for generating biospecies imprints: Many biological receptors contain metal ions that serve both structural and catalytic functions and hence determine the reactive center [13]. In this regard, metal-ligand coordination for achieving patterned orientation in functional monomer-template complex results in an MIP with substantially reduced nonspecific binding. Some key necessities for metal-coordinated imprinting are optimized polymerization conditions (i.e., the metal center does not inhibit

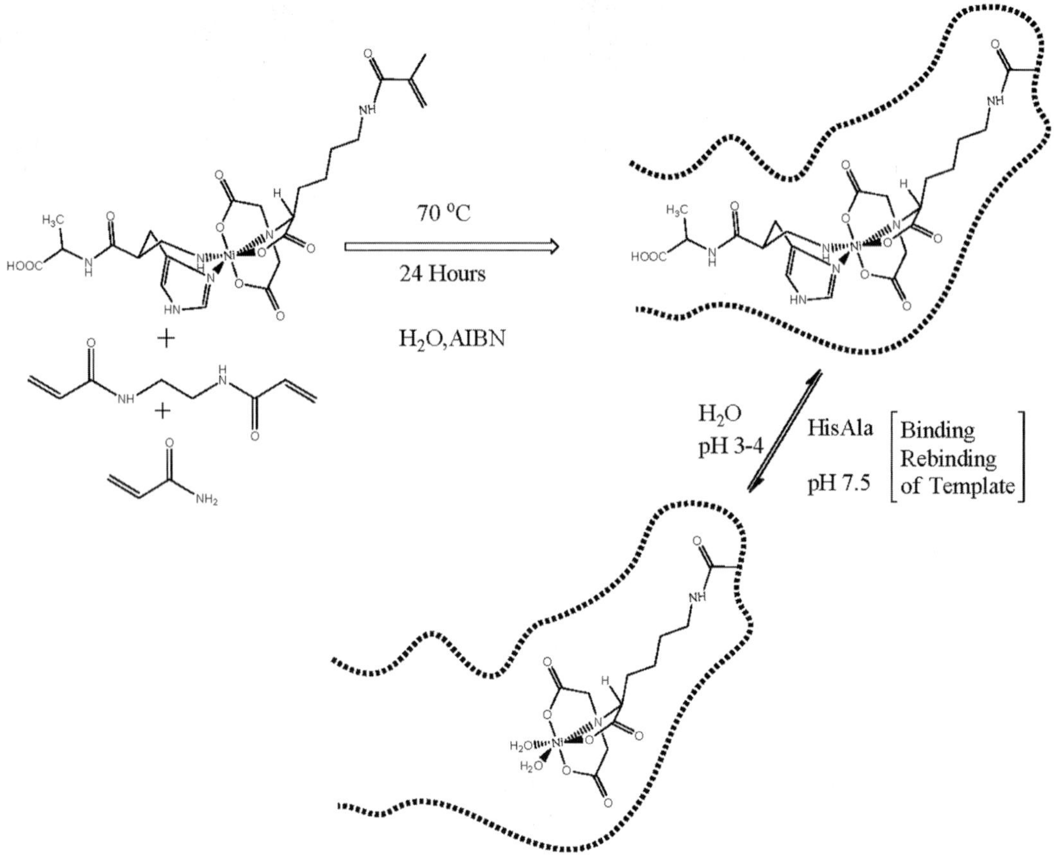

FIGURE 9.5 Imprinting of peptide; copolymerization of the (His-Ala)-Ni-NTA complex with mono- and bisacrylamides showing removal of the template peptide at pH 3-4 and rebinding at pH 7.5 via polymer containing Ni-NTA complex. *[Reprinted with permission from [18], © 2001 American Chemical Society].*

initiation or propagation steps during polymerization). Second, metal centers should possess a well-defined coordination sphere. Third, the stability of the metal–template complex during polymerization is decisive; it should be both stable enough to be preserved during templating and labile enough for easily extracting the template [13,18].

Combinatorial Imprinting: In combinatorial imprinting, functional monomers with a series of appropriate and compatible cross-linkers are combined in a suitable solvent to generate libraries of MIPs, which are optimized for a specific template via a fully automated system. This one-pot approach is used to attain better-quality recognition materials. Takeuchi et al. [19] for the first time proposed a combinatorial approach to prepare imprinted materials or artificial receptors that bear superior properties. They used an automated screening approach and designed libraries of MIPs by varying the proportions of functional monomers, cross-linking agents and templates, and other parameters [20]. Combinatorial approaches provide improved imprinting and high-throughput methodology while utilizing multivariate strategies, experimental design, and computational modeling capacities. This allows for parallel screening of a wide range of possible MIP receptors, thus substantially reducing the development time and produces optimized MIPs for desired applications.

Some other known imprinting approaches: Several other advancements in imprinting have been made, some of which shall be introduced shortly, notably, **double imprinting,** which makes use of two dissimilar templates [21]. For instance, in imprinting PAH using a binary template mixture resulted in substantially higher selectivity and sensitivity. Another imprinting approach, **epitope mapping in MIPs** [22,23], is specifically designed for recognition of proteins [23] and peptides. The main idea is to use not an entire compound as the template, but only a substructure, which results in higher cost effectiveness. According to D. F. Tai et al. [24] the foremost advantages are as follows: (1) Crystal structure and epitope mapping revealed that plenty of accessible epitopes of protein antigens can be fabricated; (2) it generates cavities that are replicas of binding sites as found in parent proteins; (3) flexible epitope-cavities are produced that can maintain the native protein conformations during inclusion/reinclusion; and finally, (4) manipulation of proteins with defined orientation. Recently, an **on/off-switchable MIP approach** has been reported. S. Li et al. [25] introduces for the first time a zipper-like on/off-switchable molecularly imprinted polymer (MIP). MIPs are based on template-imprinted polymeric networks, which can show significant switchable on/off interactions with recognition capacity toward the template. Furthermore, increasing temperature enhances the recognition ability of imprinted polymers, which is due mainly to dissociation of the interpolymer complexes, whereas these interactions are stronger at lower temperatures, thus restricting the template to recognition site for a possible interaction.

In summary, molecular imprinting has provided a platform and gained significant importance in synthesizing materials and mimicking naturally occurring systems. Some recent examples include the following: MIP consisting of methyl methacrylate (MMA) and divinylbenzene (DVB) copolymer imprinted with toluene and *p*-xylene for chemical sensing [7]; uniform MIP microspheres and nanoparticles containing chiral recognition sites to be used as binding assays and for microfludic separations [9]; characterization of functionalized QCM sensor crystal surfaces coated with MIP nanoparticles [26]; polymers imprinted with polycyclic hydrocarbons mixtures for comparing fluorescence and QCM sensors [27]; and pollen-imprinted polyurethanes as artificial recognition materials for QCM allergen sensors [28]. Imprinting molecular memory on the surface of polymer particles used as plastic/artificial antibodies used for recognition and neutralization of toxic peptides in vivo [8] and QCM sensor array for monitoring emanated terpenes as well as freshness of herbs [29]. From the preceding applications, it can be concluded that MIP technology has outstanding potential in analytical chemistry for grappling with several problems.

1.2. The Quartz Crystal Microbalance as an Example of a Mass-Sensitive Transducer

All mass-sensitive devices are based on the piezoelectric effect discovered by Pierre and Jacques Curie in 1880 [30,31]. When mechanical stress is applied on crystals, this results in electric potential on the surface possessing a polar axis without symmetry center, for example, quartz. Of course, the contrary is also true: the so-called inverse piezoelectric effect was discovered by Lippmann [32] and later confirmed by the Curie brothers [30]. Here,

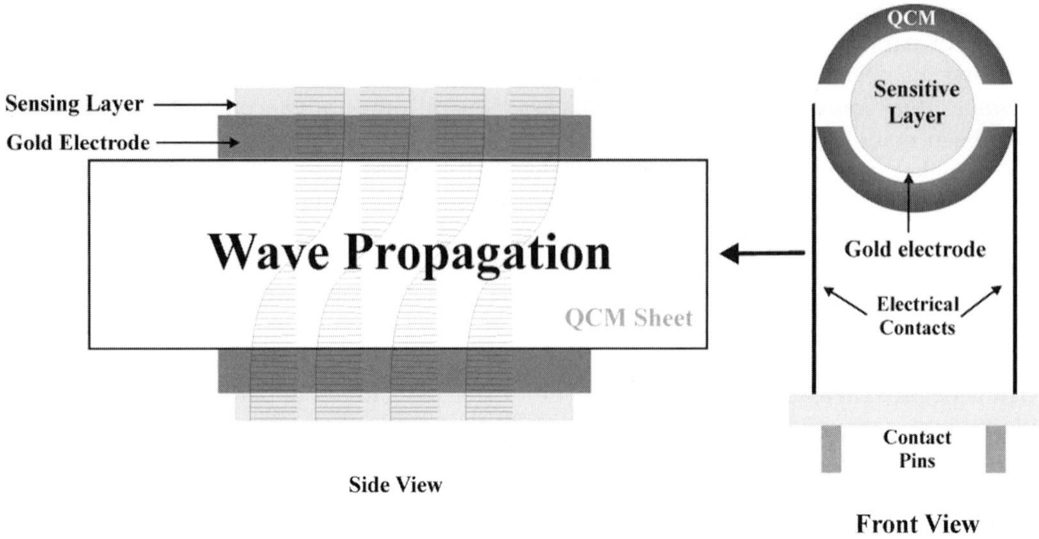

FIGURE 9.6 Schematic of a 10 MHz QCM transducer and thickness shear wave propagation.

mechanical deformation of the substrates results from applying an electric field to a crystal. The effect finds applications, for example, for the production and detection of sound, generation of high voltages, electronic frequency generation, microbalances, and ultrafine focusing of optical assemblies [33]. For chemical sensing, a range of piezoelectric substrates and cutting angles with respect to the asymmetric axis is used that defines the respective oscillation mode. The quartz crystal microbalance consists of an AT-cut quartz plate coated with two gold electrodes (one on each face) as shown in Fig. 9.6. When applying voltage to the electrodes, the two faces of the crystal move against one another in a shear motion. In the case of an alternating voltage, whose wavelength equals two times the thickness of the quartz plate, the crystal resonates, as electromechanical standing waves are generated. The displacement therefore obviously is maximized at the crystal faces, making the device sensitive to surface interactions. These transducers were originally employed in oscillator circuits used to determine the rate of metal deposition in certain vacuum systems [34]. The oscillation frequency tracks the crystal resonance and indicates mass accumulation on the device surface.

When a QCM is connected to oscillator circuit, it resonates close to its fundamental resonance frequency. A typical equivalence circuit [35] for a QCM is shown in Fig. 9.7.

The inductance L in the equivalence circuit can be correlated with the mechanically vibrating mass; the dynamic capacitance C complies with the elastic behavior of the resonator; the resonance resistance R represents the dynamic vibration losses; and the static capacitance C_0 is for the capacity between the two electrodes. A shift in the resonance frequency of the oscillating QCM is observed owing to the interaction of analyte on the surface of chemical film. This behavior can be well explained by the Sauerbrey equation [36].

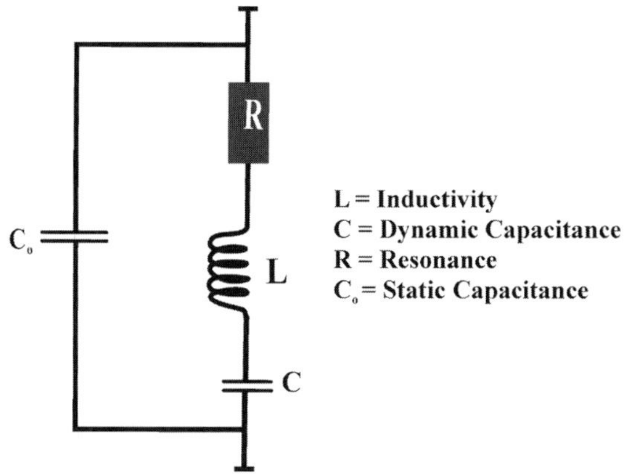

FIGURE 9.7 A typical equivalence circuit for a QCM.

$$\Delta f = -\frac{f_o^2}{K_f \cdot \rho \cdot A} \cdot \Delta m$$

where, Δf = change in frequency, Δm = change in mass, ρ = density of the sensitive layer, K_f = frequency constant of the crystal, f_o, = resonance frequency of QCM, and A = cross-sectional area of the electrode.

The equation explains a decrease in resonant frequency by increasing mass on the transducer surface. From Sauerbrey, sensitivity depends quadratically on fundamental frequency, which should therefore be as high as possible within the operational range. Strictly speaking, the initial Sauerbrey equation is only applicable for gaseous state measurements. In liquid phase, some additional parameters (e.g., viscosity) must be considered. This limitation of the Sauerbrey equation was ruled out by Kanazawa et al. [37] and modified for the viscosity term in liquid phase measurements as:

$$\Delta f = f_o^{\frac{3}{2}} \left[\frac{\rho\, \eta}{\pi\, \eta_q \rho_q}\right]^{\frac{1}{2}}$$

Hence, the change in resonant frequency Δf depends on density ρ and viscosity η of the medium. When the viscosity of the liquid medium increases, the damping also increases, which hinders sensor performance; therefore, using thin layers of polymers in a viscous medium can give better sensitivity.

In addition, QCMs have advantages such as ease of manufacturing, ruggedness in harsh environments, thermal stability, and, most importantly, excellent sensitivity upon minute mass variations on transducer surface [38]. The operational frequency range of QCM resonators is from 5–30 MHz, which corresponds to a thickness of 332 μm–42 μm, respectively. During the past few decades, ample research has been done on chemical sensors based on QCM. Chemists used its mass sensing ability in real time for measuring very small mass changes, because it is approximately 100 times more sensitive than an electronic fine balance (the latter typically being able to weigh 0.1μg) [39]. In general, acoustic sensors enable sensitive probing of the changes taking place in sensor layers coated on the electrode surface of the resonating transducers. However, effects other than mass changes, so-called nongravimetric effects, like energy dissipation, temperature, and humidity, can influence the sensor signal [40]. The foremost advantage of using the QCM-based mass-sensitive technique is as follows [33]: it exterminates the requirement of specific labeling. The working efficacy of this technique is remarkable; that is, signal transduction via the piezoelectric mechanism operates well in complex, often optically opaque, media. Furthermore, the technique is capable of detecting subtle changes in the solution—surface interface that can be due to density—viscosity changes in the solution, viscoelastic changes in the bound interfacial materials, and changes in the surface free energy. Moreover, in electrochemical sensors, supplying potential to the upper metal electrode of the electrochemical quartz crystal microbalance (EQCM) offers the user interesting ways to create or probe surface bound mass by following the ongoing electron transfer reactions. At last, the technique is relatively easy to use (i.e., user friendly), and the basic equipment is inexpensive to purchase. So we can say that these devices are miniaturized and cost-effective.

2. MOLECULARLY IMPRINTED NANOMATERIALS: RECENT TRENDS IN QCM-BASED MASS-SENSITIVE DEVICES

In recent years, there has been a substantial increase in the number of papers published on MIPs in combination with mass-sensitive devices. These can be distinguished by different parameters, such as the morphology of the recognition material (thin film versus nanoparticle), the type of analyte (pure compound versus mixture), its size (small molecule versus biospecies), and others. During the following sections some typical approaches will be discussed and summarized.

2.1. Nano-Templated Thin Films in Mass-Sensitive Devices

Depositing MIP thin films is one of the main strategies used to generate sensor layers on QCM. Usually, the coating techniques are very straightforward, namely, spin-coating or drop-coating. Both result in uniform layer heights that are ideally suited for nanosensors; experience shows that inhomogeneous films lead to increased noise in both gas-phase and liquid-phase measurements. Furthermore, rough surfaces are mainly an issue in liquid phase due to the higher viscous interactions between the sensor layer and the surrounding sample matrix. The majority of MIP thin films for mass-sensitive sensing are based on organic polymers [41] via the noncovalent imprinting approach [42,43]. The strategy leads to surprisingly high selectivity; for instance, F. Liu et al. [44] prepared permeably thin films for enantioselective discrimination of L- and D-tryptophan in citric acid buffer solution using acrylamide (AM) and trimethylolpropane as functional monomer and cross-linking agent, respectively. Recognition takes place mainly via hydrogen bonding, which due to its directed nature and rather high binding energy leads to very favorable sensor signals. Another example for this has been given by N. Tsuru et al. [45], who developed recognition cavities for Bisphenol A (BPA). MIP thin films from different compositions of copolymers—ethylene glycol dimethacrylate (EGDMA), methacrylic acid (MAA), and (EGDMA)-4-vinylpyridine (4-Vpy)—were employed for fabrication of MIP-QCM sensors leading to the selectivity pattern as

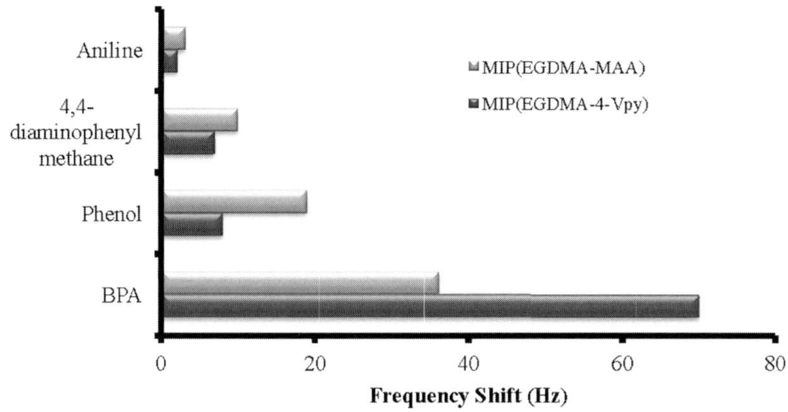

FIGURE 9.8 Sensor effect vs relative affinity of BPA towards MIP thin layers. Concentrations of BPA were 600ppb for composition 4 and 1.2 ppm for composition 1 respectively. *[Adapted from ref 45].*

summarized in Fig. 9.8. The sharply increased sensitivity of the polymer containing 4-vinylpyridine toward BPA was due to additional H-bonding in the MIP.

Going further down the size scale for molecules, other noncovalent interactions become more prominent. For instance, M. Matsuguchi et al. [46] presented MIP polymer powder from highly cross-linked methyl methacrylate (MMA)-(divinylbenzene)DVB polymer system in toluene and *p*-xylene solvents/templates. MIP powder was blended in polymethyl methacrylate (PMMA) to cast thin films for selective monitoring of toluene and *p*-xylene templates via QCM sensor. Thermogravimetric analysis revealed the high analyte sorption capacity of MIP with respect to nonimprinted films. This result can be attributed to the formation of microcavities of templates in the cross-linked polymers based on $\pi-\pi$ interactions. A similar example is the formation of polyurethane (PU) double-MIP films for polycyclic aromatic hydrocarbons (PAHs) via a noncovalent imprinting approach [27]. Binary mixtures of PAHs containing a smaller and a larger species basically enhance both sensitivity and selectivity based on $\pi-\pi$ interactions. The main reason for this is that the smaller template in this case takes over the function of a porogen, thereby substantially increasing the accessibility of the interaction sites formed within the film. Therefore a larger number of high-affinity interaction sites are available leading to substantial improvement of the sensor signals.

All previous techniques have in common that the interaction sites are distributed within the entire bulk of the thin film, which could be regarded as 3-D imprinting. Some noncovalent strategies have also led to appreciable results for 2-D imprinting [47–50], where organosilanes or thiols are usually self-assembled onto transducers in the presence of template molecules, thus producing molecular imprinted monolayers (MIMs). Hence, all recognition centers are located directly at the surface of the respective sensitive material. One practical example is given by X. Bi et al. [51], who developed MIM–QCM for real-time measurements of neonicotinoid pesticides, namely, imidacloprid and thiacloprid in celery juice. MIMs were obtained via a 2-D imprinting strategy in which template cavities were generated into monolayers of alkanethiols as represented in Fig. 9.9.

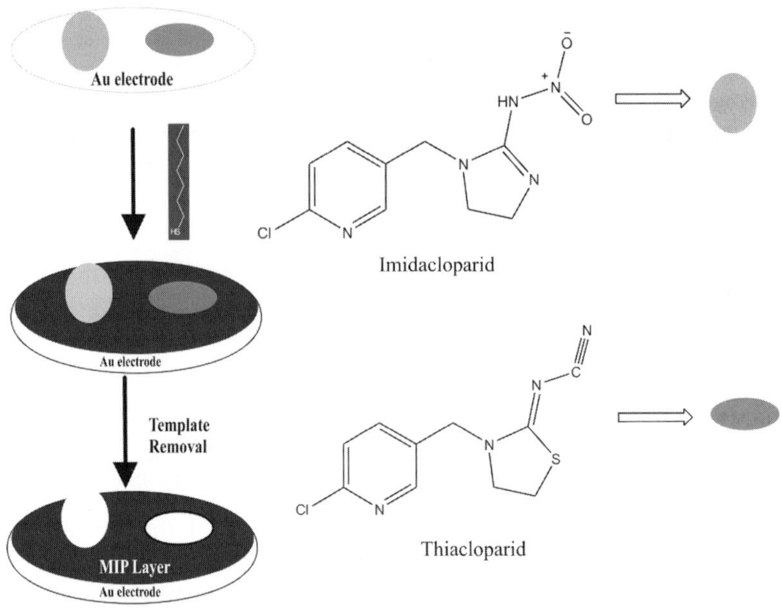

FIGURE 9.9 Schematic of 2D molecular imprinting strategy used for generating neonicotinoid-selective surface binding sites. At first templates (thiacloprid and imidacloprid) were pre-adsorbed on a gold surface followed by self-assembled monolayer of alkanethiols around the templates. The templates were removed from the surface and two different molecularly imprinted binding sites were created. *[Adapted from ref 51].*

Bi et al. reported that the detection limits for imidacloprid and thiacloprid can be improved by using long-chain alkanethiols. In this scenario, MIMs developed from hexanethiol have dissociation constants 2 to 5 times lower than those prepared from octanethiol. Hence MIM–QCM sensor developed from it quantifies 10 μM of imidacloprid and thiacloprid from celery juices, respectively, which is well beyond the 0.25 mg/kg, the EU standard for fruits and vegetables [51].

Some MIP thin films can be applied in real-life applications. R. Suedee et al. [52] used a tricholoroacetic acid (TCAA)-imprinted poly(ethyleneglycoldimethacrylate-co-4-vinylpyridine) polymer system on QCM to monitor haloacetic acid (HAA) in drinking water. Herein, the carboxylate anion and vinylpyridinium cation of MIP provided the driving force for acid–base interactions between TCAA and MIP film. Furthermore, aromatic structures of the MIP films also facilitated interactions with polarizable chlorine ions. The strategy showed fast, inexpensive, and reliable screening of HAAs with high specificity toward HAAs and a detection limit of 20–50 $\mu g l^{-1}$. That is still below its minimum allowable limit, that is, 60 $\mu g L^{-1}$. Figure 9.10(a) and (b) shows the topography of MIP film and sensor measurements, respectively.

A remarkable example of the QCM–MIP system is presented by Z. Yang et al. [53], which is utilized for detecting metal ions. MAA-trimethylolpropane trimethacrylate (TMPTM) polymer system was used for generating ion-imprinted polymer (IIP) for Cu (II) ions. The pH of the solution played a decisive role in the interaction of metal ions with MIP films. Low adsorption of Cu (II) ions at pH (2–3) gave a reduced sensor effect that reflects the abundance of hydrogen ions in solution, which preferably can interact with recognition sites rather than Cu (II) ions. This

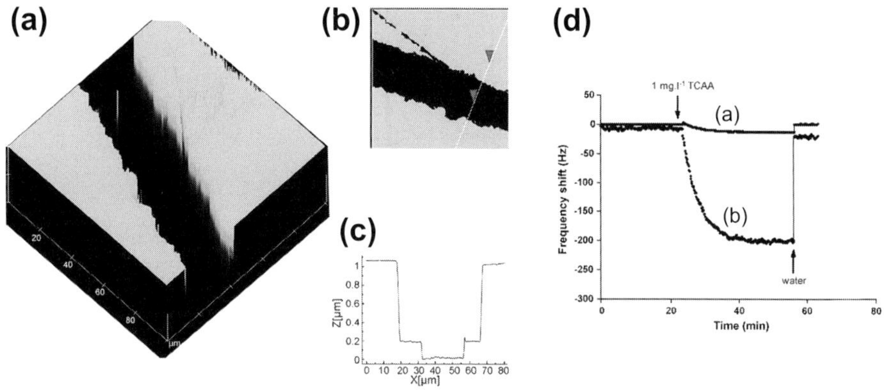

FIGURE 9.10 **(i)** (a) AFM 3D topography; (b) AFM of a cross-section of a scratch on a 20 kHz-MIP layer, and the measured depth of the polymer layer of 870 nm which is shown on (c) the plot of depth (z) vs. width (x) generated using Nanotech Electrinica WSxM scanning probe microscopy software version 3.0 Beta 8.1. Scan rate of 1000 Hz is used. **(ii)** Signal response of the QCM sensor with (a) NIP- and (b) MIP-coated electrode to TCAA at concentration level of 1 mg l^{-1}. *[Adapted from ref 52, © Elsevier 2006].*

information is useful for the desorption phenomenon of Cu (II) ions from the imprinted polymers. In contrast, at higher pH (3–5) a pronounced frequency change was observed. This change can be attributed to the deprotonation of carboxyl or hydroxyl functional groups in the imprinted polymer that becomes negatively charged at this pH. Thus, the attraction between Cu (II) ions and imprinted film is enhanced; which, first increased the recognition sites for Cu (II) ions and, second, enhanced the adsorption of Cu (II) ions. The optimal pH turned out to be six because above this level, copper starts to precipitate as the hydroxide. Such a response behavior is even more remarkable given the fact that the copper ions are much smaller than the monomer components used. This further corroborates the importance of the noncovalent interaction pattern underlying the respective recognition processes.

Furthermore, QCM sensors also allow for direct measurement of the kinetics of the underlying system. It was observed that at higher concentrations of Cu (II) ions, the sluggish sensor response can be attributed to the kinetic restriction of the Cu (II) reaction with IIP due to overloading the polymeric matrix with Cu (II) ions. Hence the original first-order reaction changes to a zero-order reaction, where the reaction rate becomes independent of the substrate concentration. The data obtained from the designed MIP-QCM sensor pertains to high-selectivity for Cu (II) ions, sharp response at optimal concentration range, reproducible sensor signals, broad linear range, that is, 0.001–50 μM for analyte concentration, and lower detection limit of 8×10^{-4}μM (S/N = 3). The success of this MIP-QCM sensor can be estimated by the feasibility of Cu (II) ions determination in wastewater [53]. A schematic of IIP fabrication on QCM electrodes and a selectivity pattern of Cu (II)-IIP-QCM are presented in Figs. 9.11(a) and (b).

Molecular imprinting is also successfully applied for development of metal chelating monomers as artificial receptors and their real-life applications via mass sensitive devices. A. Ersoz et al. [54] used a polymer system comprised of methacryloylamidohistidine-platinum (II) [MAH-Pt (II)] as metal-chelating monomer and *N-N′*-methylenebisacrylamide (*N-N′*-MBAA) as cross-linker and generated MIP films for 8-hydroxy-2′-deoxyguanosine (8-OHdG) template and determined oxidative stress in real-life samples. MIP films for QCM as well as MIP sorbent

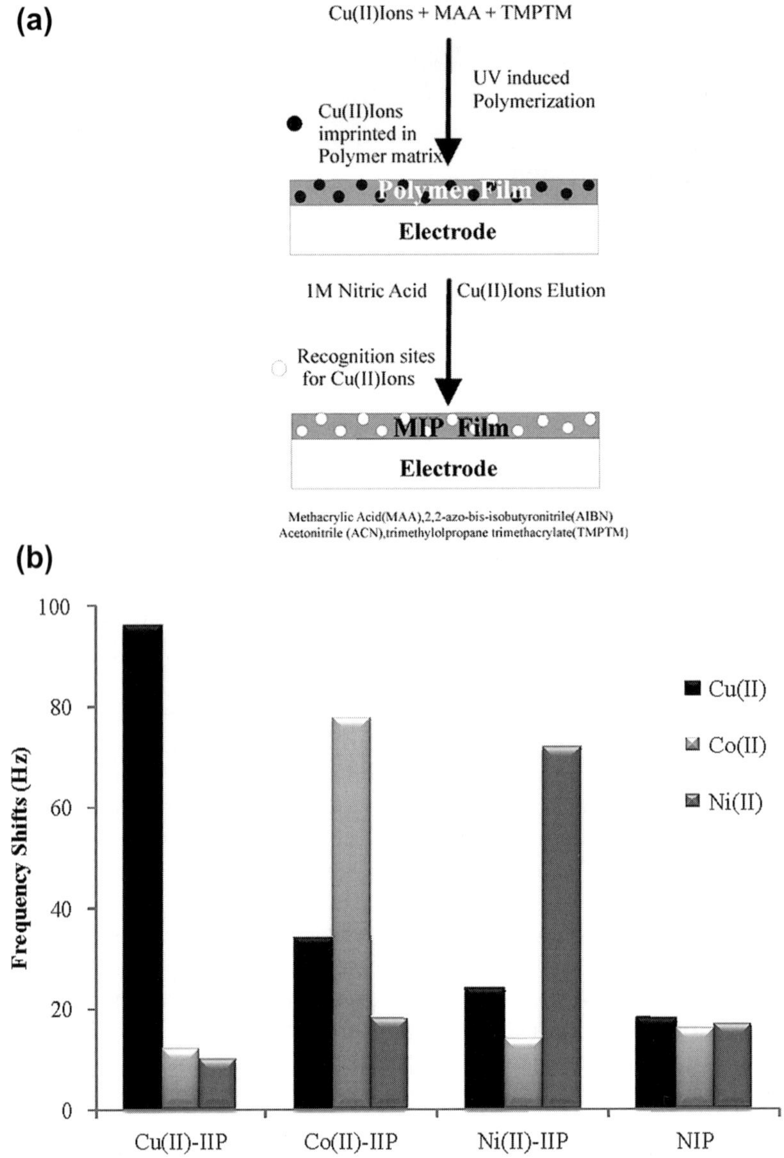

FIGURE 9.11 (a) Schematic representation of Cu (II)-IIP MIP film onto QCM resonator (b) Comparison of QCM sensor selectivity. Experimental conditions: ion concentration 0.1mM, pH 6.0 and temperature 25 °C. *[Adapted from ref 53].*

for solid phase extraction (SPE) of 8-OHdG from blood samples were generated through bulk polymerization. The ligand exchange reaction is responsible for the interaction of 8-OHdG template toward binding sites in8-OHdG-MAH-Pt (II) imprinted film. It approaches equilibrium quickly for imprinted polymer as compared to nonimprinted polymer, therefore resulting in an enhanced frequency response. The mass-sensitive data showed that high recognition sites were observed from MAH-Pt-based MIP for 8-OHdG at affinity constant K_a value ($4.85 \times 10^{10}\ M^{-1}$).

In some cases, interaction materials from nature can be incorporated directly into the polymer matrix. One example for this is the selective detection of DNA nucleobases with imprinted thin films. S. E. Diltemiz et al. [55] demonstrated MIP formation from methacryloylamidoadenine (MA-Ade)-Ethylene glycol dimethacrylate (EDMA) copolymers in dimethyl sulfoxide (DMSO) as the solvent. Template thymine performed a dual role: it was used as both a template and a trigger for noncovalent interactions between nucleobase DNA and MIP film. The thymine MIP sensor showed homogeneous binding affinity for thymine (Ka: $1.0 \times 10^5\ M^{-1}$) when estimated through the Langmuir isotherm, whereas it was heterogeneous for uracil. Similar thymine-imprinted films have also been reported for uracil, ssDNA, and ssRNA [55]. W-H. Zhou et al. [56] reported a mussel-inspired MIP-based QCM sensor for selectively determining 1,3,5-pentanetricarboxylic acid (PTA), an analogue to domoic acid. The proposed MIP strategy is straightforward, cheap, and fast, and the polymerization of monomer dopamine hydrochloride (DH) with template 1,3,5-pentanetricarboxylic acid (PTA) was carried out in ammonium per sulfate (APS) and tris-(hydroxymethyl)-aminomethane-hydrogen-chloride (Tris-HCl) buffer. The resulting polymeric matrix was deposited over QCM and placed in the dark for polymerization, and afterward the template was removed via 5% acetic acid solution in water. The MIP layer formation is described in Fig. 9.12(a). The high hydrophobicity, biocompatibility, and tunable film thickness make this MIP, polydopamine film, a prominent place in sensors. These films showed high selectivity toward domoic acid as from Fig. 9.12(d) when tested against analogous molecules (e.g., *p*-phthalic acid and *o*-phthalic acid, respectively). The MIP sensor shows excellent sensitivity and quantification limit to 5 ppb with a linear range of 0–100 ppb for domoic acid, as predicted from Fig. 9.12(c) and (d). This limit of detection obtained is lying below the maximum allowed by Canadian, American, and EU authorities.

A unique approach of macroporous protein-imprinted films on the surface of QCM has been reported by Zhou et al. [57]. This method involves the use of methyl methacrylate (MMA) as the functional monomer and trimethylolpropane-trimethacrylate (TMPTM) as a cross-linker. As porogen material, calcium carbonate nanoparticles were introduced to the polymer matrix that produces macropores, thus increasing the recognition sites and surface area of the MIP films as presented in Fig. 9.13.

Zhou et al. found that the macroporous network has no influence on the adsorption and desorption of template in MIP; rather the effect was mainly due to hydrophobic interactions. The sensor measurements corroborated better selectivity and affinity of lysozymes template in comparison to nonporous MIP films. This ultimately enhanced the mass sensitivity of the biosensor. Figure 9.14(a) and (b) depicts the surface studies of MIP and NIP films, whereas Fig. 9.14(c) and (d) shows sensitivity, selectivity of lysozyme, and cross-selectivity of lysozyme macroporous MIP toward myoglobin.

A different, very successful strategy for developing controlled thickness and tunable imprinted thin films known as electropolymerized molecular imprinting was established in the last decade [58,59]. Through this strategy, MIP films are grown adherent to electrodes; their thickness can be controlled by altering flowing charge. Thus, it provides a straightforward and facile technique, in which a direct connection between polymer and transducer surface is established and good control over polymer morphology and composition can be attained.

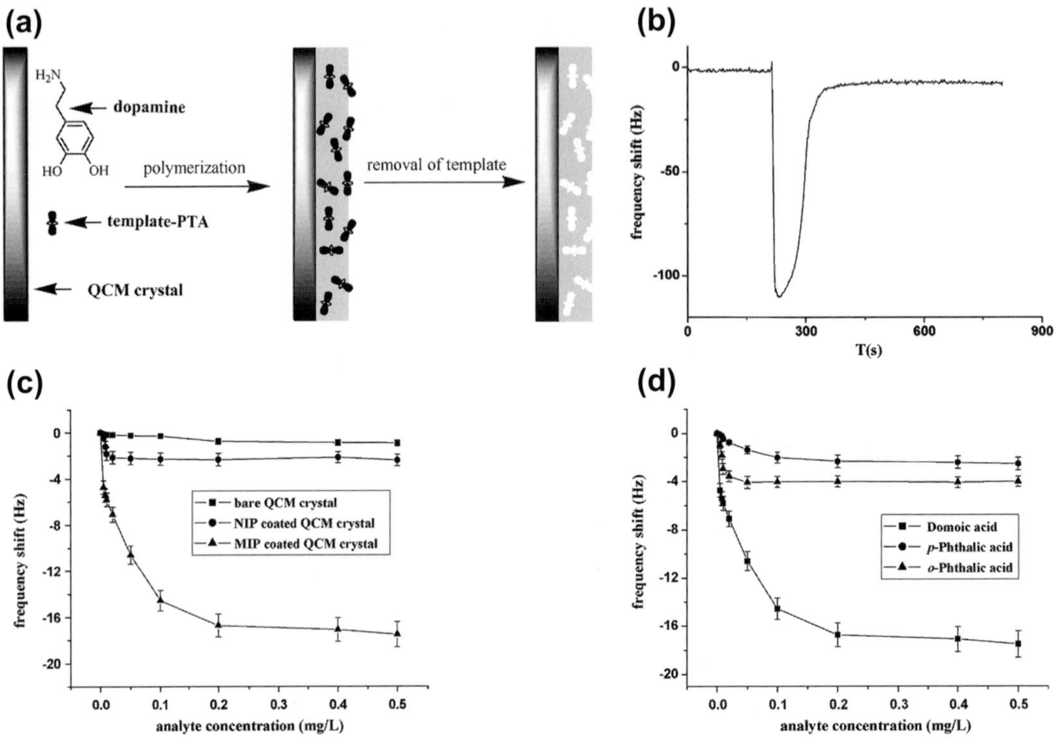

FIGURE 9.12 (a) Mussel-inspired polydopamine MIP protocol generated onto QCM. (b) Sensor response curve towards domoic acid at 0.005 mg/L (c) Change in resonant frequency for NIP, MIP coated QCMs and bare QCM to domoic acid. (d) Sensor effect of m-MIP QCM sensor towards domoic acid and its competing molecules. *[Adapted from ref 56, © Elsevier 2010].*

A variety of monomers and their resulting electrosynthesized polymers have been explored so far, showing that this imprinting polymerization is an easy route for producing biomimetic sensors for diverse classes of analytes [58,60–63]. One distinguishing example of electropolymerized molecularly imprinted polymer (E-MIP) film is a MIP bis-terthiophene dendron on quartz crystal microbalance (QCM) for folic acid template that has been presented by D.C. Apodaca et al. [64] (see Fig. 9.15).

E-MIP characteristics such as film formation, electrodeposition, and viscoelastic behavior were optimized in situ through cyclic voltammetry (CV) directly on the QCM surface. The most interesting feature of this strategy is the use of a bifunctional monomer designed to prepolymerize in the mixture together with the template. Bifunctionality implies that it has the roles of both monomer complexing agent and cross-linker. Size- and shape-specific recognition sites, similar to the folic acid template, were generated in the E-MIP and later used for its mass sensing. The pattern of recognition sites from E-MIP receptors depends on the affinity of folic acid toward terthiophene Dendron. Therefore the best suitable ratio [i.e., 1:2 (folic acid/terthiophene) in pre-polymer complex] was attained by semi-empirical AM1 quantum calculations, whereas the conformational

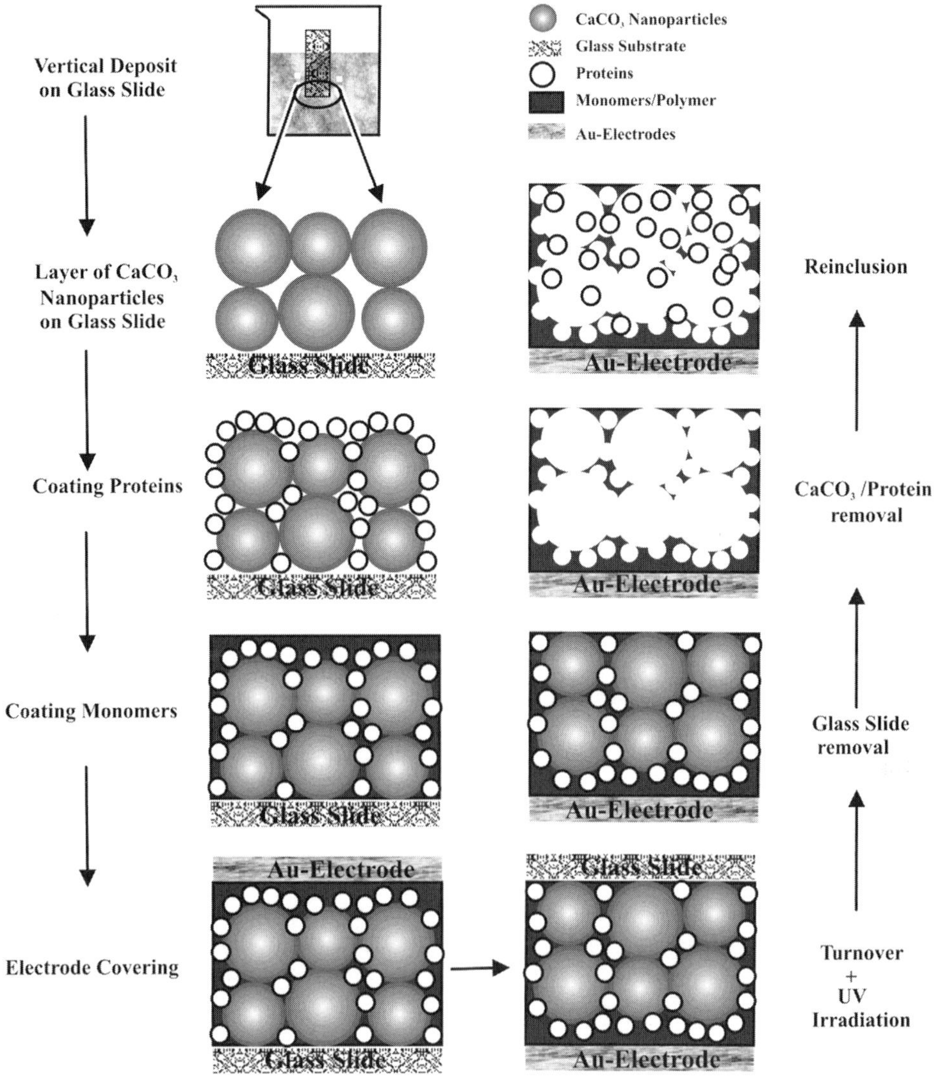

FIGURE 9.13 Fabrication of macro-porous lysozyme imprinted film. *[Adapted from ref 57, © Elsevier 2011].*

changes due to complexation and noncovalent interactions were confirmed by NMR and UV-vis spectrophotometer studies. This sensing approach generated linearity, $R^2 = 0.985$, selectivity over a concentration range of 0–100 μM folic acid with a detection limit of 15.4 μM (6.8 μg) that explicitly reflects its analytical concert, as is also shown in Fig. 9.16(a). Furthermore, Fig. 9.16(b) shows the relative cross-selectivity of the folic acid-imprinted polymer against the three competing molecules that follow this trend: pteroic acid (= 50%) > caffeine (= 41%) > theophylline (= 6%).

2.2. Multisensor Systems Based on MIP-Coated QCM

Within sensor systems, selectivity can be achieved not only by chemical ways, but also by combining a range of sensors to an array resulting in so-called electronic noses [65–69]. Mass-sensitive devices are particularly popular for this purpose [70–72]. E-noses usually consist of dozens or more different sensors used for qualitative and quantitative analysis of complex mixtures in gaseous phase [73]. In this way, they can be regarded as mimicking the natural olfaction system. Usually, in e-noses the selectivity of each sensitive layer need not be very pronounced, as any analyte mixture leads to a pattern of sensor responses which is evaluated by chemometry [74]—for example, principal component analysis (PCA) [75], partial least squares (PLS) [76], or artificial neural networks (NN) [77–79]. Due to their inherent

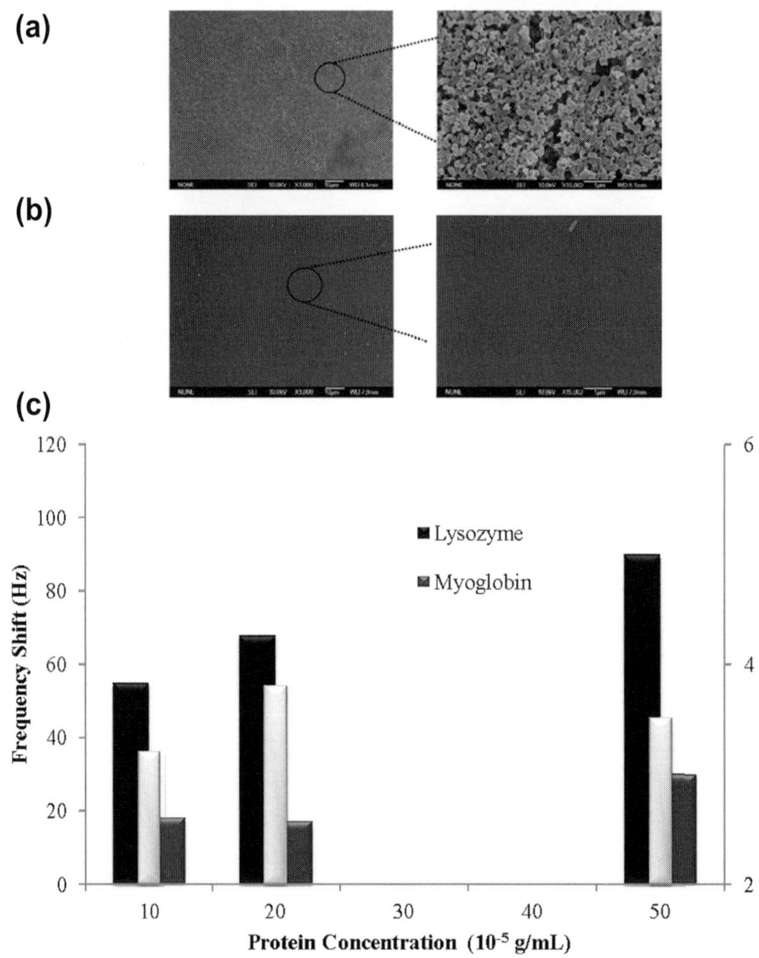

FIGURE 9.14 (a) SEM images of the porous-MIP/NIP films (b). Sensor effect from porous MIP and nonporous MIPs (c & d) The selectivity of the porous −MIP film to lysozyme and control myoglobin in a series of concentration solutions. *[Adapted from ref 57, © Elsevier 2011].*

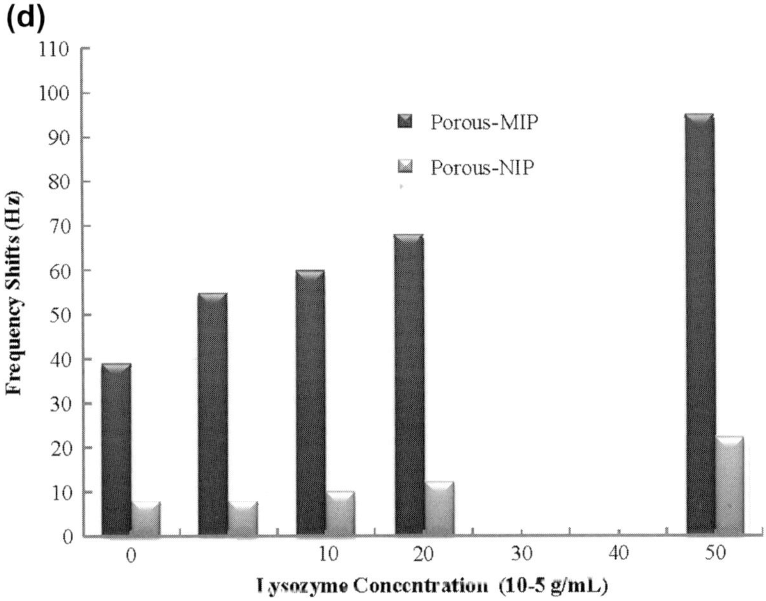

FIGURE 9.14 (*Continued*)

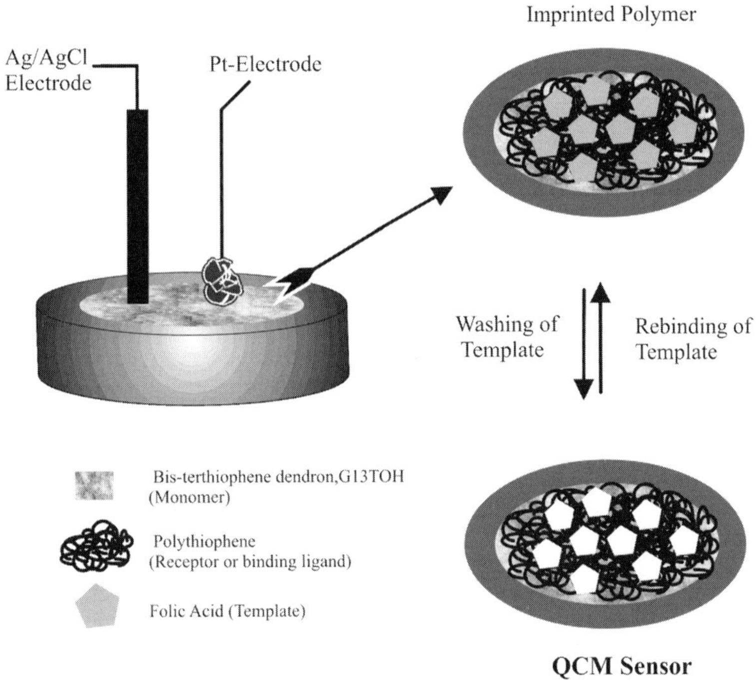

FIGURE 9.15 Schematic illustration of the fabrication of a polythiophene-based sensor for folic acid. *[Adapted from ref 64].*

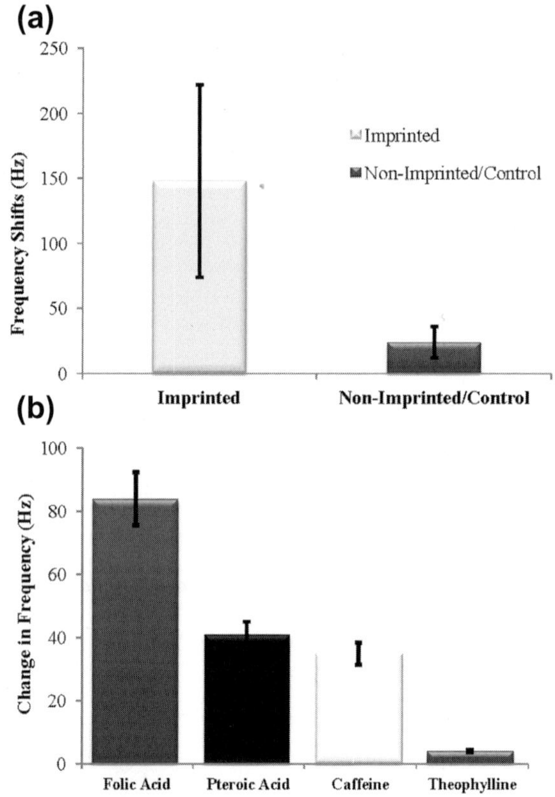

FIGURE 9.16 (a) Response of NIP and MIP sensors at 100 μM folic acid. (b) Frequency shifts of the QCM (5 MHz) upon reinclusion of competing molecules which demonstrate the relative selectivity of the imprinted polymer film towards folic acid. *[Adapted from ref 64].*

multi-analyte abilities combined with ruggedness, e-noses are generally highly promising for real-life sensor applications. Hence, a variety of sensing materials and strategies [80] have been used for their design. However, the number of e-noses combining MIP and QCM are very limited. One reason for this may be the fact that imprinting itself already aims at achieving the highest possible selectivity by the sensor material. Hence, combining it with chemometry may seem superfluous. Nonetheless, when aiming at quantitative multi-analyte detection, the strategy has proved highly successful. An example for this has been reported by F. Dickert et al. [81]: They constructed an e-nose system based on QCM sensor array using imprinted layers for online monitoring of emanations from composting. Three different polymer systems—polyvinyl alcohol (PVA), polystyrene-divinylbenzene (PS-DVB), and polyurethane (PU)—were employed for imprinting different classes of molecules such as water, alcohols, esters, and terpenes, respectively. Remarkable selectivity could be achieved: PVA cross-linked with acrylic acid layer was used as affinity material for water, thus providing a humidity sensor and hardly interacting with other analytes. The PU films imprinted with alcohols—butanol and propanol—showed better sensitivity and selectivity due to the possibility of H-bonding.

Pronounced effect from more nonpolar limonene molecule was observed due to van-der-Waals interactions with PS-DVB imprinted films, thus confirming again that PS-DVB polymer systems are highly suitable for nonpolar analytes (e.g., terpenes). Ethyl acetate bears the least pronounced functionality; therefore, it had been tested with both PU and PS MIPs to obtain complementary interaction patterns. Further progress using this strategy has been reported by P. Lieberzeit et al. [82]; a real-life application is presented for monitoring degradation products from apple, pine leaves, and grass leaves in a closed system for several days. Figure 9.17(a) presents the geometry of the QCM used, whereas Fig. 9.17(b) shows the real image of the mass-sensitive device fabricated for online measurements.

Data analysis revealed excellent agreement between the sensor data and the solvent concentrations determined by GC-MS, as can be seen in Fig. 9.18.

The next step is to simultaneously sense closely related organic vapors rather than very different ones within a complex matrix. One example is described by M. Kikuchi et al. [83], who monitored oxygenated hydrocarbons or generically terpenes by MIP-coated QCM sensor array in gas phase. The sensitive films were casted over Au electrodes of QCM from MAA-EGDMA pre-polymer solution and imprinted by immersing it into template solutions—limonene, limonene oxide, and α-pinene, respectively. The authors observed that the high content of functional monomer corresponded to the large film mass and reduced sensitivity of the MIP. Their optimized film composition (Template-MAA-EGDMA, 1:20) differentiated resembling compounds selectively. Another straightforward noncovalent imprinting approach was published by the authors of [84]: It consisted of optimizing MIP films for closely related terpene molecules and implementing them for real-life application—that is, monitoring terpene emission from odoriferous herbs. Due to the nonpolar nature of the analytes, these e-noses relied exclusively on polystyrene-based MIP. Initial experiments from sensor array detected emanated terpenes selectively below 70 ppm of the analytes. But extending this approach with chemometry for monitoring emanations of fresh and dried herbs pushed the

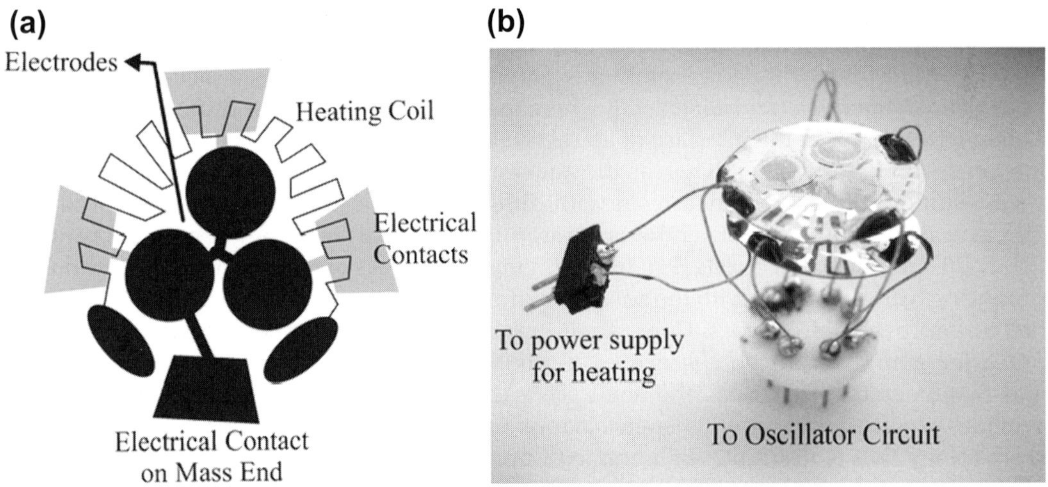

FIGURE 9.17 (a) 10 MHz QCM design: schematic representation of electrodes and heating coil, (b) actual photograph of MIP coated six electrodes QCM sensor array.

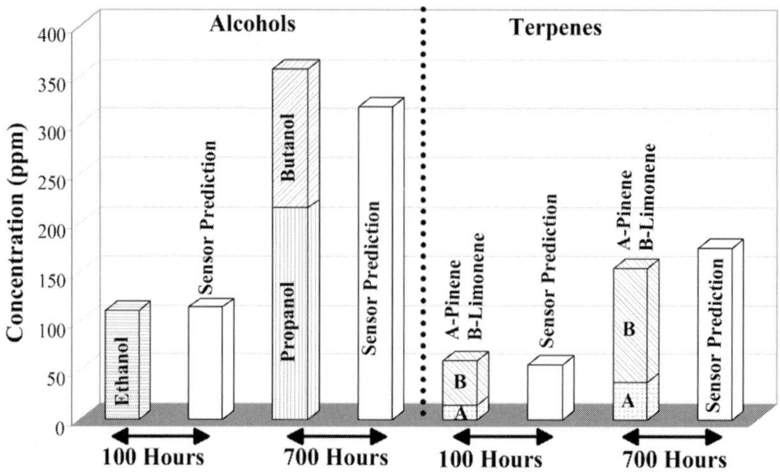

FIGURE 9.18 Comparison of QCM sensor array response with GC-MS at two selected intervals during measurements of pine composting. *(Adapted from ref [82], © Springer 2008).*

detection limit beyond 20 ppm [29]. In that strategy MIP-coated 10 MHz QCM sensor array monitored terpene emanations from selected fresh and dried herbs for several days leading to appreciable agreement between the sensor data and GC-MS, as can be seen in Fig 9.19.

2.3. Imprinted Nanoparticles

As previously mentioned, thin films have become the most widespread way to apply MIP in mass-sensitive sensing. A more recent development beyond this approach is to apply nanostructured surfaces, concretely nanoparticles. Generally, such approaches have brought enormous development in materials science and technology. Because of their size, molecularly imprinted nanoparticles (MINPs) offer more tunable characteristics as compared to bulk polymeric materials [85–90]. To date, MINPs are commonly prepared by the following synthetic strategies [91]:

(i) Emulsion polymerization [92,93], where the template molecule and the host monomers arrange themselves at the interface of an oil/water emulsion. The oil phase usually contains monomers and cross-linker, whereas the aqueous phase contains a polymerization initiator. As a result of polymerization, a resin with different particle sizes (nm range) is produced. Takagi et al. reported the first synthesis of imprinted nanoparticles; they prepared a nanoparticle resin specific for Cu^{2+} ions [94], thus proving the success of emulsion polymerization for producing MINPs [95,96]. With the passage of time, emulsion polymerization has been modified to improve its properties. **Core–shell emulsion polymerization** consists of two steps: preparing core latex in the absence of an emulsifier and then mixing it with a template and monomer that generates the shell. This leads to two improvements. First, the core provides round-shaped monodisperse nanoparticles and therefore very uniform particle morphology that is desirable for a range of applications. Second, the recognition sites are offered by the shell [91] and thus are concentrated in a rather thin layer leading to rapid responses. One example of such a synthesis is described by N. Perez et al. [92]. Initially, seed latex, that is, monodispersed core particles of 30 nm–>1μm in size from MMA, was

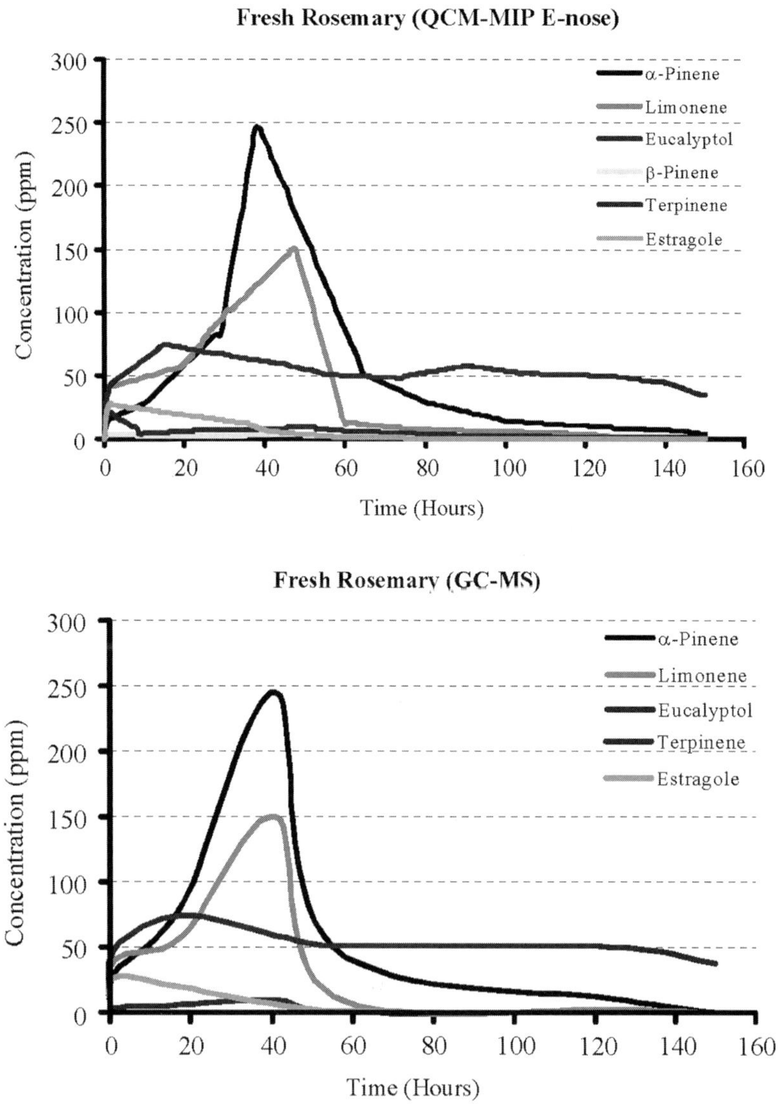

FIGURE 9.19 QCM-MIP e-nose and GC-MS comparison of emanated terpenes from fresh rosemary. *[Adapted from ref 29].*

designed and then mixed with EGDMA monomer and cholesterol as the template. The resulting nanoparticles were found to be 76 nm in diameter and to bear a surface area of 82 $m^2\ g^{-1}$. The cholesterol uptake of these MINPs was better than nonimprinted particles.

Mini-emulsion polymerization marks a step further in synthetic strategy [91,97]. In this strategy, shear forces (i.e., ultrasonification or high-pressure homogenization techniques) are employed to confine the monomer droplet size (50 to 300 nm diameter). For maintaining the steady, homogeneous, dispersed phase, a co-stabilizer is used that also inhibits diffusion, which otherwise can lead to formation of a continuous material. In addition, this also prevents the occurrence of Ostwald ripening during the polymerization process. The major difference between emulsion and mini-emulsion polymerization is the solubility of the initiator. In the former the initiator is soluble only in the dispersed phase, whereas in latter it is soluble in the continuous phase [91]. Another example in which the dimensions of the MINPs are further reduced via mini-emulsion polymerization has been presented by F. P-Capote et al. [89]. The study exposed monoclonal binding behavior of MINPs of 30–150 nm size for S-propranolol template in capillary elecrochromatographic applications [89]. An application of MINPs from this strategy has been implemented by D. Vaihinger et al. [97] for enantioselective determination of L- or D-Boc-phenylalanine anilide. Highly cross-linked polymeric nanoparticles from PMMA-co-EGDMA were synthesized in the presence of a chiral template. The physicochemical characterization and enantioselective binding of MINPs with template showed that the ratio 0.25:1 (PMMA-co-EGDMA) was most suitable for generating MINPs, with effective recognition properties ultimately presenting a portable noncovalent imprinting method for preparing synthetic affinity-based artificial receptors.

(ii) ***Precipitation polymerization:*** In precipitation polymerization [91,98], the polymeric reaction starts in a homogeneous phase, where monomers, cross-linker, and initiator are mixed in a porogenic solvent. Once polymerization has started, the growing polymer becomes insoluble and particles appear in the form of aggregates. They are then stabilized against coagulation, either sterically or by their rigid crosslinked surfaces [91]. Early developments in this scenario have been made by L.Ye et al. [98]. MINPs were prepared from an MAA-co-trimethylolpropane trimethacrylate (TMPTM) mixture containing the templates theophylline and 17-β-estradiol, respectively. This resulted in a good yield $> 85\%$, with an average diameter of 300 nm. By using a radio ligand-binding assay, the selectivity of the MINPs was observed to be extremely high, with $< 1\%$ cross-selectivity between the two templates. MINPs prepared through precipitation polymerization showed unique advantages and simplicity. K. Yang et al. [99] reported the formation of hydrophilic propranolol-imprinted nanoparticles by employing different cross-linkers to methacylic acid (MMA) in acetonitrile reflux via a simple distillation precipitation polymerization method. This does not require any surfactant to be added to the reaction mixture, which makes the strategy rather straightforward. The study revealed that this method can also provide a facile approach to prepare core-shell nanoparticles, where the core maintains imprinted binding sites and the shell offers a hydrophilic surface. Moreover, it proved possible to utilize the residual C=C bonds to immobilize fluorescent markers by Michael addition that did not affect the binding capacity of MINPs. It is also anticipated that using residual C=C bonds via similar Michael addition reaction, one can either incorporate other molecules such as amines or immobilize thiol-functionalized surfaces into MINPs and fabricate sensors for multifaceted applications [99].

Aside from these more popular approaches, several other strategies have been reported which are novel and unique but cannot be included in the preceding categories and so are discussed here. For instance, MINPs have also been prepared by combing molecular imprinting and self-assembly of block copolymers [100]. For this purpose, free radical polymerization was used to prepare di-block copolymer—poly[(tert-butyl methacrylate)-block-(2-hydroxylethyl methacrylate)] (PtBMA-b-PHEMA) functionalized by 2-acrylamido-6-carboxylbutylamidopyridine (ACAP) and cross-linked with methacryloyl. The resulting material was able to interact with 1-alkyluracil or 1-alkylthymine derivatives in chloroform to form triple hydrogen bonding complexes. By the addition of cyclohexane, the mixture yielded well-dispersed spherical micelles, imprinted with uracil or thymine within the core. Upon completion of cross-linking in the solution, recognition sites were developed that also made the micelle core-shell structure stable for the subsequent binding/rebinding process. Another exceptional and straightforward approach in which MIP films are generated on nanoparticles has been described [101]: Bifunctional (molecularly imprinted and magnetically separable) Fe_3O_4/imprinted/core-shell nanoparticles have been reported by surface-initiated atom transfer radical polymerization (ATRP), a new class of controlled living radical polymerization. Initially, a silica and then a BPA-MIP layer of controlled thickness were grown over the Fe_3O_4 nanoparticles of 300 nm size. The MIP polymer system was based on a 4-VP-co-EGDMA polymer matrix and showed exclusive recognition properties for respective analyte, that is, Bisphenol A (BPA). In the same way, C-H Lu et al. reported surface reversible addition-fragmentation chain-transfer (RAFT) polymerization techniques [101,102] for synthesis of surface-imprinted core-shell silica nanoparticles. In this strategy, trichloro(4-chloromethylphenyl)silane was immobilized on the surface of SiO_2 nanoparticles to produce chloromethylphenyl functionalized silica (silica-Cl), which was further reacted with PhC(S)SMgBr to prepare functionalized RAFT agent. Its RAFT polymerization was carried out in 2,4-dichlorophenoxyacetic acid as template and with 4-VP-EGDMA copolymer that finally led to the formation of surface-imprinted core-shell nanoparticles. In addition, D. Gao et al. [103] presented surface monomer directing highly dense imprinting for 1,2,3-trinitrotoluene (TNT) molecules at the surface of silica nanoparticles. This strategy of developing a functional monomer layer at the surface not only showed the occurrence of imprinting but also proved the insertion of template molecules into the polymer shells of silica during the polymerization step. Surface precipitation polymerization has unique advantages because of its simplicity and general applicability for different molecular imprinting systems.

Generally, combining MINPs and QCM enhances not only sensitivity, but also selectivity. On the other hand, nanoparticle inclusion to polymer matrix increases the surface area that provides more binding sites and ultimately increases the sensitivity of the underlying sensor system. One such example is reported by G. Sener et al. [104], who developed a QCM nanosensor for lysozyme detection. Imprinted nanoparticles of dimensions 50 nm were synthesized via two-phase mini-emulsion polymerization. A monolayer of MINPs was achieved by immobilizing nanoparticles on the surface of QCM by drop-coating, followed by removing the template with 1M NaCl solution. It could be shown that imprinting into nanoparticles increases the surface area of the layer and therefore the number of nanocavities, thus enhancing the amount of available binding sites, leading to a detection limit of 1.2 ng/mL lysozyme with relatively low cross-sensitivity toward competing molecule albumin. Another application of MINPs in mass sensing is presented by Z. Zeng et al. [86]. In this stratagem, peptide-imprinted nanoparticles from micro-emulsion of acrylamide, ethylene-bis-acrylamide, and tetramethylethylenediamine (TEMED) were prepared. MINPs revealed an antibody-like affinity toward the hydrophilic peptide (GFP-9: H-THGMDELYK-OH, M_w = 1092Da) prepared by inverse micro-emulsion polymerization. In order to adjust the imprint peptide at the interface of water/oil during polymerization, it was coupled to a surface with fatty acids of diverse chain lengths. Hence, complementary binding sites were created in the nanoparticles. These MINPs showed 90–900 nMol binding affinities and high specificity toward respective peptide when evaluated via QCM measurements. This approach can also

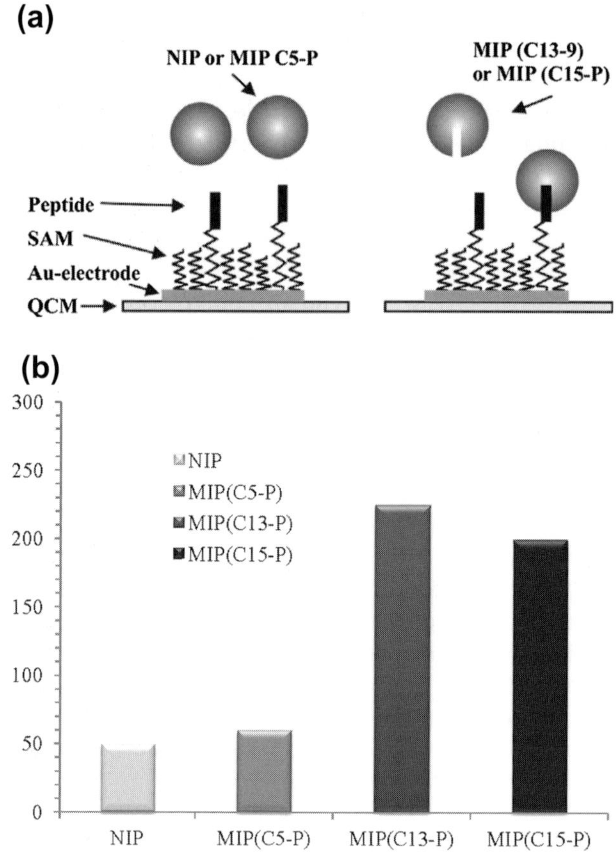

FIGURE 9.20 Interactions between nanoparticles immobilized peptide by QCM. (a) Schematic of QCM experiments for monitoring interactions between NPs (left, NIP or MIP (C5-P) and right, MIP(C13-P) or MIP(C15-P)) and GFP-9 peptide immobilized on QCM electrode. (b) Frequency shift upon injections of NIP, MIP(C5-P), MIP(C13-P), and MIP(C15-P) to QCM sensor cells with GFP-9 immobilized on the surface. Data represent the mean frequency change standard deviation (n = 3) after injection of 64 μg/mL polymeric NP solutions in GFP- 9-immobilized QCM cells. *[Adapted with permission from ref 86, © American Chemical Society 2010].*

design polymer antibodies as replicas for peptides and proteins through an epitope imprinting technique. Figure 9.20 shows the sensor development and outcome.

Besides emulsion polymerization, MINPs synthesized from precipitation polymerization are also known for their simple process. For instance, K. Reimhult et al. [105] proposed a generic assimilation of imprinted nanoparticles into QCM sensors. MINPs from MAA-co-trimethylolpropane trimethacrylate (TMPTM), AIBN cross-linker, and *(R) & (S)*-propanolol as templates were prepared. These pre-made MINPs were then functionalized onto QCM substrate by physical entrapment into a thin layer of poly(ethylene terephthalate) (PET). The deposition and distribution of the particles were controlled via micro-profilometry, and morphology studies were carried out by AFM. Through these investigations, high nanoparticle loading in a stable PET layer was achieved. SEM and dynamic light scattering studies showed that nonimprinted nanoparticles (NINPs) are two times bigger in size than MINPs. This variation in size was attributed to template molecules, which in turn, influenced nucleation and growth of particles during polymerization. Furthermore, the imprinting effect in MINPs and NINPs was confirmed by determining the ratio of radio-ligand bound toward *(R) & (S))*-propanolol, which showed that MINPs bounded more effectively toward *(S)*-propanolol than (R)-propanolol, whereas NINPs showed no binding even at higher concentrations. In contrast to this, QCM measurements with MINP layers showed moderate chiral selectivity that was assumed due to nonspecific interactions resulting from a high concentration of analytes.

Due to the appealing recognition capacity of MINPs, significant developments have been made to estimate their specific binding for biomacromolecules. In a particular case, Y. Hoshino et al. [87] developed an imprinted nanoparticle approach for peptide bee toxin melittin (Mel) with QCMs. The authors described a mild precipitation polymerization approach using acrylamides such as *N*-isopropylacrylamide (NIPAm) for imprinting biomacromolecules, which yields NPs of 10–100 nm dimensions in aqueous solution and in limited surfactant. NIPAm was used as backbone monomer in combination with acrylamide (AAm), acrylic acid (AAc), *N*-(3-aminopropyl) methacrylamide hydrochloride (APS), and *N-tert*-butylacrylamide (TBAm), as hydrogen bonding, negative-charged, positive-charged, and hydrophobic functional monomers, respectively, and *N,N′*-methylenebisacrylamide (BIS) as cross-linker. A schematic is shown in Fig. 9.21(a) where a sensor outcome is described in Fig. 9.21(b). The proposed polymer synthesis does not require organic solvent or a heating step that would be expected to cause denaturation of many biomacromolecules. Following the above strategy, a small combinatorial library of NPs was generated and tested over a QCM sensor for a wide range of concentrations.

A further application of precipitation polymerization for immobilizing MINPs in mass sensors has been presented by R. Schirhagl et al. [106], who developed a double imprinting strategy for immunoglobulins—that is, natural antibodies imprinted in polymer nanoparticles and later on fabricating plastic copies of these MINPs (via stamp method) in a pre-polymerized matrix over QCM surface. MINPs were prepared from monomer poly(vinylpyrrolidone) and methacrylic acid cross-linked with N,N′-(1,2-dihydroxyethylene)-bisacrylamide (DHEBA) in acetonitrile. Initially, the monomer solutions were pre-polymerized in the presence of template antibody, and later MINPs were precipitated in acetonitrile. After recovering and washing, they were then stamped into a polyurethane (PU) pre-polymerized layer coated on QCM electrodes. In this way replicas or "plastic copies" were fabricated which retained the biological functionality as for starting immunoglobins. This MINP encapsulation on pre-polymerized Au-electrode of QCM is described in Fig. 9.22.

The QCM measurements revealed sixfold enhanced sensor effects from plastic antibodies as compared to their natural counterparts attributed to the large surface area provided by MINPs. This subsequently offered more binding sites for templates and increased the possibility of noncovalent interactions—H-bonding or hydrophobic interactions that at the end resulted in a pronounced response. In addition, the designed QCM chemosensor was exposed to human rhinovirus 14 (HRV 14) and foot-and-mouth disease virus (FMDV) to investigate cross-sensitivities. It was observed that neither of the MINP layers responded to FMDV, as summarized in Fig. 9.23.

A more recent development in generating imprinted nanoparticles utilized folic acid as the template in a precipitation polymerization approach of an acrylate system in acetonitrile.

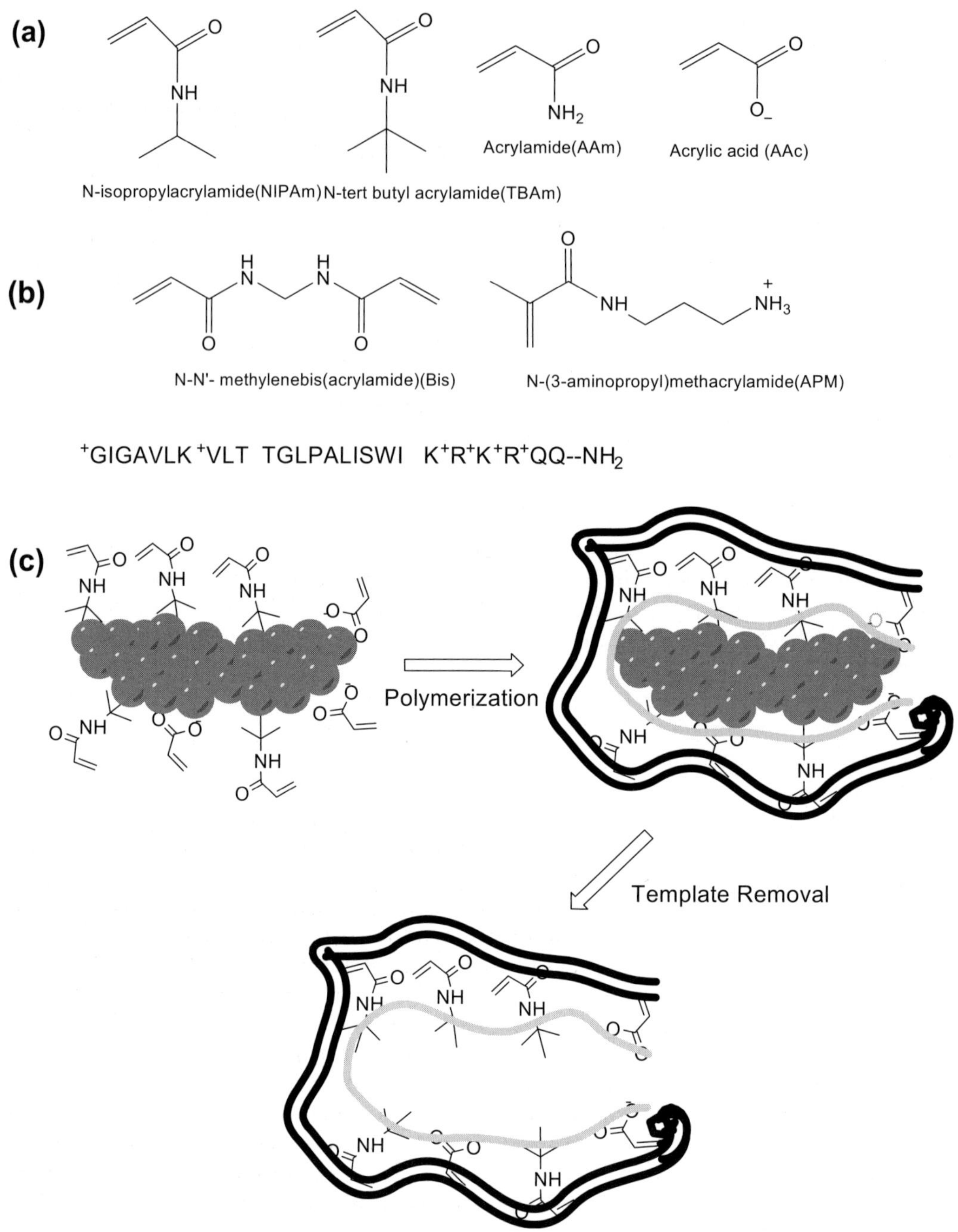

FIGURE 9.21 [i] (a) Monomers used for MINPs synthesis. (b) Amino acid sequence of Mel. (c) Schematic representation of the Mel imprinting process. *[Adapted from ref 87].*

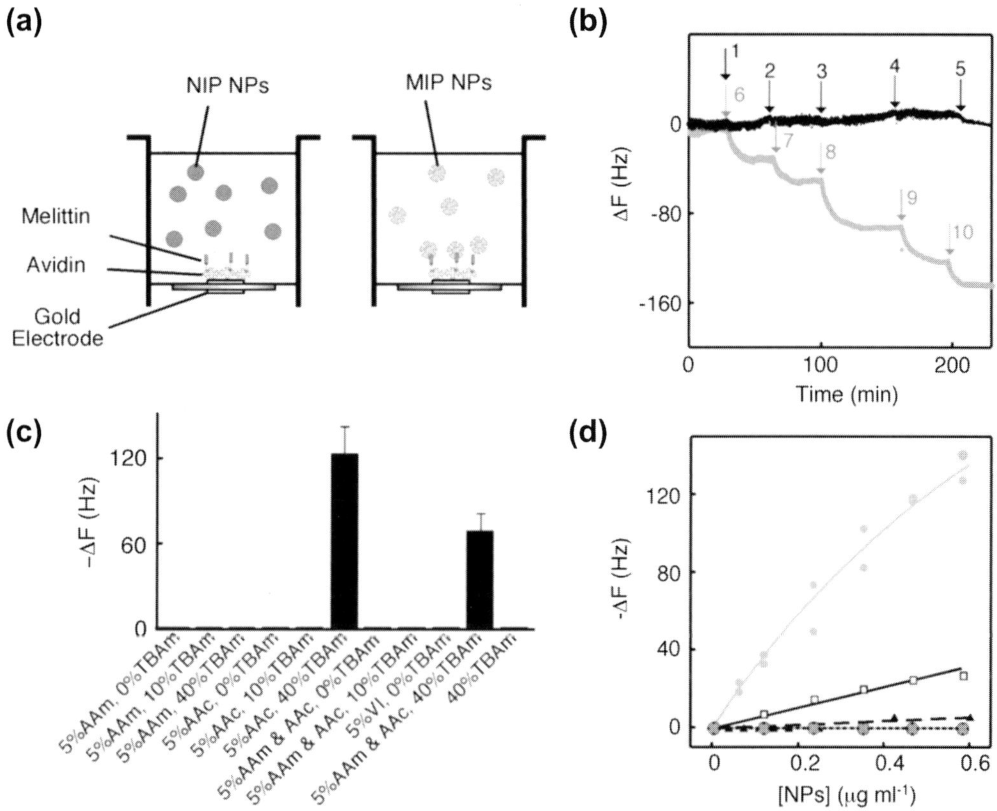

FIGURE 9.21 [ii] Interaction of polymeric NPs with Mel and plasma proteins measured via QCM in water with 6.9 μM SDS at 25 °C. (a) Schematic illustrations of interactions between NIP/MIP (left/right) NPs and Mel immobilized on QCM. (b) Time frame of frequency change (ΔF) of Mel immobilized on a 27 MHz QCM following injection of NIP 14 and MIP 9 solutions. Solutions of NIP (black line) and MIP (gray line) were injected at the time points indicated by the black and gray arrows into the QCM cell. Final concentrations after each arrow are 0.11 (1), 0.18 (2), 0.24 (3), 0.35 (4), 0.42 (5), 0.12 (6), 0.23 (7), 0.35 (8), 0.46 (9), and 0.58 (10) μg mL-1. (c) QCM screening of the library of polymer NPs for their contact with Mel (the mean frequency change [± standard deviation (n = 3)] after injection of 0.6 μg mL-1 NPs into a Mel-immobilized QCM. (d) Binding isotherms of MIP NPs to Mel (gray circles), BSA (white circles), fibrinogen (gray crosses), and γ-globulin (white squares) immobilized on the QCM surface and NIP NPs to Mel (black triangles) immobilized QCM. *[Adapted with permission from ref 87, © American Chemical Society 2008].*

The sensor measurements resulted in improved sensitivity with a detection limit of 5 pmm and appreciable selectivity against folic acid metabolites, as can be seen in Fig. 9.24(a) and (b).

Currently, the majority of MINPs reported are based on organic matrices. However, inorganic systems have also been explored [91,107,108]. One such example is presented by P.A. Lieberzeit et al. [109], who designed imprinted Titania (TiO_2) sol-gel nanoparticles of 50−300 nm dimensions imprinted with capric acid. The latter served as a model compound for oxidation products found during degradation of engine oil. Immobilized on QCMs, these MINPs revealed sensor effects that were two times larger than thin-film layers of the same thickness (see Fig. 9.25(a) and (b)).

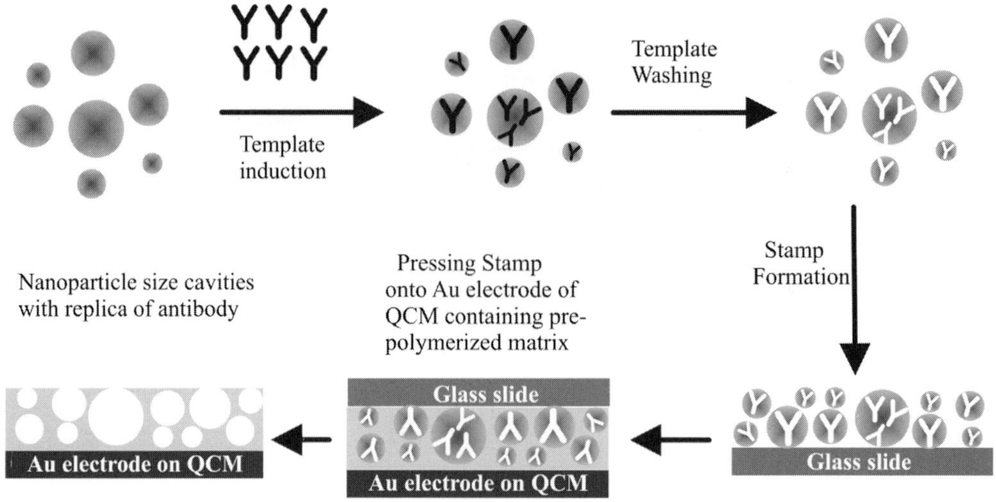

FIGURE 9.22 Imprinting of natural antibody in nanoparticles for generating artificial antibody replicas, followed by stamp formation. Finally pressing stamp into a pre-polymerized layer onto QCM electrode reproduces cavities in the size of the nanoparticles with antibody replicas on their respective surfaces. *[Adapted from reference 106].*

2.4. Bioanalyte MIP

Bioanalysis is gaining increasing importance and has generated substantial demand for easy-to-use, rapid, and selective analytical methods. The main challenges in this case are the broad variety of bioanalytes and their competitors in fairly complex systems as well as their usually low concentrations. In this context, molecularly imprinted polymers combined with mass-sensitive transducers provide a platform to solve many of these questions [110]. Biospecies often are of macromolecular size or even larger and therefore cannot be imprinted into the bulk of the polymers. The reasons for this are as follows: First, there is the slow or impossible template removal to achieve recognition sites in the polymer matrix; second, the rebinding capacity of the template in the bulk imprinted polymer is strongly hindered

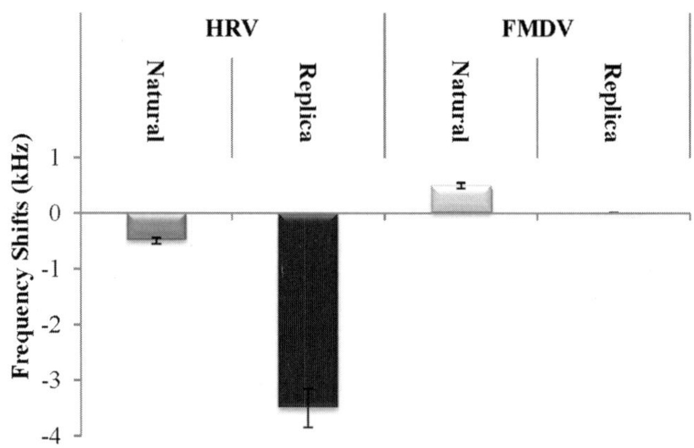

FIGURE 9.23 Sensor responses for HRV 14 and FMDV against natural immunoglobulin and its polymer mimic [106].

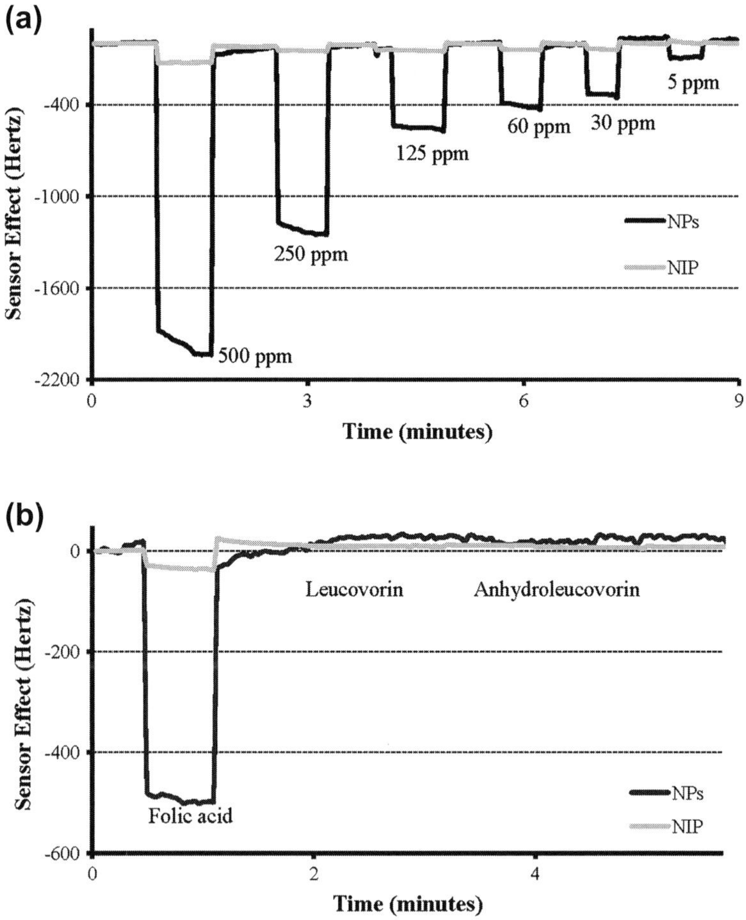

FIGURE 9.24 (a) Sensitivity curve, folic acid imprinted NPs generated from poly methacrylate bulk MIP, layer thickness 15 kHz. (b) Selectivity pattern (at 100 ppm), folic acid imprinted NPs generated from poly methacrylate bulk MIP, layer thickness 15 kHz.

due to diffusion; and third, the layer heights usually exceed the typical working range of QCM devices [111]. The solution to overcome these limitations is to confine the generation of interaction centers to the surface of the pre-polymerized polymer, which also receives increasing attention [112] in other areas of analysis, such as optical sensing, solid phase extraction , solid phase micro-extraction, and chromatographic separation [113]. A schematic model of surface imprinting [114] on a transducer is presented in Fig. 9.26.

Direct surface imprinting [115–117] not only avoids grinding and sieving of a possible bulk polymer, but it also provides sufficient recognition sites [111]. Of course, in bulk imprinting, interaction sites are distributed throughout the entire polymeric matrix. Due to the reasons mentioned above, they are not suitable for large species. However, limiting recognition to the surface does not restrict the applicability of the material: even though the number of possible interaction sites is lower, the size and mass of an individual

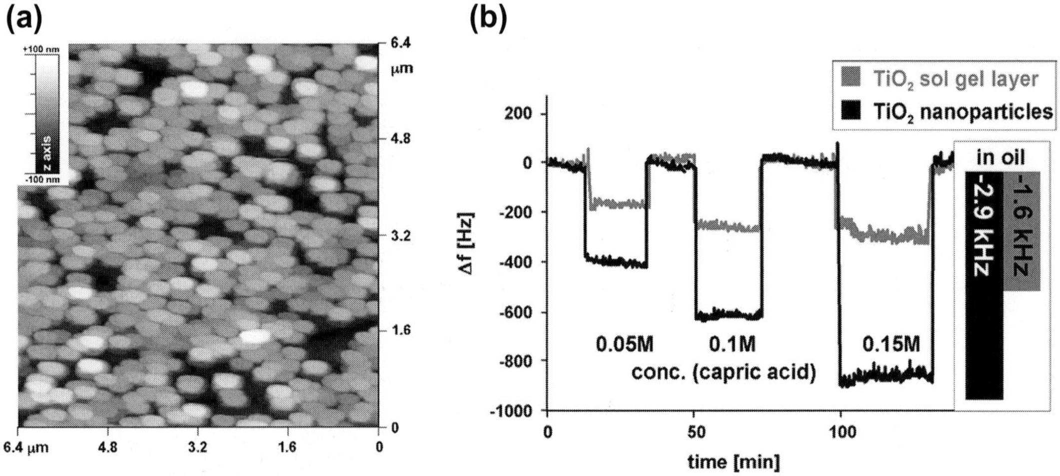

FIGURE 9.25 (a) AFM image of molecularly imprinted TiO_2 nanoparticles on a transducer Surface (b) QCM sensor signals towards different concentrations of capric acid for TiO_2 MIP thin film (grey) and nanoparticle layer (black), respectively. The insert shows the corresponding frequency shifts when changing from fresh oil to waste oil. *[Adapted from ref. 109, © Elsevier 2007].*

analyte is large enough to compensate for this apparent shortcoming. For the first time, such a strategy was reported by A. Aherne et al. [112]. A bacteria-mediated lithographic procedure was developed, where the cells in effect act as temporary protecting groups and structural templates to generate recognition cavities in the growing MIP polymer. O. Hayden et al.

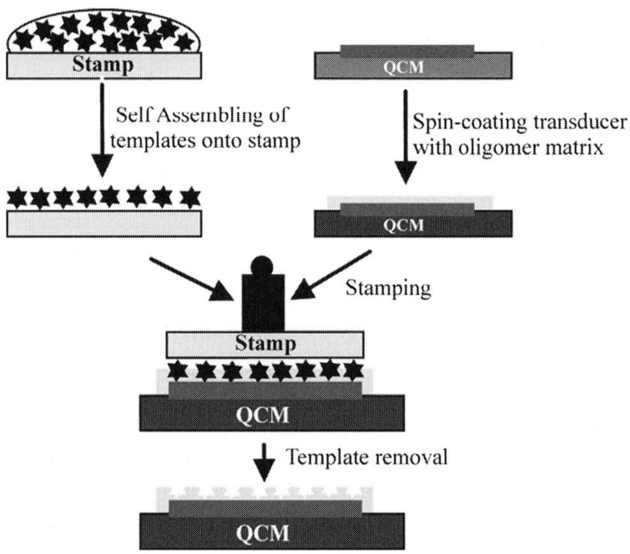

FIGURE 9.26 Schematic showing stamping strategy of imprinting bioanalytes: from self assembling to designing bioanalytes stamp followed by casting on transducer. *[Adapted from ref 114, © Wiley & Sons, 2006].*

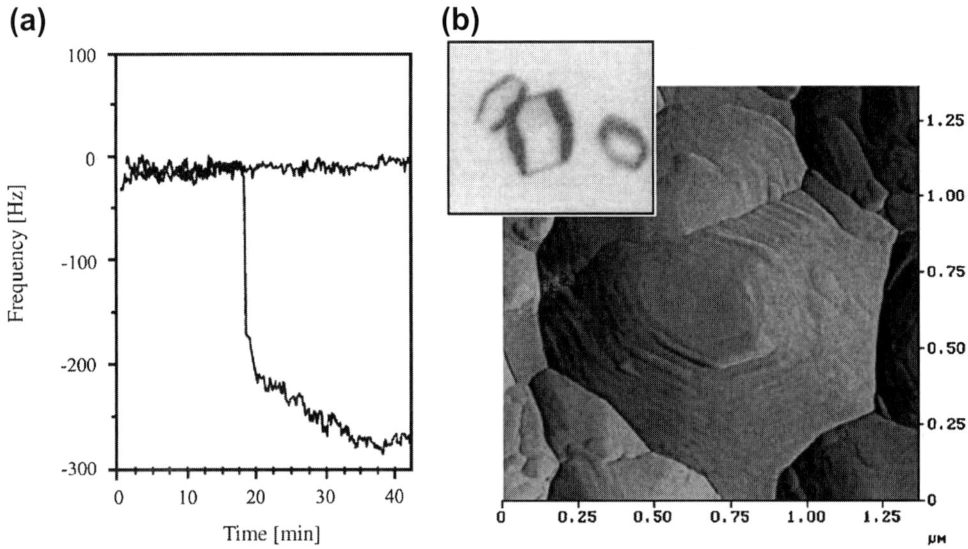

FIGURE 9.27 (a) QCM sensor response to 4 mg/ml lysozyme in PBS buffer with flow rate of 250 ml/min. NIP reference shows no sensor response. (b) Surface studies of lysozyme by AFM, onset shows crystallization in lysozyme on imprinted polymer surface from a solution of 5 mg/ml at 20°C whereas NIP shows no crystallization. *[Adapted from reference 120, © Elsevier 2003].*

[118] have presented an example for using surface MIP for bioanalyte during QCM sensing. They used cell surface-imprinted polyurethanes as artificial receptors for detecting yeasts. The template–MIP interactions in this case are based mainly on H-bonding. Prepolymerized mixtures of polyurethanes were coated on gold electrodes of QCM and stamped by the respective microorganism followed by curing. The differential measurements showed sensitivity as well as selectivity of the imprinted matrix, which could not be observed for the nonimprinted one. The respective interaction processes could also be confirmed by in-situ AFM measurements on the QCM surface [119]. In another surface imprinting approach, the same group [120] used methacrylic acid (MAA)-co-polystyrene (PS) divinylbenzene (DVB), amylase-imprinted and polyurethane systems for mass sensitive detection of lysozyme, tobacco mosaic virus (TMV), and even erythrocytes [120], respectively. Figure 9.27 shows the mass sensing effect observed for lysozyme and its surface studies via AFM.

Two strategies were followed to fabricate master stamps for yeast strains; hydrophilic cell properties were modeled by casting chemical recognition centers via polyurethane, whereas hydrophobic binding areas were addressed using polydimethylsiloxane (PDMS) [121]. Finally, thereafter an artificial yeast stamp was designed by a double imprinting strategy to generate selective polymer layers on the QCM, applying the same imprinting strategy as with native cells. Figure 9.28(a) shows the reproducible sensor effect of imprinted and nonimprinted polymers; that is, it is more selective for *S. cerevisiane* as compared to *S. bayanus*. On the other hand, Fig. 9.28(b) explains the recognition behavior of PU-imprinted QCM against native cells and their plastic replicas. The hydrophilic plastic cells showed enhanced selectivity compared to the native cells due to the possibility of H-bonding and polar effects, thus contributing more in its better recognition.

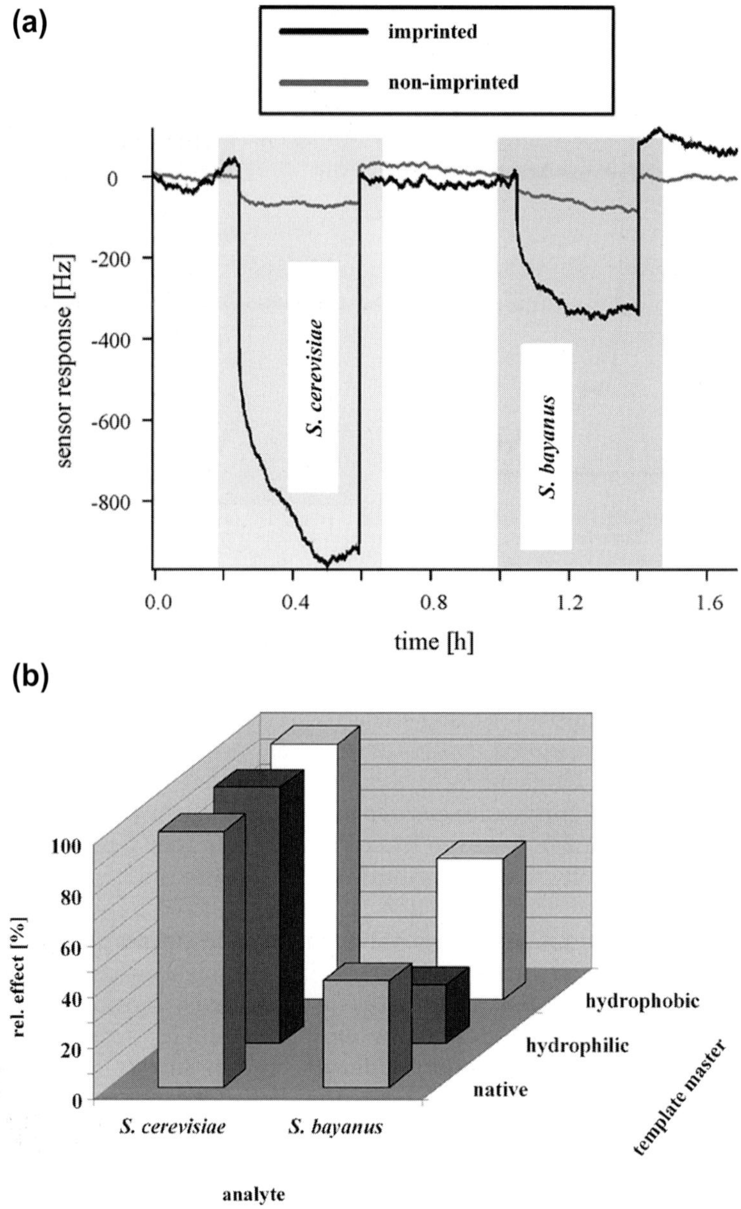

FIGURE 9.28 (a) Sensor response towards 108 particles/mL of *S. cerevisiae* and *S. bayanus* against polyurethane imprinted with plastic cells. In parallel similar unspecific effects on the non-imprinted electrode, selectivity for the template cell strain, *S. cerevisiae*, has been achieved throughout the imprinting procedure, thus leading to the larger frequency shifts (b) Sensor effect of yeast cells [*Templates: native cells of S. cerevisiae and epoxy cells generated by casting an epoxy resin in S. cerevisiae printed polyurethane (hydrophilic) and PDMS (hydrophobic)*] imprinted with polyurethane on QCMs. The cross-sensitivity pattern is normalized to the response towards *S. cerevisiae*. The "hydrophilic" plastic cell yields to enhanced selectivity in comparison to the native cells. *[Adapted from reference 121, © Elsevier 2009]*

In addition, the authors [121] also established a recognition strategy for assessing erythrocytes of different blood groups via the MIP-QCM approach based on polyvinylpyrrolidones. The resulting sensors were even selective enough in distinguishing subgroups A1 and A2. These findings obviously show that surface-imprinted polymers correctly reproduce not only the shape of the respective biospecies, but also its surface functionality. Figure 9.29(a) and b show the selectivity achieved as well as the surface topology of plastic erythrocytes.

The comparably straightforward synthetic approach makes MIP an outstandingly interesting tool in the design of rapid sensor systems for species that are currently not accessible by rapid analysis techniques. Viruses are examples of such analytes. Because of their size, they cannot be detected by optical scattering methods, but they require microbiological

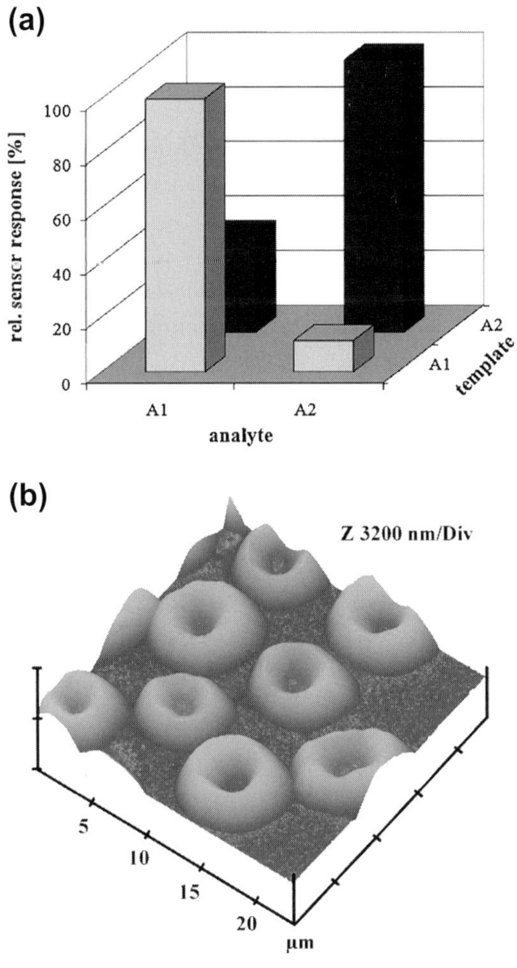

FIGURE 9.29 (a) Selectivity of MIP imprinted with erythrocytes of blood group A1 and A2 (b) Contact mode AFM studies shows erythrocytes replica having identical geometry and characteristics as of natural ones. *[Adapted from reference 121, © Elsevier 2009].*

tests. This can be overcome by surface bioimprinted QCM sensors [122] developed for monitoring viruses such as TMV with a detection range of 100 ng^{-1} to 1 mg mL^{-1} and even implemented for screening of plant sap. Polyurethane-imprinted sensor layers were used with slight modification. A blocker such as glucose, 4-aminobenzoic acid, and 4-aminophenol is introduced in order to avoid covalent linking between proteinogenic virus surface and pre-polymerized polyurethanes. However, there is more behind it: surface imprinting with HRV revealed not only sensitive detection, but also selectivity between different serotypes. Therefore, obviously not only the shape of the respective species is depicted in the polymer, but also its surface functionality. Mass sensing of viruses *[human rhinovirus (HRV, HRV-1a and HRV-16)]*, which are twice the dimension of proteins (myoglobin/hemoglobin) has been reported by generating an MIP-based artificial antibodies strategy for their recognition [114]. Self-assembly of bioanalytes onto a stamp and substantial press on pre-polymerized correspond to a geometrical replica of bioanalytes. In the end this can interact noncovalently with target molecules. Significant differentiation

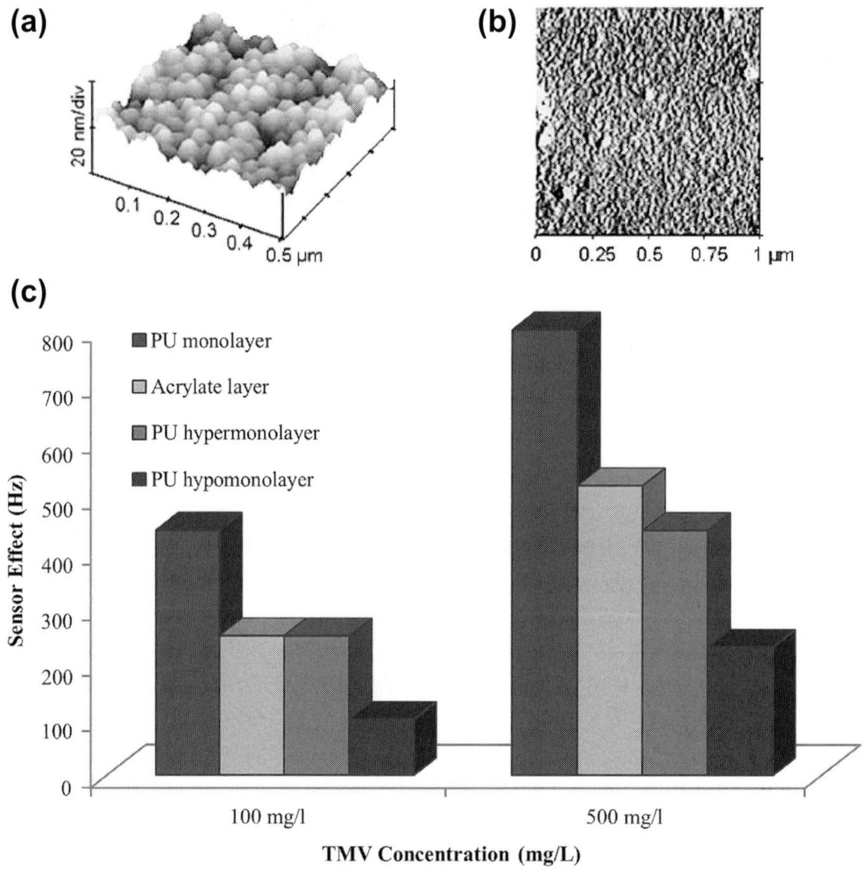

FIGURE 9.30 (a) 3D view of AFM, (b) Surface roughness and **(c)** Sensor response for four MIP sensitive layers exposed to various TMV Concentration. *[Adapted from ref 114, © Wiley & Sons, 2006].*

of alike proteins—hemoglobin/myoglobin as well as different conformations of natural neurotoxins—were predicted through use of this strategy. Figure 9.30(a), (b), and (c) show some AFM studies and sensor measurements.

While looking at enormous applications of bioanalytes in the liquid phase, chemists also explored mass sensors combined with surface imprinting for analysis of biospecies in the gaseous phase. One such example, unique in its type, is described by M. Jenik et al. [28,123]. QCM allergen sensors from pollen imprinted on the surface of polyurethane layers have been described. Polydimethylsiloxane (PDMS) stamps of brick (25 μm) and nettle (15 μm) pollens were developed and casted onto Au-electrodes of QCM already containing pre-polymerized polyurethane matrix. Imprinted pollen grains from the layer were removed by subsequent ultrasonification; however, the thickness of the MIP layer greatly affected template removal. Therefore a layer height of 500 nm was optimized. At higher layer heights,

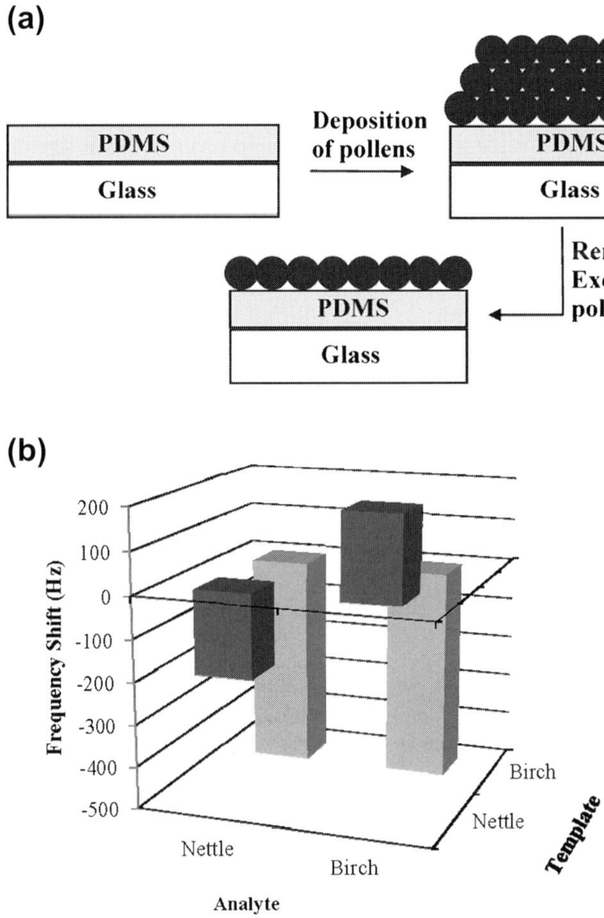

FIGURE 9.31 (a) Schematic of pollen stamp preparation: The pollen grains adhere to PDMS due to adhesive forces. Excess of pollen are blown off with pressurized air generates monolayer of stamp. (b) Selectivity pattern of *Nettle* and *Brich* towards MIP on QCM Sensor. *[Adapted from ref 28 © Springer 2009].*

noncovalent interactions such as H-bonding and $\pi-\pi$ stacking become stronger and hinder template removal. Sensor measurements were performed using a piezo actuator (oscillated at 1000 Hz) containing pollen grains and passing air stream of 10 L/min to supply airborne pollens. The sensor results showed Sauerbrey behavior due to occupation of recognition cavities with pollens and anti-Sauerbrey effect due to increases in the number of pollens in comparison to available cavities. The excess of pollen caused free motion over the QCM surface and showed positive frequency shifts. These sensor behaviors are depicted in Fig. 9.31(a) and (b).

3. CONCLUDING REMARKS AND FUTURE PERSPECTIVES

Molecular imprinting and mass-sensitive devices seem to be two components "naturally" complementing one another: the devices address mass, which is the most fundamental property an analyte can have, whereas the MIP provides the necessary selectivity in shape and functionality. Although substantial progress has been made, especially in the field of small compounds, hardly any sensors that make use of these technologies are commercially available. Besides an inherent conservativism in many application areas, such as diagnostics, issues of reproducibility and sensor standardization also have to be addressed. On the other hand, science still has to tackle a range of issues concerning synthetic techniques allowing for mass-producing MIP layers. In this regard, it is safe to assume that the next years will see a substantial increase in the amount of papers dealing with synthesizing (nano-) particle MIP that can inherently be implemented in technological processes.

References

[1] H. Breer, Anal. Bioanal. Chem. 377 (2003) 427.
[2] S.A. Goff, H.J. Klee, Science 311 (2006) 815.
[3] N. Misawa, H. Mitsuno, R. Kanzaki, S. Takeuchi, Proc. Natl. Acad. Sci. U S A 107 (2010) 15340.
[4] A.G. Mayes, K. Mosbach, TrAC, Trends Anal. Chem. 16 (1997) 321.
[5] S.A. Piletsky, N.W. Turner, P. Laitenberger, Med. Eng. & Phy. 28 (2006) 971.
[6] L. Ye, K. Mosbach, Chem. Mater. 20 (2008) 859.
[7] M. Matsuguchi, T. Uno, Sens. Actuators B 113 (2006) 94.
[8] K. Haupt, Nat. Mater. 9 (2010) 612.
[9] K. Yoshimatsu, K. Reimhult, A. Krozer, K. Mosbach, K. Sode, L. Ye, Anal. Chim. Acta. 584 (2007) 112.
[10] G. Wulff, Angew. Chem. Int. Ed. Engl. 34 (1995) 1812.
[11] G. Wulff, W. Vesper, R. Grobe-Einsler, A. Sarhan, Die Makrom, Chemie. 178 (1977) 2799.
[12] G. Wulff, S. Schauhoff, J. Organ. Chem. 56 (1991) 395.
[13] M. Yan, O. Ramstrom (Eds.), Molecularly Imprinted Materials Science and Technology, CRC Press, 2004.
[14] O. Norrlöw, M. Glad, K. Mosbach, J. Chromat. A 299 (1984) 29.
[15] K. Mosbach, Tren. Biochem. Scien. 19 (1994) 9.
[16] S.H. Cheong, S. McNiven, A. Rachkov, R. Levi, K. Yano, I. Karube, Macrom. 30 (1997) 1317.
[17] M.J. Whitcombe, M.E. Rodriguez, P. Villar, E.N. Vulfson, J. Am. Chem. Soc. 117 (1995) 7105.
[18] B.R. Hart, K.J. Shea, J. Am. Chem. Soc. 123 (2001) 2072.
[19] T. Takeuchi, D. Fukuma, J. Matsui, Anal. Chem. 71 (1998) 285.
[20] D. Batra, K.J. Shea, Curr. Opin. Chem. Biol. 7 (2003) 434.
[21] L. Dickert, P. Achatz, K. Halikias, J. Fresenius', Anal. Chem. 371 (2001) 11.
[22] D.-F. Tai, C.-Y. Lin, T.-Z. Wu, L.-K. Chen, Anal. Chem. 77 (2005) 5140.
[23] D.-F. Tai, Y.-F. Ho, C.-H. Wu, T.-C. Lin, K.-H. Lu, K.-S. Lin, Analyst 136 (2011) 2230.

[24] D.-F. Tai, M.-H. Jhang, G.-Y. Chen, S.-C. Wang, K.-H. Lu, Y.-D. Lee, H.-T. Liu, Anal. Chem. 82 (2010) 2290.
[25] S. Li, Y. Ge, S.A. Piletsky, A.P.F. Turner, Adv. Funct. Mater (2011). In press. DOI: 10.1002/adfm.201100593.
[26] K. Reimhult, K. Yoshimatsu, K. Risveden, S. Chen, L. Ye, A. Krozer, Biosens. bioelectronics 23 (2008) 1908.
[27] P.A. Lieberzeit, K. Halikias, A. Afzal, F.L. Dickert, Anal. Bioanal. Chem. 392 (2008) 1405.
[28] M. Jenik, A. Seifner, P. Lieberzeit, F.L. Dickert, Anal. Bioanal. Chem. 394 (2009) 523.
[29] N. Iqbal, G. Mustafa, A. Rehman, A. Biedermann, B. Najafi, P.A. Lieberzeit, F.L. Dickert, Sensors 10 (2010) 6361.
[30] J. Curie, P. Curie, C.R. Acad, Sci. Gen. 93 (1881) 1137.
[31] J. Curie, P. Curie, Bull. Soc. Min. 3 (1880) 903. de France.
[32] L.G. An, Chim. Phys. Ser. 24 (1881) 145.
[33] K.A. Marx, Biomacr. 4 (2003) 1099.
[34] D.S. Ballintine, R.M. White, S.J. Martin, A.J. Antonio, E.T. Zellers, G.C. Frye, H. Wohltjen, R. Stern, M. Levy, Acoustic Wave Sensor—Theory, Design, and Physico-Chemical Applications, Academic Press, San Diego, CA, USA, 1997.
[35] Y. Arnau, J.V. Montagut, Y. Garcia, Jimenez, Meas. Sci. Technol. 20 (2009). 124004/1.
[36] G. Sauerbrey, Zeitschrift für Physik A Hadrons and Nuclei 155 (1959) 206.
[37] K.K. Kanazawa, J.G. Gordon, Anal. Chem. 57 (1985) 1770.
[38] M.G. Schweyer, J.A. Hilton, J.E. Munson, J.C. Andle, Ultrasonics Symposium (1997). Proceedings, 1997 IEEE, Toronto, Ont., Canada, 1997, p. 371.
[39] H.-J. Janshoff, C. Galla, Steinem, Angew. Chem. Int. Ed. 39 (2000) 4004.
[40] R. Lucklum, P. Hauptmann, Anal. Bioanal. Chem. 384 (2006) 667.
[41] M. Ávila, M. Zougagh, Á Ríos, A. Escarpa, TrAC Tren. Anal. Chem. 27 (2008) 54.
[42] F.L. Dickert, O. Hayden, TrAC Trends in Anal. Chem. 18 (1999) 192.
[43] G. Bunte, M. Heil, D. Röseling, J. Hürttlen, H. Pontius, H. Krause, Prop. Explos. Pyrotech. 34 (2009) 245.
[44] X. Liu, S.-C. Liu, H.S.-O. Ng, Chan, Sens. Actuators B B113 (2006) 234.
[45] N. Tsuru, M. Kikuchi, H. Kawaguchi, S. Shiratori, Thin. Solid Films 499 (2006) 380.
[46] M. Matsuguchi, T. Uno, Sens. Actuators B B113 (2006) 94.
[47] V.M. Mirsky, T. Hirsch, S.A. Piletsky, O.S. Wolfbeis, Angew. Chem. Intern. Edit. 38 (1999) 1108.
[48] W. Senaratne, L. Andruzzi, C.K. Ober, Biomacromol. 6 (2005) 2427.
[49] R. Voicu, K. Faid, A.A. Farah, F. Bensebaa, R. Barjovanu, C. Py, Y. Tao, Langmuir 23 (2007) 5452.
[50] F.S. Kim, G. Ren, S.A. Jenekhe, Chem. Mater. 23 (2010) 682.
[51] X. Bi, K.-L. Yang, Anal. Chem. 81 (2009) 527.
[52] R. Suedee, W. Intakong, F.L. Dickert, Talanta 70 (2006) 194.
[53] Z.-p. Yang, C.-j . Zhang, Sens. Actuators B B142 (2009) 210.
[54] S.E. Ersoz, A.A. Diltemiz, A. Ozcan, R. Denizli, Say, Biosens. Bioelectron. 24 (2008) 742.
[55] S.E. Diltemiz, D. Hur, A. Ersoz, A. Denizli, R. Say, Biosens. Bioelectron. 25 (2009) 599.
[56] W.H. Zhou, S.F. Tang, Q.H. Yao, F.R. Chen, H.H. Yang, X.R. Wang, Biosens. bioelectron. 26 (2010) 585.
[57] D. Zhou, T. Guo, Y. Yang, Z. Zhang, Sens. Actuators B 153 (2011) 96.
[58] I. Malitesta, P.G. Losito, Zambonin, Anal. Chem. 71 (1999) 1366.
[59] T.L. Panasyuk, V.M. Mirsky, S.A. Piletsky, O.S. Wolfbeis, Analy. Chem. 71 (1999) 4609.
[60] L. Özcan, Y. Sahin, Sens. Actuators B 127 (2007) 362.
[61] M. Riskin, R. Tel-Vered, I. Willner, Advan. mater. 22 (2010) 1387.
[62] Y. Liu, Q.-J. Song, L. Wang, Microchem. 91 (2009) 222.
[63] M.R. Aghaei, M. Milani Hosseini, Najafi, Electrochim. Act. 55 (2010) 1503.
[64] D.C. Apodaca, R.B. Pernites, R.R. Ponnapati, F.R. Del Mundo, R.C. Advincula, ACS appl. mater. interf. 3 (2011) 191.
[65] F. De Cesare, E. Di Mattia, S. Pantalei, E. Zampetti, V. Vinciguerra, F. Canganella, A. Macagnano, Soil Biol. Biochem. 43 (2011) 2094.
[66] A. Tuantranont, P. Wisitsora-at, K. Sritongkham, Jaruwongrungsee, Anal. Chim. Acta 687 (2011) 114.
[67] E.H. Oh, H.S. Song, T.H. Park, Enzy. Microb. Technol. 48 (2011) 427.
[68] M. Peris, L. Escuder-Gilabert, Anal. Chim. Acta 638 (2009) 1.
[69] A.D. Wilson, M. Baietto, Sensors 9 (2009) 5099.
[70] T. Alizadeh, S. Zeynali, Sens. Actuators B 129 (2008) 412.
[71] C.-D. Kohl, in: R. Waser (Ed.), Nanoelectronics and Information Technology: Advanced Electronic Materials and Novel Devices, Wiley-VCH, Berlin, 2003, p. 851.

[72] M. Garcia, M. Fernandez, J. Fontecha, J. Lozano, J. Santos, M. Aleixandre, I. Sayago, J. Gutierrez, M. Horrillo, Talanta 68 (2006) 1162.
[73] D.B. Hibbert, Electroanalysis 10 (1998) 1077.
[74] F.L. Dickert, P.A. Lieberzeit, in: C. Steinem, A. Janshoff (Eds.), Piezoelectric Sensors (Springer Series on Chemical Sensors and Biosensors), Springer, Berlin Heidelberg, 2007, p. 173.
[75] Y. Yin, X. Tian, Sens. Actuators B. 124 (2007) 393.
[76] L. Moreno i Codinachs, J.P. Kloock, M.J. Schoening, A. Baldi, A. Ipatov, A. Bratov, C. Jimenez-Jorquera, Analyst (Cambridge, U. K.) 133 (2008) 1440.
[77] S. Borah, E. Hines, M. Leeson, D. Iliescu, M. Bhuyan, J. Gardner, Sens. Instrument. Food Quality and Safety 2 (2008) 7.
[78] N. Rock, U. Barsan, Weimar, Chem. Rev. 108 (2008) 705.
[79] M. Penza, G. Cassano, Sens. Actuators B 89 (2003) 269.
[80] P.A. Lieberzeit, B. Najafi, F.L. Dickert, in: S.G. Pietro Vincenzini (Ed.), Advances in science and technology, CIMTEC, Sicily, Italy, 2008, p. 103.
[81] F.L. Dickert, P.A. Lieberzeit, P. Achatz, C. Palfinger, M. Fassnauer, E. Schmid, W. Werther, G. Horner, Analyst 129 (2004) 432.
[82] P.A. Lieberzeit, A. Rehman, B. Najafi, F.L. Dickert, Anal. Bioanal. Chem. 391 (2008) 2897.
[83] M. Kikuchi, N. Tsuru, S. Shiratori, Sci. Technol. Adv. Mater. 7 (2006) 156.
[84] P.A. Lieberzeit, A. Rehman, N. Iqbal, B. Najafi, F.L. Dickert, Monatsh. Chem. 140 (2009) 947.
[85] I.S. Chronakis, A. Jakob, B. Hagstrom, L. Ye, Langmuir 22 (2006) 8960.
[86] Z. Zeng, Y. Hoshino, A. Rodriguez, H. Yoo, K.J. Shea, ACS Nano 4 (2010) 199.
[87] Y. Hoshino, T. Kodama, Y. Okahata, K.J. Shea, J. Am. Chem. Soc. 130 (2008) 15242.
[88] X. Yu, Y. Li, K. Kalantar-zadeh, Sens. Actuators B 136 (2009) 1.
[89] L. Priego-Capote, S. Ye, S.A. Shakil, S. Shamsi, Nilsson, Anal. Chem. 80 (2008) 2881.
[90] L. Ye, K. Yoshimatsu, D. Kolodziej, J. Da Cruz Francisco, E.S. Dey, J. Appl. Polym. Sci. 102 (2006) 2863.
[91] K.E. Geckeler, H. Nishide, Advanced Nanomaterials, Wiley-VCH, Weinheim (2010).
[92] N. Pérez, M.J. Whitcombe, E.N. Vulfson, J. Appl. Polym. Sci. 77 (2000) 1851.
[93] S.R. Carter, S. Rimmer, IEE Proc Nanobiotechnol. 152 (2005) 169.
[94] H. Kido, T. Miyajima, K. Tsukagoshim, I. Maeda, M. Takagi, Analytical Sci. 8 (1992) 749.
[95] M. Murata, S. Hijiya, M. Maeda, M. Takagi, Bull. Chem. Soc. Japan 69 (1996) 637.
[96] M. Yoshida, K. Uezu, M. Goto, S. Furusaki, M. Takagi, Chem. Lett. 27 (1998) 925.
[97] K. Vaihinger, I. Landfester Kräuter, H. Brunner, G.E.M. Tovar, Macrom. Chem. Phys. 203 (2002) 1965.
[98] L. Ye, P.A.G. Cormack, K. Mosbach, Anal. Comm. 36 (1999) 35.
[99] K. Yang, M.M. Berg, C. Zhao, L. Ye, Macrom. 42 (2009) 8739.
[100] Z. Li, J. Ding, M. Day, Y. Tao, Macrom. 39 (2006) 2629.
[101] C.-H. Lu, Y. Wang, Y. Li, H.-H. Yang, X. Chen, X.-R. Wang, J. Mater. Chem. 19 (2009) 1077.
[102] C.-H. Lu, W.-H. Zhou, B. Han, H.-H. Yang, X. Chen, X.-R. Wang, Anal. Chem. 79 (2007) 5457.
[103] Z. Gao, M. Zhang, C. Wu, G. Xie, D. Guan, Wang, J. Am. Chem. Soc. 129 (2007) 7859.
[104] G. Sener, E. Ozgur, E. Yilmaz, L. Uzun, R. Say, A. Denizli, Biosens. Bioelectron 26 (2010) 815.
[105] K. Reimhult, K. Yoshimatsu, K. Risveden, S. Chen, L. Ye, A. Krozer, Biosens. Bioelectron 23 (2008) 1908.
[106] R. Schirhagl, P.A. Lieberzeit, D. Blaas, F.L. Dickert, Advan. Mater. 22 (2010) 2078.
[107] P.A. Lieberzeit, A. Afzal, G. Glanzing, F.L. Dickert, Anal. Bioanal. Chem. 389 (2007) 441.
[108] Y. Hu, Z. Zhang, H. Zhang, L. Luo, S. Yao, Surf. Interf. Analy. (2011). DOI: 10.1002/sia.3807.
[109] P.A. Lieberzeit, A. Afzal, A. Rehman, F.L. Dickert, Sens. Actuators, B B127 (2007) 132.
[110] P.A. Lieberzeit, F.L. Dickert, Anal. Bioanal. Chem. 391 (2008) 1629.
[111] H. Tsunemori, K. Araki, K. Uezu, M. Goto, S. Furusaki, Bioseparation 10 (2001) 315.
[112] C. Aherne, M.J. Alexander, N. Payne, E.N. Perez, Vulfson, J. Am. Chem. Soc. 118 (1996) 8771.
[113] W.-Z. Xu, W. Zhou, L.-H. Bian, W.-H. Huang, X.-Y. Wu, J. Sep. Sci. 34 (2011) 1746.
[114] O. Hayden, P.A. Lieberzeit, D. Blaas, F.L. Dickert, Adv. Funct. Mater. 16 (2006) 1269.
[115] G. Wulff, T. Gross, R. Schönfeld, T. Schrader, C. Kirsten, Molecular and Ionic Recognition with Imprinted Polymers (ACS Symposium Series, Vol. 703), American Chemical Society, 1998, p. 10.
[116] B. Sellergren, Molecular and Ionic Recognition with Imprinted Polymers (ACS Symposium Series, Vol. 703), American Chemical Society, 1998, p. 49.

[117] H. Zheng, X. Du, J. Phy, Chem. B 113 (2009) 11330.
[118] O. Hayden, F.L. Dickert, Advan. Mater. 13 (2001) 1480.
[119] O. Hayden, R. Bindeus, F.L. Dickert, Meas. Sci. Technol. 14 (2003) 1876.
[120] O. Hayden, R. Bindeus, C. Haderspock, K.-J. Mann, B. Wirl, F.L. Dickert, Sens. Actuators B B91 (2003) 316.
[121] M. Jenik, A. Seifner, S. Krassnig, K. Seidler, P.A. Lieberzeit, F.L. Dickert, C. Jungbauer, Biosens. Bioelectron. 25 (2009) 9.
[122] F.L. Dickert, O. Hayden, R. Bindeus, K.-J. Mann, D. Blaas, E. Waigmann, Anal. Bioanal. Chem. 378 (2004) 1929.
[123] F.L. Dickert, O. Hayden, P.A. Lieberzeit, C. Palfinger, D. Pickert, U. Wolff, G. Scholl, Sens. Actuators B 95 (2003) 20.

CHAPTER

10

Luminescent Optical Sensors Based on Nanoscale Molecularly Imprinted Polymers

Meiping Zhao, Yibin Liu, Dongqin Zhao, Xiaocui Zhu, Shanshan Wang

Beijing National Laboratory for Molecular Sciences, MOE Key Laboratory of Bioorganic Chemistry and Molecular Engineering, College of Chemistry and Molecular Engineering, Peking University, Beijing, China

OUTLINE

1. INTRODUCTION

Molecular imprinting is a useful approach to prepare tailor-made receptors for separation, catalytic reaction, and chemical sensors [1,2]. Compared with natural biological molecules such as proteins and nucleic acids, molecularly imprinted polymers (MIPs) offer distinct advantages in ease of preparation, low cost, and high stability. In combination with nanotechnology, MIPs prepared at nanoscale further provide high specific surface area and better

Molecularly Imprinted Sensors.
DOI: 10.1016/B978-0-444-56331-6.00010-4

accessibility to the binding sites [3–7], which are especially beneficial for developing chemical sensors.

Luminescent optical sensors are a sensing method that transforms the analyte-receptor binding events into measurable light signals, such as fluorescent signals, chemiluminescent signals, and colorimetric signals. The advantages of luminescent optical sensors include high sensitivity, flexibility, environmental stability, ease of miniaturization, inexpensiveness, and nondestructive analyte analysis [3].

There are generally four signal transduction ways to rebind MIPs to the target analytes [3,8]. (1) *A special optical property of the template itself.* To avoid a possible high background due to residual template molecules in the polymer matrix, the MIP can be prepared with a nonfluorescent analyte analogue. For the analyte that does not display desired optical properties, a labeled template or analogue derivative can also be used for a displacement or competitive assay [9–11]. (2) *Incorporation of a responsive chromophore* or *fluorophore into the polymer matrix* [12–16]. When the analytes bind to the imprinted cavities, the microenvironment (e.g., polarity, pH) around the fluoro/luminophore is altered, resulting in quenching or enhancement of the fluorescence or energy transfer [17–21]. (3) *Interfacing the MIP with a transducer element.* The MIP can either be synthesized in situ at the transducer surface [22,23] or preformed, followed by immobilization on the surface of the transducer [24]. (4) *Signals generated by MIP catalytic reactions.* For example, MIP can catalyze some derivatization reactions and produce spectroscopically active species, which may serve an efficient transduction way for the analyte [25]. Employing a suitable optical signal transduction way is greatly beneficial for discrimination between different analytes and thus extends the scope of MIP-based sensors to applications where cross-reactivities or nonspecific interactions would cause problems.

2. FLUORESCENT SENSORS BASED ON MIP WITH DIFFERENT NANOSTRUCTURES

Quantum dots (QDs) are semiconductor nanocrystals ranging from 2–10 nm in diameter. They provide narrow, symmetric, and tunable emission spectra and show higher resistance to photobleaching than organic fluorescent dyes [26–29]. MIP-embedded CdSe/ZnS QDs or CdSe@SiO2@MIP core-shell nanoparticles have all been extensively studied for chemical sensing of various templates.

Lin et al. prepared MIP-embedded CdSe/ZnS QDs for uracil and caffeine using 4-vinylpyridine and ethylene glycol dimethacrylic acid as the functional monomer and cross-linker, respectively [30,31]. The obtained QD/MIPs provided positive fluorescent responses after rebinding with the templates. Stringer et al. developed a MIP sensor for highly explosive compounds by cross-linking fluorescent QDs to the MIP particles [32]. The system was capable of detecting 2,4-dinitrotoluene (DNT) as well as 2,4,6-trinitrotoluene (TNT) in solution with high sensitivity. The lower limits of detection for DNT and TNT were 30.1 and 40.7 μM, respectively. The response time for TNT was only 1 min. A molecularly imprinted fluorescent conjugated polymer material with an intrinsic capability for signal transduction was also synthesized for the detection of TNT and related nitroaromatic compounds. [33]

Diltemiz et al. developed a CdS QD nanosensor for guanosine and its analogues by attaching a new metal-chelating monomer methacryloylamidohistidine-platinium to CdS QDs. [34] The guanosine can simultaneously chelate to Pt(II) metal ion and fit into the shape-selective cavity on the surface of the MIP.

Another MIP-based fluorescence nanosensor was developed by Li et al. [35] by anchoring the MIP layer on the surface of silica nanospheres embedded CdSe QDs via a surface molecular imprinting process. The uniform core-shell lambda-cyhalothrin (LC)-imprinted silica nanospheres (CdSe@SiO2@MIP) are highly photostable. The fluorescence intensity reduces proportionally with the increasing concentrations of LC in the range of 0.1−1000 μM. The method is highly selective and sensitive, with a detection limit of 3.6 μg L^{-1} for LC in water.

Zhang et al. synthesized a QD-based MIP-coated composite for selective recognition of cytochrome c (Cyt) by sol-gel reaction [36]. The fluorescence of MIP-coated QDs was strongly quenched by the template. Under optimum conditions, the linear range for Cyt is from 0.97 μM to 24 μM, and the detection limit is 0.41 μM. Similar QD/MIPs composites were prepared by Lee and co-workers [37] for several protein biomarkers using phase inversion of poly(ethylene-co-vinylalcohol) solutions. QD fluorescence quenching was observed on the binding of analytes to composite MIPs in a concentration-dependent manner. The composite MIP particles were used for the quantitative detection of amylase, lipase, and lysozyme in real saliva samples.

Medina-Castillo et al. prepared a highly magnetic MIP optical sensor for pyrene [38]. Since the optical signal of pyrene in pMIP was not affected by the presence of other luminescent polycyclic aromatic hydrocarbons (PAHs) such as acenaphthene, fluorene, anthracene, benzo[a]pyrene, and benzo[a]anthracene, the target compound can be selectively measured with a detection limit of 7 ng mL^{-1}.

Yang et al. [39] monitored the ligand−receptor interactions by using fluorescent silica nanoparticles (FNPs) labeled ligand as ligands and MIP as artificial receptors. The FNPs had a uniform size of 120 nm and showed strong fluorescence intensity, with excitation and emission peaks at 289 nm and 570 nm, respectively. L-phenylalaninamide was employed as a test ligand and the template for MIP. By using 4,4′-azobis (4-cyanovaleric acid) as the grafting initiator, a MIP receptor membrane was firmly grafted on the support layer by layer. The specific affinity between artificial MIP receptor and ligand conjugated to the FNPs was successfully monitored.

Valero-Navarro et al. synthesized a polyurethane-based magnetic-MIP for selective detection of 1-naphthylamine (1-NA) in drinking water [40]. They first incorporated the magnetite-coated-oleic acid nanoparticles (γ-Fe_3O_4-OA) into a lipophilic polymeric matrix poly-methyl methacrylate-co-ethylene dimethacrylate. Then they encapsulated these magnetic seeds into the MIP structure by precipitation polymerization. The mag-MIP showed a well-organized structure with magnetite in the material and the imprinted polymer coating the magnetic core. Then a magnetic optical sensor MIP was developed by using an optical fiber coupled with a magnetic separator. The hybrid material showed high selectivity for 1-NA and allowed the detection of 1-NA in water even in the presence of four structurally related compounds (2-naphthylamine, 1-naphthol, 2-naphthol, and 1-naphthalenemethylamine), with a low limit of detection of 18 ng mL^{-1}.

A fluorescent MIP capable of picking up glucose from aqueous media was reported by Manju et al. [41]. The fluorescence intensity of the polymer film was found to decrease linearly with the increasing concentration of glucose, facilitating its use as a glucose sensing element. The detection limit of glucose was observed to be 10 μg/mL.

Jenkins et al. [42] developed a portable miniaturized lanthanide europium (III)-based MIP sensor for the detection of the hydrolysis product of the nerve agent soman in water. The imprinted polymer was coated onto the tip of a multimode optical fiber, delivering the excitation light and collecting back the emission from the formation of the hydrolyzed soman-europium (III) polymer in the presence of the analyte. The lanthanide-based MIP sensor achieved a low detection limit of 7 ppt with the response time less than 10 min.

A fiber-optic chemical sensor for cocaine was developed based on a MIP containing a fluorescein moiety as the signaling group [43]. The fluorescent MIP was formed and covalently attached to the distal end of an optical fiber. The sensor exhibited an increase in fluorescence intensity in response to cocaine in the concentration range of 0–500 μM in aqueous acetonitrile mixtures, with good reproducibility over 24 h.

A fluorescent sol-gel MIP sensor for fluorene was obtained by using a fluorene analogue functionalized silane [44]. The template was removed by cleavage of the carbamate linkage, leaving an amine group for attachment with the environmentally sensitive fluorescent probe 7-nitrobenz-2-oxa-1,3-diazole (NBD). Fluorene binding alters the local polarity around the NBD probe and induces a change in fluorescence intensity. The sensing films deposited on glass slides exhibited rapid response (<60 s) and low detection limit (<10 ppt) to fluorene. The MIP sensor had higher selectivity for fluorene than the structurally related compounds such as naphthalene, fluoranthene, and anthracene. A limitation of the sensor is the lack of reversibility following fluorene binding.

Güney et al. developed an MIP membrane fluorescent sensor for Hg^{2+} detection [45]. The polymers were synthesized with 4-vinylpyridine as functional monomer and Hg^{2+} as template. 9-Vinylcarbazole was used as both a complex-forming agent and a fluorescence probe. The template was ion-bonded to pyridine, and carbazole groups in the polymer membrane formed within a semicylindrical Teflon mold. After removal of the template by acid treatment, the MIP membranes could be used to directly determine Hg^{2+} concentration in aqueous solutions by monitoring of the fluorescence intensity of the carbazole groups quenched upon complex formation with metal ions. Recently, a simple chemical etching approach to generate "ion-imprinted" sites on the surface of CdTe QDs was also developed as a fluorescence turn-on probe for Cd^{2+} [46].

Tan et al. [47] reported an ion-imprinted mesoporous silica-based fluorescence turn-on sensor array for discrimination of Zn^{2+} and Cd^{2+}. The fluorescent functional monomer contained an 8-hydroxyquinoline moiety, which was covalently anchored in the inorganic mesoporous silica matrix. The binding of metal ions to the imprinting site was directly transformed into fluorescence signals.

Tao et al. [48] proposed an innovative sensor strategy for selective and quantitative determination of proteins by introducing luminescent reporter molecules to the protein-imprinted xerogels to transduce the binding event into an appropriate signal. The method achieved a limit of detection of about 2 pg/mL.

Sánchez-Barragán et al. [49] proposed a novel concept for optosensing by introducing a heavy-atom effect in the MIP sensing system. The polymer allows one to perform room-temperature phosphorescence (RTP) transduction of the analyte. The noncovalent MIP was synthesized using tetraiodobisphenol A as one of the polymeric precursors and fluoranthene as template. Once recognized by the MIP, the iodide included in the polymeric structure induced efficient RTP emission from the analyte in the presence of an oxygen scavenger. The developed optosensing system has demonstrated a high specificity for fluoranthene against other polycyclic aromatic hydrocarbons. The detection limit for the target molecule was 35 ng/L. The synthesized sensing material showed good stability and reusability.

3. CHEMLUMINESCENT SENSORS BASED ON MIP WITH DIFFERENT NANOSTRUCTURES

MIP-capped Mn-doped ZnS QDs was synthesized for 4-nitrophenol (4-NP) and applied to the chemiluminescence (CL) system using a KIO_4-H_2O_2 system [50]. The authors believe that the MIP-capped Mn-doped ZnS QDs play a luminophor role in the process of the CL reaction. H_2O_2 directly reacts with $NaIO_4$ to generate $\bullet O_2^-$ radicals, which may produce energy-rich precursors of excited molecules $(O_2)_2^*$. The excited molecules $(O_2)_2^*$ transfer their energy to the QDs and result in a strong light emission. In the presence of the template quencher (4-NP), which was adsorbed on the surface of MIP-capped Mn-doped ZnS QDs, the fluorescence emission was quenched. The CL intensity decreased almost linearly with the increasing concentrations of 4-NP. The application combines the good selectivity of MIP with the high sensitivity of CL. The linear range of the CL system is from 0.1 to 40 μM, and the detection limit can reach 76 nM for 4-NP in the water.

Another positively responsive 96-well microplate CL sensor for determination of deltamethrin was developed by Ge et al. through layer-by-layer self-assembly of CdTe QDs and deltamethrin-imprinted polymers [51]. The water-soluble thioglycolic acid (TGA)-capped CdTe QDs were prepared using NaHTe as the Te precursor. The chemiluminescence intensity had a linear relationship against the concentration of deltamethrin over the range of 0.053–46.5 μg mL^{-1} with a lower detection limit of 0.018 μg mL^{-1}. The chemiluminescent reaction was considered to be initiated by the reaction between TGA and dissolved oxygen, which may produce superoxide radical reaction intermediate. The superoxide ions can easily donate one electron and lead to the injection of an electron from superoxide ion into the 1Se quantum-confined orbital of CdTe QDs. Meanwhile, in the presence of HCl, deltamethrin, and $KMnO_4$, the Cl^- can be converted into hydrochloric radical (Cl*). The formed Cl* radical should be able to inject a hole in the 1Sh quantum-confined orbital of the CdTe QDs at the pH values used. Then the exciton CL reaction arises accompanying the light emission. When the deltamethrin was added and adsorbed on the MIP, the CL intensity increased almost linearly with the increasing concentrations of deltamethrin.

Chang et al. developed a MIP-CL method for detection of proline using precipitation polymerization with hydroxyproline (a structural analogue) as the template [52]. The MIPs were immobilized in microtiter plates (96 wells) for selective adsorption of the analyte (dansyl-proline). After washing, the bound fraction was quantified based on peroxyoxalate chemiluminescence reaction enhanced by imidazole. Under the optimum conditions, the relative CL intensity was linear in relationship to the proline concentration in the range from 1 to 40 μmol L^{-1}, and the detection limit was found to be 0.3 μmol L^{-1}.

Thongchai et al. incorporated a MIP preconcentration step into a CL-microfluidic system for the determination of chloramphenicol in honey samples [53]. The MIP was prepared by using chloramphenicol as the template, diethylaminoethyl methacrylate as the function monomer, ethylene glycol dimethacrylate as the cross-linking monomer, 2,2′-dimethoxy-2-phenylacetophenone as the free radical initiator, and toluene and dodecanol as the solvent. The MIP was preloaded into a 10 mm long, 2 mm wide, and 150 μm deep channel in a planar glass microfluidic device. When the sample containing chloramphenicol was introduced into the microfluidic device, it was first preconcentrated on the MIP and then detected by an enhancement effect on the chemiluminescence reaction of tris(2,2′-bipyridyl) ruthenium(II) with cerium(IV) sulphate in sulphuric acid. A micro-syringe pump was used to pump the reagents. The CL intensity had a linear relationship with the concentration of chloramphenicol in the range from 1.55×10^{-4} to 3.09×10^{-3} $\mu mol\ L^{-1}$, and the detection limit was 7.46×10^{-6} $\mu mol\ L^{-1}$.

Wan et al. synthesized amoxicillin-imprinted polymer with methacrylic acid as the functional monomer and ethylene glycol dimethacrylate as the cross-linker [54]. 3-(3′-nitrophenyl)-5(2′-sulfonylphenylazo)-rhodanine was synthesized as CL reagent. A MIP-CL sensor for the determination of amoxicillin was developed based on the CL reaction of amoxicillin with potassium permanganate in an acidic medium. The sensor displayed excellent selectivity and high sensitivity. The linear response range was from 5.0 ng mL^{-1} to 1.0 μg mL^{-1} and the detection limit was 1.3 ng mL^{-1}. The sensor was applied to the determination of amoxicillin in urine samples with satisfactory results.

A triazophos (TAP)-MIP was synthesized by the precipitation polymerization method [55]. The adsorbed TAP strongly enhances the CL of luminol-H_2O_2. An online enrichment and detection method was then developed for TAP. The change in CL intensity is linearly proportional to the concentration of TAP in the range from 1.0 μM to ~4.0 nM, and the detection limit is 2.5 nM, which is about one order of magnitude lower than that of the conventional CL assay. The method was successfully applied to the determination of TAP in vegetable samples.

4. COLORIMETRIC SENSORS BASED ON MIP WITH DIFFERENT NANOSTRUCTURES

A colorimetric test-system for phenols in water was developed by Sergeyeva et al. based on free-standing MIP membranes [56]. The MIP membranes were obtained by copolymerization of a functional monomer (e.g., itaconic acid, 2-acrylamido-2-methyl-1-propanesulfonic acid, acrylamide, or methacrylic acid) with triethyleneglycoldimethacrylate and oligourethaneacrylate. Phenol molecules were selectively adsorbed by the MIP membranes and quantified using color reaction with 4-aminoantipyrine. The intensity of the membranes' staining was proportional to the phenol concentration in the analyzed sample. The optimized colorimetric test system demonstrated a low detection limit (50 nM), a wide working range (50 nM−0.5 mM), and an excellent storage stability at room temperature (12 months).

A fast and direct competitive bio-mimetic enzyme-linked immunosorbent assay (BELISA) method was developed by Wang et al. for the determination of methimazole in urine sample based on a molecularly imprinted film as an artificial antibody [57]. The imprinted film was directly synthesized on the well surface of polystyrene 96-well plate and showed antibody-like binding ability, rapid adsorption speed, and high stability. The BELISA method had a good selectivity for methimazole over other close structural analogues. The IC_{50} (calculated as the concentration giving 50% inhibition of color development) and the detection limit were observed to be 70 ± 4 μg L^{-1} and 0.9 ± 0.04 μg L^{-1}, respectively. The method was applied to the determination of methimazole in spiked urine sample with satisfactory recoveries, and the imprinted film could be reused 20 times without loss of sensitivity.

Hong and co-workers developed an on-chip MIP biosensor for rapid detection of anesthetic propofol optically at 655 nm after the reaction with color reagent [58]. The detection limit of propofol is 0.25 ppm, with a linear detection range from 0.25 to 10 ppm.

5. OTHER MIP OPTICAL SENSORS

Surface-enhanced Raman scattering (SERS) in combination with MIP measurement allows detection of an analyte based on the specific vibrational spectra characteristic of the compound adsorbed on the MIP. The interrogation of MIP layers by SERS was first reported by Wulff and co-workers [59]. Recently, Kantarovich et al. [60] developed a MIP-based SERS biochip. The MIP solution droplets were printed using a pipet or a nano fountain pen on SERS-active surfaces. The results have shown that SERS can be used to monitor MIP-based sensors on microscale. Later, they [61,62] fabricated a chemical nanosensor based on composite MIP particles and SERS for (S)-propranolol with a detection limit of 0.1 μM. The molecularly imprinted coreshell nanocomposite particles were synthesized by precipitation and seeded emulsion polymerization. The layer of gold colloids located between the polymeric core and the MIP shell enhanced the signals by several orders of magnitude in a single molecularly imprinted composite particle by contrast of the conventional Raman measurements on plain MIP particles and on bulk propranolol.

Yang et al. reported a method for the ultra-trace detection of TNT on p-aminothiophenol-functionalized silver nanoparticles coated on silver molybdate nanowires based on SERS [63]. The system presents optimal imprint molecule contours, with the p,p′-dimercaptoazobenzene (DMAB) forming imprint molecule sites and cross-linking the silver nanoparticles deposited on the silver molybdate nanowires. Anchoring of the TNT analyte at these SERS "hot spots" leads to a pronounced intensification of its Raman emission due to p-donor-acceptor interactions between the p-acceptor TNT and the p-donor DMAB. The detection limit for TNT was as low as 10^{-12} M. With this new type of SERS substrate, the silver/silver molybdate nanowires complex can yield new silver nanoparticles during the detection process, which makes the Raman signals very stable.

Holthoff et al. reported on a MIPs-SERS sensor for the detection of the explosive TNT [64]. Micron thick films of sol-gel-derived xerogels were deposited on a SERS-active surface as the sensing layer. Xerogels were molecularly imprinted for TNT using noncovalent interactions with the polymer matrix, and the rebinding of the TNT resulted in unique SERS bands. The MIP-SERS sensor showed reversible and stable response to TNT and achieved a detection limit of 3 μM.

By combination of colloidal-crystal and molecular-imprinting techniques, Hu et al. prepared two imprinted photonic-hydrogel films with theophylline and (1R,2S)-(-)-ephedrine as template molecules, respectively [65]. The imprinted photonic polymer consists of a three-dimensional (3D), highly-ordered and interconnected macroporous array with a thin hydrogel wall, where nanocavities complementary to analytes are distributed. The polymer can quickly and directly report the molecular recognition event based on Bragg diffractive shifts due to the lattice change of their 3-D ordered macroporous arrays upon rebinding to the target molecules. The method was highly sensitive and specific, which provided a rapid and handy approach for stimulant assay and drug analysis in athletic sports.

Gratings are prepared by using microtransfer molding based on SiO_2/Si molds. MIP 2D-DGs exhibit 2D optical diffraction patterns, and the refractive index of the micropatterned MIP material was estimated to be 1.486 via solvent index matching experiments. Target-molecule detection was demonstrated based on the changes in diffraction efficiency.

Zhao et al. report a new type of suspension array for the multiplex label-free detection of biomolecules [67]. The microcarriers of their suspension array are MIP beads with photonic crystal structure, which not only provide diffraction peaks for encoding but also convert the slight physicochemical response signals to the obvious changes of optical signals. This technique combines the advantages of suspension arrays, molecular imprinting, and photonic crystal sensors.

6. CONCLUSION

Optical sensors developed on the basis of MIP at nanoscale exhibit excellent performance in comparison with their conventional-sized counterparts, typically represented by the thorough removal of the template, fast response time, and easiness of combination with other functionalities. Recently published work has shown many encouraging results of various MIP nanostructures, including nanoparticles, nanowires, and nanofilms, for miscellaneous templates and applications. However, some remaining key issues still need to be addressed in the development of nano-MIP-based optical sensors.

The most urgent issue is how to further enhance the sensitivity of the MIP optosensing system. A possible solution is to incorporate efficient signal amplification steps. Another critical problem is to further reduce the nonspecific binding of undesired components by the polymer matrix. To address this problem, functional monomers need to be designed more precisely. Catalytic sensors are often superior to the sensors solely dependent on the binding event itself. Detection of the product from the analyte conversion may increase the sensitivity and avoid interference by nonspecific binding. The reports on such exploration are still very rare. Another important trend in this field goes toward miniaturization, high-throughput, and multisensing. On-chip MIP biosensors have shown the advantages of compact size, high sensitivity, high selectivity, low cost, and fast response. Integration of the latest advanced microfabrication and signal processing techniques may significantly improve the performance. Nano-MIP-based sensors have great potential for wide applications in clinical tests, environmental monitoring, and other analytical areas.

References

[1] G. Wulff, Angew. Chem. Int. Ed. 34 (1995) 1812.
[2] A.G. Mayes, K. Mosbach, Trends Anal. Chem. 16 (1997) 321.

[3] O.Y.F. Henry, D.C. Cullen, S.A. Anal, Bioanal. Chem. 382 (2005) 947.
[4] K. Haupt, Anal. Chem. 75 (2003) 376A.
[5] Y. Ge, A.P.F. Turner, Chem. Eur. J. 15 (2009) 8100.
[6] K. Flavin, M.M. Resmini, Anal. Bioanal. Chem. 393 (2009) 437.
[7] S. Tokonami, H. Shiigi, T. Nagaoka, Anal. Chim. Acta 641 (2009) 7.
[8] K. Haupt, K. Mosbach, Chem. Rev. 100 (2000) 2495.
[9] K. Haupt, A.G. Mayes, K. Mosbach, Anal. Chem. 70 (1998) 3936.
[10] R. Levi, S. McNiven, S.A. Piletsky, S.H. Cheong, K. Yano, I. Karube, Anal. Chem. 69 (1997) 2017.
[11] J.L. Suarez-Rodriguez, M.E. Diaz-Garcia, Biosens. Bioelectron. 16 (2001) 955.
[12] P. Turkewitsch, B. Wandelt, G.D. Darling, W.S. Powell, Anal. Chem. 70 (1998) 2025.
[13] Y. Liao, W. Wang, B.H. Wang, Bioorg. Chem. 27 (1999) 463.
[14] J. Matsui, Y. Tachibana, T. Takeuchi, Anal. Commun. 35 (1998) 225.
[15] W. Wang, S.H. Gao, B.H. Wang, Org. Lett. 8 (1999) 1209.
[16] A.L. Jenkins, O.M. Uy, G.M. Murray, Anal. Chem. 71 (1999) 373.
[17] F.L. Dickert, O. Hayden, Trends Anal. Chem. 18 (1999) 192.
[18] F.L. Dickert, M. Tortschanoff, W.E. Bulst, G. Fischerauer, Anal. Chem. 71 (1999) 4559.
[19] S. Al-Kindy, R. Badia, J.L. Suarez-Rodriguez, M.E. Diaz-Garcia, Crit. Rev. Anal. Chem. 30 (2000) 291.
[20] A. Ellwanger, C. Berggren, S. Bayoudh, C. Crecenzi, L. Karlsson, P.K. Owens, K. Ensing, P. Cormack, D. Sherrington, B. Sellergren, Analyst 126 (2001) 784.
[21] A. Salinas-Castillo, I. Sanchez-Barragan, J.M. Costa-Fernandez, R. Pereiro, A. Ballesteros, J.M. Gonzalez, A. Segura-Carretero, A. Fernandez-Gutierrez, A. Sanz-Medel, Chem. Commun. 25 (2005) 3224.
[22] K. Haupt, K. Noworyta, W. Kutner, Anal. Commun. 36 (1999) 391.
[23] A. Kugimiya, T. Takeuchi, Electroanalysis 11 (1999) 1158.
[24] D. Kriz, O. Ramstrum, A. Svensson, K. Mosbach, Anal. Chem. 67 (1995) 2142.
[25] C.F. Chow, M.H.W. Lam, M.K.P. Leung, Anal. Chim. Acta. 466 (2002) 17.
[26] L. Basabe-Desmonts, D.N. Reinhoudt, M. Crego-Calama, Chem. Soc. Rev. 36 (2007) 993.
[27] T. Jin, F. Fujii, H. Sakata, M. Tamura, M. Kinjo, Chem. Commun. 34 (2005) 4300.
[28] T. Jin, F. Fujii, H. Sakata, M. Tamura, M. Kinjo, Chem. Commun. 22 (2005) 2829.
[29] K. Konishi, T. Hiratani, Angew. Chem. Int. Ed. 45 (2006) 5191.
[30] C.I. Lin, A.K. Joseph, C.K. Chang, Y.D. Lee, J. Chromatogr. A 1027 (2004) 259.
[31] C.I. Lin, A.K. Joseph, C.K. Chang, Y.D. Lee, Biosens. Bioelectron. 20 (2004) 127.
[32] R.C. Stringer, S. Gangopadhyay, S. Grant, Anal. Chem. 82 (2010) 4015.
[33] J.H. Li, C.E. Kendig, E.E. Nesterov, J. Am. Chem. Soc. 129 (2007) 15911.
[34] S.E. Diltemiz, R. Say, S. Buyuktiryaki, D. Hur, A. Denizli, A. Ersoz, Talanta 75 (2008) 890.
[35] H.B. Li, Y.L. Li, J. Cheng, Chem. Mater. 22 (2010) 2451.
[36] W. Zhang, X.W. He, Y. Chen, W.Y. Li, Y.K. Zhang, Biosens. Bioelectron. 26 (2011) 2553.
[37] M.H. Lee, Y.C. Chen, M.H. Ho, H.Y. Lin, Anal Bioanal Chem 397 (2010) 1457.
[38] A.L. Medina-Castillo, G. Mistlberger, J.F. Fernandez-Sanchez, A. Segura-Carretero, I. Klimant, A. Fernandez-Gutierrez, Macromolecules 43 (2010) 55.
[39] L.Z. Yang, Z.L. Chen, Anal. Lett 44 (2011) 687.
[40] A. Valero-Navarro, A.L. Medina-Castillo, J.F. Fernandez-Sanchez, A. Fernández-Gutiérrez, Biosens. Bioelectron. 26 (2011) 4520.
[41] S. Manju, P.R. Hari, K. Sreenivasan, Biosens. Bioelectron. 26 (2010) 894.
[42] A.L. Jenkins, O.M. Uy, G.M. Murray, Anal. Chem. 71 (1999) 373.
[43] T.H. Nguyen, T. Sun, K.T.V. Grattan, S.A. Hardwick, Proc. of SPIE 7653 (2010). 76531V-1−5.
[44] C.A. Carlson, J.A. Lloyd, S.L. Dean, N.R. Walker, P.L. Edmiston, Anal. Chem. 78 (2006) 3537.
[45] O. Güney, F.C. Cebeci, J. Appl. Polym. Sci. 117 (2010) 2373.
[46] P. Wu, X.P. Yan, Chem. Commun. 46 (2010) 7046.
[47] J. Tan, H.F. Wang, X.P. Yan, Biosens. Bioelectron. 24 (2009) 3316.
[48] Z.Y. Tao, E.C. Tehan, R.M. Bukowski, Y. Tang, E.L. Shughart, W.G. Holthoff, A.N. Cartwright, A.H. Titus, F.V. Bright, Anal. Chimi. Acta 564 (2006) 59.
[49] I. Sanchez-Barragan, J.M. Costa-Fernandez, R. Pereiro, A. Sanz-Medel, A. Salinas, A. Segura, A. Fernandez-Gutierrez, A. Ballesteros, J.M. Gonzalez, Anal. Chem. 77 (2005) 7005.
[50] J.X. Liu, H. Chen, Z. Lin, J.M. Lin, Anal. Chem. 82 (2010) 7380.

[51] S.G. Ge, C.G. Zhang, F. Yu, M. Yan, J.H. Yu, Sensor. Actuat B 156 (2011) 222.
[52] P.P. Chang, Z.J. Zhang, C.Y. Yang, Anal. Chim. Acta 666 (2010) 70.
[53] W. Thongchai, B. Liawruangath, S. Liawruangrath, G. Greenway, Talanta 82 (2010) 560.
[54] F. Wan, J. Yu, P. Dai, S. Ge, Anal. Lett 43 (2010) 1033.
[55] C.G. Xie, H.K. Zhou, S. Gao, H.F. Li, Microchim Acta 171 (2010) 355.
[56] T.A. Sergeyeva, L.A. Gorbach, O.A. Slinchenko, L.A. Goncharova, O.V. Piletska, O.O. Brovko, L.M. Sergeeva, G.V. Elska, Mater. Sci. Eng. C 30 (2010) 431.
[57] J.P. Wang, W.W. Tang, G.Z. Fang, M.F. Pan, S. Wang, J. Chin. Chem. Soc. 58 (2011) 1.
[58] C.C. Hong, P.H. Chang, C.C. Lin, C.L. Hong, Biosens. Bioelectron. 25 (2010) 2058.
[59] S. Kostrewa, M. Emgenbroich, D. Klockow, G. Wulff, Macromol. Chem. Phys. 204 (2003) 481.
[60] K. Kantarovich, I. Tsarfati, L.A. Gheber, K. Haupt, I. Bar, Anal. Chem. 81 (2009) 5686.
[61] M. Bompart, Y.D. Wilde, K. Haupt, Adv. Mater. 22 (2010) 1.
[62] K. Kantarovich, I. Tsarfati, L.A. Gheber, K. Haupt, I. Bara, Biosens. Bioelectron 26 (2010) 809.
[63] L.B. Yang, L. Ma, G.Y. Chen, J.H. Liu, Z.Q. Tian, Chem. Eur. J. 16 (2010) 12683.
[64] E.L. Holthoff, D.N. Stratis-Cullum, M.E. Hankus, Sensors 11 (2011) 2700.
[65] X.B. Hu, G.T. Li, M.H. Li, J. Huang, Y. Li, Y.B. Gao, Y.H. Zhang, Adv. Funct. Mater. 18 (2008) 575.
[66] C.A. Barrios, C. Zhenhe, F. Navarro-Villoslada, D. López-Romero, M.C. Moreno-Bondi, Biosens. Bioelectron 26 (2011) 2801.
[67] Y.J. Zhao, X.W. Zhao, J. Hu, J. Li, W.Y. Xu, Z.Z. Gu, Angew. Chem. Int. Ed. 48 (2009) 7350.

CHAPTER

11

Molecularly Imprinted Polymer-Based Potentiometric Sensors for the Determination of Drugs in Pharmaceutical, Biological, and Environmental Samples

Mehran Javanbakht[1], *Behrouz Akbari-adergani*[2]

[1]Department of Chemistry, Amirkabir University of Technology, Tehran, Iran

[2]Food & Drug Laboratory Research Center, Food & Drug Department, Ministry of Health and Medical Education, Tehran, Iran

OUTLINE

Molecularly Imprinted Sensors.
DOI: 10.1016/B978-0-444-56331-6.00011-6

1. INTRODUCTION

The application of potentiometric sensors has become a well-established routine analytical technique in many fields, including clinical and environmental analysis, physiology, and process control. Over the past two decades, imprinted polymers have attracted broad interest from scientists appointed in electrochemical sensor development [1–3]. This attention can be clarified by the serious potential advantages of using molecularly imprinted polymers (MIPs) in place of natural receptors and enzymes such as their better stability, low cost, high selectivity, and easy preparation. Also, MIPs are much more stable to organic solvents, pH, high temperature, and pressure than their biological counterparts. Moreover, the cost of producing these materials is low, and they can be stored in the dry state at room temperature for long periods of time. In molecular imprinting, highly cross-linked polymers can be utilized to form three-dimensional cavities with binding sites, mimicking the specificity of biological molecules, through a course in which functional monomers, such as methacrylic acid (MA), and cross-linkers—for instance, trimethylolpropane trimethacrylate (TRIM) or ethylene glycol dimethacrylate (EGDMA)—are copolymerized in the presence of the selected analyte as template. After polymerization, the template is removed to give binding sites complementary to the template structure. These polymers are often referred to as artificial antibodies.

MIPs possess a high potential as synthetic receptors in numerous applications such as solid phase extraction [4–7], liquid chromatography [8] assays and sensor technology [9–12], and drug delivery systems [13,14]. Recent reviews have been dedicated to the application of MIPs in the sensor development [2–4,10,11,15–22]. Biomimetic sensors using MIPs offer some interesting advantages to typical biological-based sensors, such as those using antibodies or enzymes. The molecular imprinting technique is a powerful method for preparing artificial recognition sites with predetermined selectivity for a wide range of target molecules.

In contrast with the relatively increasing number of MIP reports on electrochemical sensors with electrochemical transduction of capacitive, conductometric, amperometric, and voltammetric, it is surprising that in spite of the relatively simple transduction of the potentiometric signal, only a very few MIP-based sensors have been reported using a potentiometric transducer [21,23].

In particular, the areas of environmental monitoring and food and drug analysis require analytical tools capable of discriminating chemicals with high molecular specificity, considering complex biological fluids and increasing numbers of complex environmental contaminants. Furthermore, environmental, food, drug and process safety control issues favor the application of online in-situ analytical methods with high molecular selectivity. Molecular imprinting has a great variety of applications because of the theoretical lack of restrictions on the shape, size, or chemical nature of the imprinted molecule. The possibility of tailor-made, highly selective artificial receptors makes these synthetic materials ideal recognition elements in electrochemical sensors [24,25].

MIPs play an advantageous role when associated with potentiometric sensors. These electrodes use membranes to enable the selective recognition of a specific ion by transferring it (selectively) across the interface between the sample and membrane phase. This transfer generates a potential difference that is a measure of the activity of the transferred ion.

There are various MIP-based potentiometric sensors that are described by dispersing MIP particles in plasticizer and embedding in polyvinylchloride (PVC) matrix [26,27], forming a glassy membrane [28], assembling the template on the polar surface of indium tin oxide (ITO) glass plate [29], depositing a MIP polymeric film on the gate surface of an ion-sensitive field-effect transistor (ISFET) [30], and preparing the chemically modified carbon paste electrodes (CMCPEs) by mixing graphite/binder paste and MIP as modifier [12,31].

This chapter first provides an overview of the principle of potentiometric sensors and MIPs as recognition elements and then offers a survey of the different methods in MIP-based sensor fabrication and biomedical applications of the potentiometric devices in the determination of drugs in pharmaceuticals, biological fluids, and environmental samples.

2. BASIC PRINCIPLES OF POTENTIOMETRIC SENSORS

In chemical sensors, a chemical signal is created upon the binding of the analyte to the recognition element. A transducer then translates this signal into a quantifiable output signal. The same general principle applies if a MIP is used as the recognition element. A classification of the recognition elements used in sensors can be made on the basis of their physico-chemical/biological/biomimetic origin and catalytic/noncatalytic nature.

Biomimetic recognition elements cover all recognition elements based on host–guest molecular recognition. Host–guest chemistry has been used to develop receptors for simple or complex molecules. MIPs acting as a carrier provide a means to enhance the selectivity of ion-selective electrodes (ISEs) membranes. These materials are designed to carry out specific host–guest interactions between template and carrier. Once the imprinted templates are washed out from the polymeric matrix, they leave behind binding sites with specific shape and functional group complementarity in the polymeric network. These sites have the ability to rebind the analyte, generating an emf variation.

Electrochemical devices usually monitor the current at a fixed voltage (amperometry), the voltage at zero current (potentiometry), or measure conductivity or impedance changes. A potentiometric sensor operates under conditions of zero (or near-zero) current flow and measures the difference in potential between the working electrode and a reference electrode.

Potentiometric sensors are the most important of this class of transducer, in which the cell voltage is measured as a function of the activity of the analyte in the sample.

The essential part of a MIP-based potentiometric sensor is the MIP membrane that is commonly placed between two aqueous phases, that is, the sample and inner solutions that contain an analyte ion. The membrane may be a glass, a crystalline solid, carbon paste, or a liquid. The potential difference across the membrane is measured with two reference electrodes positioned in the respective aqueous phases: Reference electrode 2 || sample solution |MIP membrane | inner solution || Reference electrode 1 in which phase boundaries are indicated by vertical lines and the two parallel lines mark a liquid–liquid interface.

In MIP-based potentiometric sensors, by incorporation of the imprinted polymer as the active ingredient in a membrane of an ISE, new electrochemical transduction by chemical recognition is obtained. Until now, several mechanisms have been suggested for the response mechanism of these types of electrodes. One of these mechanisms is based on the phase boundary theory [32]. Typically, the membrane potential is divided into three separate potential contributions: the phase boundary potentials at both interfaces and the diffusion potential within the ion-selective membrane. However, the membrane's internal diffusion potential is zero, if there is no concentration gradient within the membrane. Under equilibrium conditions, the measured potential (or emf), E, can be expressed by the Nernst equation. The Nernst equation provides the fundamental principle of all potentiometric transducers. A Nernstian response is expected in a potentiometric cell when the liquid junction potential can be neglected and the inner filling solution of the electrode is kept constant [33]. According to this equation, the potential changes are logarithmically proportional to the specific ion activity:

$$E = E_I^0 + \frac{RT}{z_I F} \ln a_I \tag{11.1}$$

where z_I is the charge of the analyte ion and a_I is its activity in the sample solution. R, T, and F are the universal gas constant, the absolute temperature, and the Faraday constant, respectively. It is assumed that the interfacial ion transfer and the complexation processes are relatively fast and, consequently, the equilibrium is held at the interface.

For the development of MIP sensors, it is important that the creation of membrane potential between sample and inner filling solutions do not require the template to be extracted from the membrane. This is an advantage because extraction of the template to leave recognition sites ready for binding is a cause for uncertainty at the determination or a sensitivity-limiting factor [11]. Another unique feature of potentiometry is that the species do not have to diffuse through the membrane, so that there is no size restriction on the template compound.

3. SYNTHESIS OF MIPs AS RECOGNITION ELEMENTS

3.1. Optimization of MIP Formulation

Despite their moderately long history, the development of highly selective imprinted polymers is not an easy task. Several variables, such as amount and nature of monomer, cross-linker, and solvent, affect the final characteristics of the obtained materials in terms of capacity, affinity, and selectivity for the target analyte. Thus, by achieving an optimum combination of cross-linker and functional monomer, nonspecific binding should be able to be minimized. Generally, proper molar ratios of functional monomer to template are very important to enhance the specific affinity of polymers and number of MIPs recognition sites. High ratios of functional monomer to template result in a high nonspecific affinity, while low ratios produce fewer complexations due to insufficient functional groups [34]. For example, four molar ratios of the monomer methacrylic acid (MAA) to the template (metoclopramide) of 4:1, 9:1, 23:1, and 34:1 were used in the synthesis of metoclopramide-molecularly imprinted polymers [5]. The optimum ratio of functional monomer to template for the specific rebinding of metoclopramide was 9:1 (Table 11.1), which had the best specific affinity and the highest recovery of 97.5%, while that of the corresponding nonimprinted polymers (NIPs) was low at 31.0%.

The process of producing a molecular-sensing polymer includes the following steps: (1) choice and preparation of proper agents (monomer, cross-linker, progen) (2) synthesis of the cross-linked copolymers, (3) washing of the polymers for removing the template (4) optimizing of the polymer for molecular selectivity, and (5) use of the polymers in the construction of MIP-based membrane sensors.

3.2. Characterization of Molecularly Imprinted Polymers

The imprinted polymer particles can be characterized by Fourier transform infrared spectroscopy (FT-IR). Usually there are O-H and C=O stretching vibration in the leached MIP, and these peaks in the unleached MIP are shifted to lower frequencies

TABLE 11.1 Effect of Composition of Metoclopramide MIP Particles on the Specific Rebinding

MIP	MAA (mmol)	MCP (mmol)	EGDMA (mmol)	AIBN (mmol)	Rebinding (%) (mean ± SD)[a]
MIP1	0.90	0.217	9.90	0.347	61.5 (± 3.1)
MIP2	2.04	0.217	9.90	0.347	97.5 (± 3.4)
MIP3	5.04	0.217	9.90	0.347	55.2 (± 2.9)
MIP4	7.57	0.217	9.90	0.347	40.9 (± 3.3)
MIP5	2.04	0.217	6.88	0.347	38.0 (± 2.8)
MIP6	2.04	0.217	20.6	0.347	60.3 (± 3.1)
MIP7	2.04	0.217	27.5	0.347	70.6 (± 3.3)
NIP1	0.9	-	9.90	0.347	27.8 (± 1.5)
NIP2	2.04	-	9.90	0.347	31.0 (± 1.4)
NIP3	5.04	-	9.90	0.347	28.5 (± 1.8)
NIP4	7.57	-	9.90	0.347	29.0 (± 1.5)

[a]*Average of five determinations.*

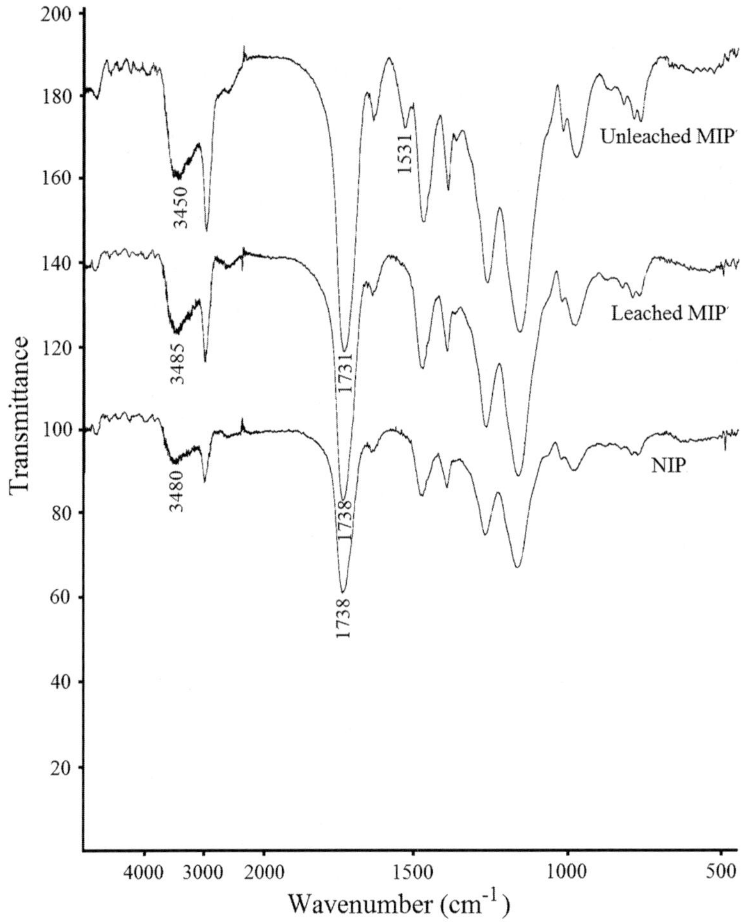

FIGURE 11.1 FT-IR spectra of unleached and leached dipyridamole-imprinted polymers.

[4–6,12,14,32–37]. This displacement toward lower frequencies is due to the hydrogen interaction of O-H and C=O groups with the template. Figure 11.1 shows the FT-IR spectra of unleached and leached dipyridamole-imprinted polymers.

MIPs with well-controlled physical forms in different size ranges are highly desirable. For example, MIP nanospheres are very suitable for use in developing binding assays. Scanning electron microscopy (SEM) is employed to determine the shape and surface morphology of the produced polymer particles [14,36,37]. In this method, the polymer particles are sputter coated with gold prior to the SEM measurement. Figure 11.2 shows the SEM images of carbamazepine-imprinted polymer nanospheres and corresponding blank polymer nanoparticles. The uniformly sized nanospheres were obtained in acetonitrile as progen and by an appropriate amount of template/functional monomer (MAA)/cross-linker agent (TRIM) (molar ratio of 1:6:8) during the precipitation polymerization.

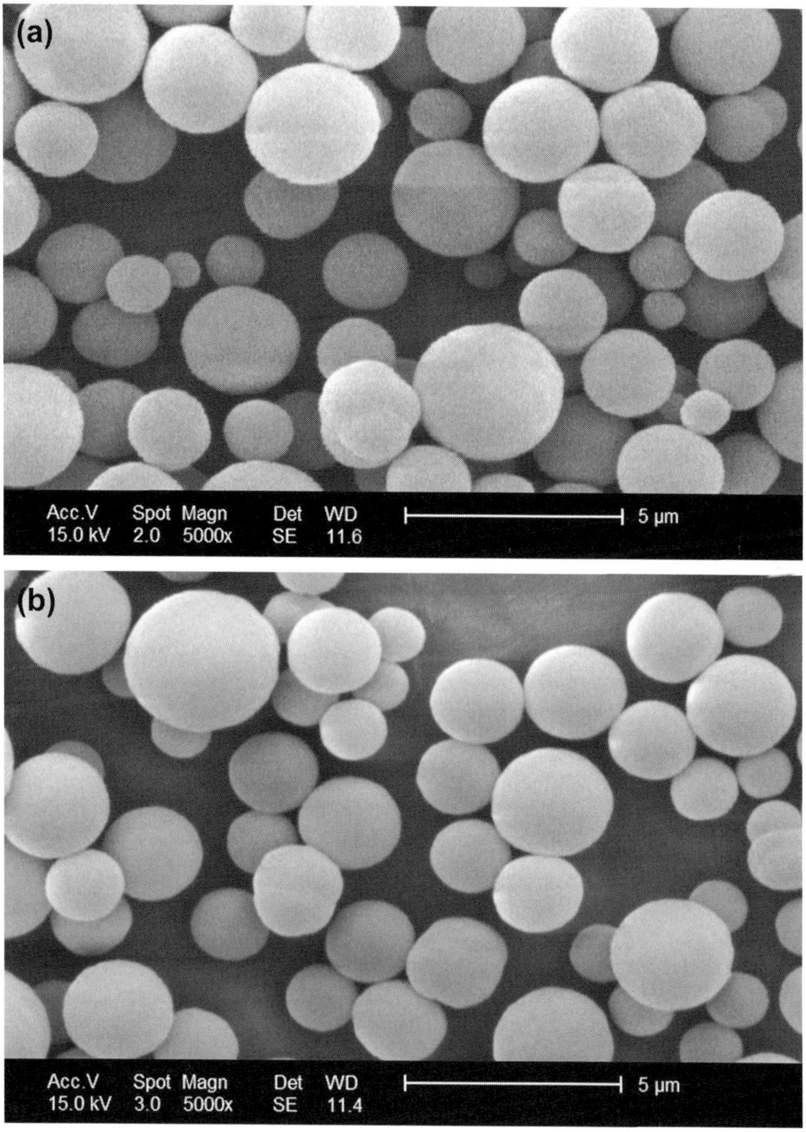

FIGURE 11.2 Scanning electron micrographs of carbamazepine molecularly imprinted nanospheres (a) and corresponding NIPs (b) prepared by precipitation polymerization.

The particle size and size distribution of polymer particles can be measured by photon correlation spectroscopy. Figure 11.3 shows a very narrow particle size distribution for carbamazepine-imprinted polymer nanospheres prepared in acetonitrile. The polydispersity index obtained was 0.063 for MIP and 0.083 for corresponding nonimprinted polymer (NIP) [14].

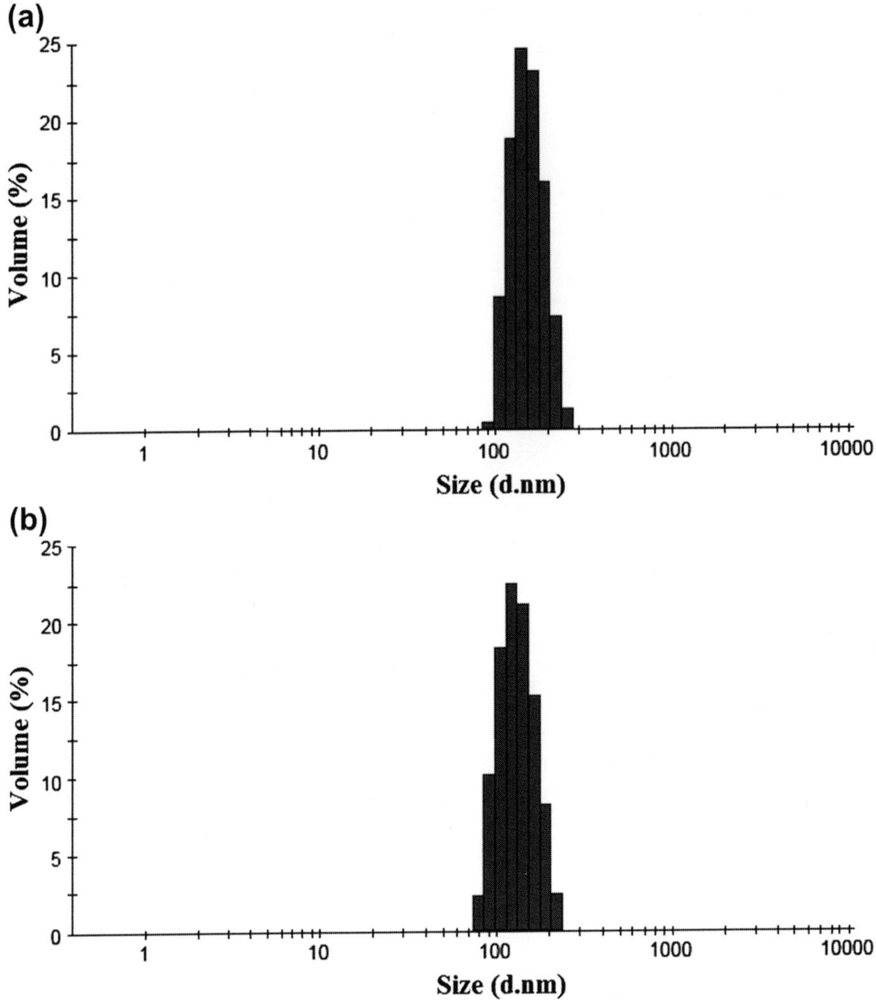

FIGURE 11.3 Particle-size distribution of carbamazepine-imprinted polymer nanospheres (a) and corresponding blank polymer nanoparticles (b) measured by photon correlation spectroscopy.

The specific surface area, pore volume, and average pore diameter of polymer particles are obtained by the Brunauer–Emmett–Teller (BET) method [14,37]. Table 11.2 shows the texture properties (specific surface area, pore volume, and average pore diameter) of the carbamazepine-imprinted polymer particles and corresponding NIP, measured by nitrogen adsorption porosimetry [14]. These data indicate that different porogens had significant effects on the porosity of the polymer particles. In addition, Javanbakht et al. showed that upon extraction with a good solvent, only the specific surface of the imprinted polymer particles can be changed significantly [37].

The thermal properties of the leached and unleached MIPs and blank polymer can be characterized by thermogravimetric analysis (TGA) and differential scanning calorimetry

TABLE 11.2 Results of BET Analysis for Carbamazepine-Imprinted Polymers and Control Polymer Prepared by Different Porogens

Polymer	Polymerization solvent	Specific surface area (m^2/g)	Pore volume (mL/g)	Average pore diameter (nm)
MIP1	Acetonitrile	242	1.00	16
NIP1	Acetonitrile	268	0.95	14
MIP2	Chloroform	18	0.10	22
MIP3	Acetonitrile/Chloroform (1:1, v/v)	121	1.05	35
MIP4	Toluene	6	0.03	20

(DSC). As a rule, an endothermic transition was observed for unleached MIP, which can be explained as the melting of template load in polymer [14]. Figure 11.4 depicts the TGA plots of the unleached and leached hydroxyzine-MIP particles [12]. Regarding the unleached MIP particles, TGA revealed two decomposition states: one mass loss between 100 °C and 180 °C (10% weight loss), assigned to the decomposition of the free monomer and the cross-linker, and one starting at 190 °C, related to the hydroxyzine dihydrochloride decomposition as the melting point of hydroxyzine dihydrochloride, which is approximately 193 °C. These observations also indicated that the rigidity of the unleached and leached MIP particles is further than blank materials: the former exhibits decomposition above 300 °C, whereas the latter starts its decomposition at approximately 250 °C onward.

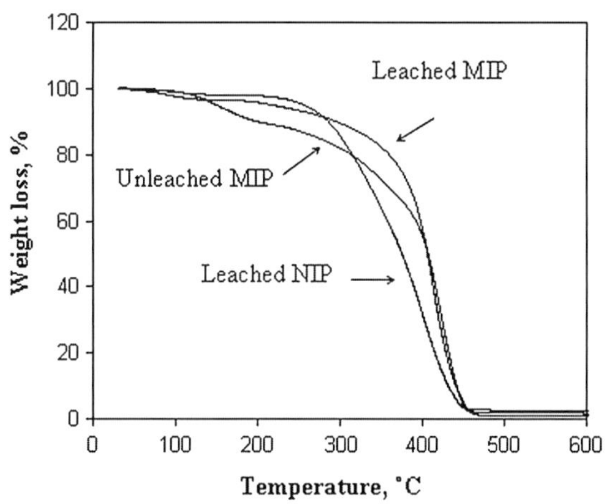

FIGURE 11.4 TGA plots of the leached NIP, and unleached and leached hydroxyzine-imprinted polymer particles.

3.3. Batch Rebinding Experiments

The selectivity of a MIP results from the existence of specific cavities designed for the target analytes. Therefore, the various approaches used to evaluate the presence of these cavities of the MIP usually consist in the study of the interactions between the template or the target analyte and the MIP. Furthermore, these studies are commonly carried out in parallel on a NIP that is obtained by using the same procedure of polymerization in the absence of the template. The NIP possesses the same chemical properties as the MIP but without possessing any specific cavities. The existence of specific cavities can be assessed by various methods, such as binding experiments and chromatographic measurements.

Equilibrium batch rebinding is one of the most general methods for evaluating the presence of recognition cavities [4,5,36]. A well-known amount of template is introduced in a vial with a given amount of imprinted or nonimprinted polymers. Once the system has reached equilibrium, the amount of free template in the solution is measured to determine the amount of adsorbed template. This amount is compared to the one bound on the NIP, the number of cavities being correlated to the difference between the amounts adsorbed on both polymeric sorbents.

4. MIP-BASED POTENTIOMETRIC SENSORS

There are a variety of combinations of recognition elements and signal transducers. Potentiometric sensors, in which an electrode is used as the transduction element for monitoring the voltage at zero current, represent an important subclass of electrochemical sensors. The immobilization of MIPs on the transducer surface is an important feature in the design of MIP-based potentiometric sensors.

4.1. Plasticized PVC Membrane Sensors

Entrapment of MIP particles into PVC membranes has often been used for electrochemical transducers. These prepared materials are applied onto the transducer surface. In a typical manner, the sensing membrane was prepared by mixing PVC powder and templated polymer particles with a plasticizer. The mixture was stirred until the PVC was well moistened, and then the mixture was dispersed in THF. The resulting mixture was adequately mixed with a magnetic stirrer and then was poured onto a glass plate. The solvent was evaporated slowly at room temperature until a solid membrane with about 0.3 mm thickness was formed. A desired piece of the membrane was cut and then was attached to an end of a Pyrex tube using a viscous solution of PVC in THF as an adhesive. The resulting sensor was then filled with an internal solution of the analyte and conditioned for a proper period of time (see Fig. 11.5 for a schematic diagram).

The first imprinted polymer ISEs were prepared for calcium and magnesium ions by Mosbach's group [38]. The monomer used in the fabrication of the electrode was a neutral ionophore N,N′-dimethyl-N,N′-bis(4-vinylphenyl)-3-oxapentadiamide. This was chosen as a Ca^{2+}-selective ionophore to produce a polymer cavity that would resemble a polyether site, with imprinting being the means to acquire size selection. The membranes were prepared

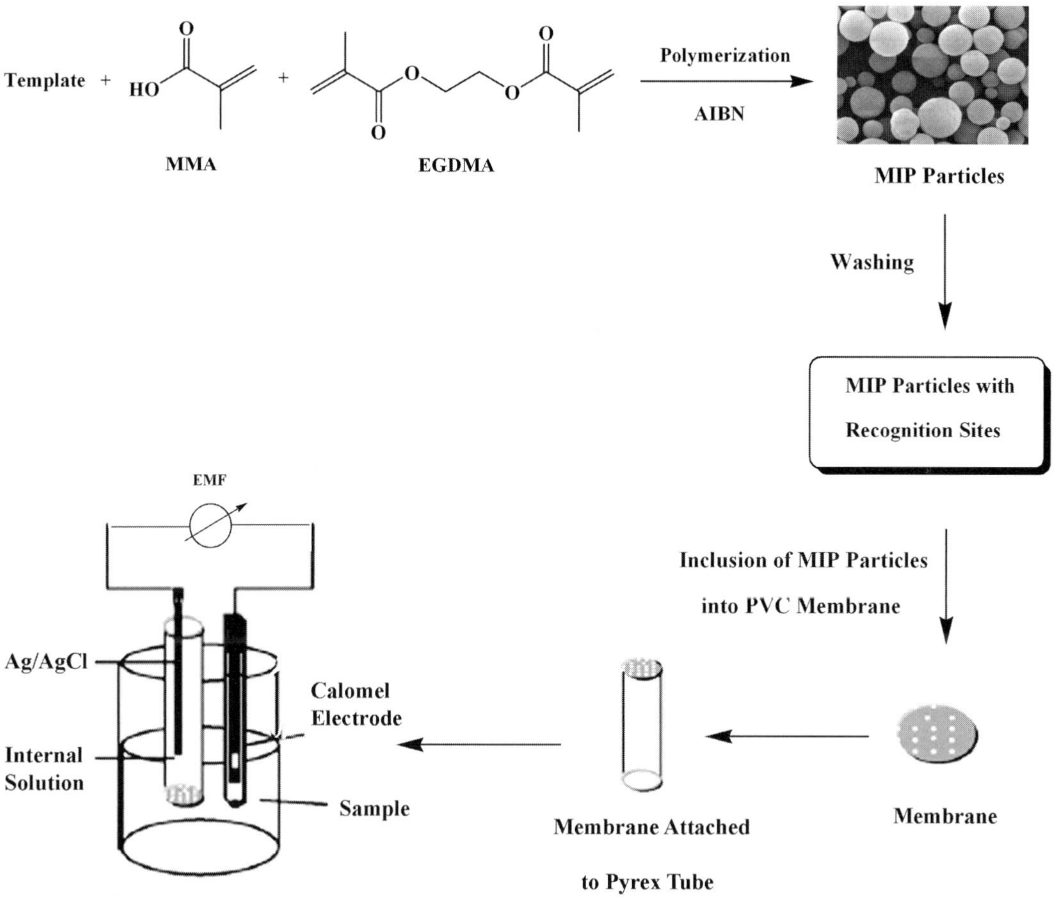

FIGURE 11.5 Schematic diagram for fabrication of a MIP-based PVC membrane sensor.

using a mixture of 32.1 wt% PVC, 65.3 wt% bis(2-ethylhexyl) sebacate, 1.85 wt% imprinted polymer, and 0.7 wt% potassium tetrakis (p-chlorophenyl)borate.

T. Prasada Rao and co-workers developed a biomimetic potentiometric sensor by dispersing the atrazine-imprinted polymer particles in di-*n*-octyl phthalate plasticizer and then embedding in a polyvinyl chloride matrix [39]. The sensor responds to atrazine in the pH range 2.5–3.0 over a working range of 0.0001–10 mM, with a detection limit of 0.5 μM. They also reported a biomimetic potentiometric sensor for the recognition of methylphosphonic acid (MPA) [40]. This involves the preparation of MPA-imprinted polymer particles and removal of the template by Soxhlet extraction. Subsequently, the leached MIP particles were dispersed in 2-nitrophenyloctyl ether (as a plasticizer) and embedded in polyvinyl chloride matrix. The sensor responds to MPA in the concentration range 5×10^{-8} to 1×10^{-4} and 1×10^{-3} to 1×10^{-1} M, with a detection limit of 5×10^{-8} M.

A plasticized PVC membrane sensor based on MIPs for recognition and determination of levamisole hydrochloride was developed by Sadeghi et al. [41]. Characteristics of the proposed sensor were evaluated by measuring the potential response to levamisole hydrochloride in the activity range from 2.5 μM to 100 mM, with a near Nernstian response of 57.0 mV/decade and a limit of detection of 1.0 μM. The response time of the sensor was short (<15 s), and the membrane could be used for 4 months. The applicability of the sensor was tested in potentiometric titration of levamisole hydrochloride in pure drug and tablet pharmaceutical form.

Qin and co-workers described a polymeric membrane ion-selective electrode for determination of melamine [42]. The membrane electrode shows a near-Nernstian response (54 mV/decade) to the protonated melamine over the concentration range of 5.0×10^{-6} to 1.0×10^{-2} mol L^{-1}. The electrode exhibits a short response time of 16 s and can be stable for more than 2 months. Combined with the flow analysis system, the potentiometric sensor has been successfully applied to the determination of melamine in milk samples.

M. Goreti, F. Sales, and co-workers reported a new potentiometric membrane sensor with cylindrical configuration for tetracycline based on the use of MIP material consisting of 2-vinylpyridine as a functional monomer in a plasticized PVC membrane [43]. The sensor exhibited a significantly enhanced response to tetracycline over the concentration range $1.59 \times 10^{-5} - 1.0 \times 10^{-3}$ mol/L at pH 3–5 with a detection limit of 1.29×10^{-5} mol/L. The response was near-Nernstian, with average slopes of 63.9 mV/decade. The possibility of applying the proposed sensor to tetracycline determination in spiked biological fluid samples was demonstrated. This teamwork also reported a 3-D-mirror synthetic receptor for ciprofloxacin host–guest interactions and potentiometric transduction [44]. The host cavity was shaped on a polymeric surface assembled with methacrylic acid or 2-vinyl pyridine monomers by radical polymerization. Molecularly imprinted particles were dispersed in 2-nitrophenyl octyl ether and entrapped in a poly(vinyl chloride) matrix. The sensors exhibited a near-Nernstian response in steady-state evaluations. Slopes and detection limits ranged from 26.8 to 50.0 mV/decade and 1.0×10^{-5} to 2.7×10^{-5} mol/L, respectively. Good selectivity was observed for trimethoprim, enrofloxacin, tetracycline, cysteine, galactose, hydroxylamine, creatinine, ammonium chloride, sucrose, glucose, sulphamerazine, and sulfadiazine. The sensors were applied to the determination of ciprofloxacin concentrations in fish and in pharmaceuticals. Similar studies show a biomimetic sensor for the potentiometric transduction of oxytetracycline [45]. The artificial host was imprinted in methacrylic acid and/or acrylamide-based polymers. Different amounts of MIPs were dispersed in different plasticizing solvents and entrapped in a poly(vinyl chloride) matrix. These sensors exhibited a near-Nernstian response in steady-state evaluations; slopes and detection limits ranged from 42–63 mV/decade and 2.5–31.3 μg/mL, respectively. These sensors were independent from the pH of test solutions within 2.0–5.0. Good selectivity was observed toward glycine, ciprofloxacin, creatinine, acid nalidixic, sulfadiazine, cysteine, hydroxylamine, and lactose. In flowing media, the biomimetic sensors presented RSD of ±0.7%, sensitivity of 65 mV/decade, linear range from 5.0×10^{-5} to 1.0×10^{-2} mol/L, and detection limit of 19.8 μg/mL.

Comparable work shows biomimetic sensors for chlorpromazine [46]. Slopes and detection limits ranged from 51–67 mV/decade and 0.46–3.9 μg/mL, respectively, in steady-state conditions. Sensors were independent from the pH of test solutions within 2.0–5.5. Good selectivity was observed toward oxytetracycline, doxytetracycline, ciprofloxacin, enrofloxacin, nalidixic acid, sulfadiazine, trimethoprim, glycine, hydroxylamine, cysteine, and creatinine.

4.2. Self-Assembled Monolayer-Based Sensors

In a typical self-assembled monolayer (SAM) approach reported by M. Rafailovich [47], the gold-coated silicon chip was used after cleaning with de-ionized water and dried with pure nitrogen gas. Biomolecules such as proteins were dissolved in de-ionized water, and the thiol was dissolved in acetic acid. Then the mixture was made by blending them in the 19:1 (water: acetic acid) volume ratio. In this way, a good solubility was obtained for thiol and no precipitation of proteins occurred. Then the Au coated plate was immersed into the solution for at least 2 h to co-absorb the protein and thiol. While the thiol molecules can be tightly attached to the electrode surface through a sulfur—metal bond, the proteins are adsorbed onto the gold surface mostly through hydrophobic interactions and electrostatic forces, in the absence of strong chemical bindings. Cavities that are complementary with the template proteins can then be created in the SAM matrix by removing the protein molecules through repetitive rinsing with de-ionized water. Due to the unique character, the so-prepared electrode is expected to have a higher affinity to adsorb the same kind of template protein molecules than the other "guest" molecules. The electrode was then dried at room temperature overnight before electrochemical measurements were made. This technique was applied for detection of globular proteins such as myoglobin and hemoglobin with parts per million accuracy.

4.3. Sensor Based on In-Situ Polymerization

In-situ polymerization is the best immobilization procedure and consists of in-situ electrosynthesis of MIPs at the transducer surface. As a result, imprinted polymers prepared by an in-situ technique do not require any consequent treatment of the resultant material, except washing to extract the template, and they can be used directly in their applications.

In-situ polymerization has the potential advantage of integrating the immobilization step into an automatic mass-production process.

A polymeric membrane-based ion-selective electrode for the determination of l-ascorbic acid was developed by D. Tonelli et al. [48]. The sensor was fabricated by modifying glassy carbon electrodes with molecularly imprinted polypyrrole (PPy) synthesized by electropolymerization of the monomer in the presence of ascorbate. The molecularly imprinted polymer improves the performance of the sensor, especially if the PPy film has been submitted to an oxidation treatment. The PPy-MIPox electrode shows near-Nernstian response (approx. -57 mV/decade) to the ascorbate over the concentration range between 5×10^{-6} mol/L and 2×10^{-3} mol/L, and it can be used for more than 3 weeks. The potentiometric sensor has been applied to the determination of ascorbate in food and pharmaceutical samples (with recoveries of about 100%).

4.4. Sensor Based on ITO Thin Film

A potentiometric chemosensor for determination of dipicolinic acid (2,6-pyridinedicarboxylic acid, DPA) was developed based on the surface imprinting technique coupled with a nanoscale transducer, an indium tin oxide (ITO)-coated glass plate by Y. Zhou [49]. The DPA sensor could recognize DPA from 3,5-pyridinedicarboxylic acid. Potentiometric measurements demonstrated detection of DPA in a concentration range of 1.5×10^{-6} to

0.0194 M. In this technique, DPA (template) and octadecyltrichlorosilane were co-adsorbed on the polar solid surface of the ITO glass plate from a $CHCl_3/CCl_4$ solution. Then, the electrode was rinsed with 30 × 1 mL $CHCl_3$.

4.5. Sensor Based on Surface Coating

Surface coating (spin and spray coating) is another method of applying and immobilizing MIPs onto a transducer surface. Although normally used when MIPs are integrated with optical transducers [1], it is also possible to apply this surface-coating method with electrochemical transducers. Use of the surface-coating technique allows production of a thin film of MIPs.

4.6. Sensors Based on Ion-Selective Field-Effect Transistor

Potentiometric devices can also include the chemically sensitive field-effect transistors (FETs), which are semiconductor devices that respond to the surface electric gradient or charge at the gate electrode. A chemical reaction at the surface of the silicon chip will shift the surface potential and hence affects the current, which allows the rate of the reaction to be monitored. This type of device can be used for monitoring practically any type of charged template molecules.

A biomimetic sensor for cAMP was fabricated in combination with an ion-sensitive field-effect transistor (ISFET) as a transducer and a cAMP-imprinted polymer as a molecular recognition material [50]. The cAMP-imprinted polymer was prepared using 1-allyl-2-thiourea. The imprinted polymer membrane was coated on the ISFET electrode, and the response of the sensor was evaluated. The cAMP-imprinted polymer showed high binding ability to and selectivity for cAMP in aqueous media. The linear relationship was obtained from 0.1 to 1.0 mmol/L cAMP. For polymer membrane coating onto ISFET electrode, first, the reaction mixture was degassed and dropped onto the ISFET electrode, and covered with micro cover glass coated with trimethylchlorosilane. Then, the polymerization was initiated by UV light irradiation at room temperature for 60 min under a nitrogen atmosphere. The blank polymer-coated ISFET sensor was prepared without template molecule, cAMP, and was used as a control. The ISFET electrodes were washed, first with 1% NH_3 aq, followed by distilled water, and then dried in air.

4.7. Chemically Modified Carbon Paste Electrodes

Another interesting technique is the preparation of chemically modified carbon paste electrodes (CMCPEs). The CMCPEs can be prepared by thoroughly mixing graphite powder, paraffin oil, and MIP particles. The electrode body was fabricated from a glass tube. After mixture homogenization, the paste was packed carefully into the tube tip to avoid possible air gaps, often enhancing the electrode resistance. A copper wire was inserted into the opposite end to establish an electrical contact. The external electrode surface was smoothed with soft paper. The surface was renewed by scraping out the old surface and replacing the carbon paste.

In 2008, Javanbakht et al. reported the first CPE, based on a noncovalent imprinted polymer, for the recognition and determination of hydroxyzine in tablets and biological fluids [12].

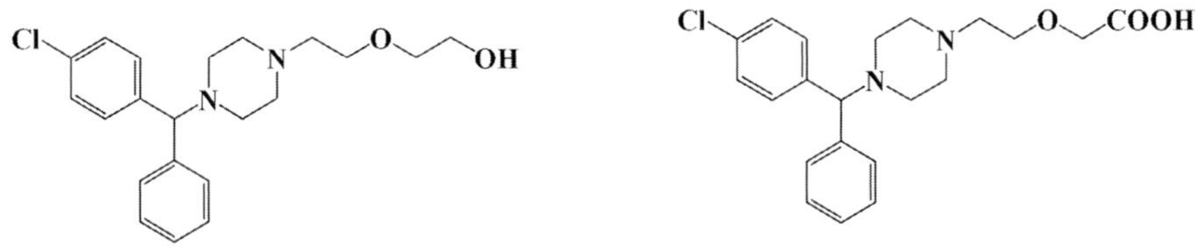

Hydroxyzine

Cetirizine

FIGURE 11.6 Chemical structures of hydroxyzine and cetirizine.

Hydroxyzine (see Fig. 11.6), (*RS*)-2-[2-((4-chlorophenyl)phenylmethyl)-piperazine-1-yl]ethoxy ethanol, is a piperazine antihistamine. It diminishes the main action of histamine by competitive reversible blockade histamine receptors sites H1 in the tissues. The MIPs were synthesized by precipitation polymerization, using hydroxyzine dihydrochloride as template molecule, methacrylic acid (MAA) as a functional monomer, and ethylene glycol dimethacrylate (EGDMA) as a cross linking agent. The schematic representation of the imprinting and the removal of hydroxyzine from the imprinted polymer are shown in Fig. 11.7. The fresh modified electrode surfaces were conditioned by exposure to a 1.0×10^{-4} M hydroxyzine

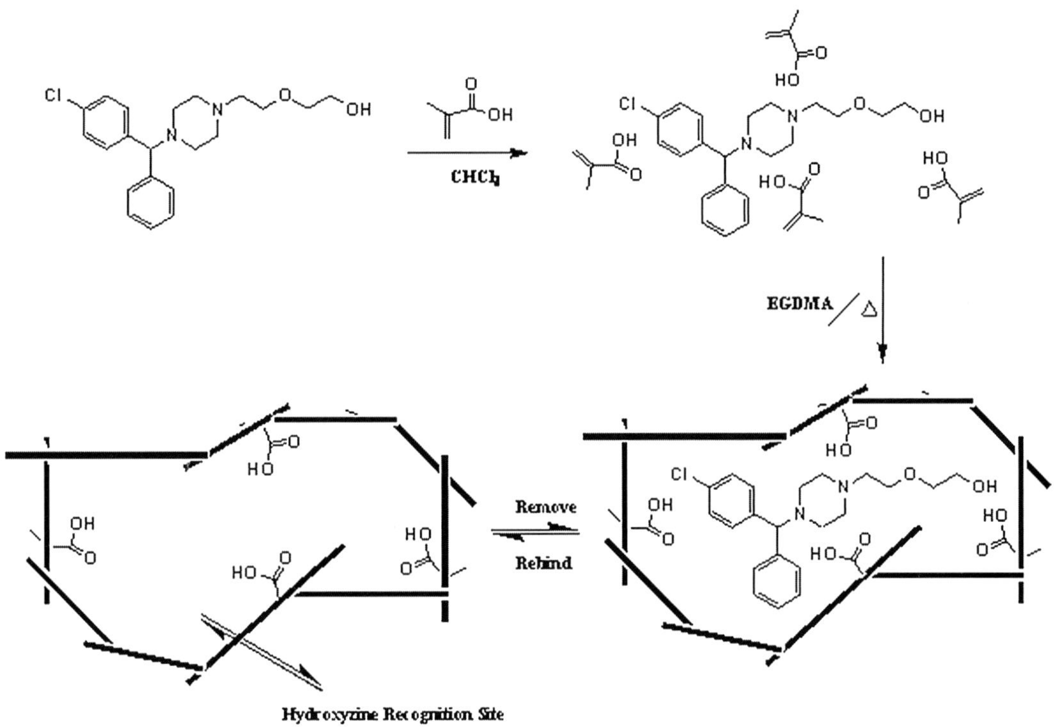

FIGURE 11.7 Schematic representation of the hydroxyzine MIP synthesis.

dihydrochloride solution at the pH value of 2.0, adjusted with a 0.01 M phosphate buffer for 36 h. The electrodes were then rinsed with de-ionized water for 30 s. After electrode conditioning, the sensors showed high selectivity and sensitive response to the template in the aqueous system.

In this study, the solution pH displayed a great effect on the MIP sensor response owing to partial hydroxyzine deprotonation in the sample solutions. Consequently, when the electrodes are employed, buffered solutions at the pH value of 2.0 must be used.

In a comparable work, a biomimetic potentiometric sensor was fabricated for the recognition and determination of cetirizine [31]. Cetirizine (2-[2-[4[(4-chlorophenol)phenylmethyl]-1-piperazinyl]ethyloxy] acetic acid dihydrochloride) (Fig. 11.6) is an active metabolite of hydroxyzine, a first-generation H1-receptor antagonist. The pK_a values of cetirizine dihydrocholride as a basic compound with three ionozable groups—for the carboxylic group, the tertiary amine group, and the nitrogen heterocyclic—are 2.2, 2.9, and 8.0, respectively.

In this study, the solution pH also displayed a great effect on the MIP sensor response owing to partial cetirizine deprotonation in the sample solutions. Subsequently, when this electrode is employed for the measurement of pharmaceutical samples, the pH parameter should be adjusted at the constant pH value of 2.0.

5. MIP-BASED SENSORS FOR THE DETERMINATION OF DRUGS IN BIOLOGICAL AND PHARMACEUTICAL SAMPLES

Analysis of pharmaceutical compounds can be performed in biological matrices such as blood, plasma, or serum samples, and this process usually requires a simple sample pretreatment. Generally, it is necessary to clean up the sample before electrochemical analysis with MIP sensors or even before the chromatographic separation that can be commonly achieved by solid phase extraction (SPE). The most commonly used procedure involves solvent extraction of blood or urine. In this case the choice of best solvent is crucial in order to eliminate the interference of naturally occurring electroactive compounds of the biological matrix. Pretreatment of these samples with SPE has many advantages because it is the most popular clean-up technique because of factors such as convenience, cost, time saving, and simplicity. However, classical SPE sorbents, used for the treatment of biological samples, mainly cause nonspecific hydrophobic interactions that lead to the co-extraction of interfering compounds. Many classic pretreatment methods for biological samples can be used before using electrochemical sensor-based MIPs. Recently, molecularly imprinted solid phase extraction (MISPE) as the sample clean-up technique was developed for determining various drugs such as dextromethorphan, metochloprmide, bromhexine, and tramadol in plasma, serum, or urine samples [4–6,36]. MIP-based solid phase extraction cartridges can allow a straightforward and fast pretreatment due to their specific recognition properties.

Undoubtedly the blood is the most frequently sampled biological fluid that is collected for therapeutic purposes. A number of factors govern the actual portion of the blood to be analyzed for drug content. Available analytical techniques and the component that correlates the respective concentration with the pharmacologic effect of the drug are two main factors that affect the experimental sample. The separation of the compound from its biological matrix may be performed by using several procedures. Ultracentrifugation or filtration methods may be employed to remove the interference of molecules of high molecular weight, such as proteins, which may inhibit the electrode response. A precipitation of the proteins using an adequate precipitating agent may also be performed. In such cases, care must be taken to avoid the loss of drugs, some of which are strongly bound to the proteins. In clinical laboratories, three types of samples can be derived from venous blood sample: whole blood, serum, and plasma. Urine is the other biological fluid that is collected mainly for following drugs in unmetabolized form or even metabolized form in overdose cases.

In the analysis of whole blood, cellular components of the blood (composed primarily of erythrocytes) must be lysed so that their content can form a homogeneous mixture with the fluid portion of the blood. The first step in the treatment of blood samples is to discard the red cells from the serum to prevent the presence of hemoglobin or its degradation products, which may interfere in subsequent analysis. The cells can be lysed by using a sonicator or by freezing the sample for a period of time followed by sonication. In some difficult cases and complex matrices, it is helpful to use HPLC for better separations. This technique, associated with ultraviolet, fluorescence, or even electrochemical detection, has been shown to be useful and powerful in measuring low levels of drugs in blood and other biological fluids.

The use of plasma and its products has evolved over a period of four decades. For preparation of plasma, the fibrinogen must be removed in the process of clotting. Heparin is the most commonly used anticoagulant additive that is added to the blood to retard clot formation. The term *fresh frozen plasma (FFP)* refers to the liquid portion of human blood that has been frozen and preserved after a blood donation. FFP can refer to the fluid portion of one unit of human blood that has been centrifuged, separated, and frozen solid at −18 °C (−0 °F), or colder, within eight hours of collection. The use of FFP has increased tenfold during the years 2000–2010, and it can be very important in the detection of drugs by MIP sensors and in clinical diagnostics. Serum is a portion of whole blood that actually includes all proteins not used in the blood-clotting process. It is used in numerous diagnostic tests, as well as in blood typing, and because of its simpler matrix it can be used in many electrochemical sensors.

5.1. Analysis of Drugs in Urine Samples

Urine samples often are diluted with water or a buffer of the appropriate pH prior to the sample addition. In some cases, acid hydrolysis (for basic compounds) or base hydrolysis (for acidic compounds) is used to ensure that the compounds of interest are freely solvated in the urine sample. Usually a strong acid (e.g., concentrated HCl) or base (e.g., 10 M KOH) is added to the urine. The urine is heated for 15–20 minutes, then cooled and diluted with a buffer, and the pH is adjusted appropriately prior to the electrochemical potentiometric analysis. Enzymatic hydrolysis that frees bound compounds or drugs may also be used. Liang and co-workers used only a buffer solution for pretreatment of urine samples and found recoveries between 96.1 and 103.0% for determination of caffeine in human urine [51].

A typical solvent extraction method for the preparation of spiked urine samples was reported for hydroxyzine potentiometric sensor [12]. In this approach, 1 mL of hydroxyzine dihydrochloride aqueous solution was transferred in to a 5-ml volumetric flask, and then the solution was diluted with urine and vortexed. Next the solution was adjusted to pH 10 by the concentrated sodium hydroxide solution. For the determination of hydroxyzine, 2 mL dichloromethane was added to one milliliter of urine samples and vortexed. The mixture was centrifuged at 6000 rpm to separate the aqueous and organic layers. After removal of the organic layer, the extraction was repeated on the residual aqueous layer. The dichloromethane layers were pooled and dried under a gentle stream of nitrogen. After drying, samples were reconstituted with 5 ml of HCl and vortexed for 1 min. Then the analysis was followed up as indicated in the general analytical procedure.

5.2. Analysis of Drugs in Serum and Plasma Samples

In many cases, analytes such as drugs may be protein-bound, which reduces recoveries. To disrupt protein binding in these biological fluids, we can use one of the following methods for potentiometric analysis of drugs with MIP-based sensors:

1. Shift the pH of the sample to extremes ($pH<3$ or $pH>9$) with acids or bases in the concentration range of 0.1 M or greater. Use the resulting supernatant as the sample for biological sensor.
2. Precipitate the proteins using a polar solvent such as acetonitrile, methanol, or acetone (two parts solvent per one part biological fluid is typical). After mixing and centrifugation, remove the supernatant and dilute with water or an aqueous buffer for subsequent analysis.
3. To precipitate proteins, treat the biological fluid with acids or inorganic salts, such as formic acid, perchloric acid, trichloroacetic acid, ammonium sulfate, sodium sulfate, or zinc sulfate. The pH of the resulting supernatant may be adjusted prior to inserting sensor electrodes.
4. Sonicate the biological fluid for 15 minutes, add water or buffer, centrifuge, and use the supernatant for subsequent potentiometric analysis.
5. An organic solvent such as dichloromethane was added, vortexed for a few minutes, and centrifuged the obtained mixture for the separation of the aqueous and organic layers. The dichloromethane layers were pooled and dried under a gentle stream of nitrogen. After drying, samples were reconstituted with an acid and buffered and the analysis was followed [12,31].

5.3. Analysis of Drugs in Pharmaceuticals

Tablets or capsules taken orally remain one of the most effective means of treatment available. Before any potentiometric analysis with an MIP-based sensor, it is necessary to convert these formulations into a solution that is free from any particle. For this procedure, usually the contents of ten tablets or capsules are weighed and finely powdered in a small dish. Then an accurately weighed portion of the powder, equivalent to one tablet or capsule, is dissolved in buffer solution with the pH in acidic range and filtered into a volumetric flask. The solutions are completed to the mark with appropriate buffer. An aliquot of each solution is transferred to a 50-ml beaker, and the drug sensor in conjunction with a double junction reference electrode is immersed and compared with the calibration graph [52].

In the assays of hydroxyzine and cetirizine in tablets [12,31], twenty tablets were weighed, crushed, and mixed in a mortar and pestle. A portion of powder equivalent to the weight of one tablet was accurately weighed into volumetric flasks, and phosphate buffer was added to each flask. The volumetric flasks were sonicated to effect complete dissolution of the hydroxyzine or cetirizine, and the solutions were then made up to volume with phosphate buffer. Suitable aliquots of solution were filtered through a 0.45 μm millipore filter, rejecting the first portion of the filtrate. The desired concentrations were obtained by accurate dilution with phosphate buffer.

6. MIP-BASED SENSORS FOR THE DETERMINATION OF POLLUTANTS IN THE ENVIRONMENTAL SAMPLES

Trace determination of persistent organic pollutants (POPs) in complex matrices requires a sample pretreatment. Centrifugation, filtration, liquid–liquid extraction, dilution in buffers, Soxhlet extraction with acids or solvents, and even solid phase extraction, especially with MIPs, are the most common pretreatment methods used for environmental samples such as natural waters and soil and plant samples. In addition to ensuring proper pH of the sample, other sample pretreatment criteria must be considered. The following researches describe how some difficult sample matrices should be pretreated before being applied to potentiometric analysis. Jenkins and co-workers developed an MIP-based sensor for pesticides and insecticides in water. In this work, the direct imprinting of nonhydrolyzed organophosphates, including pesticides and insecticides, is described. The detection limit for this sensor for detection of some organophosphates in environmental waters was less than 10 parts per trillion (ppt), with long linear dynamic ranges (ppt to ppm) and response times of less than 15 min [53]. T. Prasada Rao and co-workers reported a biomimetic potentiometric sensor for determination of atrazine in ground waters. This sensor had a response time of about 2 min and high selectivity with respect to other important classes of pesticides and herbicides [39]. They also reported a biomimetic potentiometric sensor for the determination of methylphosphonic acid (MPA), the degradation product of nerve agents sarin, soman, VX, and so on [40]. The selectivity of the sensor has been tested with respect to chemical analogues such as phosphoric acid, sodium dihydrogen phosphate, organophosphorous pesticides, and triazine herbicides. The utility of the sensor was tested for field monitoring of MPA in spiked groundwater.

7. ANALYTICAL PERFORMANCE

The properties of an ion-selective electrode are characterized by parameters such as linearity, limit of detection (LOD), limit of quantification (LOQ), sensitivity, selectivity, reproducibility, accuracy, robustness, response time, and lifetime.

7.1. Linearity and Detection Limits

Perhaps the most surprising and innovative research direction in potentiometric sensors is the dramatic improvement of the lower detection limit. Each potentiometric sensor has a lower and upper detection limit where the response starts to distance significantly from

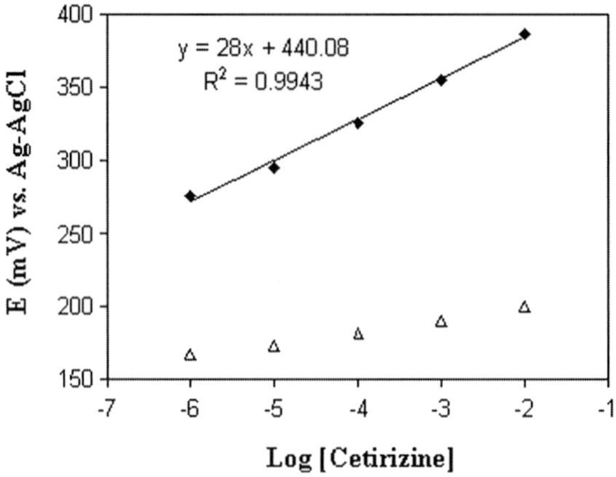

FIGURE 11.8 Calibration curve for a cetirizine potentiometric sensor.

a Nernstian electrode slope. Generally, they fall into activity ranges where the electrode starts to lose sensitivity to the target ion. According to the IUPAC recommendation, the detection limit is defined by the cross-section of the two extrapolated linear calibration curves [54]. The linearity of a potentiometric sensor is defined as the activity ratio of upper and lower detection limit and approximately corresponds to the range where the electrode responds according to the Nernst equation.

Figure 11.8 exhibits a Nernstian slope of 28.0 mV/decade in a cetirizine concentration range between 1.0×10^{-6} mol/L and 1.0×10^{-2} mol/L, with a lower detection limit of 7.0×10^{-7} mol/L [31]. The carbon paste electrode with the nonimprinted polymer presented a nonlinear potentiometric response, and the slope was very low (approximately 9.2 mV/decade).

Table 11.3 summarizes the MIP-based potentiometric sensors reported in the literature and includes the studies of their linearity and detection system.

7.2. Sensitivity

Providing that some conditions are gotten together, the theoretical slope (sensitivity) of the calibration curve of an ISE toward a given singly charged ion, at 25 °C, should be 59.2/z mV/decade, as formulated by the Nernst equation. The ISE membranes must have a sufficient perm-selectivity. For the following two situations, a non-Nernstian may be seen:

1. Co-extraction of the counter ions, which leads to considerable ion pairing, reduces slopes.
2. Non-Nernstian behaviors may also happen, if an ISE is exposed to ions that complex by the ionophore.

Surfactants and proteins, and other species that may be present together with the target ions in pharmaceuticals and biological fluids, can also interfere with the normal ion-extraction process resulting in nontheoretical (non-Nerstian) behaviors.

It has been demonstrated that the membrane composition has a significant effect on the sensitivity of potentiometric sensors. Table 11.4 displays the slopes, linear ranges, and LODs of the resulting potentiometric calibration curves for the hydroxyzine ion-selective

TABLE 11.3 Construction, Recognition System, and Linear Range of Different Potentiometric Electrodes Based on MIP

Analyte	Sensor construction	Recognition system	Linear range (mol/L)	Ref.
Dipicolinic acid	Template and OTS were co-adsorbed on the polar solid surface of the ITO glass plate.	ITO-coated glass plate	1.5×10^{-6} to 1.9×10^{-2}	49
Atrazine	A glassy membrane is obtained from polymerization of a reagent mixture containing MAA, EGDMA, template, and AIBN.	A glassy membrane	3×10^{-5} to 1×10^{-3}	28
Atrazine	By dispersing the atrazine-imprinted polymer particles in the DOP plasticizer and, then, embedding in the PVC matrix.	A PVC membrane	1×10^{-7} to 1×10^{-2}	39
Methylphosphonic acid	By casting a membrane after dispersing the MPA-imprinted polymer particles in NPOE and embedding in the PVC matrix.	A PVC membrane	5×10^{-8} to 1×10^{-4} and 1×10^{-3} to 1×10^{-1}	40
Methylphosphonic acid	By surface imprinting technique coupled with a nanoscale transducer, ITO.	ITO-coated glass plate	1×10^{-2} to 1×10^{-1}	30
Pinacolyl methylphosphonate	By dispersing the PMP-imprinted polymer materials and the suspension in PVC.	A PVC membrane	4×10^{-8} to 1×10^{-3}	26
Levamisole	By dispersing the levamisol-imprinted polymer particles in the DBS plasticizer and, then, embedding in the PVC matrix.	A PVC membrane	2.5×10^{-6} to 1.0×10^{-1}	41
cAMP	The imprinted polymer membrane was coated on the ISFET electrode	ISFET	5×10^{-6} to 1.2×10^{-2}	50
Hydroxyzine	A CPE is prepared by mixing graphite powder, paraffin oil, and hydroxyzin-imprinted polymer.	Carbon paste	1.0×10^{-6} to 1.0×10^{-1}	12

Abbreviations: OTS, octadecyltrichlorosilane; ODS, octadecylsiloxane; ITO, indium tin oxide; DBS, sebacic acid dibutyl ester; PVC, polyvinylchloride; DOP, dioctyl phthalate; o-NPPE, o-nitrophenyl octyl ether; PMP, pinacolyl methylphosphonate; N-CBZ-Asp, N-cabobenzoxy-aspartic acid.

TABLE 11.4 Composition of the MIP-Based Hydroxyzine CPEs and Their Potentiometic Response Characteristics, Measured in Phosphate Buffer Solutions (pH 2.0)

Electrode	Composition%			Slope (mV decade^{-1})	Linear range (M)	LOD (M)
	MIP	Graphite powder	Paraffin oil			
CPE1	0[a]	64	21	~8 (±2.0)	-	5.0×10^{-6}
CPE2	5	71	24	10.7(±1.9)	1.0×10^{-5} to 1.0×10^{-1}	7.0×10^{-6}
CPE4	7	70	23	12.9(±1.8)	5.0×10^{-6} to 1.0×10^{-1}	3.0×10^{-6}
CPE5	10	68	22	21.8(±1.5)	1.0×10^{-5} to 5.0×10^{-2}	7.0×10^{-6}
CPE6	15	64	21	29.4(±1.0)	1.0×10^{-6} to 1.0×10^{-1}	7.0×10^{-7}

[a]*Containing 15% NIP.*

electrodes, obtained in buffered solutions at the pH value of 2.0 [12]. Clearly, an increase at the MIP level in the CPE caused a slope increase of the calibration curve. The MIP-modified electrode with the graphite powder/paraffin oil/MIP percentage ratio of 64%/21%/15% was selected as the one with the optimal membrane ingredient composition. The MIP-modified electrodes exhibited Nernstian response (29.4 ± 1.0 mV/decade) in the concentration range 1.0×10^{-6} to 1.0×10^{-1} mol/L for hydroxyzine.

7.3. Selectivity

The potentiometric selectivity coefficients can be determined by different procedures, namely, the so-called fixed interference method (FIM), the separate solution method (SSM), and the matched potential method (MPM). The coefficients describe the preference of the suggested electrode for an interfering ion, X, with reference to the analyte ion. The IUPAC recommended the use of SSM and FIM, two different procedures to determine the Nicolskii coefficients of ISEs [54]. The SSM involves the measurement of two separate solutions, each containing a salt of the determined ion only. The Nicolskii coefficient is then calculated from the two observed emf values. In the FIM, an entire calibration curve is measured for the analyte ion in a constant interfering ion background, a_j(BG). The linear response curve of the electrode as a function of the analyte ion activity is extrapolated until, at the lower detection limit a_I(DL), it intersects with the observed potential for the background alone. The FIM coefficient is calculated from the resulting the lower detection limit to the interfering ion activity ratio (Eq. 11.2):

$$K_{I,X}^{MPM} = \left[\frac{\frac{a_1(DL)}{z_i}}{a_j^{z_j}} \right] \tag{11.2}$$

The MPM was introduced in the mid-1980s by Gadzekpo and Christian to provide a selectivity formalism that would give empirically more meaningful results [55]. According to the MPM method, the specified activity (concentration) of the primary ions is added to

TABLE 11.5 Potentiometric Selectivity Coefficient Values, $K^{Pot}_{Hyd,X}$, for a Hydroxyzine Potentiometric Sensor

Interfering, X	Log $K^{MPM}_{Hyd,X}$	Log $K^{FIM}_{Hyd,X}$	Interfering, X	Log $K^{MPM}_{Hyd,X}$	Log $K^{FIM}_{Hyd,X}$
Cetrizine	−1.8	−1.9	Pyrrole	−4.2	−4.3
Promethazine	−3.9	−4.1	Aniline	−4.2	−4.2
Salbutamol sulfate	−4.1	−4.2	KNO_3	−3.3	−3.5
Piperazine	−3.3	−3.3	$NaNO_3$	−4.8	−4.1
Metochlorpramide	−4.1	−4.4	NaCl	−4.8	−5.0
Pipyridine	−3.4	−3.5	$Ca(NO_3)_2$	−3.5	−3.6
$N(CH_2CH_3)_3Cl$	−3.5	−3.7	$Mg(NO_3)_2$	−4.0	−4.2
Terazosine	−4.5	−4.6			

a reference solution, and the potential is measured. In another experiment, the interfering ions (X) are successively added to an identical reference solution, until the measured potential matches that obtained before the addition of the primary ions. The MPM selectivity coefficient, $K^{MPM}_{I,X}$, is then given by the ratio of resulting primary ion activity (concentration) to the interfering ion activity increases in the two experiments (Eq. 11.3):

$$K^{MPM}_{I,X} = \Delta a_1/\Delta a_X \tag{11.3}$$

The MPM and FIM selectivity coefficients for the hydroxyzine ion-selective electrode at the constant pH value of 2.0 are listed in Table 11.5 [12]. As is obvious from this table, when the MIP sensor is applied to determine hydroxyzine, all the other substances (except for cetirizine) hardly interfere with the determination. In most cases, the selectivity coefficients are on the order of 5.0×10^{-4} and lower. The MPM selectivity sequence of the employed MIP for different drugs and organic materials (Fig. 11.9) approximately obeys the order: cetirizine > piperazine > pipyridine > tri-ethyl ammonium chloride > promethazine > salbutamol ≈ metochlorpramide > pyrrole ≈ aniline. As a result, the MIP molecular recognition is based both on the template molecular structure (shape) and on the interactions between the print molecule and the imprinted polymer. Cetirizine is of a similar chemical structure, in which an acetic acid ($-CH_2-COOH$) moiety presents instead of the ethanol ($-CH_2-CH_2-OH$) moiety in the aliphatic chain. This behavior confirmed that this molecule could interfere in the hydroxyzine detection.

7.4. Response Time

In previous IUPAC recommendations, response time is the average time required for the electrodes to reach a potential response within ±1 mV of the final equilibrium value, or has reached 90% of the last value, after successive immersions in a series of target solutions, each having a 10-fold concentration difference [54]. More recently, it has been extended to be able to treat drifting systems as well. In this case, the second time instant is defined as the one at which the emf/time slope (dE/dt) becomes equal to a limiting value [56]. Fig. 11.10 shows the

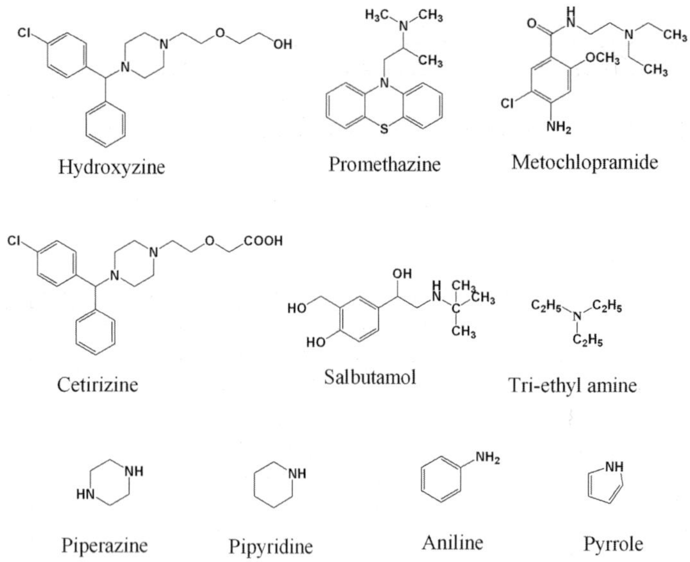

FIGURE 11.9 Structure of the investigated drugs and organic materials as interferences in MIP-based hydroxyzine sensor.

resulting potential-time responses for the hydroxyzine potentiometric sensor, obtained upon changing the hydroxyzine concentration from 1.0×10^{-3} mol/L to 1.0×10^{-2} mol/L and from 1.0×10^{-2} mol/L to 1.0×10^{-3} M. It is evident that the potentiometric response of the electrode is rapid (approximately 15 s) and reversible, although the time needed to reach the equilibrium value for the case of high-to-low sample concentration is longer than that of the low-to-high sample concentration [12].

Rafailovich et al. introduced a hemoglobin-imprinted thiol modified electrode with the response time of 2–10 min [47].

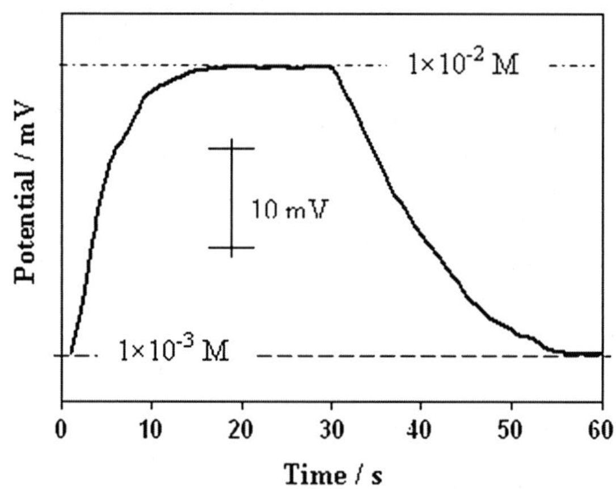

FIGURE 11.10 Dynamic response of the MIP-based hydroxyzine sensor for low-to-high (1.0×10^{-3} mol/L to 1.0×10^{-2} mol/L, left part) and high-to-low (1.0×10^{-2} mol/L to 1.0×10^{-3} mol/L, right part) step concentrations.

7.5. Lifetime

The lifetime of the MIP-based electrode can be studied by periodically recalibrating the potentiometric response to the target drug in the standard solutions. After the conditioning step, the electrode was repeatedly calibrated three times every month, and any change in the electrode performance was recorded.

The average lifetime for most of the reported PVC-plasticized membrane electrodes is in the range of 1–2 months [39,40,42–46]. After this time, a slow deterioration of selectivity and response will be observed. It is well established that the loss of plasticizer or ionic site from the polymeric film due to leaching into the sample is a primary reason for the limited lifetimes of ISEs [33]. From this point, the chemically modified electrodes were found to have significant advantages over their corresponding plasticized PVC membranes. Since these electrodes did not contain any special plasticizer or membrane solvent, they were more durable and less toxic than the plasticized PVC membranes to be applied to ion sensors.

MIP-modified electrodes for hydroxyzine and cetirizine demonstrated the successful fabrication and durable lifetime, and so no significant change in the electrodes' performance during 5 months was observed [12,31].

7.6. Accuracy and Precision

Usually, the accuracy of the measurements by means of the potentiometric sensors is checked by calculating the recovery of a known analyte concentration.

The hydroxizine potentiometric sensor was applied to assay of hydroxizine in spiked human serum and urine samples. The results of the recovery studies are listed in Table 11.6. The recoveries of the methods are in the range of 95 to 110% and 96 to 109% for the spiked serum and urine samples, respectively [12].

For the evaluation of reproducibility, the analytical method must be repeated several times. The precision of the procedure can be reported in terms of relative standard deviation. The analytical performance of the hydroxizine potentiometric sensor was evaluated with five repeated potentiometric measurements of the 1.0×10^{-5} M hydroxizine solution in different laboratories. The precision of the described procedure in terms of relative standard deviation was 6.5% [12].

TABLE 11.6 Recovery Results for the Hydroxyzine Potentiometric Sensor in Urine and Serum Samples

Amount added (M)	Urine		Serum	
	Amount found[a] (M)	Recovery%	Amount found[a] (M)	Recovery%
5.0×10^{-6}	$5.40(\pm0.34) \times 10^{-6}$	109	$5.51(\pm0.24) \times10^{-5}$	111
1.0×10^{-5}	$1.06(\pm0.06) \times 10^{-5}$	104	–	–
5.0×10^{-5}	–	–	$5.06(\pm0.05) \times10^{-5}$	106
1.0×10^{-4}	$1.02(\pm0.09) \times 10^{-4}$	103	$1.4(\pm0.37) \times 10^{-4}$	107
5.0×10^{-4}	–	–	$4.83(\pm0.19) \times 10^{-4}$	94

[a] *Average of three determinations.*

7.7. Robustness/Ruggedness

Robustness is defined as the persistence of a system's characteristic behavior under perturbations or conditions of uncertainty. In other hands, robustness is a measure of its capacity to remain unaffected by small but deliberate variations in the analytical procedure parameters. A robust chemosensor capable of specifically sensing the presence of chloropropanols in complex sample matrices with adequate sensitivity was reported by K.P.L. Mitch and coworkers [57]. This MIP-based sensor showed good selectivity for 3-chloro-1,2-propanediol (3-MCPD) since both 3-MCPD and 1,3-DCP often coexist as contaminants in epoxy resins used in the paper industry and in food products.

Acknowledgments

Financial support provided by the Amirkabir University of Technology (Tehran, Iran) is acknowledged.

References

[1] K. Haupt, K. Mosbach, Chem. Rev. 100 (2000) 2495.
[2] M.C. Blanco-Lopez, M.J. Lobo-Castanon, A.J. Miranda-Ordieres, P. Tunon-Blanco, Trends Anal. Chem. 23 (2004) 36.
[3] A. Merkoci, S. Alegret, Trends Anal. Chem. 21 (2002) 717.
[4] M. Javanbakht, H. Namjumanesh, B. Akbari-Adergani, Talanta 80 (2009) 133.
[5] M. Javanbakht, N. Shaabani, B. Akbari-Adergani, J. Chromatogr. B 877 (2009) 2537.
[6] M. Javanbakht, A.M. Attaran, M.H. Namjumanesh, M. Esfandyari-Manesh, B. Akbari-adergani, J. Chromatogr. B 878 (2010) 1700.
[7] L.I. Andersson, J. Chromatogr. B 739 (2000) 163.
[8] L. Zhenying, D. Hongquan, H. Yanping, L. Zhaosheng, Prog. Chem. 20 (2008) 747.
[9] B. Pejcic, R. De Marco, Electrochim. Acta 51 (2006) 6217.
[10] M.C. Blanco-Lopez, S. Gutierrez-Fernandez, M.J. Lobo-Castanon, A.J. Miranda-Ordiers, P. Tunon-Blanco, Anal. Bioanal. Chem. 378 (2004) 1922.
[11] B.R. Arnold, A.C. Euler, A.L. Jenkins, O.M. Uy, G.M. Murray, John Hopkins Apl. Tech. Digest 20 (1999) 190.
[12] S. Javanbakht, A. Eynollahi Fard, M. Mohammadi, M.R. Abdouss, P. Ganjali, L. Norouzi Safaraliee, Anal. Chim. Acta 612 (2008) 65.
[13] B. Sellergren, C.J. Allender, Adv. drug delivery rev. 57 (2005) 1733.
[14] M. Esfandyari-Manesh, M. Javanbakht, F. Atyabi, A. Badiei, R. Dinarvand, J. Appl. Polym. Sci. 121 (2011) 1118.
[15] R. Levi, S. McNiven, S.-H. Cheong, K. Yano, I. Karube, Anal. Chem. 69 (1997).
[16] E.L. Holthoff, F.V. Bright, Anal. Chim. Acta 594 (2007) 147.
[17] M. Avila, M. Zougagh, A. Rios, A. Escarpa, Trends Anal. Chem. 27 (2008) 54.
[18] T. Prasada Rao, R. Kala, S. Daniel, Anal. Chim. Acta 578 (2006) 105.
[19] S.A. Piletsky, A.P.F. Turner, Electroanalysis 14 (2002) 317.
[20] G.M. Murray, A.L. Jenkins, A. Bzhelyansky, O.M. Uy, Johns Hopkins Apl. Tech. Digest. 18 (1997) 464.
[21] T. Prasada Rao, K. Prasad, R. Kala, J.M. Gladis, Crit. Rev. Anal. Chem. 37 (2007) 191.
[22] T. Prasada Rao, R. Kala, Talanta 76 (2008) 485.
[23] D. Kriz, R.J. Ansell, in: B. Sellergren, F. Lanza (Eds.), Molecularly Imprinted Polymers, Elsevier, New York, 2001, pp. 417–440.
[24] K. Haupt, Analyst (Cambridge, U.K.) 126 (2001) 747.
[25] N. Kirsch, J.P. Hart, D.J. Bird, R.W. Luxton, D.V. McCalley, Analyst 126 (2001) 1936.
[26] N. Mostafa, M. Rasoul, Drug Testing and Analysis 3 (2011) 132.
[27] V. Vishnuvardhan, K.P. Prathish, G.R.K. Naidu, T. Rao, Electrochim. Acta 52 (2007) 6922.
[28] K.P.G. D'Agostino, G. Alberti, R. Biesuz, M. Pesavento, Biosens. Bioelectron 22 (2006) 145.
[29] X. Xu, G. Zhou, H. Li, Q. Liu, S. Zhang, J. Kong, Talanta 78 (2008) 26.

[30] M. Lahav, A.B. Kharitonov, O. Katz, T. Kunitake, I. Willner, Anal. Chem. 73 (2001) 720.
[31] S. Javanbakht, M. Eynollahi Fard, A. Abdouss, M.R. Mohammadi, P. Ganjali, L. Norouzi Safaraliee, Electroanalysis 20 (2008) 2023.
[32] E. Bakker, P. Buhlmann, E. Pretsch, Chem. Rev. 97 (1997) 3083.
[33] E. Lindner, R.P. Buck, Anal. Chem. 72 (2000) 336A.
[34] Rachkov, N. Minour, J. Chromatogr. A 889 (2000) 111.
[35] M. Javanbakht, N. Shaabani, M. Abdouss, M.R. Ganjali, A. Mohammadi, P. Norouzi, Cur. Pharm. Analysis 5 (2009) 269.
[36] M.M. Moein, M. Javanbakht, B. Akbari-adergani, J. Chromatogr. B 879 (2011) 777.
[37] M. Javanbakht, S. Mohammadi, M. Esfandyari-Manesh, M. Abdouss, J. Appl. Polym. Sci. 121 (2011) 1118.
[38] T. Rosatzin, L.I. Andersson, W. Simon, K. Mosbach, J. Chem. Soc. Perkin Trans. 2 (1991) 1261.
[39] K. Prasad, K.P. Prathish, J. Mary Gladis, G.R.K. Naidu, T. Prasada Rao, Sens. and Actuators B 123 (2007) 65.
[40] K.P. Prathish, K. Prasad, T. Prasada Rao, M.V.S. Suryanarayana, Talanta 71 (2007) 1976.
[41] S. Sadeghi, F. Fathi, J. Abbasifar, Sens. and Actuators B 122 (2007) 158.
[42] R. Liang, R. Zhang, W. Qin, Sens. and Actuators B 141 (2009) 544.
[43] T.C.M. Felismina, R.L.G. Joana, L.U.A. Vera, H.K. Ayman, G.F.S. Maria, Anal. Methods 2 (2010) 2039.
[44] H.M.V. Oliveira, F.T.C. Moreira, M.G.F. Sales, Electrochim. Acta 56 (2011) 2017.
[45] F.T.C. Moreira, A.H. Kamel, J.R.L. Guerreiro, M.G.F. Sales, Biosens. Bioelectron. 26 (2010) 566.
[46] F.T.C. Moreira, M.G.F. Sales, Mat. Sci. & Engin. C 31 (2011) 1121.
[47] Y. Wang, Y. Zhou, J. Sokolov, B. Rigas, K. Levon, M. Rafailovich, Biosens. Bioelectron. 24 (2008) 162.
[48] D. Tonelli, B. Ballarin, L. Guadagnini, A. Mignani, E. Scavetta, Electrochim. Acta, in press.
[49] Y. Zhou, B. Yu, K. Levon, Biosens. Bioelectron 20 (2005) 1851.
[50] A. Kugimiya, K. Kohara, Mat. Sci. & Engin. C 29 (2009) 959.
[51] C. Liang, H. Peng, X. Bao, L. Nie, S. Yao, Analyst 124 (1999) 1781.
[52] T.S. Ma, S.S.M. Hassan, Organic Analysis Using Ion-Selective Electrodes, vol. 1, Academic Press, London, 1982.
[53] A.L. Jenkins, R. Yin, J.L. Jensen, Analyst 126 (2001) 798.
[54] G.G. Guilbault, R.A. Durst, M.S. Frant, H. Freiser, E.H. Hansen, T.S. Light, E. Pungor, G. Rechnitz, N.M. Rice, T.J. Rohm, W. Simon, J.D.R. Thomas, Pure Appl. Chem. 48 (1976) 127.
[55] V.P.Y. Gadzekpo, G.D. Christian, Anal. Chim. Acta 164 (1984) 279.
[56] R.P. Buck, E. Lindner, E. Pure, Appl. Chem. 66 (1994) 2527.
[57] K.P.L. Mitch, K.W.C. Brenda, H.W.L. Michael, Anal. Chim. Acta 491 (2003) 145.

CHAPTER

12

Conductive Polymers for Plastic Electronics

Iva Chianella[1,*], *Konstantinos Karras*[1], *Dhana Lakshmi*[1], *Michael J Whitcombe*[1], *Silvia Marson*[2]

[1]Cranfield Health, Vincent Building, Cranfield University, Cranfield, MK43 0AL, UK

[2]School of Applied Science, Cranfield University, UK

[*]Corresponding Author: Cranfield Health, Vincent Building, Cranfield University, Cranfield, MK43 0AL, UK

OUTLINE

1. INTRODUCTION

Since the beginning of the last century, polymers have experienced a constant increase in their applications, perhaps more than any other type of material. The success seen by polymers is mainly due to important advantages such as their wide range of properties and their almost endless flexibility in processability [45]. In general, polymers have been used in the past because of their attractive chemical, mechanical, and electrically insulating properties [52]. This was true until the late 1970s, when scientists discovered that an organic polymer

Molecularly Imprinted Sensors.
DOI: 10.1016/B978-0-444-56331-6.00012-8

like polyacetylene could be rendered highly conductive when oxidized or reduced, a process named doping [57]. Since then, scientists have devoted great efforts in the attempt to synthesize successfully conductive polymers, which combine the processing and mechanical properties of conventional polymers with the electrical and optical properties of metals.

Today the use of conductive polymers in commercial products is still limited, due to the partial success achieved in producing materials with high conductivity and real plastic characteristics. Once better conductive plastics are developed, the potential applications could be endless ranging from organic bioelectronics in nanomedicine [58] to plastic electronic components for sensors and biosensors—also in the form of field-effect transistors and integrated circuits—to photovoltaic devices such as solar cells and to organic-based light emitting devices [7,52]. The main challenge encountered when attempting to use conductive polymers as plastic electronics is their limited processability using the methods normally utilized by the polymer industry (e.g., injection molding, extrusion). The difficulty in trying to process conductive polymers using such standard methods arises from the fact that these materials are not intrinsically thermoplastic [16]. Over the last two to three decades scientists have tried several different approaches to produce thermoplastic polymers with high conductivity. In this chapter, after a brief history of conductive polymers, these approaches are reviewed, with particular emphasis on polyaniline as one of the most promising conductive materials. A section that describes injection molding and micro-injection molding and highlights the thermoplastic characteristics required by the material used for the process is also included.

2. CONDUCTIVE POLYMERS—A BACKGROUND

The intrinsic ability of certain polymers to conduct electricity has been known for some time; it was discovered as far back as 1862 when Letheby produced a partly conductive material by the anodic oxidation of aniline in a sulfuric acid solution, which in all likelihood was polyaniline [23]. Polypyrrole was discovered in the 1960s, but little was understood about the mechanism of its electrical conduction, and this knowledge was essentially lost [2,17]. An inorganic and explosive polymer, polysulfur-nitride $(SN)_x$, was discovered in the early 1970s and was found to have superconductive properties at very low temperatures [23].

Polyacetylene was the first conductive polymer to be fully explored and understood and led to an increasing interest in this field. While Natta and colleagues were the first to prepare this polymer in 1958, their work was not widely recognized [56]. A new synthetic procedure developed by Shirakawa allowed the controlled production of nearly all—*cis*-polyacetylene or all-*trans*-polyacetylene pictured in Fig. 12.1, though these materials still possessed relatively modest conductivity in the order of 10^{-1} to 10^{-7} S/m [23].

In 1976 Shirakawa began a collaboration with Professors Heeger and MacDiarmid, who had extensive experience with $(SN)_x$ polymers. They used their knowledge regarding the doping of these materials to further explore the doping of polyacetylene, with some preliminary work having already been done by Shirakawa. An iodine doped all-*trans*-polyacetylene was produced with a conductivity of 3000 S/m, a vast improvement over the undoped form. Further experimentation with both the *cis* and *trans* forms led to the creation of doped polyacetylene with conductivity up to eleven orders of magnitude larger than the undoped state [23]. These findings fostered a heightened interest in conducting polymers and the many

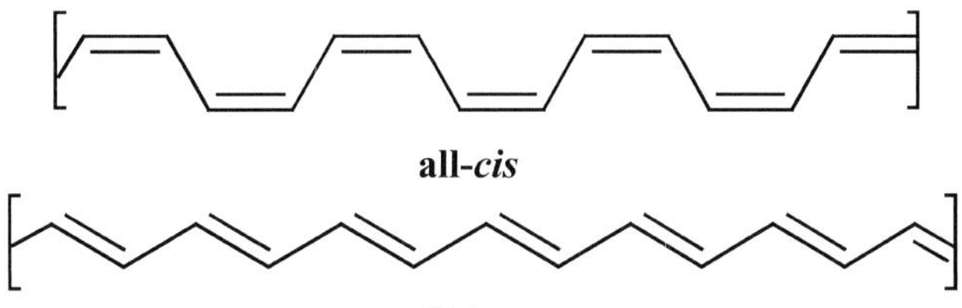

FIGURE 12.1 Molecular structure of all—cis- and all-trans-polyacetylene.

potential applications in which they could be employed, leading to Shirakawa, Heeger, and MacDiarmid being awarded the Nobel Prize in Chemistry in 2000.

The basis of conductivity in conducting polymers, also known as conjugated polymers, is due to the presence of π-conjugated bonds located along the carbon backbone, which allows electrons to move freely along the polymer chain. When a polymer is conjugated, it can often be represented as a continuous network of adjacent unsaturated carbon atoms in the sp^2 hybridized state. The carbon atom possesses the electronic structure $1s^2 2s^2 2p^2$. In conventional polymers, the C-atoms are sp^3 hybridized, and each of them forms four σ-bonds. In such nonconjugated polymers, the electronic structure of atom chains contains only σ-bonds, with a large electron energy band gap that renders these materials electrically insulating. On the other hand, in conjugated polymers the C-atoms are sp^2 hybridized and form 3 σ-bonds with a remaining electron in a p_z atomic orbital, which establishes π-bonds with the p_z-orbital of the nearest neighbor. It is the weak nature of these π-bonds that leads to the state of electron delocalization, forming a one-dimensional band within the material [52]. This delocalization occurs as the lone electron in a repeat unit is attracted to the nuclei in the adjacent units. In addition, as the p_z orbital, which is involved in the π-bonds, contains only one electron, the band can be considered to be only half filled and therefore metallic in nature [8]. In their ground state, the charge mobility is limited, and the conductivity of the material is therefore in the semiconductive range. It is by the doping process that the conductivity can be increased by several orders of magnitude. In fact, due the presence of the delocalized π-electron system, the polymer chains can be easily oxidized or reduced, usually by charge transfer with dopant species (chemical doping). The most common form of doping is p-doping whereby the chain is chemically oxidized (one electron removed) using counterions, creating a positively charged "hole" that can act as charge carrier in the polymer. Other forms of doping include electrochemical, photochemical, and interfacial doping, where the charge is changed without the use of counterions [23]. In Fig. 12.2 a scheme illustrating the available mechanisms of doping with the correspondent possible applications is reported.

With appropriate doping the conductivity can be increased even more than 12 orders of magnitude transforming the material from an insulator, through a semiconductor, and finally a metal [8].

Many of the conducting polymers that have been developed and comprehensively investigated so far include polypyrrole, polythiophene, and its many derivatives, polyphenylene-vinylene as well as polyaniline.

As can be seen in Fig. 12.2, conductive polymers have the potential to be employed as materials in a wide range of applications as electronic components: namely, electrodes, antistatic EMI shielding, photovoltaic devices, and so on. However, as already mentioned, the

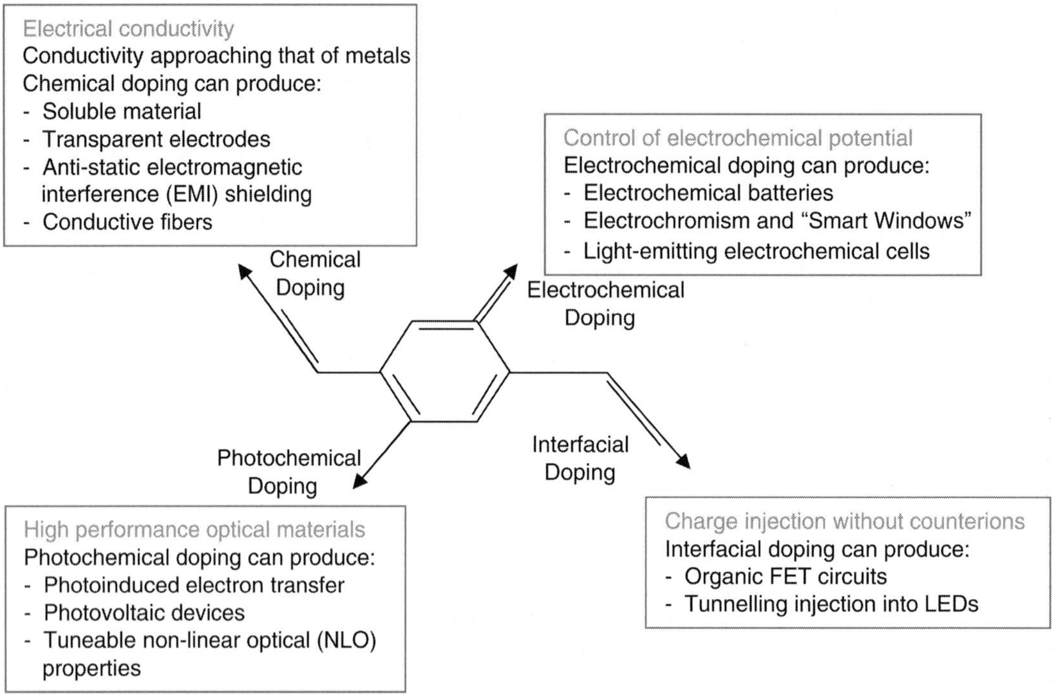

FIGURE 12.2 Doping mechanisms and related applications. *(Adapted from Heeger, 2001).*

commercial success of conjugated polymers has been hampered by their lack of processability, due to their very poor solubility in common solvents, limited infusibility, as well as their lack of thermoplastic behavior.

Manufacturers producing electronic equipment, like many other types of industries, are under constant pressure to decrease the size of equipment, increase manufacturing volume, and maintain profitability. This demand has led to production techniques that facilitate the increasing complexity of fabricated parts made in great quantities, while simultaneously decreasing costs. One field that has made great strides in achieving these goals is that of injection molding, which allows incredibly complex plastic parts of both very small and large dimensions to be inexpensively made in large volumes. A subset of this technique, called micro-injection molding, has more recently sprung forth from this field, allowing even smaller, more intricate plastic parts to be made using conventional thermoplastic polymers. A brief section describing injection molding, with emphasis on micro-injection molding, is reported in the next section.

3. INJECTION AND MICRO-INJECTION MOLDING (μIM)

Injection molding is a well-known and widely used technique for the high-volume manufacture of plastic parts. Technological advances have necessitated a need for continued miniaturization of components, as more functions need to be included in devices that seem to be perpetually decreasing in size and increasing in complexity. Micro-injection molding (μIM) is the technology that makes use of suitable processes and machines for the manufacture of miniaturized components with features in the microscale. Many advantages of injection molding, including the suitability to high-volume manufacturing, are transferred to μIM. As such, miniaturized polymer parts, even if constructed from high-end moldable materials such as conductive polymers based on polyaniline, polypyrrole, and polythiophenes, could be used in applications requiring inexpensive, disposable components [22]. Other advantages of μIM include the extensive range of thermoplastic materials available, the possibility of complete automation with decreased cycle times, as well as the ability to accurately control dimensions. In addition, much of the theoretical and practical knowledge gained from the many years that injection molding has been utilized can be directly applied to μIM [3]. Micro-injection molding now appears to have become one of the leading solutions for the medium- to large-volume production of polymer-based miniaturized components [19,20].

Similarly to conventional injection molding, μIM is the process of transferring a thermoplastic material in the form of granules from a hopper into a heated barrel so that it becomes molten and soft. The soft material is then forced at high speed and high pressure inside a mold cavity containing features with dimensions in the micrometer range (or larger features with tolerances in the micrometer range) [51]. The softened material is then maintained at a temperature typically below—but very close—to the glass transition temperature (Tg) of the polymer. Once in the mold the polymer is subjected to a holding pressure to compensate for the material shrinkage occurring when passing below the Tg. After sufficient time, the material freezes into the mold and gets ejected, and the cycle is repeated. A typical cycle lasts between a few seconds and a few minutes [3,20].

The characteristics that make μIM appealing over conventional injection molding for the manufacture of conductive miniaturized polymer parts are as follows:

Capability of manipulating small amounts of material (shot weight smaller than 0.1g).
Better process control thanks to micromolding machine design that suits the manufacture of high-precision, small-dimension parts.
Very accurate shape replication and good dimensional control.

Small residence time of the polymer in the melting unit, which allows using easily-degrading polymers. However the production of miniaturized parts by μIM imposes a number of requirements on the polymer that will need to withstand harsh processing conditions without degrading or producing defective parts:

High temperatures during the plasticization phase. A better replication of the micromold features is achieved by decreasing the polymer viscosity. However, this is obtained by increasing the melting temperature during the plasticization phase. This condition might result in polymer degradation within the plasticating unit.
High shear rates. The molten polymer travels into narrow runners and gates (sub-millimeter dimensions) to fill the micrometer-sized cavities and as such is subject to high shear rates (10^5–10^6 s^{-1}). Molten polymers tend to exhibit an increase in viscosity at increased shear rates, which might result in polymer degradation while being forced into the mold cavities.

High holding pressures. Following injection, a holding pressure is applied to the polymer in the feeding system to counteract its shrinkage and produce completely filled parts. High holding pressures can result in excessive internal stresses and polymer degradation.

Rapid cooling. This is a phenomenon caused by the high relative contact surface area of the polymer with the micromold cavity. The rapid cooling caused by the contact of the flowing polymer with the colder cavity wall, in combination with high shear rates, leads to a skin forming around the flowing polymer that might result in defective parts [34].

Demolding. The demolding of miniaturized parts can be particularly challenging when the diameter of the ejection pins is comparable to the overall dimension of the parts (mm to sub-mm). If the pressure applied by the ejection pins is too large, the parts can be subject to high mechanical stresses, which might result in parts deformation and even polymer discoloration [35].

The thermoplastic polymers commonly used in micro-injection molding are special grades (high melt flow index) of the traditionally nonconducting or insulating polymers. The development of the relatively new field of conducting polymers in conjunction with the use of micro-injection molding as the forming technology may result in the development of new, inexpensive, and easily mass-produced products such as components for plastic electronics. One of the most prominent strategies that have been sought so far to achieve melt processability of conducting polymer has been the formation of polymer blends containing a conductive polymer and a conventional thermoplastic one. Due to their easy preparation and environmental stability, polyaniline, polypyrrole, and polythiophenes are the conductive materials more commonly used for the preparation of polymer blends [16]. In the next section a comprehensive review of the polymer blends containing polyaniline as conductive polymer is reported.

4. POLYANILINE-BASED POLYMER BLENDS

Among the group of intrinsically conducting polymers, polyaniline (PANI) is known to have a combination of favorable properties: sensitivity to protonation leading to ease of doping, environmental stability, high electrical conductivity, low cost, ease of synthesis, as well as the ability to be easily dispersed in a thermoplastic polymer matrix [54]. As such it will be the main focus of this review. PANI has a very long history, first being made by Runge in 1834 [18]. When in its nonconducting form, it can exist in a wide continuum of oxidation states, ranging from the completely reduced form (leucoemeraldine) to the completely oxidized form (perniganiline), with the emeraldine base situated between them [44]. The intermediate states of oxidation are termed protoemeraldine and nigraniline [26]. The most commonly used form of polyaniline is the emeraldine form, in which the imine sites can be readily protonated, leading to a marked insulator-conductor transition as positive charges are added, while the number of π electrons remains unchanged. The formation of an emeraldine salt in this manner leads to a significant change in the properties of the polymer with new optical and paramagnetic characteristics as well as the development of electrical conductivity [44]. When a conducting polymer is used to form a blend with a thermoplastic polymer, thermal stability is a property of great importance as the conducting polymer must remain stable during the melting process, when exposed to the high temperatures, which are usually several degrees higher than the thermoplastic polymer's melting point or *Tg*. The emeraldine base form of polyaniline has high thermal stability, enduring temperatures up to 420 °C before undergoing substantial loss of material, whereas the thermal stability of doped forms is significantly decreased by several degrees depending on the dopant utilized.

The decomposition of the doped forms occurs in two stages: the first after the loss of the counter anion and the second following the breakdown of the polymer's backbone. Doping with hydrochloric acid and sulfuric acid leads to reduced thermal stability compared to doping with methane sulfonic acid [27]. Doping with *p*-toluene sulfonic acid has also been demonstrated to produce thermally stable forms of polyaniline and polypyrrole [54]. One of the major problems encountered with the use of PANI is its relative insolubility in most organic solvents, more so when it is in its doped forms. Initial methods aimed at improving the solubility and therefore the processability of PANI involved dissolving the polymer in substances such as N-methylpyrrolidone (NMP), pyrrolidine, tripropylamine, or in concentrated sulfuric or other strong acids. With specific regard to melt processing, the use of highly concentrated acids is not feasible due to the increased degree of equipment corrosion [60,67]. Despite these discoveries, there was still great difficulty in achieving blends of PANI and common thermoplastic polymers with good electrical conductivity due to the inability to dissolve the doped forms of PANI in common nonpolar or moderately polar organic solvents [67].

4.1. The Counterion-induced Processability Approach

In 1992, in the attempt to overcome the difficulty of creating blends of PANI and common thermoplastic polymers soluble in organic solvents, Cao and colleagues demonstrated that by utilizing a procedure termed counterion-induced processability, they were able to simultaneously create a doped form of PANI that was also soluble in commonly used solvents. The process utilized functionalized protonic acids, which are conventionally denoted as H^+ (M^- -R), where H^+ M^- is a protonic acid group and R is an organic group. The proton interacts with the PANI imine nitrogens, converting the emeraldine base into the conducting salt form, while the counterion species (M^- -R) contains an "R" functional group that is specifically chosen so as to impart solubility in common solvents.

This process is well demonstrated by an acid commonly used for this purpose, dodecylbenzenesulfonic acid (DBSA), where the counterion anion species M^- is SO_3^- and the R functional group is a dodecylbenzene [9,72]. The dodecylbenzene group with its long alkyl chain imparts solubility in common solvents such as toluene, decalin, chloroform, and xylenes, thereby rendering PANI compatible with bulk polymers, which contain similar molecular structures, such as polyolefins. By utilizing specific functional groups such as camphor sulfonic acid (CSA), *o*-anisidine-5-, *p*-chlorobenzene, *p*-toluene sulfonic acid (*p*TSA), or 4-nitrotoluene-2-, solubility in a wide range of solvents like CH_3Cl, formic acid, and dimethyl sulfoxide (DMSO) can be imparted to PANI, so as to allow compatible mixing with a chosen bulk polymer [9,72]. Cao and co-workers demonstrated this concept by using CSA to dope PANI forming PANI-CSA complexes, which are soluble in *meta*-cresol [9]. As *meta*-cresol is an excellent solvent for poly (methyl methacrylate) (PMMA), this allowed the formation of polyblends of PANI and PMMA. These materials displayed good electrical conductivity of 1 S/cm even in blends, where the conductive component constituted less than 2% of the total weight fraction. It was thought that this indicated an atypical morphology, where connected pathways could be formed even when extremely low fractions of the PANI-CSA complex were added, as displayed in Fig. 12.3 [9,67].

Transmission electron microscopy (TEM) was performed on a variety of these PANI-CSA-PMMA films (arrowed in Fig. 12.3) containing different concentrations of the PANI-CSA complex. The micrographs confirmed that an interconnected fibrillar network of PANI-CSA had formed with clear interspersed voids. Of note is that despite the fact that the networks did become less dense, the interconnected mesh pattern remained clearly visible as the concentration decreased from 28% down to 1.9%. At 0.96%, connections were present, but were more tenuous, and it was determined that this was approximately the concentration at which conducting pathways start forming macroscopically in these blends. This characteristic lends itself to a smooth increase in electrical conductivity—an attribute that was in stark contrast to polymer blends up to this point in time. The prevailing belief involving polymer blends was that as conducting particles were dispersed in insulating host polymers, the electrical conductivity was seen to increase suddenly at a volume fraction (v) equivalent to the percolation threshold ($v \approx 0.16$ (16%) to 0.3 (30%)—dependent on the size and aspect ratio of the particles used), which was predicted for spherical conducting particles in a nonconductive matrix.

Commonly used blends up to this point suffered from troubles in controlling the customarily sharp percolation behavior, difficulty in processing homogeneous conducting materials, as well as problems with sloughing (e.g., carbon black) and excessive cost (though not in the case of carbon black) [54,60,67]. While a lower percolation threshold (6–10%) for electrical conductivity had been achieved by Shacklette and co-workers, who reported the dispersion

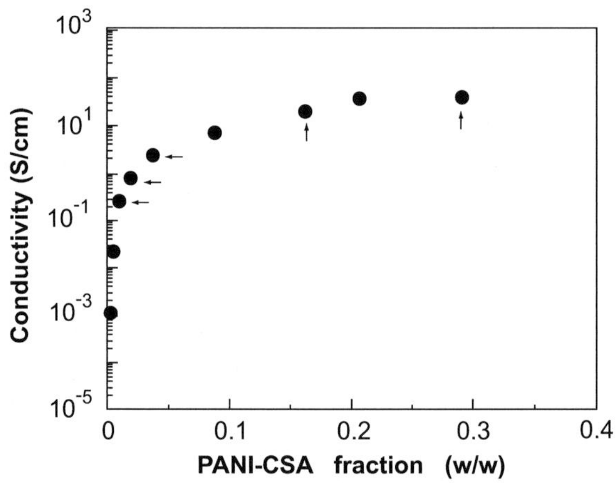

FIGURE 12.3 Electrical conductivity of solution processed PANI-CSA-PMMA polyblends as a function of weight fraction (*f*) of PANI-CSA complex. The arrows in the figure indicate the concentrations of polyblends that were investigated using TEM. *[Reprinted from Yang et al., 1993, Morphology of conductive, solution-processed blends of polyaniline and poly (methyl methacrylate).* Synthetic Metals, *53, 293–301. Copyright (1993), with permission from Elsevier].*

of doped PANI (Versicon™) within polar thermoplastic matrix polymers such as polycaprolactone and poly (ethyleneterephthalate glycol), it was determined that this was due to the tendency of the spherical PANI particles to partially reaggregate following dispersion, forming chain-like structures causing the earlier onset of conductivity [54]. With many of the polyblends based on the use of counterion-induced processability, there is no evidence of a sharp percolation threshold, and the onset of conductivity occurs at a point at least one order of magnitude lower than the accepted theoretical percolation threshold [67].

Counterion-induced processability has been used to create polyblends of PANI with several different polymers such as nylon, polyethylene (PE), polycarbonate (PC), polystyrene (PS), thermoplastic polyurethane (TPU), poly(vinyl acetate) (PVA), polypropylene (PP), poly(vinyl chloride) (PVC), poly(vinyl chloride) nitrile rubber blend (PVC-NBR), polysulfone (PSU), acrylonitrile-butadiene-styrene copolymer (ABS), and polystyrene-*block*-poly(ethylene-*ran*-butylene)-*block*-polystyrene (SEBS). All of these yielded similar findings with regard to electrical conductivity, albeit with a difference in the rate of increase of electrical conductivity depending largely on the functionalized protonic acid used [6,9,32,53,59]. An example of note was the formation of blends of PANI-DBSA with PE, where electrical conductivity was observed to increase at a slower rate relative to the weight percentage of conductive polymer added, as compared to the previous example of PANI-CSA-PMMA. In this case, gel processing was used to make blends from dilute solutions, after which the films were drawn, facilitating partial orientation of the PANI complex within the PE.

When considering the use of these blends in a variety of applications, a property that is of major significance is their stability under a wide range of conditions. A number of stability tests were performed by Davies and colleagues on a 50% by weight PANI-DBSA in a plasticized chlorinated polyethylene (CPE) blend. The change in conductivity over 70 days with the material exposed to a temperature of 70 °C was minimal, indicating significant stability over an extended period of time. In addition, tests of exposure to UV light, ozone, as well as exposure to 70 °C, while the material was immersed in water, were also performed, all of which indicated adequate stability [14].

Another technique, called thermal doping, allows the doping of PANI and the addition of alkyl groups to the PANI backbone without the need for adjuvant solvents. In thermal doping, DBSA and PANI are manually mixed together in a weight ratio of DBSA:PANI of 3:1 and then mixed in a proper apparatus at a high temperature (140 °C) for 5 minutes. As the DBSA is used in excess, a significant amount of DBSA remains unbound, forming an ionically conducting DBSA paste into which the PANI particles are dispersed. Once heated, the paste becomes a semisolid material, due to a phase transition occurring at the point where the PANI $(DBSA)_x$ salt complex is formed, coinciding with a sharp irreversible rise in electrical conductivity with values ranging from 1 to 20 S/cm. If DBSA is added in large excess, a plasticized complex is formed [24,72]. PANI-DBSA formed through thermal doping, termed Neste Complex (Neste Oy—Finland), has been successfully blended with several different nonpolar matrix polymers such as high-density polyethylene (HDPE), low-density polyethylene (LDPE), polystyrene (PS), polyvinyl chloride (PVC), and polypropylene (PP). The electrical conductivity of these blends was observed to increase smoothly and at a low level of wt% of PANI-DBSA, as can be seen in Fig. 12.4.

It was determined that by using additional plasticizers, developed subsequently by Neste Oy (Finland), the resulting polymer blends could be processed using many processing

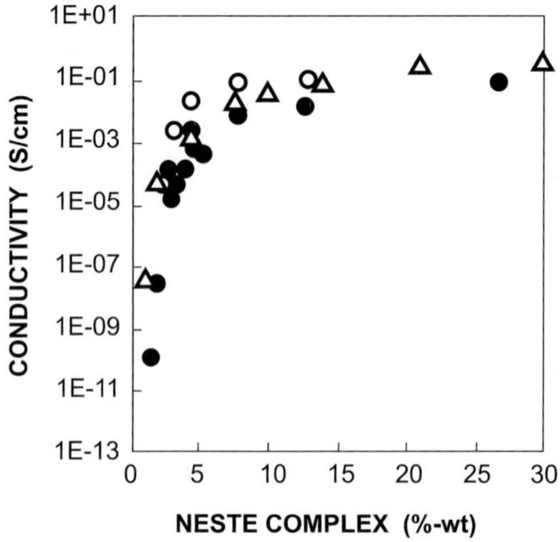

FIGURE 12.4 Electrical conductivity of Neste Complex melt blended with HDPE (•), LDPE (o) and PP (△). *Reprinted from Ikkala et al., 1995, Counterion-induced processability of polyaniline: Conducting melt processable polymer blends,* Synthetic Metals, *69, 97–100. Copyright (1995), with permission from Elsevier.*

methods such as injection molding, compression molding, and blow molding [60]. The technique behind the creation of the infusible PANIPOL™ formulations was similar to injection molding and involved an extrusion process, whereby PANI salts, the plasticizers, and any other possible additives were fed into an extruder at a temperature of 100–180 °C. They were then left in the equipment to solidify into strands of conducting material, which were then cooled and made into pellets. These pellets can then be formed into composites or blends with conventional thermoplastic polymers and undergo further processing into chosen products using conventional melt-processing techniques. The characteristic of the final blends and specifically the conductivity depends on many parameters such as the weight fraction of PANIPOL™, the level of plasticization, the processing temperature and the viscosity ratio of PANIPOL™, and the thermoplastic polymer at the selected processing temperature.

A different approach was used by Ruckenstein and co-workers to achieve conductive thermoplastic blends. They used an inverted concentrated emulsion pathway. Oxidative polymerization of aniline was performed by adding and stirring a solution of the oxidizing agent ammonium persulfate (APS) and the doping agent (HCL) into a concentrated emulsion composed of the host polymer and aniline dispersed in an ionic surfactant solution (sodium dodecylsulfonate) (SDS) [49,50]. Coprecipitation of the polyaniline formed and the host polymer was then performed, creating the polymer blend. Using this method, blends of PANI with polystyrene (PS), poly (alkyl methacrylates) including PMMA, poly(ethyl methacrylate) (PEMA), poly(butyl methacrylate) (PBMA) and combinations thereof, poly(methyl methacrylate-co-butyl acrylate) latex particles, and even rubber were produced. Depending on several factors such as the mole ratio of oxidant to aniline, concentration of HCL, and type of solvent used, conductivity from 0.1 up to 5 S/cm was achieved and the percolation threshold was reached with a small-volume fraction of PANI. In addition, dependent on the host polymer and processing route utilized, a variety of mechanical attributes ranging from strong and brittle to soft and flexible could be engineered into the materials without much change in electrical conductivity [49,50,67].

A similar route was taken at a later date by Haba and co-workers, though in this case DBSA was used, serving as both surfactant and dopant. Initially, aniline and DBSA were mixed in a stochiometric ratio to yield insoluble anilinium-DBSA complex particles. Polymerization of the aniline was achieved by the addition of APS at a temperature of 0 °C for 5 hours. Stable PANI–DBSA dispersions were created, which had electrical conductivity ranging from 3.5×10^{-3} S/cm up to 0.7 S/cm when in powdered form, depending on the duration of polymerization [21].

Rubinger and colleagues have used sulfonated polystyrene for the in-situ polymerization of aniline. The use of this compound can facilitate both the doping of PANI and the formation of an easily processable electrically conducting composite. The electrical conductivity of the products varied according to the sulfonation degree of the polystyrene, with the best values being achieved with the highest degree (30% sulfonation). This composition formed a homogeneous mixture, with electrical conductivity adequate for use in electronic devices [48].

Progress in facilitating PANI melt processability was also made through use of phosphoric acid diesters. These compounds have sufficient acidity to serve as doping agents as well as concurrently improving solubility due to their hydrophobic alkyl chains. In addition, they have well-documented polymer plasticizing properties. In the first examples of this approach, two dialkyl esters were used, namely, bis-(methylpropyl) hydrogen phosphate (DiBHP) and bis-(2-ethylhexyl) hydrogen phosphate (DioHP) [43]. Doping of already formed PANI was achieved either by treatment with a solution of the diesters dissolved in an appropriate solvent (toluene, THF or *m*-cresol, decalin, chlorinated hydrocarbons) or by mechanical mixing of the PANI and neat diesters. Blends with thermoplastic polymers [e.g., polystyrene, PMMA, and poly(acrylonitrile-butadiene-styrene)] can be obtained by solution casting at different temperatures, with those done at 160 °C displaying the useful electrical conductivity characteristics inherent to counterion-induced processing. Unfortunately, the thermal stability of these ester doped PANIs is limited; degradation occurred at temperatures greater than 140 °C, and more so with the DioHP diester [43]. Thermal stability was significantly improved through the use of Aryl diesters of phosphoric acid formed from diphenyl hydrogen phosphate, which allowed melt processing at temperatures up to 200 °C [29]. Other phosphoric acid esters used include the monoester 3-pentadecylphenylphosphoric acid (PDPPA), 3-pentadecylphenyl(bis)phosphoric acid [PDP(bis)PA], and monocardanylphosphoric acid (MCPA), which form complexes with PANI that are highly soluble in nonpolar and weakly polar organic solvents, with resultant heavily plasticized blends exhibiting thermal stability at temperatures above 200 °C [41,42].

The same research group also used sulfonic acid derivatives of 3-pentadecyl phenol (PDP), which also possess long, flexible *n*-alkyl tails, imparting high solubility in commonly used solvents, and in addition they have plastification properties, allowing the doped PANI to be melt processed. The PANI complexes demonstrated thermal stability up to 200 °C, and electrical conductivity of free-standing flexible films was as high as 65 S/cm [40].

Polyblends of PANI polyacid complexes with polyurethane (PU) were formed by Yang and Lee [66] and blended with two types of water-soluble polyurethane. Aniline was dissolved in a solution containing polystyrensulfonic acid (PSSA), polyacrylic acid (PAA), and poly (2-acrylamido-2-methyl-1-propanesulfonic acid) (PAMPSA) to which APS was added with continuous rapid stirring. Unreacted aniline monomer was removed, and the product was dried, then re-dissolved in an aqueous solution and blended with two types of water-based PU (PU1 and PU2) in varying ratios. The electrical conductivity of the blends was most affected by the type of PU, used with an order of magnitude difference between the best values recorded for each, these being 0.5 S/cm and 5 S/cm for PU1 and PU2, respectively. The difference in electrical conductivity was attributed to the presence of aromatic groups in the structure of PU2, whereas PU1 consisted only of aliphatic groups. Most likely, the aromatic–aromatic interactions between PU2 and PANI improved their compatibility. This was further clarified by observation of the blends using TEM, with the PU2 blends displaying a smooth, fully interconnected network structure, whereas the PU1 blends were far more globular in nature, with only some connections between PANI particles [66].

Liao and co-workers used a sequential polymerization technique to form an interpenetrating polymer network (IPN) of PANI-DBSA and maleimide-terminated PU (UBMI). An IPN was defined by IUPAC in 2007 as "a polymer comprising two or more networks, which are at least partially interlaced on a molecular scale, but not covalently bonded to each other and cannot be separated unless chemical bonds are broken. Note: a mixture of two or more preformed polymer networks is not an IPN." Liao's group also observed large differences dependent on the type of PU utilized. Polyether UBMI demonstrated better compatibility with PANI-DBSA than polyester UBMI, leading to increased electrical conductivity and superior mechanical properties. This was confirmed by SEM, which demonstrated a high degree of roughness within the polyester blend indicating a heterogeneous morphology, whereas a smooth and homogeneous morphology was observed with the polyether blend. Scanning tunnel microscopy demonstrated a discontinuous distribution of the PANI-DBSA network within the polyester UBMI due to phase separation, whereas a continuous interpenetrating network of PANI-DBSA had formed with the polyether UBMI [32].

A further example of incompatibility forming an obstacle in polymer blends was provided by Zhang and colleagues, who attempted to further improve on the gains made by counterion-induced processing by creating PANI containing ternary polymer blends. Initially, PANI-CSA was blended with PMMA, and in a separate experiment PANI-CSA was blended with CPA (a copolymer composed of caprolactam and decylamide in a 1:1 ratio). Using a melt-processing technique, each blend was then mixed with low-density polyethylene (LDPE) in a rheometer at a temperature of 160 °C. Through the use of transmission electron microscopy (TEM) and differential scanning calorimetry (DSC), it was determined that phase separation of the individual components had occurred due to incompatibility. The microscopic analysis highlighted a cloud-like form and faint fibrillar nature in the PANI-CSA/CPA/LDPE and PANI-CSA/PMMA/LDPE blends, respectively, with DSC demonstrating no change in the melting points of the individual components. In addition, electrical conductivity measurements were relatively low at 8×10^{-7} S/cm and 1.5×10^{-7} S/cm for the PMMA and CPA containing blends, respectively [71].

More recently, "compatibilizers" have been developed to aid the compatibility of the separate components in polymer blends. An example of this technique is the use of a block copolymer of styrene and butadiene (SBS) to help the formation of blends of PS with PANI, which had been doped with DBSA and PSS (polystyrene sulfonic acid). It was determined that there was improved interaction between individual components, as observed in morphological characterization of the material. This indicated that the PANI-DBSA-PSS particles were more extensively and favorably distributed, leading to corresponding increases in electrical conductivity [36].

Bae and co-workers used an in-situ polymerization technique to form PANI/PS blends. Of note were their observations on the differences in PANI structure when polymerization was performed at different temperatures, namely, 0 °C and 15 °C. Using field emission scanning microscopy (FESEM), it was observed that use of the lower temperature led to a more interconnected network, while the higher temperature caused the formation of a more granular matrix composed of smaller particles. As one would expect, this led to a significant difference in electrical conductivity from 1.81 S/cm to 7.86 S/cm for the samples created at higher and lower temperatures, respectively. In addition, it is believed that lower temperatures are advantageous when forming PANI through oxidative polymerization due to the exothermic nature of the process [4].

Del Castillo-Castro and co-workers developed a technique to improve compatibility between PANI chains and PS host polymers, through the use of poly(styrene-metal acrylate) ionomers. Poly(styrene-*co*-potassium acrylate) (PSAK), poly(styrene-*co*-sodium acrylate) (PSANa), and poly(styrene-*co*-lithium acrylate) (PSALi) were synthesized by emulsion polymerization of styrene and the specific metal acrylate in varying ratios of styrene to metal acrylate. Composites were formed by the addition of one of the ionomers to a solution of aniline in DBSA, followed by oxidative polymerization of the aniline using APS under a N_2 atmosphere at 5 °C. The potassium acrylate containing blends had the highest electrical conductivity of all the composites, and when mixed with poly(n-butyl methacrylate) and formed into films exhibited higher conductivity than films made from the same components, but lacking the ionomer. IR spectroscopy suggested that a reaction takes place between the acid groups of the ionomer and the amine groups of PANI, creating amine salt groups, capable of imparting higher thermal stability to the composites [15].

Wang and colleagues have employed a mixture of an organic acid (*p*-toluene sulfonic acid) (*p*-TSA) and an inorganic one (phosphoric acid) (PA) to dope TPU (thermoplastic polyurethane)/PANI blends using in-situ polymerization. The combination of acids was chosen on the premise that *p*-TSA would improve processability, while PA would enhance conductivity. A variety of combinations were produced by altering the percentage of aniline and the ratio of the acids added. The blend with the highest electrical conductivity (8.3 S/cm) had 20% (by weight) of aniline added with a ratio of 2:1 (by molar ratio) of *p*-TSA:PA. At higher percentages of aniline, electrical conductivity decreased; the authors report that the likely reasons were that at the 20% level of aniline, there was an optimal degree of both aromatic—aromatic interactions of aniline with the hard segments (aromatic diisocyanate) of the TPU and hydrogen bonds with the soft segments (ether oxygen groups) of TPU, leading to in-situ formation of PANI within the TPU network structure. Most likely, at higher levels of aniline there was a greater tendency for the aniline monomers and oligomers to react with each other rather than with TPU, forming PANI that was removed during post—treatments [62]. The composites formed were soluble in commonly used solvents such as dimethylformamide (DMF). Of further interest were the results of the DSC analysis performed on the blends formed with 20% aniline. DSC showed a characteristic sharp endothermic peak at 165 °C, thought to be due to polymer crystal melting (see Fig. 12.5). This could be attributed either to the melting of polyether

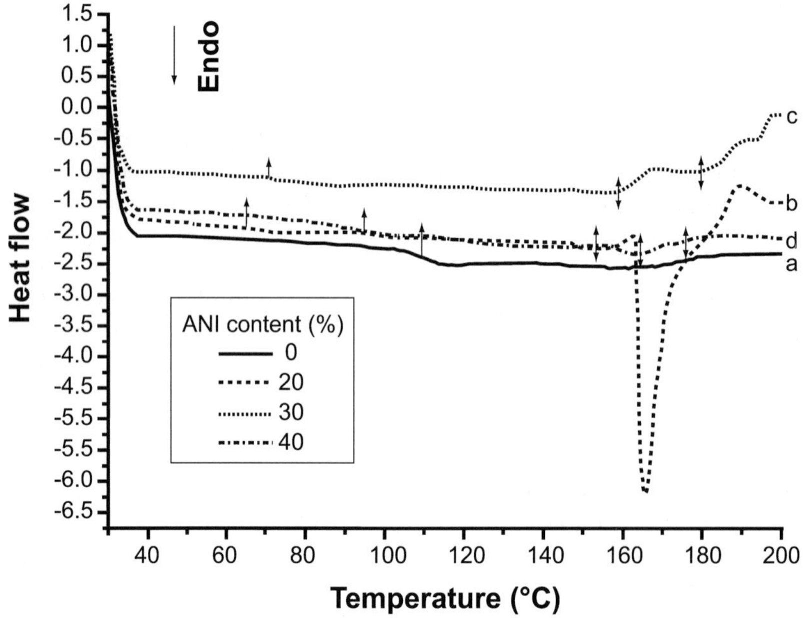

FIGURE 12.5 DSC curves of the in-situ TPU/PANI composites with varying aniline contents. *Reprinted with permission from Wang et al., 2010,* Journal of Applied Polymer Science, *115, 1886, Copyright 2010, John Wiley and Sons.*

crystals that may have formed in the TPU or to the melting of PANI oligomer crystals trapped within the network structure. If so, the latter would be the first finding of this kind and could hint at the possibility of preparing PANI that would be as processable as any other regular thermoplastic polymer [62].

Creation of polymer blends to improve the solubility and processability of PANI is not the only approach used so far by scientists. Alternative methods involving the chemical modification and copolymerization of PANI are being explored in attempts to render it more soluble and therefore processable. Many of the examples that have been reported in the literature are reviewed in the next section.

4.2. Modifications of PANI by Various Methods: Substitution and Copolymerization

PANI is not very easy to manipulate due to the presence of aromatic rings, interchain hydrogen bonds, and charge delocalization effect. Therefore, one of the simpler ways to modify PANI is by chemical substitution to make it more processable and soluble in various solvents. Some of the methods used in recent years are ring substitution, *N*-alkylation, and protonation with functionalized protonic acids, such as dodecylbenzene sulfonic acid and camphorsulfonic acid.

Alkyl substituents on the polymer backbone have considerably improved the solubility of PANI in doped and undoped forms in common organic solvents and even in water. A few researchers have tried to make colloidal forms of this substituted PANI by using polymeric surfactants [1].

Electrophilic substitution on the aromatic rings of PANI has enhanced to some extent its solubility and processability, although the presence of substituents also has the effect of reducing PANI's electrical conductivity. Therefore, attention has been given to sulfonation to make PANI more soluble and processable for blend applications. Yue and Epstein in 1990 [70] tried to dissolve the emeraldine base of PANI in a fuming sulfuric acid solution and extracted a green powdered product, which was further precipitated using methanol and acetone. The resulting sulfonated PANI possessed sulfonation of 50% of the aromatic rings and was termed SPAN (sulfonated PANI). It was highly soluble in basic solutions, and the measured electrical conductivity was in the order of 0.1 S/cm and independent of pH up to a value of 7. This level of conductivity was achieved without the need for further doping and was therefore termed self-doped.

Additional studies were made by Wei and Epstein in 1995 in order to improve the degree of sulfonation by further derivatizing SPAN (highly sulfonated polyaniline or HSPAN) using an electrophilic substitution reaction. Higher sulfonation of the aromatic rings (75%) was achieved by using the leucoemeraldine base form dissolved in fuming sulfuric acid. Solubility in alkaline solutions was improved by 1.5 times, and electrical conductivity also increased by one order of magnitude to 1 S/cm, this being independent of pH throughout the pH range. It was thought that the improvement in degree of sulfonation was a result of the use of the fully reduced form of polyaniline, which by possessing the highest electron density of all the PANI forms, allowed more complete sulfonation and improved yield in the subsequent electrophilic ring substitution reaction (Wei and Epstein, 1995). Despite this result, the general trend is a decrease in conductivity with an increase in sulfonation degree, with electrical conductivity ranging from 5 S/cm for unsulfonated PANI to 0.02 S/cm for 45% sulfonated PANI. Only SPAN is the exception, by possessing a sulfonation degree greater than 40%, and by being completely soluble in dilute alkaline solutions while also maintaining its self-doped behavior [61].

Electrophilic substitution reaction was also used to sulfonate poly (N-methylaniline) (PNMANI) using fuming sulfuric acid, which led to a polymer, termed SPNMANIes, containing both sulfonate and tertiary amine groups [5]. The same researchers have also reported different methods for functionalization of PANI and making its copolymers. The other methods used by Barbero and colleagues include nucleophilic addition and coupling with diazonium salts. Figure 12.6 describes some of the methods used to prepare SPNMANIes employing electrophilic substitution and nucleophilic addition reactions. The nucleophilic addition was performed by adding the sulfonate groups to a stirred solution of the emeraldine base PNMANI in a saturated solution of hydrogen sulfite (HSO_3^-), while pure O_2 was bubbled through the solution, forming the product SPNMANIna. Other nucleophilic additions have also been demonstrated by the authors, all occurring by addition to the quinonimine units in the oxidized form of PANI. These techniques facilitate simple modification of PANI by adding functional groups directly to the backbone, improving solubility and miscibility with commonly used polymers such as poly(methyl-methacrylate). Some of these methods can be chemically or electrochemically controlled, allowing a quantitative tailoring of the modification.

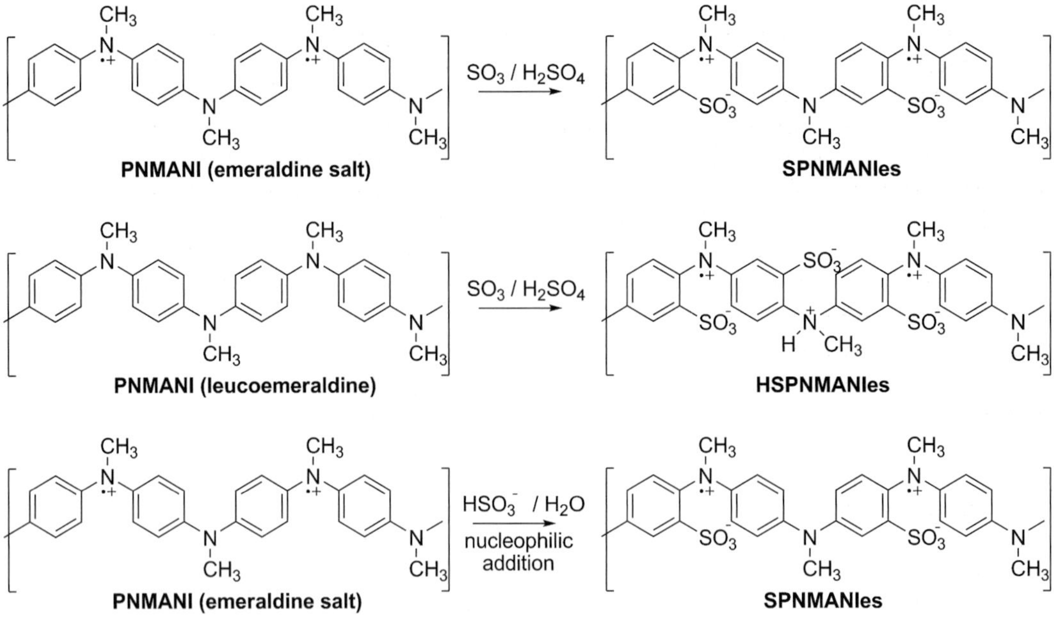

FIGURE 12.6 Alternative routes of sulfonated poly (*N*-methylaniline) synthesis. Top and middle rows—electrophilic substitution reactions. Bottom row—nucleophilic addition reaction. *Reprinted from Barbero et al., 2004, Novel synthetic methods to produce functionalized conducting polymers I. Polyanilines,* Electrochimica Acta, *49, 3671. Copyright 2004, with permission from Elsevier.*

Chen and Hwang in 1994 demonstrated that sulfonation also involves substitution at the nitrogen position rather than on the aromatic ring. Emeraldine base was modified by exposure to 1,3-propanesulfone to form poly(aniline-*co*-*N*-propanesulfonic acid-aniline) (PAPSAH), a self-doped PANI derivative with conductivity of 0.015 S/cm that was soluble in water.

Lin and Chen produced a completely water-soluble self-doped sulfonated PANI, poly (aniline-*co*-*N*-benzoylsulfonic aniline) (PABSAH) by treating PANI with *o*-sulfobenzoic anhydride. Conductivity of PABSAH was 4.7×10^{-4} S/cm without need of external doping or inorganic acid in the purification step. Being highly soluble in water, stable solutions of PABSAH and many other conventional polymers soluble in water can be formed, for example, poly (vinyl alcohol), polyacrylamide, poly (acrylic acid), and hydroxycellulose [33]. As opposed to sulfonated polyaniline, once the polymer has been already formed, techniques have been developed to use sulfonated aniline monomer derivatives in the polymerization process. Metanilic acid (*m*-anilinesulfonic acid) is one of such monomers and was polymerized using the same approach used to make PANI [55]. The polymer that was formed had a low molecular weight and poor electrical conductivity (3×10^{-6} S/cm) thought to be due to steric hindrance as well as the low reactivity of the monomer due to the highly deactivating influence of the electron-withdrawing properties of the sulfonic group. It was felt that the addition of an electron donating group to the monomer would improve reactivity. 2-Methoxyaniline-5-sulfonic acid was chosen for this purpose due to its electron donating methoxy group [55]. Polymerization using the monomer with the additional methoxy group led to formation of a higher molecular weight polymer with electrical conductivity of 0.04 S/cm at room temperature. Even though there was an improvement, the low conductivity of this sulfonated PANI was thought to be due to a greater twist of the aromatic rings as well as a larger interchain separation due to the added methoxy groups and greater number of sulfate groups [30,55].

A new form of sulfonated PANI was synthesized by 4-aminodiphenylamine-2-sulfonic acid. The resulting PANI was self-doped and soluble in aqueous alkaline solutions but had electrical conductivity in the semiconducting range (1.3×10^{-5} S/cm) [47].

The low electrical conductivity was thought to be due to a higher level of disorder within the polymer structure evidenced by EPR (electron paramagnetic resonance) spectroscopy, which demonstrated identical g-values but a wider peak to peak (line-width) size when compared to analogous sulfonated polyanilines [47].

Self-doping sulfonated PANI can also be formed by utilizing copolymerization of two monomers when one or both contain a sulfonate group. This has been performed with aniline and metanilic acid using $(NH_4)_2S_2O_8$ (ammonium peroxidisulfate) or Fe $(ClO4)_3$ $9H_2O$ (Iron (III) perchlorate) as the oxidizing agent. The copolymer formed had a 50% sulfonation degree and higher thermal stability than polyaniline doped with Iron (III) perchlorate [37]. This process of copolymerization allows the production of tailor-made polymers with precisely designed properties. As evident in the example just mentioned, a common method to produce PANI involves copolymerization of aniline with monomers that are similar to aniline, except that one of the aromatic ring hydrogens is substituted with another functional group, imparting new characteristics such as improved solubility.

It is well-known that studies of copolymerization could lead to knowledge of the reactivities of monomers and their relationship with the chemical structure of the monomers and, therefore, to a better understanding of the mechanism of polymerization. Copolymerization also greatly increases the ability of polymer scientists to tailor-make a material with specifically desired properties. It is for these reasons that scientists are trying to make various copolymers of aniline with substituted anilines to improve solubility and processability. Substituted aniline derivatives can be formed by substitution reaction. For examples, *o*-, *m*-, or *p*-toluidine can be formed when one hydrogen on the aromatic ring is substituted with a methyl group, and *o*-, *m*-, or *p*-anisidine can be obtained when a methoxyl group is substituted [64]. Several polyaniline derivatives based on these alkyl and alkoxyl ring-substituted anilines have been formed, which have very different solubility and electronic characteristics as demonstrated by poly(*o*-ethoxyaniline), which was found to soluble in water. In 1990 Wei and colleagues were able to successfully form true copolymers of aniline with *o*- and *m*-toluidine, using chemical and electrochemical methods. The chemical copolymerization utilized APS as the doping agent for aniline and various molar fractions of the specific toluidine in a 1M HCl solution. Electrochemical copolymerization involved the use of cyclic potential sweep techniques to deposit copolymer films onto a platinum working electrode that had been placed in a 1M HCL solution containing the monomers in varying ratios. The resulting copolymers, namely, poly (aniline-*co*-*o*-toluidine) and poly (aniline-*co*-*m*-toluidine), were compared to homopolymers of the monomers used, that is, polyaniline, poly (*o*-toluidine), and poly (*m*-toluidine), which were prepared using the same conditions. IR spectroscopy demonstrated a marked similarity between the spectra of the copolymers and the polytoluidine spectra, whereas there were distinct differences when compared to the polyaniline spectra. This was thought to indicate the presence of a high proportion of toluidine units in the copolymers. NMR spectroscopy also demonstrated similar spectra for both the toluidine-based copolymers and homopolymers, whereas in the spectra obtained for polyaniline there was no signal in the region assigned to methyl protons. The composition of the copolymers could therefore be determined by the ratio of the peak areas of the aromatics to the methyl protons.

The final ratio of the monomers was compared to the ratio of monomers initially added, and it was determined that the toluidine monomers were significantly more reactive than the aniline monomers, as evidenced by the fact that even a relatively small percentage of toluidine in the initial reaction formed a high content in the copolymer. Cyclic voltammetry was performed on the materials formed by electrochemical and chemical polymerization, so as to confirm that true copolymers rather than a mixture of two homopolymers had been formed. All the $E_{1/2}$ values for the copolymers were between the values obtained for the polyaniline and polytoluidine homopolymers, and single peaks were demonstrated for each copolymer material rather than the two that would be expected had a mixture of homopolymers been formed. The cyclic voltammograms of the copolymers formed through chemical or electrochemical polymerization were essentially identical, indicating that both processes led to the formation of true copolymers. The electrical conductivity of the copolymers was between the values obtained for the homopolymers, polyaniline and polytoluidine yielding 10 S/cm and 0.1 S/cm, respectively, and was dependent on the proportion of toluidine added [64].

Pandey and co-workers used a chemical synthetic method to form the copolymer poly (aniline-*co*-*o*-anisidine), by oxidation using APS of a 1M HCl solution containing equimolar concentrations of aniline and anisidine. The respective homopolymers, namely, polyaniline and poly (*o*-anisidine), were also made using the same method for comparative characterization. UV and IR spectroscopy as well as DSC were performed on the various samples, as well as being compared to composite blends of aniline and anisidine in a 1:1 ratio—all providing evidence that a random copolymer had been formed. Poly (aniline-*co*-*o*-anisidine) was found to be soluble in a variety of commonly used solvents, including acetone, chloroform, THF, DMF, and DMSO, with the best level of solubility being achieved in N-methylpyrrolidone (NMP). An electrical conductivity of 1.2×10^{-2} S/cm was observed for the doped form of the copolymer [39]. Similarly, poly (aniline-*co*-anthranilic acid) was synthesized through APS oxidation of a 1M HCl solution containing either a 2:1, 1:1, or 1:2 molar ratio of aniline and anthranilic acid. These were compared to homopolymers of aniline and anthranilic acid prepared in the same way [12]. It was observed that as increasing amounts of anthanilic acid were added to the polymerization mixture, a reduction in the yield of the corresponding copolymer occurred, this was thought to be due to lower reactivity of the anthranilic acid monomer caused by the electron-withdrawing properties of its carboxylate group combined with its effects on steric hindrance. Elemental analysis and X Ray Photoelectron spectroscopy were perfomed so as to characterize the materials formed. The existence and amount of carboxylate group could be determined using XPS, thereby indicating the amount of anthranilic acid in the copolymer. Self-doping of the polymer by the carboxylate group was shown to occur, though this was far less effective than the doping with HCl. Electrical conductivity was of the order of 4 S/cm for the copolymer containing the least anthranilic acid, and it decreased sharply when more anthranilic acid was present to values of 7.0×10^{-3} S/cm and 2.0×10^{-5} S/cm. DSC findings indicated the occurrence of intra- and intermolecular hydrogen bonding, which was made possible by the carboxylate groups of anthranilic acid. This type of bonding usually leads to higher thermal stability, but thermogravimetric analysis (TGA) failed to show any decrease in polymer decomposition compared to the homopolymers. It was determined that the hydrogen bonding is dependent on the copolymer being in the doped state, but as the HCl is eluted at relatively lower temperature, any favorable interactions are lost once the higher temperatures are reached [12]. Using chemical synthesis and electrochemical methods, several other copolymers have been formed using aniline as one of the monomers for example, poly (aniline-*co*-*N*-(4-sulfophenyl) aniline) (PAPSA), poly (aniline-*co*-*N*-butylaniline), and poly (aniline-*co*-*o*-bromoaniline). Examples of terpolymers made of three monomers, namely, aniline, 2-pyridylamine, and 2,3-xylidine, have also been reported [31,38,46,68].

A water-soluble self-doped copolymer possessing pendant aniline dimers and sulfonic acid groups was reported by Yin and Ruckenstein by grafting polyaniline onto a water-soluble polymer. The copolymer poly (aniline-*co*-2-acrylamido-2-methyl-1-propanesulfonic acid) (PAMPANI) was formed in several stages, first being the formation of APMA (*N*-4-anilinophenyl methacrylamide) through a catalytic aminolysis reaction involving *p*−aminodiphenylamine and methyl methacrylate as well as dibutylin serving as a catalyst and phenothiazine acting as a polymerization retardant. Then AMP (2-acrylamido-2-methyl-1-propanesulfonic acid) underwent oxidative polymerization using APS for 1 hour in a water-based surfactant-free emulsion at 70 °C under a nitrogen atmosphere. At 1 hour the APMA was added in a drop-wise fashion, and the reaction allowed continuing for 4.5 hours, forming the product poly (AMP-*co*-APMA).The graft polymerization of aniline onto the poly (AMP-*co*-APMA) was then performed by the drop-wise addition of a poly (AMP-*co*-APMA) and APS solution into a 1M HCL solution containing the aniline. The product obtained was undoped with NH_4OH to form PAMPANI-NH_4, which was converted to PAMPANI using an H^+ ion exchange resin, exchanging NH_4^+ for H^+. The three major steps involved in the production are highlighted in Fig. 12.7 [69].

FT-IR (Fourier transform infrared spectroscopy) of PAMPANI-Na showed an absorption peak corresponding to the -CO-N< (tertiary amide) stretching vibration, indicating that a copolymer of aniline and poly (AMP-*co*-APMA) had formed. In addition, a shift in specific frequencies similar to those observed when emeraldine is doped with sulfonic acid was observed, indicating that PAMPANI is in a self-doped state. NMR (nuclear magnetic resonance) data were in line with the IR results and confirm the structure reported in Fig. 12.7. Electrical conductivity using four-point testing of a dried thin film of the material yielded a value of approximately 0.8 S/cm [69].

A significant step forward in the production of a truly thermoplastic, intrinsically conducting polymer was made by Carone and colleagues, when they developed a method of grafting PANI and SPAN (sulfonated PANI) onto the backbone of a carboxylated segment of the conventional thermoplastic polymer polyurethane (PEUA); the binding occurred between the free amine units of PANI or SPAN and the carboxyl group in the "hard" unit of the PU, to form amide groups and PEUAPAN and PEUASPAN, respectively. PANI in the emeraldine base and sulfonated polyaniline were synthesized by accepted methods, and polyether urethane acid (PEUA) was synthesized in a modified two-step process: (1) the formation of a pre-polymer from poly (propylene oxide) (PPO) and methylene-bis-4-4′-phenyl isocyanate (MDI); and (2) the extension of the pre-polymer through use of dihydroxymethylpropionic acid (DHMPA). The grafting procedure was performed by

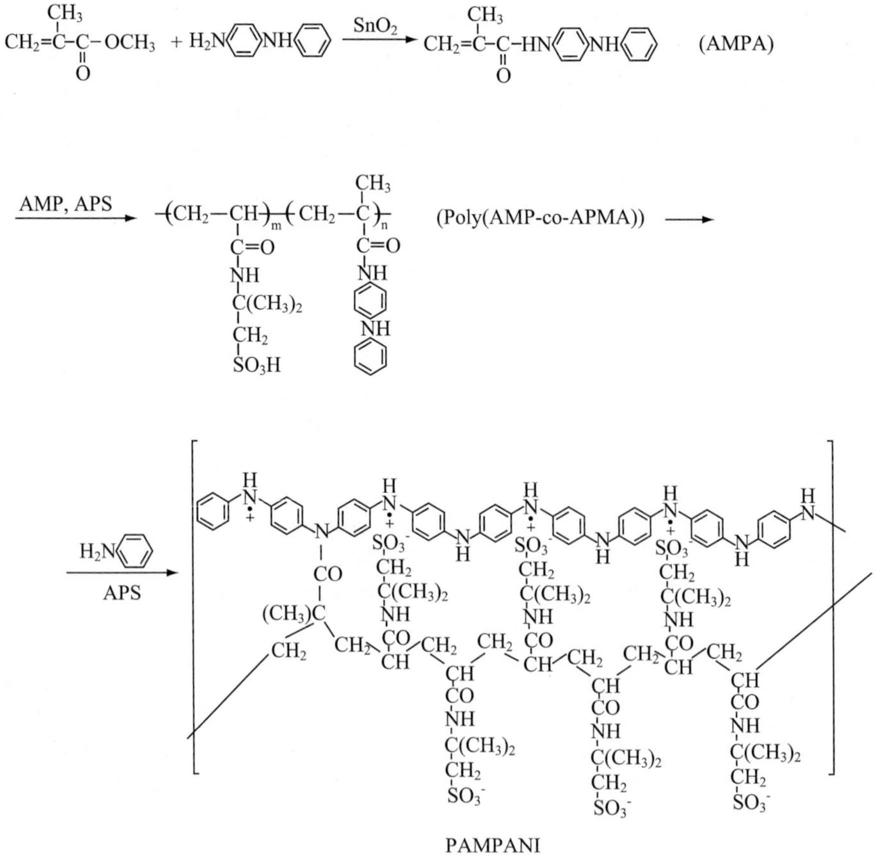

FIGURE 12.7 Schematic representation of synthesis of PAMPANI, poly (aniline-*co*-2-acrylamido-2-methyl-1-propanesulfonic acid). Top row: synthesis of AMPA (*N*-4-anilinophenyl methacrylamide). Middle row: Copolymerization with AMP (2-acrylamido-2-methyl-1-propanesulfonic acid) forming poly (AMP-*co*-AMPA). Bottom row: Graft polymerization of aniline onto poly (AMP-*co*-AMPA) forming PAMPANI. *Reprinted with permission from Yin and Ruckenstein 2000,* Macromolecules, *33, 1129. Copyright 2000, American Chemical Society.*

carbodiimide (specifically) dicyclohexylcarbodiimide and *N*-hydroxysuccinimide (NHS) in an NMP solution at 0 °C. High molecular weight PANI, insoluble in NMP, was removed by centrifugation, and the product PEUAPAN was precipitated. Unreacted products and unwanted by-products were removed by Soxhlet extraction. PEUASPAN was created using the same process [10]. ^{1}H and ^{13}C NMR was performed on the amidated and unamidated polyurethanes, with results indicating amidation of PEUAPAN and PEUASPAN polymers, with a large peak for the carboxylic group and a smaller, well-defined peak for the amidic carbon atom. By calculation of the aromatic ^{1}H signal, it was also established that there were an average of three aniline rings per PEUA repetitive unit in the PEUAPAN and one for every repetitive unit in the PEUASPAN polymer. This finding led the authors to infer that the amidation degree of the PEUAPAN may have been of the order of 60% and 20% in the PEUASPAN. Measurement by UV spectroscopy showed a peak at 600 nm, specific to the quinoid rings, which allowed determination of the amount of PANI or SPAN units per PU repetitive unit. The results were in agreement with the previous findings. FT-IR investigation confirmed the higher amount of emeraldine compared to SPAN in the respective samples.

Thermogravimetric analysis (TGA) of all the samples was made. The first weight loss of emeraldine base and SPAN occurred between 25 °C and 200 °C, due to water or solvent evaporation, with further weight loss of SPAN occurring from 200 °C to 320 °C, when the sulfonic acid groups are lost, and both samples losing weight after 400 °C due to polymer decomposition. Pure polyether urethane acid PEUA had two distinct points of weight loss at 280 °C and 350 °C due to decarboxylation and polymer decomposition, respectively. Decarboxylation was not seen with PEUAPAN due to the higher amidation degree, and with PEUASPAN the two processes of decarboxylation and desulfonation combined into one larger weight loss from 200 °C to 330 °C. The degree of maximum decomposition occurred at 380 °C for both PEUAPAN and PEUASPAN, a value marginally lower than that for PEUA, though their residual weight was higher due to the greater stability of aromatic rings compared to the other constituents in the polymers [10]. Figure 12.8 shows the thermogravimetric curves for PEUA and for the grafted copolymers PEUAPAN and PEUASPAN.

DSC analysis of the PEUA and the amidated PEUAs was performed, with all samples showing a similarly positioned soft segment glass transition. The greatest difference in the thermal behavior of the samples was made evident by the heat capacity change (ΔC_p) at the glass transition temperature (T_g) (see Fig. 12.9). It was concluded that the inclusion of PANI intensified the hard-soft phase segregation characteristics within the polymer due to the strong tendency for PANI chains to aggregate, the effect of which could be seen in the lowered ΔC_p and slight decrease in T_g values (see the thermal analysis data in Fig. 12.9).

Dynamic mechanical analysis carried out confirmed the DSC findings but also indicated a decrease in the viscous component for PEUAPAN, thought to be due to increased stiffness

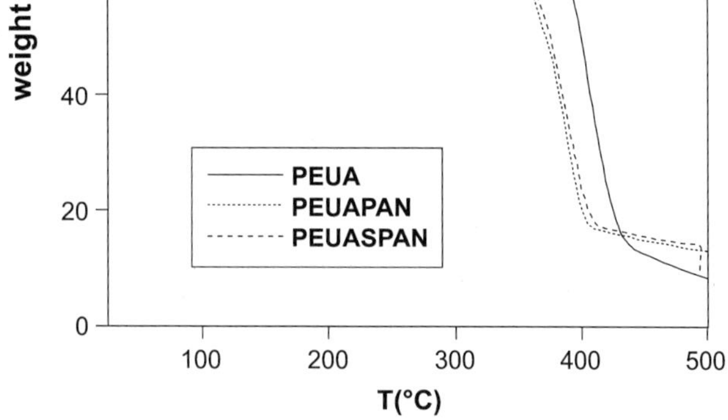

FIGURE 12.8 Thermogravimetric curves for PEUA and the grafted copolymers PEUAPAN and PEUASPAN. *Reprinted with permission from Carone et al., 2002b,* Journal of Applied Polymer Science, *86, 1259. Copyright 2002, John Wiley and Sons.*

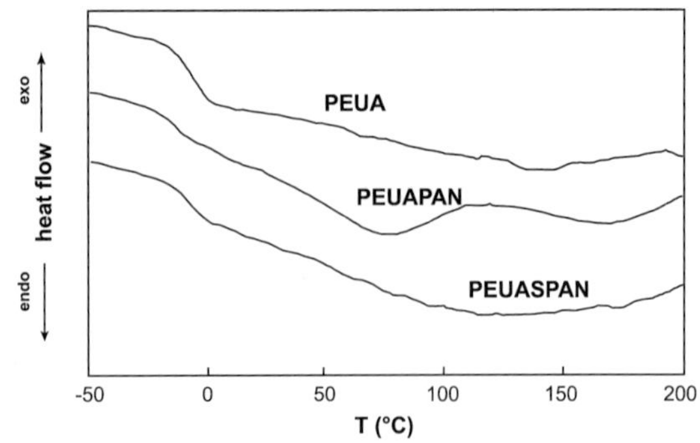

Thermal Analysis Data of PEUA, PEUAPAN and PEUASPAN

Sample	T_g (°C)	C_p (J g^{-1} K^{-1})
PEUA	-7	0.45
PEUAPAN	-13	0.16
PEUASPAN	-8	0.24

FIGURE 12.9 DSC thermograms (graph) and thermal analysis data for PEUA, PEUAPAN, and PEUASPAN. *Reprinted with permission from Carone et al., 2002b,* Journal of Applied Polymer Science, *86, 1259. Copyright 2002, John Wiley and Sons.*

as a result of the linkage of the conductive polymer to the PU backbone. Electrical conductivity was measured using the four-probe technique, yielding values of 6.1×10^{-3} S/cm and 1.5×10^{-3} S/cm for PEUAPAN and PEUASPAN, respectively.

4.3. PANI with Additional Functionalities

We have already discussed the attempts to improve the processability of PANI, including the introduction of sulfonic acid groups, to produce self-doped water-soluble polymers and the blending with polyvinyl alcohol or anionic polymers such as polyacrylic acid or polystyrene sulfonate in Section 4.2. Incorporation of PANI into composites with polyacrylamide has also been used to enhance polymer stability and to modify the mechanical properties of the material. The major problems with blending such materials, however, involve the possibility of phase separation.

In this direction, many researchers have tried to incorporate different functionalities into aniline molecules and have made copolymers by various means. When it comes to functionalize the aniline moiety, at Cranfield, our group recently synthesized two new monomers, which have polyaniline as core molecule and methacrylamide and dithiocarbamate moieties as additional functionality. Two of such monomers are *N*-phenylethylene diamine methacrylamide (NPEDMA) [28] (Fig. 12.10) and *N,N*-diethyldicarbamoyl ethyl amido ethyl) aniline (NDDEAEA) [25]. These new monomers have the potential to produce melt-processable material once copolymerized with other conventional monomers.

NPEDMA [28] possesses both an aniline and a methacrylamide group, and either or both of these groups can be polymerized by several methods, namely, free radical polymerization, chemical oxidative methods, as well as electrochemical polymerization to create polyaniline or poly(methacrylamide) forms. In addition, cross-linked polymers can also be produced when a cross-linker such as ethylene glycol dimethacrylate is integrated into the structure. The polymers created using these methods have high electrical conductivity (6.2 S/cm), on the order of polyaniline (6.9 S/cm), which was produced for comparison using the same methods. Even higher measurements of electrical conductivity were achieved in the cases where small amounts of a cross-linker were added (12.6 S/cm with 5% of EGDMA). These

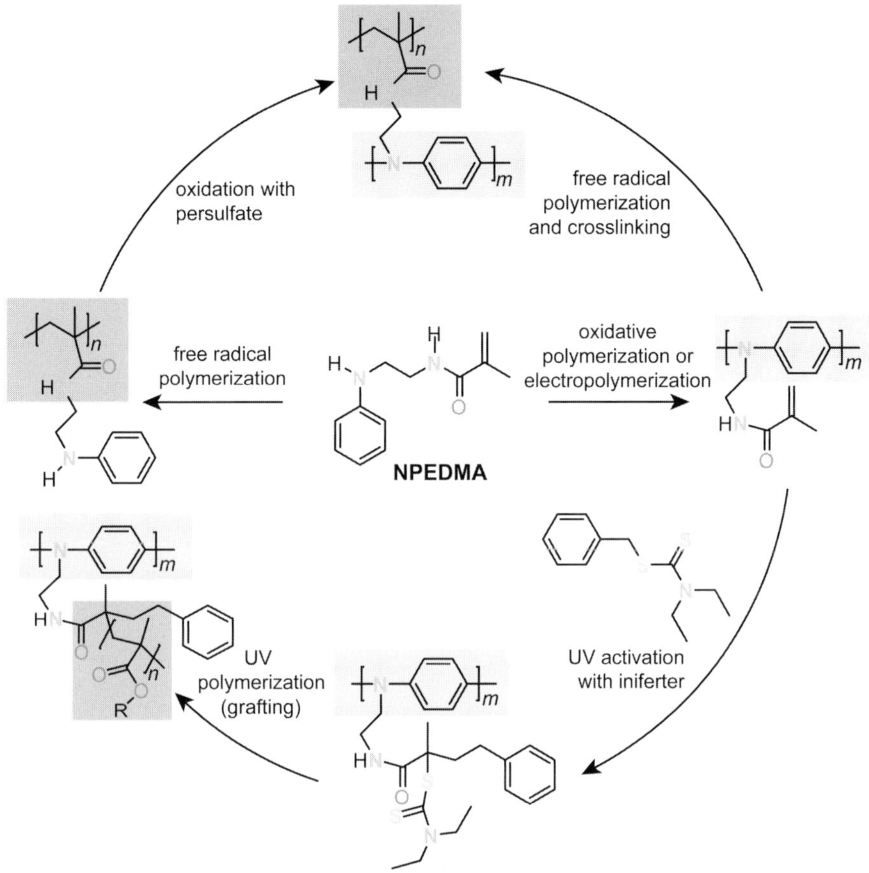

FIGURE 12.10 Orthogonal ways of polymerization shown by NPEDMA monomer. *Lakshmi et al., 2009.* Chemical Communication, *reproduced by permission of the Royal Society of Chemistry.*

results were surprising, with the expectation being that an aniline monomer possessing an N-substituted nonconjugated side chain would have lower conductivity. This seemingly paradoxical finding may be partially explained by the use of cross-linking, leading to improved proximity and enhanced interaction between polymer chains with possible effects on crystallinity and orientation of the PANI chains, supporting the movement of charge carriers through the polymer system. One of the main advantages of NPEDMA appears to be its versatility, with a variety of polymerization techniques available and many routes through which functionality can be added (see Fig. 12.10). This has led to a wide range of potential applications where NPEDMA may be employed, including electronic equipment, sensors, electrochromic devices, antistatic shields, protective coatings, photolithography, molecularly imprinted polymers, and biosensors. This type of functional monomers provides a better option for making blends and processing applications.

Another aniline-based monomer, which was synthesized at Cranfield, is *N*-(*N,N*-diethyl-dicarbamoyl ethyl amido ethyl) aniline (NDDEAEA) (Fig. 12.11).

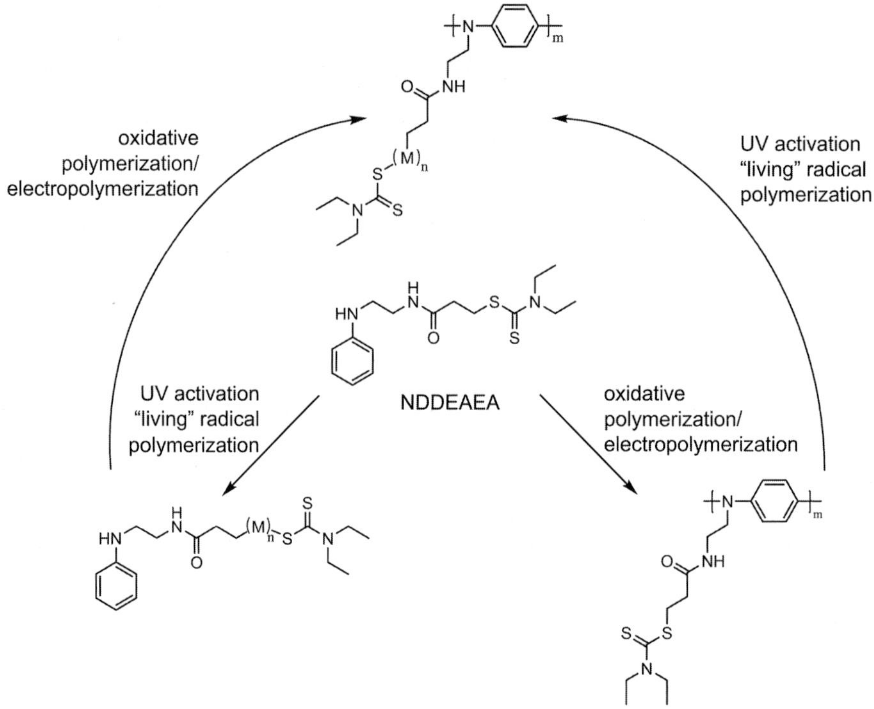

FIGURE 12.11 Schematic representation of various ways of polymerization using NDDEAEA monomer.

Conjugated poly (NDDEAEA), consisting of N-substituted polyaniline (PANI) backbones with dithiocarbamate ester pendant groups, can be synthesized using chemical and electrochemical polymerization. UV grafting can be used for grafting other polymers over poly-NDDEAEA via iniferter activation. This monomer can be used in making blends or thermoplastic via surface confined UV grafting. Initial investigations on the use of this monomer to produce thermoplastic and conductive polymers are currently being performed at Cranfield. After chemical polymerization of a mixture of aniline and NDDEAEA using APS in acidic conditions, the resulting polymer, soluble in organic solvents, was dissolved in DMF and the iniferter groups were UV grafted with methyl methacrylate. Initial results are demonstrating that although the resulting polymers still did not demonstrate a truly thermoplastic behavior by DSC analysis, electrode conductivity as high as 18 S/cm was measured with a four-point probe. This high conductivity combined with the improved solubility in organic solvent makes these novel materials encouraging candidates for application in plastics electronics.

References

[1] S.P. Armes, M. Aldissi, S. Agnew, S. Gottesfeld, Synthesis and characterization of aqueous colloidal dispersions of poly (vinyl alcohol)/polyaniline particles, Mol. Cryst. Liq. Cryst. 190 (1990) 63–74.

[2] D. Ateh, H. Navsaria, P. Vadgama, Polypyrrole-based conducting polymers and interactions with biological tissues, J. R. Soc. Interface 3 (2006) 741–752.

[3] U.M. Attia, S. Marson, J.R. Alcock, Micro-injection moulding of polymer microfluidic devices, Microfluid. Nanofluid. 7 (2009) 1–28.

[4] W.J. Bae, W.H. Jo, Y.H. Park, Preparation of polystyrene/polyaniline blends by in situ polymerization technique and their morphology and electrical property, Synth. Met. 132 (2003) 239–244.

[5] C. Barbero, H.J. Salavagionee, D.F. Acevedo, D.E. Grumelli, F. Garay, G.A. Planes, G.M. Morales, M.C. Miras, Novel synthetic methods to produce functionalized conducting polymers I. Polyanilines, Electrochim. Acta 49 (2004) 3671–3686.

[6] G.M.O. Barra, R.R. Matins, K.A. Kafer, R. Paniago, C.T. Vasques, A.T.N. Pires, Thermoplastic elastomer/polyaniline blends: Evaluation of mechanical and electromechanical properties, Polym. Test. 27 (2008) 886–892.

[7] S. Bhadra, N.K. Singha, D. Khastgir, Semiconductive composite form ethylene 1-octene copolymer and polyaniline coated nylon 6: Studies on mechanical, thermal, processability, electrical, and EMI shielding properties, Polym. Eng. Sci. 48 (2008) 995–1006.

[8] D. Bott, "Electrically conducting polymers," Phys. Technol. 16 (1985) 121–126.

[9] Y. Cao, P. Smith, A.J. Heeger, Counter-ion induced processibility of conducting polyaniline and of conducting polyblends of polyaniline in bulk polymers, Synth. Met. 48 (1992) 91–97.

[10] E. Carone, L. D'ilario, A. Martinelli, New conducting thermoplastic elastomers. I. Synthesis and chemical characterization, J. Appl. Polym. Sci. 83 (2002a) 857–867.

[11] E. Carone, L. D'ilario, A. Martinelli, New conductive thermoplastic elastomers. Part II. Physical and chemical-physical characterization, J. Appl. Polym. Sci. 86 (2002b) 1259–1264.

[12] H.S.O. Chan, S.C. Ng, W.S. Sim, K.L. Tan, B.T.G. Tan, Preparation and characterization of electrically conducting copolymers of aniline and anthranilic acid: Evidence for self-doping by x-ray photoelectron spectroscopy, Macromolecules 25 (1992) 6029–6034.

[13] S.A. Chen, G.W. Hwang, Synthesis of water-soluble self-acid-doped polyaniline, J. Am. Chem. Soc. 116 (1994) 7939–7940.

[14] S.J. Davies, T.G. Ryan, C.J. Wilde, G. Beyer, Processable forms of conductive polyaniline, Synth. Met. 69 (1995) 209–210.

[15] T. Del Castillo-Castro, M.M. Castillo-Ortega, I. Villarreal, F. Brown, H. Grijalva, M. Perez-Tello, S.M. Nuno-Donlucas, J.E. Puig, Synthesis and characterization of composites of DBSA-doped polyaniline and polystyrene-based ionomers, Composites Part A 38 (2007) 639–645.

[16] M.A. De Paoli, W.A. Gazotti, Conductive polymer blends: Preparation, properties and applications, Macromol. Symp. 189 (2002) 83–103.

[17] J.M. Fonner, L. Forciniti, H. Nguyen, J.D. Byrne, Y.F. Kou, J. Syeda-Nawaz, C.E. Schmidt, Biocompatibility implications of polypyrrole synthesis techniques, Biomed. Mater. 3 (2008) 034124.

[18] E.M. Geniès, A. Boyle, M. Lapkowski, C. Tsintavis, Polyaniline: A historical survey, Synth. Met. 36 (1990) 139–182.

[19] J. Giboz, T. Copponnex, P. Mélé, Microinjection molding of thermoplastic polymers: A review, J. Micromech. Microeng. 17 (2007) R96–R109.

[20] J. Giboz, T. Copponne, P. Mélé, Microinjection molding of thermoplastic polymers: Morphological comparison with conventional injection molding, J. Micromech. Microeng. 19 (2009) 25023.

[21] Y. Haba, E. Segal, M. Narkis, G.I. Titelman, A. Siegmann, Polymerization of aniline in the presence of DBSA in an aqueous dispersion, Synth. Met. 106 (1999) 59–66.

[22] M. Heckele, W.K. Schomburg, Review on micro molding of thermoplastic polymers, J. Micromech. Microeng. 14 (2004) R1–R14.

[23] A.J. Heeger, Nobel Lecture: Semiconducting and metallic polymers: The fourth generation of polymeric materials, Rev. Mod. Phys. 73 (2001) 681–700.

[24] O.T. Ikkala, J. Laakso, K. Väkiparta, E. Virtanen, H. Ruohonen, H. Järvinen, T. Taka, P. Passiniemi, Counter-ion induced processibility of polyaniline: Conducting melt processible polymer blends, Synth. Met. 69 (1995) 97–100.

[25] P.K. Ivanova-Mitseva, V. Fragkou, D. Lakshmi, M.J. Whitcombe, F. Davis, A. Guerreiro, J.A. Crayston, D.K. Ivanova, P.A. Mitsev, E.V. Piletska, S.A. Piletsky, Conjugated polymers with pendant iniferter groups—Versatile materials for grafting, Macromolecules 44 (2011) 1856–1865.

[26] M.B. Keivani, K. Zare, M. Aghaie, H. Aghaie, M. Monajjemi, Synthesis of nano conducting polymer based polyaniline and its composite: Mechanical properties, conductivity and thermal studies, E-J. Chem. 7 (2010) 105–110.

CHAPTER

13

Molecularly Imprinted Sol-Gel Sensors

Zi-Hui Meng

School of Chemical and Environmental Engineering, Beijing Institute of Technology, Beijing, China

OUTLINE

1. INTRODUCTION

Most of the sensor devices based on imprinted polymers are either optical or electrochemical. The routine method of molecularly imprinted polymers (MIPs) preparation includes polymerization of organic functional monomers, and the majority of MIPs are based on organic acrylate or acrylic-type polymer. A severe limitation of these types of MIPs is that the full potential of MIPs as a specific binding element is not exploited, since the organic polymers are usually relatively thick, diffusion of analyte into the polymer is slow, and the recognition sites lack direct electrical or optical contact with the transducer.

Sol-gel technology is being increasingly used for the development of optical sensors and biosensors, due to its simplicity and versatility [1,2]. Sol-gel as a solid matrix for chemical and biochemical sensor has the following advantages:

1. Compared to organic polymers, sol-gel material is chemically, photochemically, thermally, and mechanically stable; therefore, it is convenient for applications in harsh environments.

Molecularly Imprinted Sensors.
DOI: 10.1016/B978-0-444-56331-6.00013-X

2. Sol-gel material is optically transparent (down to 250 nm), which is good for spectrophotometric and spectrofluorimetric measurements.
3. Sol-gel material is formed at low temperature and under mild chemical conditions, allowing the entrapment of molecules with poor thermal and chemical stability (e.g., proteins and enzymes).
4. Sol-gel can be cast as various forms including monoliths, thin films, fibers, and powders. This allows various configurations of sensing elements for different applications.

More recently, the construction of MIPs using sol-gel technology has drawn much attention. Molecularly imprinted sol-gel (MISG) systems are a family of inorganic polymers that have backbones consisting of rigid macropores with specific recognition of the template, these macropores in the polymer allow small inorganic and organic molecules to penetrate freely, and the rigid physical structure can maintain its original properties when being treated by strong acid, strong base, or heat. Compared with acrylic polymer-based MIPs, sol-gel-based MIPs can easily control the thickness, porosity, and surface area, while the selectivity and diffusion are comparable or even better, making the sol-gel method suitable for the construction of electrochemical and optical sensors. Marx et al. compared the performance of propranolol-imprinted MIPs based on acrylic polymers and sol-gel, respectively. The acrylic system was found to have high uptake toward propranolol, but this was accompanied by a lower specificity. The sol-gel system has lower capacity, but remarkably it lowers nonspecific binding. The uptake kinetics of the acrylic polymer is significantly slower than the sol-gel polymer, reaching saturation after 10 h, relative to <1 h for the sol-gel polymer [3]. Therefore, the combination of the molecular imprinting with sol-gel technique is ideally propitious for constructing sensing devices [4–7].

2. PREPARATION OF MOLECULARLY IMPRINTED SOL-GEL (MISG)

Sol-gel is an optically transparent glass-like material produced by the hydrolysis and polycondensation of organometallic compounds at low temperature. The organometallic compounds commonly used are silicon alkoxides, for example, tetramethoxysilane (TMOS) and tetraethyl orthosilicate (TEOS). The porous matrix of sol-gel contains interconnected "bottleneck-like" pores (or so-called cavities and cages) formed by a three-dimensional network, and the pores are usually in the size range of 1.5 to 10 nm. Usually, large sensing agents can be securely trapped, but small analytes can diffuse readily into and out of the pores of sol-gel matrix. In the typical procedure, the precursor is mixed with water and a co-solvent (usually ethanol or methanol), yielding a homogeneous sol. Hydrolysis and polycondensation can be accelerated by employing an appropriate acid or base catalyst. As the reactions proceed, gradual increase of the solution's viscosity occurs and a rigid, porous, interconnected gel is formed. After aging and drying at room temperature, a xerogel is obtained, which may be further densified at high temperatures if a nonporous glass is intended. The physicochemical properties of the obtained gel depend on the parameters of the process. Factors such as the type of precursor, the pH, the

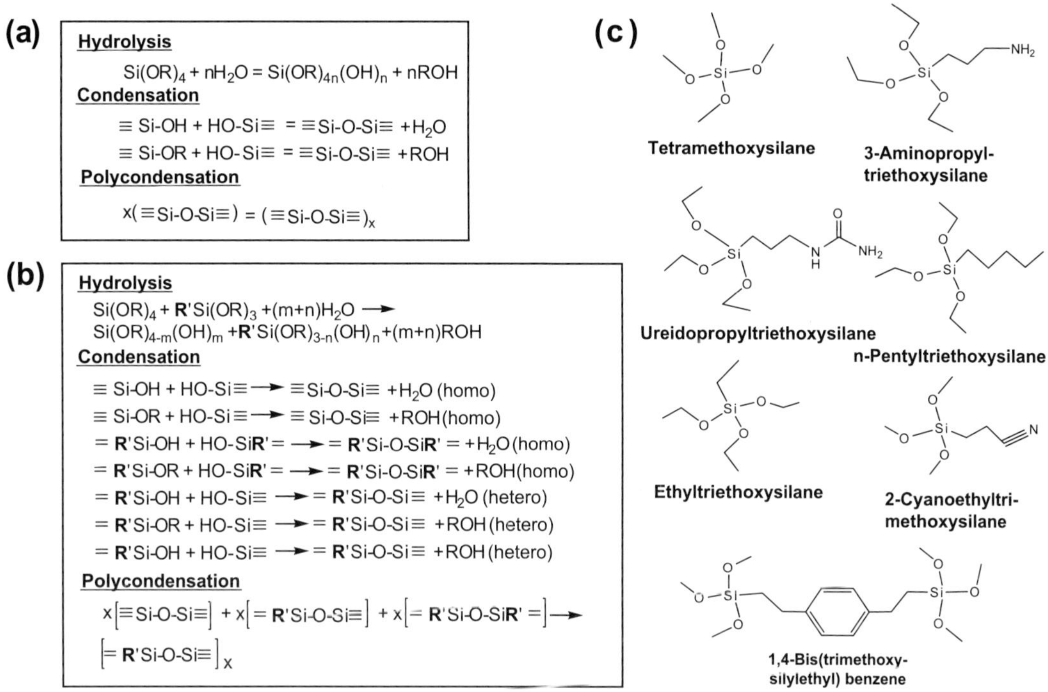

FIGURE 13.1 Sol-gel processing. (a) For a tetraalkoxysilane. (b) For the co-hydrolysis and co-condensation of a tetraalkoxysilane and an alkyl-trialkoxysilane to form a class II ORMOSIL. (c) Representative compilation of silane-based alkoxide precursors that can be used alone or in combination to form xerogels [6].

nature and concentration of the catalyst, the H_2O:Si molar ratio (R), the type of co-solvent, temperature, method, and extension of drying, the presence of doping substances, or even the chemical nature of the gelation vessel, may have a strong influence on the sol-gel material characteristics: porosity, surface area, refractive index, thickness, and mechanical properties. Considering silica alkoxydes as starting materials, the sol-gel process can be represented by Fig. 13.1, in which R is an alkyl group. A wider variety of precursors, including those based on Ti, Ce, V, Zr, and Hf as well as Si, can be obtained from several commercial sources (Sigma–Aldrich, Gelest) [6]. Sol-gel can be cast into thin films, fibers, or monoliths while it is still in the solution phase. The methods used to dry the gel have a dramatic influence on the properties of final material. Aging and drying process are accompanied by the loss of solvent (alcohol, water), which increases the strength and decreases porosity. When drying is performed under ambient conditions, the resulting composite is called a xerogel. An aerogel is formed when the sol is processed under supercritical conditions. Xerogels have lower surface areas, smaller pore sizes, and higher densities in comparison to aerogels.

Dickey's work was the first documented demonstration of MISG. In his studies, an imprinted silica matrix was formed from an acidified silica solution with homologues of

TABLE 13.1 MISG reported

Template	Precursor	Reference
Cu(II), Ni(II), and Zn(II)	1,2-bis(triethoxysilyl)ethane (BTSE), N-[3-(trimethoxysilyl)propyl]ethylenediamine	[9]
dopamine	phenyltrimethoxysilane (PTMOS), methyltrimethoxysilane (MTMS)	[10]
	aminopropyltriethoxysilane (APTES)	[11]
DDT	his(trimethoxysilylethyl)benzene (BTEB)	[12]
estazolam	APTES, PTMOS	[13]
bovine hemoglobin	TEOS, APTES	[14]
triazolam	γ-aminopropyltriethoxysilane, PTMOS	[15]
bisphenol A	TEOS	[16]
tetracycline	allyltriethoxysilane (AtEOS), TEOS	[17]
2,4-dinitrotoluene	PTMOS	[18]
theophylline	3-aminopropyltrimethoxysilane (APTMS), TEOS	[19]
bovine serum albumin (BSA) and lysozyme (Lyz)	MTMS, γ-methacryloxypropyltrimethoxysilane	[20,21]
caffeine	TEOS, MTEOS, PTEOS	[22]
	TEOS with poly{(methacrylamide)-co-(vinyl trimethoxysilane)}, poly{(methacrylic acid)-co-(vinyl trimethoxysilane)}	[23]
creatinine	TEOS	[24]
	2-acrylamido-2-methylpropane-sulfonic acid, TEOS	[25]
cholesterol	(cholesteryl propylcarbamate)triethoxysilane, tetraethyl orthosilicate	[26]
4-(4-propyloxyphenylazo) benzoic acid	$Ti(O{-}^{n}Bu)_4$	[27,28]
bilirubin	$Ti(O{-}^{n}Bu)_4$	[29]
propranolol, 1,1′-bi-naphthol, and 2-(4-isobutylphenyl)-propionic acid	$Ti(O{-}^{n}Bu)_4$	[30]
bensulfuron-methyl	$Ti(O{-}^{n}Bu)_4$	[31]

methyl orange; after dye extraction, methyl orange was found to adsorb better to the imprinted gel than to a blank, nontemplated gel [8]. Parts of MISG reported until now are listed in Table 13.1.

Although most MISGs are based on silica alkoxydes, the development of sol-gel imprinting technique was limited by the performance of the silica matrix.

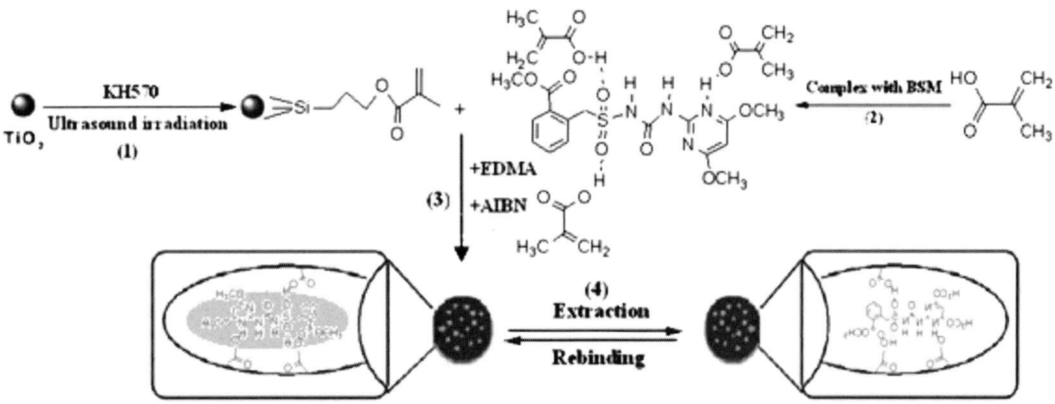

FIGURE 13.2 Scheme for the synthesis of organic–TiO_2 hybrid MIP nanosphere. (1) KH570-modified-TiO_2 nanoparticles, (2) host–guest complexes, (3) precipitation polymerization, (4) removal of template [31].

Organic–inorganic hybrid materials based on TiO_2 prepared by the sol-gel process have attracted increasing attention in recent years. Robustness and flexibility of the ultrathin titania layer is demonstrated by the formation of interconnected titania hollow structures. Developing the sol-gel imprinting technique based on monomers except for silica matrix is necessary [32–34]. The major advantages of molecularly imprinted hybrid polymers, in contrast to pure organic polymers, are larger surface area and more ordered imprinted cavities with improved mechanical and thermal stability. Yao et al. developed a simple and efficient method to prepare organic–TiO_2 hybrid MIP nanospheres [31]. Bensulfuron-methyl (BSM) molecules were imprinted in the organic–inorganic network structures using noncovalent interactions. The BSM-imprinted hybrid MIP nanospheres were obtained through a step-by-step polymerization (Fig. 13.2) and have a more irregular shape and smaller average size compared to organic-only MIP nanospheres. Furthermore, faster adsorption kinetics was achieved, which made it promising in chemical sensor applications. Chen et al. also reported a MISG based on TiO_2, in which α-cyclodextrin (CD) was selected as the template molecule and doped into TiO_2 in a sol-gel reaction. The molecularly imprinted TiO_2 sol was spin-coated on a glass slide, and appropriate template cavities in the TiO_2 sol-gel material were formed after the template molecules were removed. This modified glass slide can be used to recognize α-CD from a sample solution containing equal amounts of α-, β-, and γ-CD [35].

Acid catalyzed sol-gel process of TEOS with poly{(methacrylamide)-co-(vinyl trimethoxysilane)} and poly{(methacrylic acid)-co-(vinyl trimethoxysilane)} in the presence of caffeine resulted in the formation of highly transparent monoliths (Fig. 13.3). The ionic and non-specific adsorptions, which are considered to be the main disadvantages of the MIP, are prevented to a considerable extent by the end capping of surface silanol groups [23].

A TEOS sol-gel matrix is imprinted with creatinine (Fig. 13.4). To characterize and confirm the MISG, SEM and solid state ^{29}Si NMR spectra were analyzed [24]. The specific uptake of creatinine by the MISG was investigated by carrying out adsorption experiments in

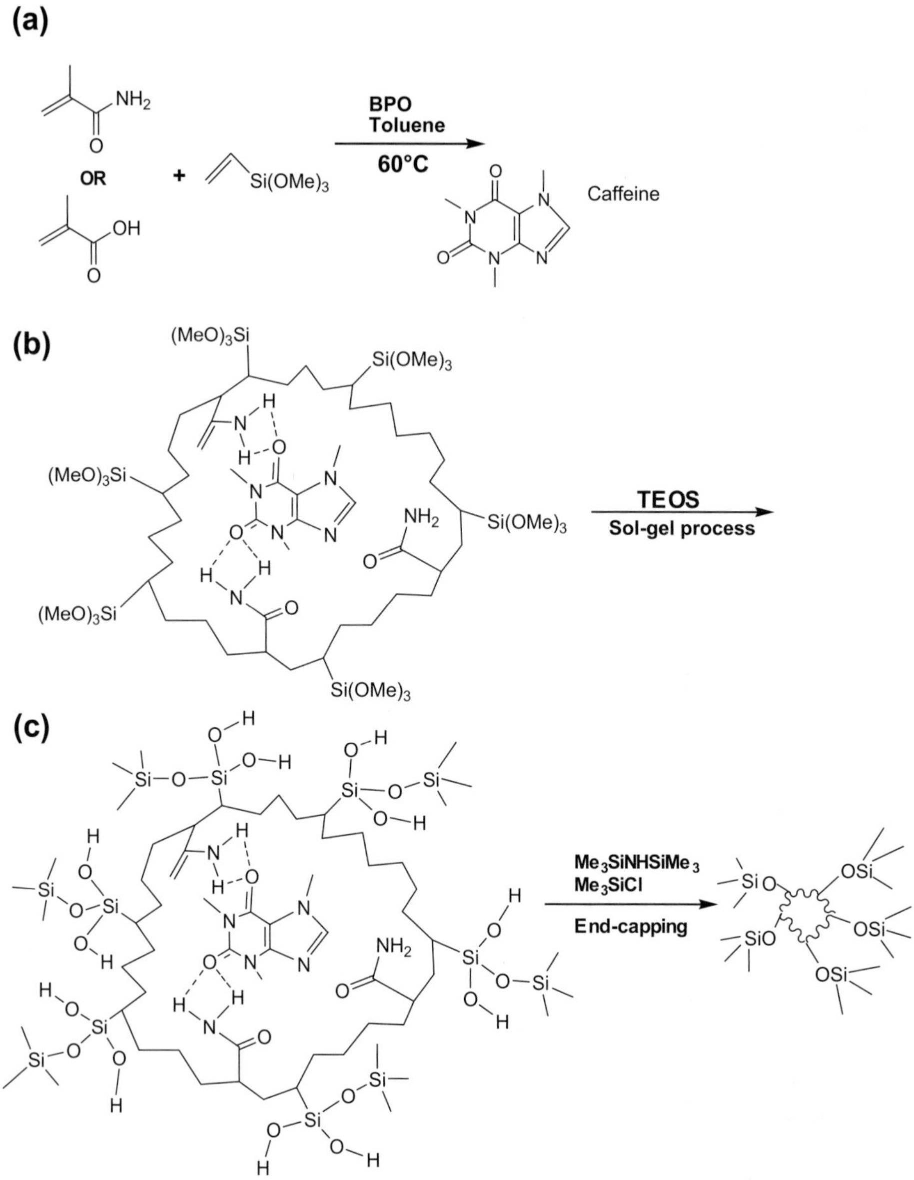

FIGURE 13.3 Scheme for the synthesis of organic–inorganic hybrid MIP for caffeine [23].

creatinine mixture solutions. *N*-Hydroxysuccinimide and 2-pyrrolidinone as creatinine's analogues and creatine as the metabolite in the presence of serum were used to form various mixture solutions for the binding task. Creatinine MISG can also be synthesized from TEOS and an organic functional monomer, 2-acrylamido-2-methylpropane-sulfonic acid (AMPS) [25].

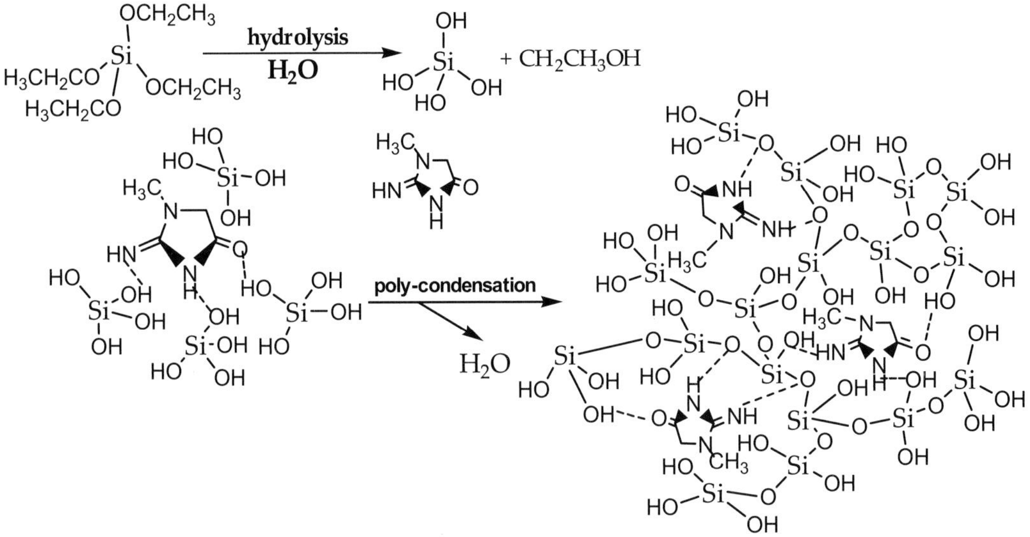

FIGURE 13.4 The preparation scheme of the creatinine imprinted poly(tetraethoxysilanol) [24].

MISG for cholesterol was prepared by using (cholesteryl propylcarbamate)triethoxysilane and TEOS as the hydrophobic monomer and cross-linker, respectively (Fig. 13.5). The MISG had a larger pore volume when formation was catalyzed at a higher pH, with the pore diameters being 3.5 nm and 2 nm, respectively, for materials formed at high and low pH values. Both of these pore volumes were greater than those found for the respective control polymers formed without template. However, only the MISG formed at low pH values showed an imprinting effect. Compared with other methods explored, the sol-gel procedure gave only a small amount of nonspecific binding for both the MISG and NISG when synthesized at low pH values [26].

Lofgreen et al. prepared a molecularly imprinted mesoporous organosilica (MIMO) using a semicovalent imprinting technique [16]. A thermally reversible covalent bond was used to link bisphenol A (BPA) template molecule to a functional alkoxysilane monomer at two points to generate a covalently bound imprint precursor. This precursor was incorporated into a cross-linked periodic mesoporous silica matrix via a typical acid-catalyzed triblock copolymer-templated, sol-gel synthesis (Fig. 13.6). After thermal treatment, the imprinted material removed more than 90% of appropriately sized bisphenol species from water, yet showed significantly lower binding for both smaller and larger molecules containing phenol moieties.

Synthetic conditions affect the properties of MISG for serine significantly; the influence of hydrophobicity, presence of amino groups, additives, water/ethanol ratio, and type of catalyst (basic or acid) were optimized using a Plackett-Burman design in 12 blocks. Imprinted and nonimprinted material was studied by competitive and noncompetitive protocols. Using the Plackett-Burman statistical design, it is possible to distinguish critical variables from a large number of suspected contributing parameters for the response variable in use.

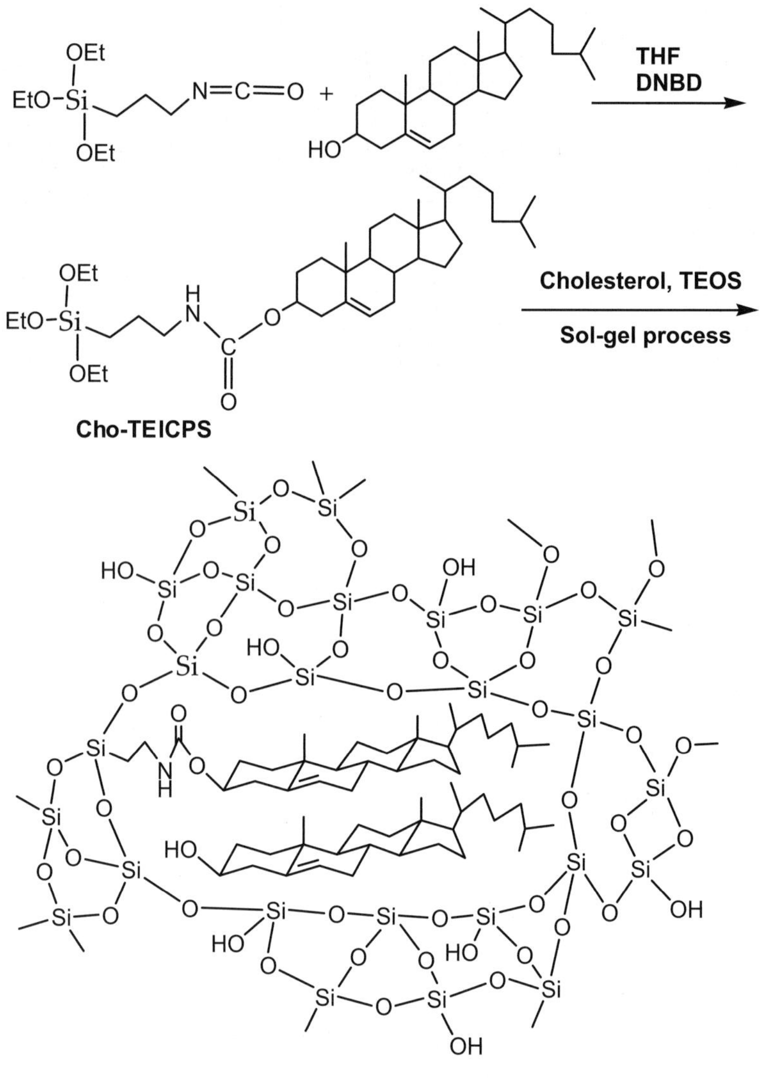

FIGURE 13.5 Preparation of cholesterol MISG [26].

Therefore, this design is extremely useful in preliminary studies to identify factors with a significant effect that can be fixed or eliminated in further studies [36]. Molecular dynamic (MD) simulations of different pre-gelification mixtures representing intermediate stages of the sol-gel process were set up to gain insight into the molecular imprinting process in xerogels, namely, to assess the template gel affinity and template self-aggregation. The physical plausibility of the parameterization was checked, confirming the reliability of the simulations. The simulation results, expressed mainly by the radial distribution functions and respective coordination numbers, showed that the affinity of the template molecule,

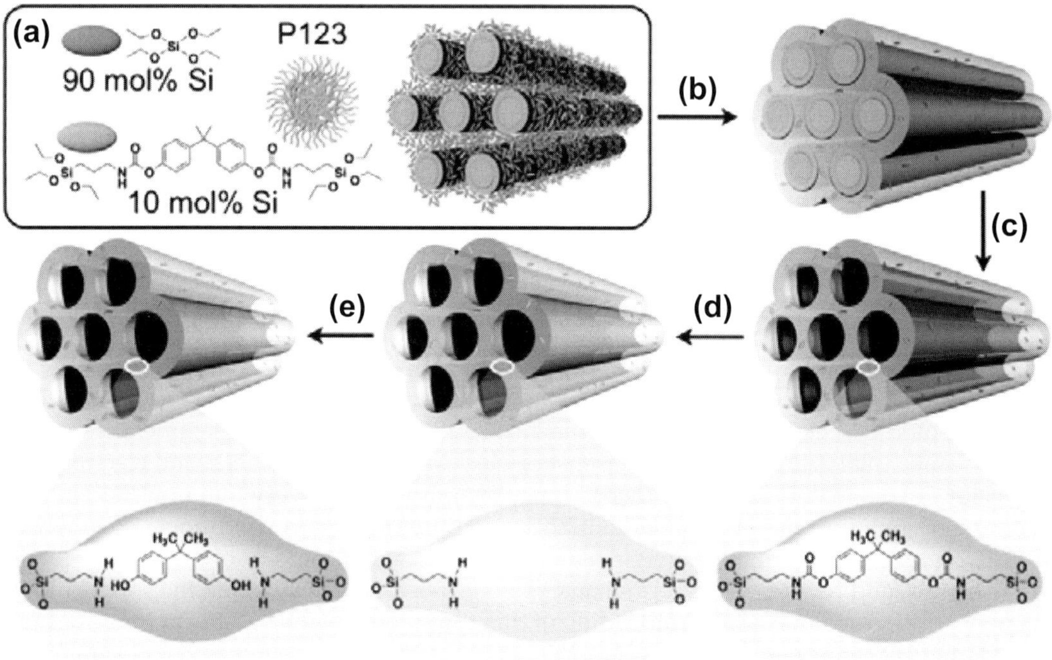

FIGURE 13.6 Synthesis of molecularly imprinted mesoporous organosilica (MIMO), imprint removal to yield MIMO-ir, and interaction of a target molecule with the imprint site: (a) the mixture of TEOS and BPAP assembles around and between the hexagonal close-packed core-shell micelles of P123 in acidic aqueous media; (b) stirring at room temperature for 24 h followed by quiescent curing at 80 °C for 24 h; (c) P123 template removal by Soxhlet extraction with ethanol for 20 h; (d) thermal cleavage of the imprint by heating in wet DMSO for 5 h; (e) sequestration of an appropriately sized target bisphenol molecule by hydrogen bonding between phenols and amines [16].

damascenone (a hydrophobic species), for the gel backbone would not be attained without the tested functional group, phenylaminopropyl. The affinity, related to the capability to trap the template within the gel network, was derived mostly from the hydrophobic interaction. It was also inferred from MD simulations that lower water content (methanol-richer mixtures) would facilitate a better dispersion of both the functional group and the template within the final gel, therefore favoring the imprinting process. Congruence between simulated and experimental results demonstrates the ability of MD simulations to provide information regarding the fine aspects of molecular interactions in pre-gelification mixtures for imprinting [37]. Since an optimal template-functional monomer interaction is a necessary condition for successful imprinting, the choice of the best monomers is of the greatest importance, and computational method may constitute an expeditious and reliable screening tool. Molecular modeling was applied to the design of MISG [38], and the requisite level of theory for the selection of suitable alkoxysilane functional monomers was determined. A comparative study, applied to the design of an MIP for β-damascenone, involving different levels of theory, basis set superposition error (BSSE), correct ion and basis set augmentation, and also semi-empirical methods, was performed. The computation results suggest that the use of the 3-21G basis set concomitantly with a method for BSSE correction represents a good compromise between theory level and computation time for the successful screening of functional monomers.

Microwave processing makes it possible to drive reactions uniformly throughout the matrix rapidly and is a useful technique to prepare sol-gels based on TiO_2, SiO_2, and SnO_2. Thus a novel microwave-assisted sol-gel method for molecular imprinting of silica microspheres was developed by Feng et al. [39], methyl orange and ethyl orange dye molecules were used as template molecules respectively. The method had the advantage of being able to quickly prepare the imprinting matrix with good reproducibility and stability. The microspheres obtained via microwave irradiation have smooth surfaces and good dispersion.

3. MISG USED IN ELECTROCHEMICAL SENSORS

Sol-gel-modified electrodes have found many applications, for example, in electrocatalysis, enzymatic catalysis, sensors and biosensors, electorchemiluminescence, and electroactive films [40]. Sol-gel materials inherently combine the ability to form very thin films and matrix porosity. In addition, the ease of sol-gel fabrication, mild reaction conditions, commercial availability of a wide variety of functional monomers, physical rigidity of the matrix, chemical inertness, and resistance to thermal and solvent stresses have all made the sol-gel methodology attractive for molecular imprinting of thin films and a perfect platform for designing various electrochemical applications. A novel, sensitive, molecularly imprinted electrochemical sensor was constructed for the selective detection of chlorogenic acid (CGA) by deposition of a MISG film, prepared by the sol-gel process, onto an Au bare electrode surface (Fig. 13.7). Initially, a (3-mercaptopropyl)siloxane layer was formed on the Au bare surface, followed by a siloxane layer obtained from the acid-catalyzed hydrolysis/condensation of a solution constituted by TEOS, PTEOS, APTMS, and CGA, as a molecular template. The MISG/Au sensor was tested in a solution of the CGA template and other similar molecules. This electrode displayed excellent selectivity toward CGA when compared with structurally similar molecules. Under optimized experimental conditions, the peak current response of the sensor for CGA was linear from 5.0×10^{-7} M to 1.4×10^{-5} M, and the detection limit was 1.48×10^{-7} M. The MISG/Au sensor was successfully applied for the determination of CGA in coffee and tea samples [41].

Cyclodextrin (CD) hosts and imprinting effects were combined to fabricate binding sites within a TiO_2 ultrathin layer (Fig. 13.8), and the electrode surface was modified with ultrathin TiO_2 gel film containing a 2:1 complex of β-CD and BPA [42]. The resulting BPA-imprinted TiO_2/β-CD film showed specific and sensitive detection of BPA, as confirmed by quartz crystal microbalance (QCM) and cyclic surface-polarization impedance (cSPI) measurements. The high selectivity for BPA relative to structurally related guest molecules was estimated to be 1.6–3.4 at the guest concentration of 2×10^{-7} M, and was considered to arise from the synergic effect of the binding site.

Titania sol-gel layers imprinted with capric acid have been used as synthetic receptors for highly sensitive detection of oxidized products resulting from degradation of automotive engine oil [43]. These layers have been applied as sensitive coating material on shear transverse wave (STW) resonators of frequencies ranging from 100 MHz to 430 MHz. The relatively small size of STW resonators, that is, about 2 mm for 430 MHz, makes these devices

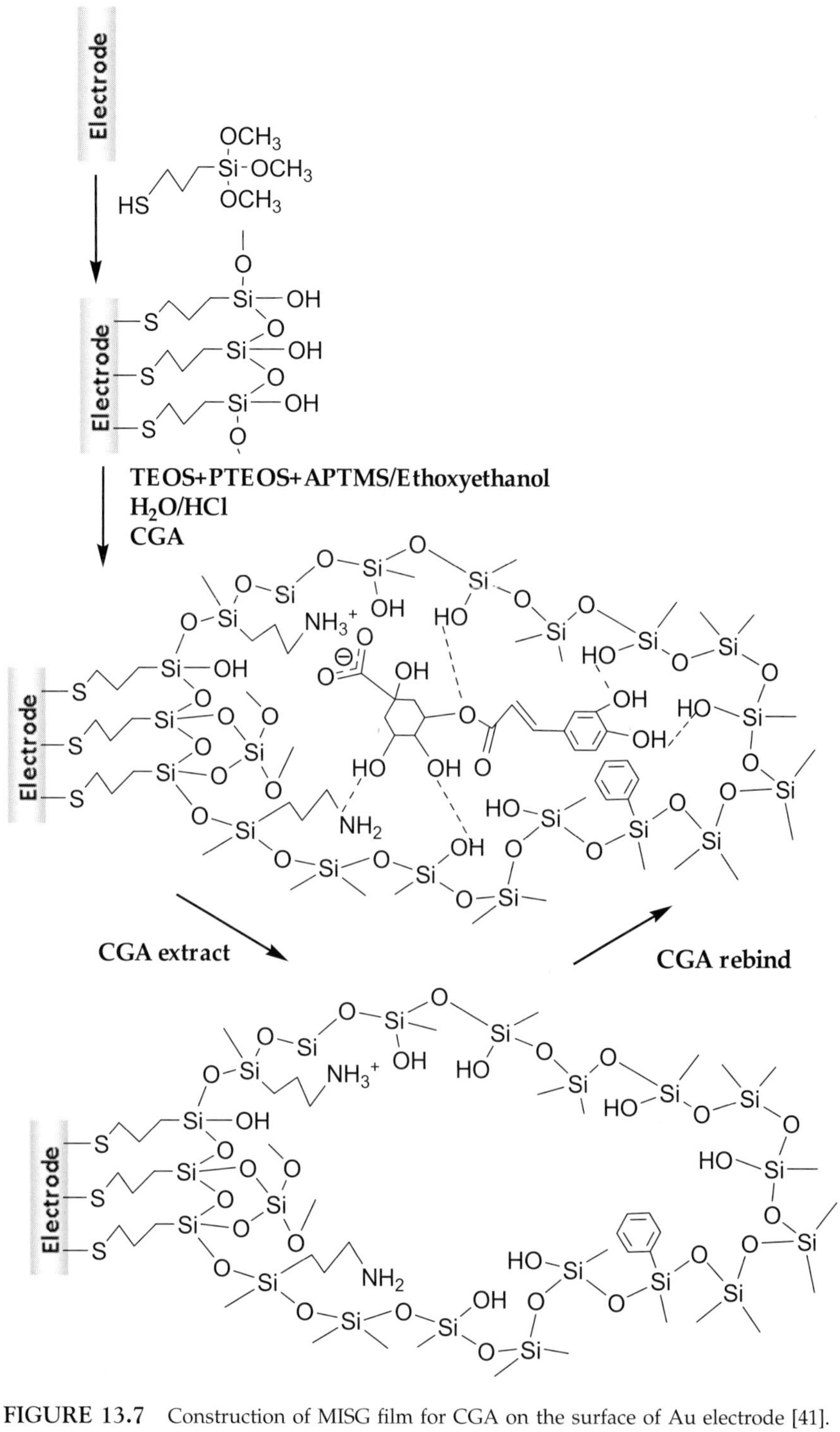

FIGURE 13.7 Construction of MISG film for CGA on the surface of Au electrode [41].

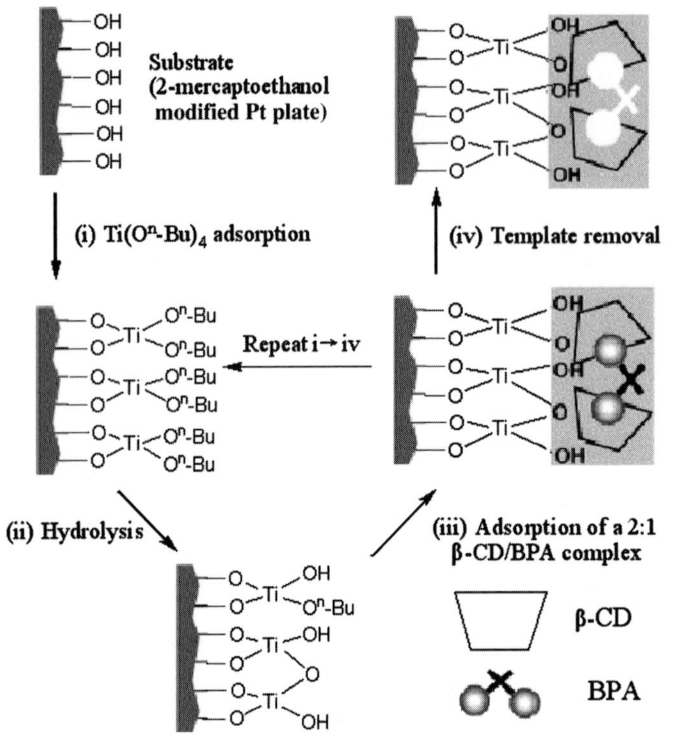

FIGURE 13.8 Molecular imprinting via surface modification with a 2:1 β-CD/BPA complex [42].

extremely useful while considering the concept of miniaturization. It has been proved experimentally that by increasing the fundamental resonance frequency of these devices, a very high sensor response—22 kHz up to 460 kHz—can be generated. The thin titania layers coated on a transducer surface provide excellent diffusion pathways to oxidized products of waste engine oil for selective and reversible re-inclusion, that is, recovery time of 30 min (Fig. 13.9). Molecular recognition sites for mercury ions were imprinted in TiO_2 film using stable ground-state complex of 1-amino-8-naphtol-3,6-disodium sulfonate (ANDS) and mercury ions as template [44]. The complex ratio between mercury ions and ANDS was estimated to be 2:1. Compared with the controlled and pure TiO_2 electrodes, the imprinted electrode revealed selectivity toward the imprinted ions. Linearity for mercury ions detection was obtained, and the regression equation was I_p (μA) $= 4.29 \times 10^{-7} + 19.40$ $[Hg^{2+}]$ with a detection limit of 3.06×10^{-9} M. The imprinted electrode could be used for more than 1 month. Recoveries were calculated at both high and low concentrations, with a mean recovery of 99.6%.

Introduction of polyaniline (PAN), β-CD or multiwalled carbon nanotubes (MWNTs) nanocomposite materials can improve the overall surface area without substantially changing the size of the electrode and enhancing the rate of electron transfer. A sensitive and selective electrochemical sensor based on a polyaniline-modified carbon electrode for the determination of L-phenylalanine has been proposed by utilizing β-CD incorporated

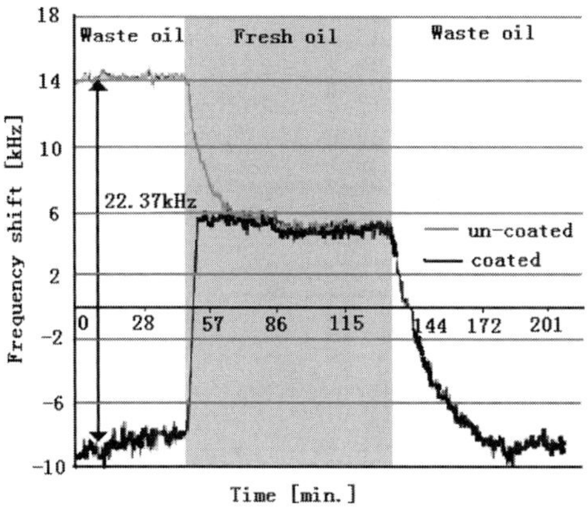

FIGURE 13.9 Sensor response of 266 h old oil sample with coated (50 nm TiO_2 layer) and uncoated device on 100 MHz STW [43].

MWNT and imprinted sol-gel film (Fig. 13.10). The electrochemical behavior of the sensor toward L-phenylalanine was investigated by CV, DPV, and amperometric *i*–*t* curve. The responsive mechanism of the imprinted sensor for L-phenylalanine was based on the inclusion interaction of β-CD and molecular recognition of the imprinted film for L-phenylalanine. A linearity was obtained covering the concentration range from 5.0×10^{-7} to 1.0×10^{-4} M with a detection limit of 1.0×10^{-9} M. With excellent sensitivity, selectivity, stability, reproducibility, and recovery, the electrochemical imprinted sensor can detect L-phenylalanine in blood plasma samples successfully [45].

Surface electrodeposition is a simple and rapid method to deposit a uniform MISG layer, with good adherence to the transducer surface. Moreover, it allows easy control of the film thickness by varying the amount of circulated charge. By using layer-by-layer assembly technology [46], a novel MIP/sol-gel/MWNT/Au electrode specific to clindamycin was fabricated by electrodepositing an imprinted sol-gel solution onto the surface of an Au electrode decorated by a MWNT layer (Fig. 13.11). This sensor exhibits convenient, low-cost, excellent selectivity and sensitivity to clindamycin, and was successfully employed to detect clindamycin in human urine samples. The excellent performance of the MIP/solgel/MWNT/Au electrode toward clindamycin was ascribed to the MWNT functional monolayer with subtle electronic properties and huge specific surface area and the porous imprinted sol-gel film with plentiful selective binding sites. Under the selected analytical conditions, the linearity for clindamycin detection was from 5.0×10^{-7} M to 8.0×10^{-5} M with the detection limit of 2.44×10^{-8} M. The MWNT functionalized imprinted sensor, with high selectivity, sensitivity, and speed, met the requirements of clindamycin detection, and it could be potentially applied in analysis of real samples. MWNT combined with MISG can also be used for the direct detection of *L*-histidine [47]. The sensor was fabricated onto an ITO electrode via stepwise modification of MWNTs and a thin film of MISG. The introduced MWNTs exhibited noticeable enhancement on the sensitivity of the MISG sensor. Meanwhile, the molecularly imprinted film displayed high sensitivity and excellent selectivity for the target molecule *L*-histidine. A linearity from 2.0 μM to 1.0 mM for the detection of *L*-histidine

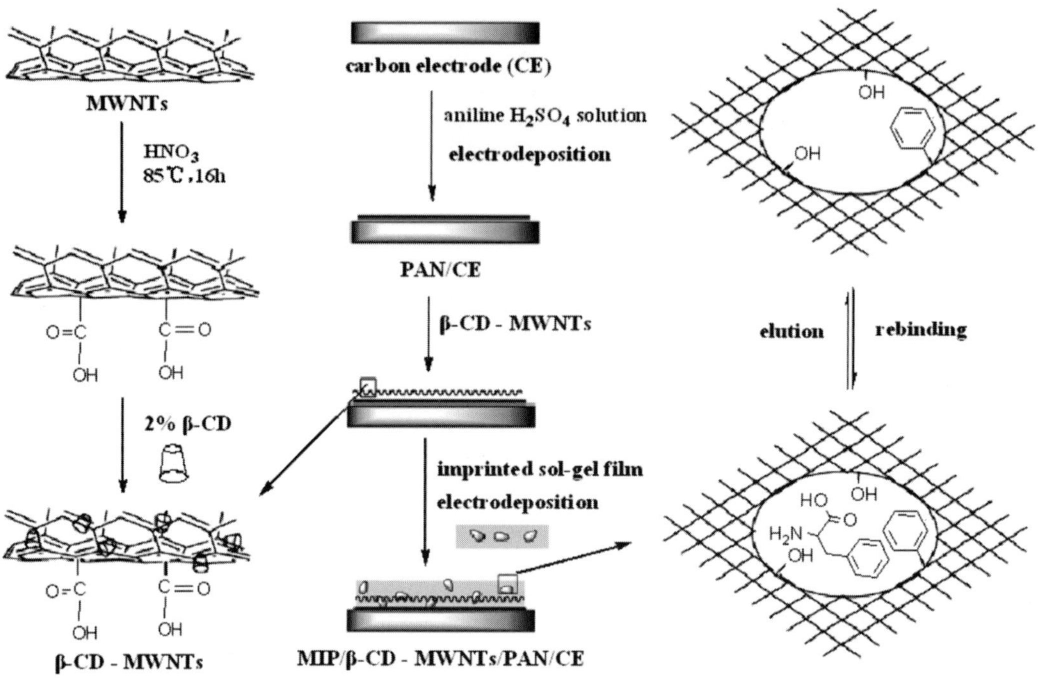

FIGURE 13.10 Illustration of the preparation process of the MIP/β-CD–MWNTs/PAN/CE [45].

was observed with the detection limit of 5.8×10^{-9} M. This imprinted electrochemical sensor was successfully employed to detect *L*-histidine in human blood serum.

Functionalized polyurethane and sol-gel material with amino group as ligand were synthesized to enhance the selectivity toward metal ions [48]. Tetraethylammonium picrate was used for structuring the sol-gel materials with no functionality. The recognition pattern of these polymers is based on ionic sizes of picrate salt. The difference in frequency shifts obtained by QCMs between imprinted and nonimprinted layers indicates the interaction properties in these synthesized coatings. All of these polymer layers are highly structured and can be varied in their chemical properties for sensing purposes. A four-channel sensor has been prepared on 10-MHz QCM by a special sieve designed for screen printing (Fig. 13.12). This tetraelectrode mass-sensitive sensor will allow the sensing of metal ions in a diverse analyte environment, that is, ionic and neutral, considering one electrode as reference. In this way, the concept of reliable, miniaturized, and universally adaptive sensor devices can be realized.

Molecular recognition sites for the nucleotides adenosine 5′-monophosphate, guanosine 5′-monophosphate, cytosine 5′-monophosphate, and uridine 5′-monophosphate are imprinted in an acrylamide-acrylamidephenylboronic acid copolymer membrane [49]. The imprinted membranes are assembled on piezoelectric Au quartz crystals or Au electrodes via electropolymerization or on the gate surface of an ISFET device by radical polymerization. The imprinted membranes reveal selectivity toward the imprinted nucleotide, and the association of the respective nucleotides with the recognition sites is transduced by the

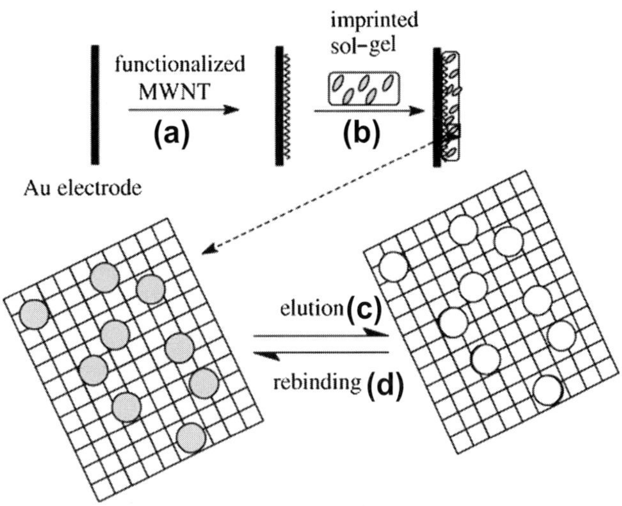

FIGURE 13.11 Preparation procedure of the MIP/sol-gel/MWNT/Au electrode: (a) modification with functionalized MWNT, (b) electrodeposition, (c) removal of the template and (d) rebinding of the template [46].

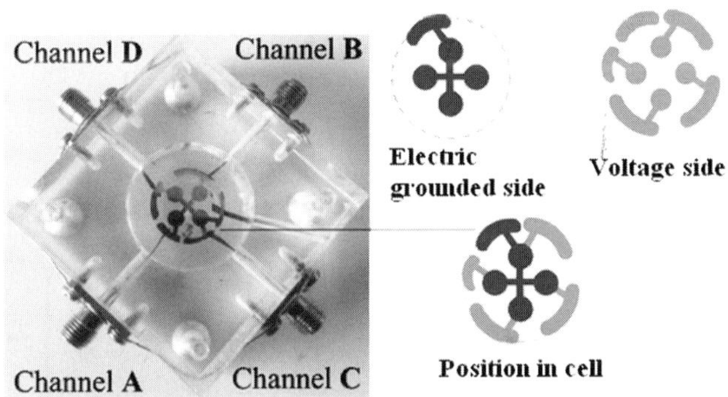

FIGURE 13.12 Measuring cell with tetraelectrode QCM [48].

following: (i) microgravimettic, quartz crystal microbalance (QCM) measurements; (ii) Faradaic impedance analyses; and (iii) potentiometric responses of the ISFET devices. Selective sensing of the respective monosaccharides can also be accomplished in the presence of the imprinted membrane-functionalized ISFET devices. (R)- or (S)-2-Methylferrocene carboxylic acids, (R)-1 or (S)-1, (R)- or (S)-2-phenylbutanoic acid, (R)-2 or (S)-2, and (R)- or (S)-2-propanoic acid, (R)-3 or (S)-3, were imprinted in thin TiO2 films on the gate surface of ISFET devices [50]. The imprinting is performed by hydrolyzing the respective carboxylate Ti(IV) butoxide complex on the gate surface, followed by washing off the acid from the resulting TiO2 film (Fig 13.13). The imprinted sites reveal chiroselectivity only toward the sensing of the imprinted enantiomer. The chiral recognition sites reveal not only chiroselectivity but also chirospecificity, and, for example, the (R)-2-imprinted film is active in the sensing of (R)-2, but insensitive toward the sensing of (R)-2-phenylpropanoic acid, (R)-3, which exhibits a similar chirality. The chiroselectivity and chirospecificity of the resulting imprinted films are attributed to the need to align and fit the respective substrates in precise molecular

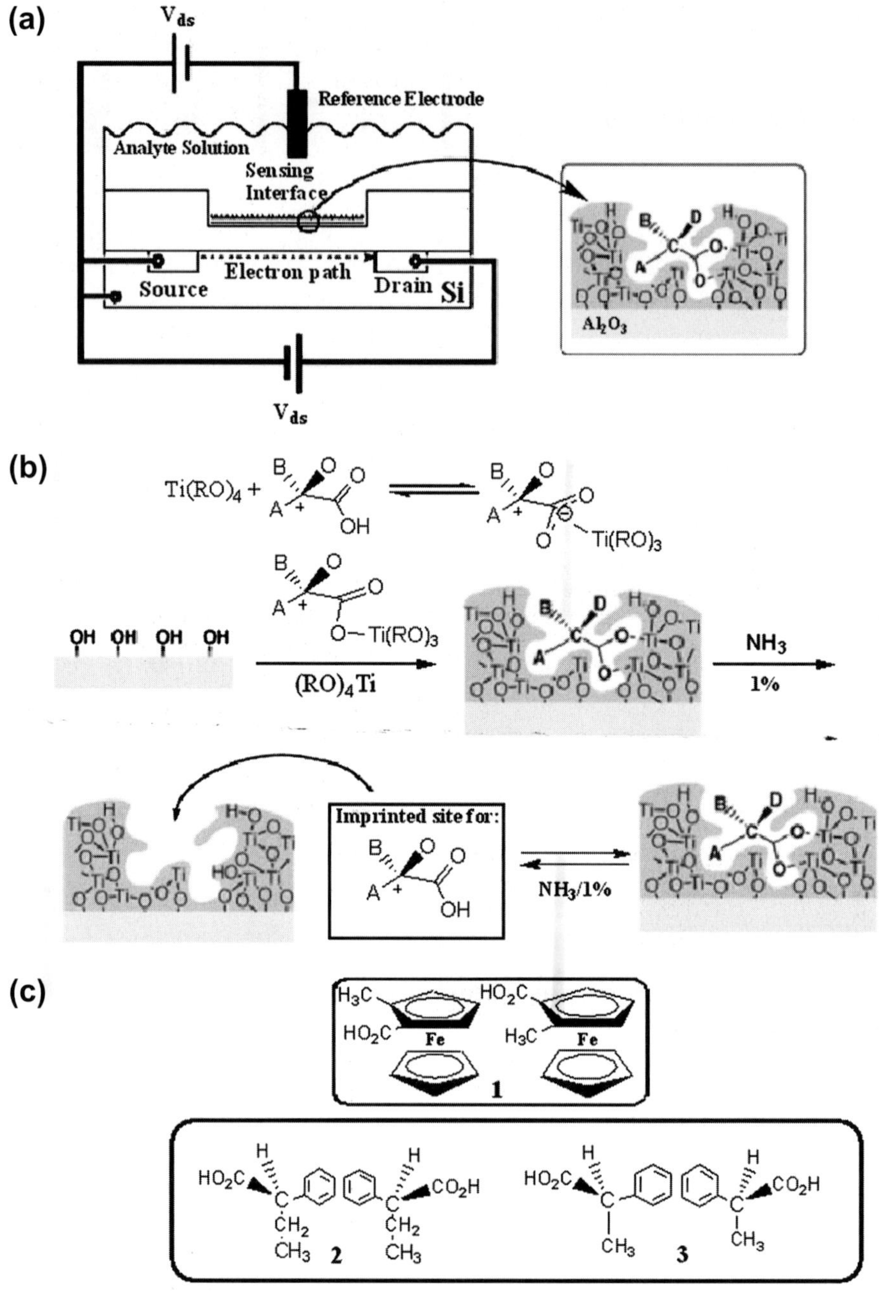

FIGURE 13.13 (a) Schematic configuration of the molecular imprinted ISFET device. (b) Preparation of molecular imprinted sites for carboxylic acids in a TiO_2 film acting as the sensing interface on the ISFET gate [50].

contours generated in the cross-linked TiO_2 films upon the imprinting process. MISG film with enantioselectivity can also be fabricated on ITO electrodes [51].

The chiral selectivity recognition was detected using two enantiomer pairs: D- and L-3, 4-dihydroxyphenylalanine (D- and L-dopa) and (R)- and (S)-N,N′-dimethylferrocenylethylamine [(R)-Fc and (S)-Fc]. Chiral selection properties were measured by CV and SWV. For both template molecules, very good chiral recognition was revealed by electrochemical measurement. The nonspecific adsorption measured for reference nonimprinted films was negligible (less than 5%). Dopa-imprinted films revealed both high sensitivity, by the detection of 0.2 ppb concentration, and excellent selectivity, when challenged with a series of catechol derivatives. Fc-imprinted films were able to detect 2 ppm of the target molecule, with very good enantioselectivity and low nonspecific adsorption. The SiO_2 gate of an ISFET can be functionalized with a TiO_2 film that includes imprinted molecular sites for 4-chlorophenoxy acetic acid, (**1**), or 2,4-dichlorophenoxy acetic acid, (**2**). The functionalized devices that include the imprinted interfaces reveal an impressive selectivity in the sensing of the imprinted substrates Na^+-**1** or Na^+-**2**. The detection limit for Na^+-**1** is $(5 \pm 2) \times 10^{-4}$ M, which corresponds to 38 $mV \cdot dec^{-1}$ in the concentration range of 0.5 to 6 mM. The detection limit for the analysis of Na^+-**2** is $(1.0 \pm 0.2) \times 10^{-5}$ M, which corresponds to 28 $mV \cdot dec^{-1}$ in the concentration range 0.1–9.0 mM. The equilibration time of the devices is 5 min [50,52]. Gao et al. developed an ITO electrode using MISG as a recognition element for dopamine(DA) [11]. Aminopropyl-derivatized organosilane aminopropyltriethoxysilane (APTES) was first self-assembled on an ITO electrode, and then imprinted sol was spin-coated on the modified surface (Fig. 13.14). APTES, which can interact with DA through hydrogen bonds,

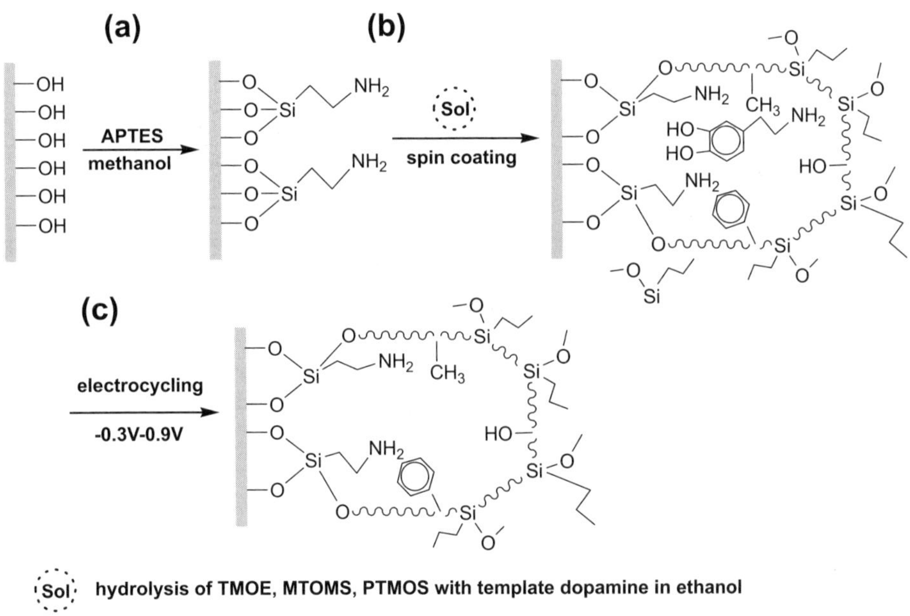

FIGURE 13.14 Schematic representation of hypothetical DA-templated three-dimensional sol-gel imprinting film based biomimetic sensor: (a) Surface modification of an ITO electrode with APTES. (b) Spin-coating of DA-imprinted sol on the functionalized surface. (c) Removal of DA from the MIP film [11].

brought more binding sites located closely to the surface of the ITO electrode, thus making the prepared sensor more sensitive for DA detection. Potential scanning is presented to extract DA from the modified film, therefore, DA can rapidly and completely leach out. The affinity and selectivity of the resulting biomimetic sensor were characterized using CV. It exhibited an increased affinity for DA over that of structurally related molecules; the anodic current for DA oxidation depended on the concentration of DA in the linear range from 2×10^{-6} M to 0.8×10^{-3} M with a correlation coefficient of 0.9927. In contrast, DA-templated film prepared under identical conditions on a bare ITO showed obviously lower response toward dopamine in solution.

MISG casted on glassy carbon electrodes was used for parathion detection in aqueous solutions [53]. The binding of parathion to the imprinted films in the liquid phase was investigated by steady-state experiments with analysis by GC-FPD and CV. The imprinted films showed high selectivity toward parathion in comparison to similar organophosphates. Another sensor for the determination of parathion based on *p-tert*-butylcalix[6]-1,4-crown-4 as functional monomer was fabricated by the sol-gel method using molecularly imprinted technology (Fig. 13.15) [54]. The electrochemical behavior of parathion at the imprinted *p-tert*-butylcalix[6]arene-1,4-crown-4 sol-gel film sensor was characterized by CV, linear sweep voltammetry, chronoamperometry, and alternating current impedance spectroscopy. A fast response of parathion can be obtained after being incubated in 0.1 M phosphonate buffer solution which contain appropriate amount of parathion for 20 min. A linear response over parathion concentration in the range of 5.0×10^{-9} to 1.0×10^{-4} M was exhibited with a detection limit of 1.0×10^{-9} M. This imprinted film sensor has been applied in determining parathion in a real sample, and the results were consistent with those obtained by HPLC. The same technology was used to detect *O,O*-dimethyl-(2,4-dichlorophenoxyacetoxyl)(3′-nitrophenyl) methinephosphonate (Phi-NO_2) [55]. The functional monomer, *p-tert*-butylcalix[6]arene, possesses plentiful hydroxyl groups for the formation of inorganic–organic hybrid film through covalent bonding with Si–OH, and it can overcome the bleeding effect of the functional monomer for the removal of template, resulting in better stability and reproducibility. Otherwise, hydrophobic cavities of

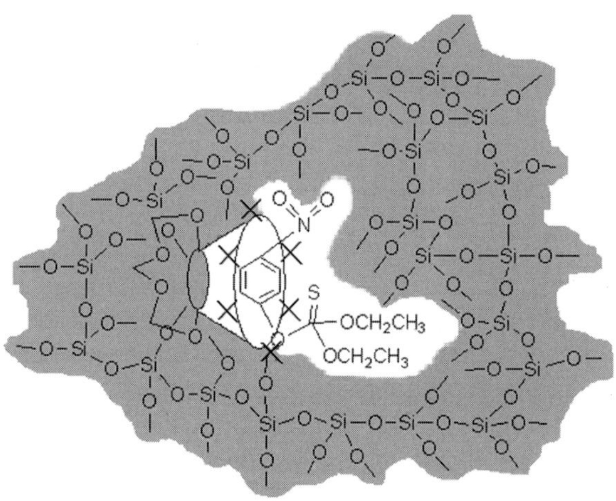

FIGURE 13.15 Schematic illustration of a selective site for parathion in the imprinted *p-tert*-butylcalix[6]-1,4-crown-4 sol-gel film [54].

p-tert-butylcalix[6]arene provide a hydrophobic environment for the interaction between Phi-NO_2 and functional monomer by hydrophobic interactions/$\pi-\pi$ bonding, allowing the rebinding state to be rapidly obtained in 30 min. The sensor responds linearly to Phi-NO_2 over the concentration range of 2.0×10^{-5} to 1.0×10^{-8} M, and the detection limit is 1.0×10^{-9} M. This sensor provides an efficient way of eliminating interferences from coexisting substances in the solution. This sensor exhibits a high degree of sensitivity and selectivity; it can be used for the detection of Phi-NO_2 in cabbage samples and can lead to a generation of a new type of sensors for Phi-NO_2.

A hybrid sol-gel material formed by acid hydrolysis of a mixture of TEOS and PTEOS as functional monomers was imprinted by using tyramine and DA as template molecules for the purpose of sensing [56]. Imprinted materials were spin-coated as thin films on the surface of glassy carbon electrodes and then were characterized using CV. Imprinted films showed higher affinities toward their template molecules compared to other structurally similar molecules, especially for tyramine-imprinted film. With the exception of tyramine and norepinephrine, the interference level did not exceed 5% for all compounds studied for DA-imprinted films. Tyramine-imprinted films, however, showed a high affinity for tyramine with DA 40% interference. The sensing surface lifetime extended to 2 weeks, with a decay in response signal that ranged from 22% to 40% and 60% for DA, tyramine, and norepinephrine, respectively. The standard deviation from the mean of measurements for the repeated experiments is 7.4%.

A MIP brush grafted to sol-gel film on graphite electrode is reported for the selective and sensitive analysis of barbituric acid (BA) in aqueous, blood plasma, and urine samples (Fig. 13.16) [57]. The modified electrode was preanodized at +1.6 V (vs. saturated calomel electrode), where encapsulated BA involved hydrophobically induced hydrogen bondings, in MIP cavities exposed at the film/solution interface, at pH 7.0. The differential pulse, cathodic stripping voltammetry (DPCSV) technique was employed to investigate the binding performance of the sol-gel-modified imprinted polymer brush, which yielded a linear response in the range of 4.95–100.00 $\mu g\ mL^{-1}$ of BA, with a detection limit of 1.6 $\mu g\ mL^{-1}$ ($S/N = 3$).

MISG individually imprinted with either Ura or 5FU was also brush "grafted to" the surface of sol-gel immobilized on cost-effective, homemade solid-phase microextraction (SPME) fibers (Fig. 13.17), and was used in combination with a voltammetric sensor duly modified with the same MISG [58]. This combination provided up to 10- and 8.4-fold preconcentrations of Ura and 5FU, respectively, which was more than sufficient for achieving stringent detection limits in the primitive diagnosis of uracil disorders and fluoropyrimidine toxicity in DPD-deficient patients. The proposed method permits the assessment of Ura and 5FU plasma concentrations with detection limits of 0.0245 and 0.0484 ng mL^{-1} respectively, without any problems of nonspecific false-positives and cross-reactivities in complicated matrices of biological samples.

Selective and quantitative recognition of folic acid can be achieved using a preanodized sol-gel-coated pencil graphite (grade 2B) electrode with imprinted polymer immobilized to its exterior surface [59]. The MIP capable of binding specifically folic acid was prepared by the stoichiometric imprinting process (template/monomer molar ratio, 1:3), creating multiple binding sites within the cross-linked hyperbranched polymer. Dendrimer-like chains were obtained by an "initiator-fragment incorporation radical polymerization"

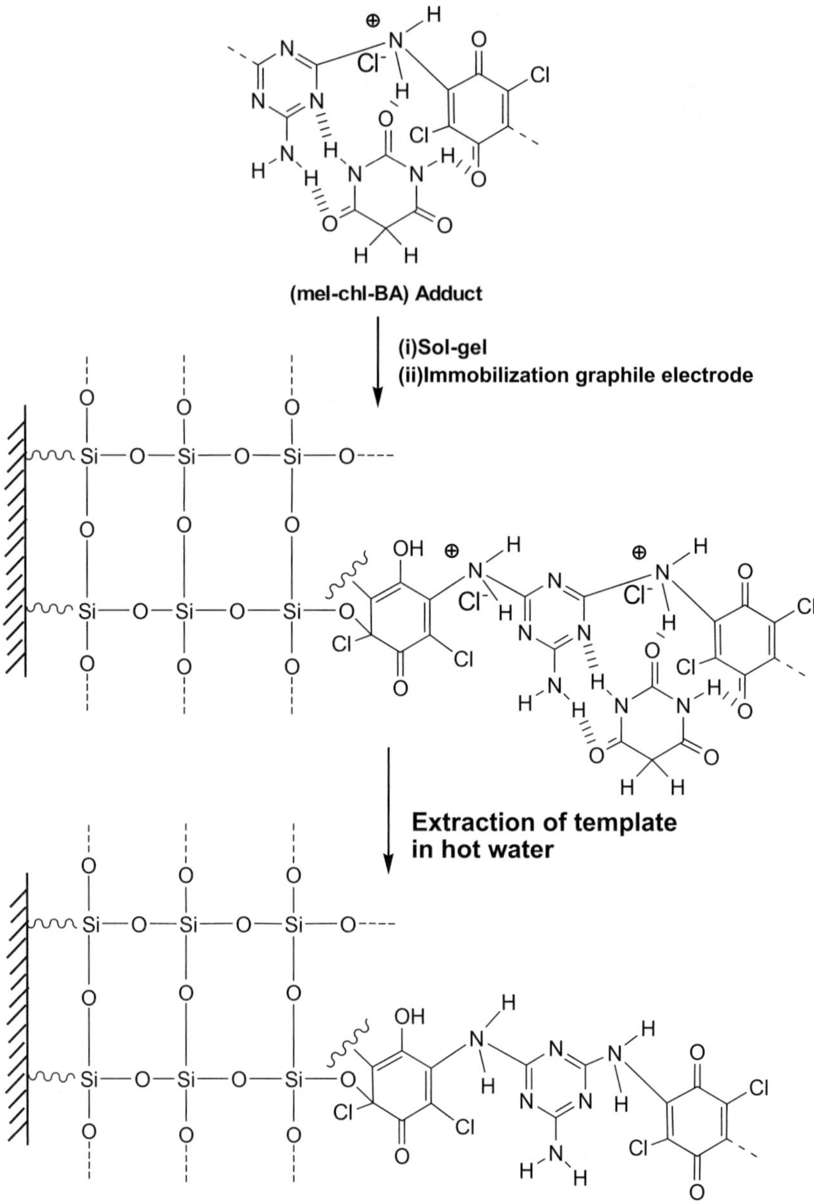

FIGURE 13.16 Schematic representation of the preparation of a brush MIP grafted sol-gel film on graphite electrode [57].

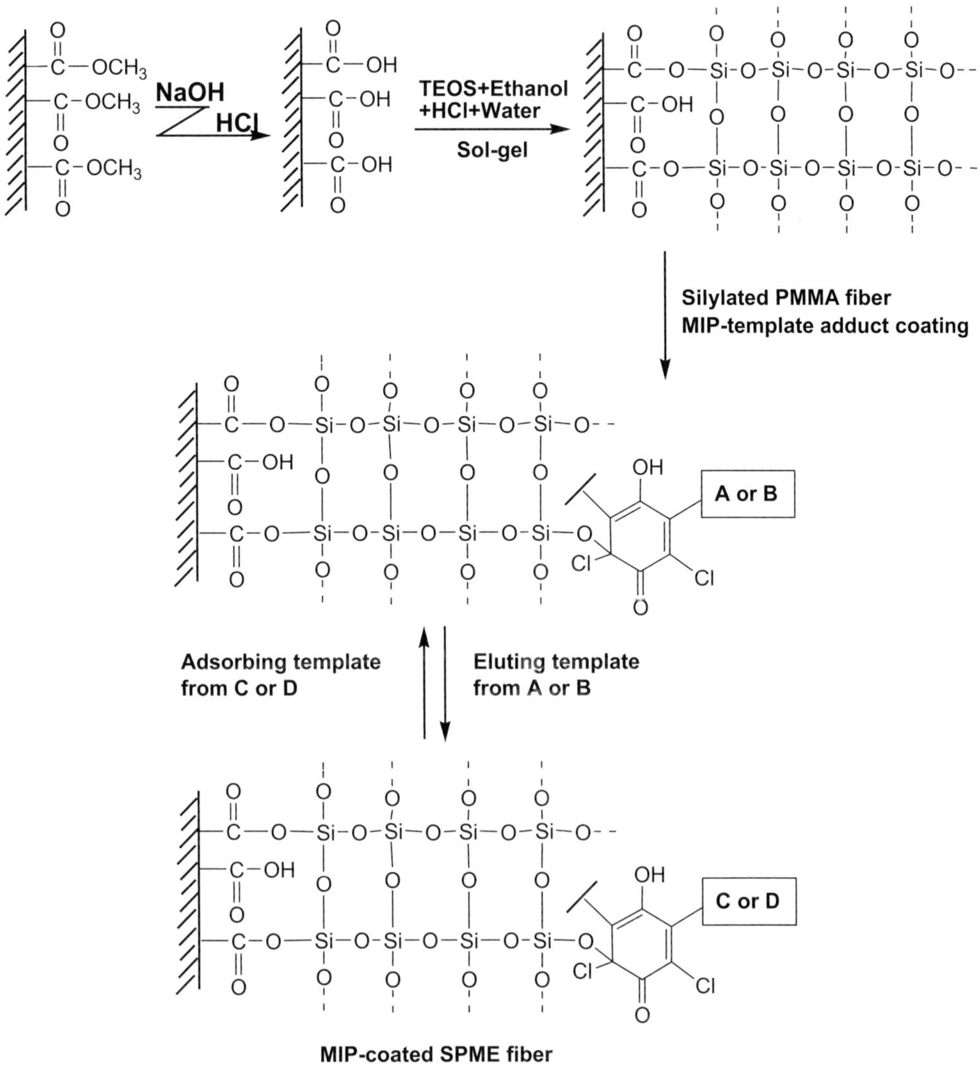

FIGURE 13.17 Schematic representation of MIP-coated SPME fiber preparation [58].

technique involving a new trifunctional monomer, 2,4,6-trisacrylamido-1,3,5-triazine (Fig. 13.18). During the preconcentration step at +0.8 V (with respect to Ag/AgCl), the analyte recapture at pH 2.5 in aqueous environment simultaneously involved mixed hydrophobically driven hydrogen bondings and ionic interactions with pteridine substructure and purely hydrogen bonding interactions with glutamic acid residue of folic acid. The encapsulated analyte was instantly oxidized and then cathodically stripped off responding differential pulse cathodic stripping voltammetric signal. The folic acid was selectively

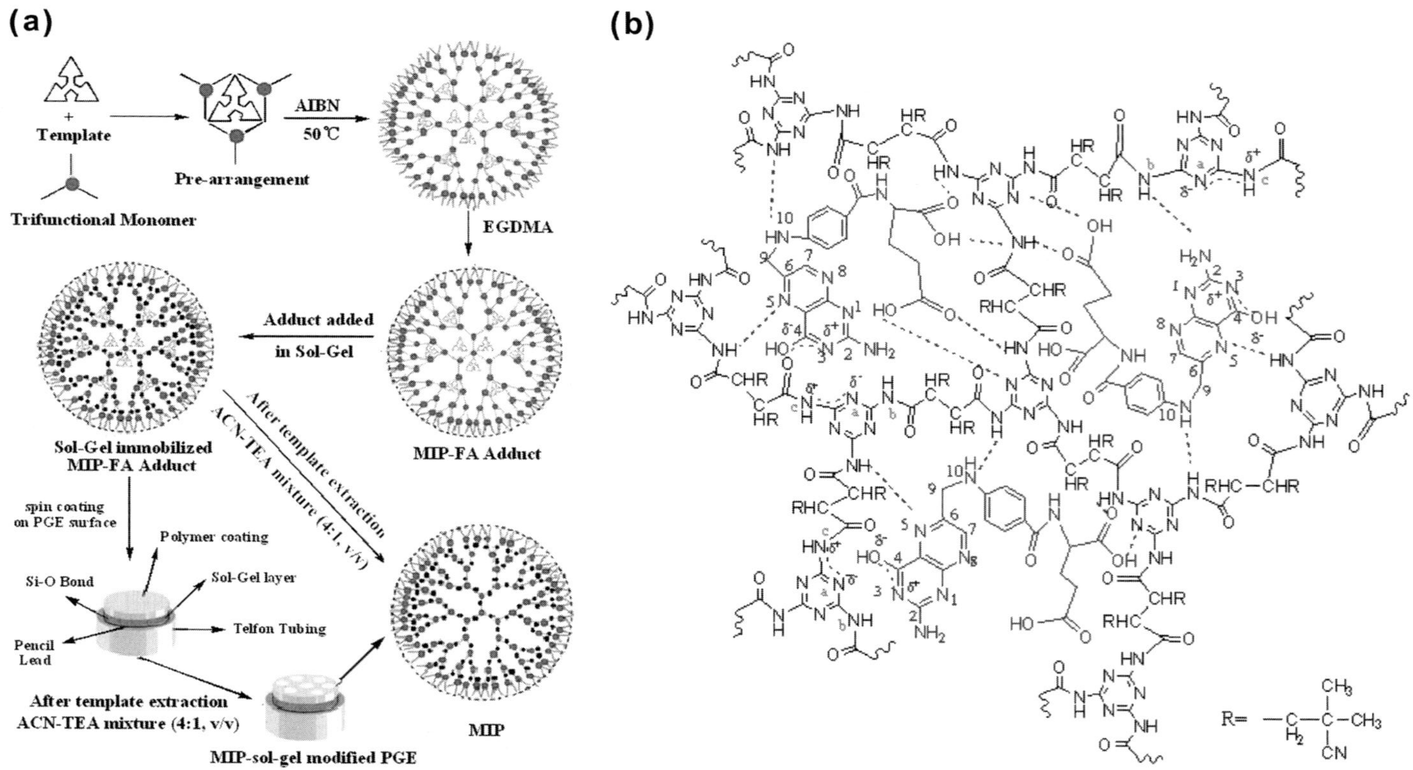

FIGURE 13.18 (a) Schematic representation of the preparation of the MIP–sol-gel coated PGE. The symbol (●) represents silane binding as Si–O–C–OH at various carbonyl centers of MIP. (b) Recognition mechanism of template (FA) bindings within MIP cavities via noncovalent interactions showing multimolecular imprinting in hyperbranched (dendrimer-like) polymer [59].

detected with a LOD of 0.002 μg mL^{-1}, without any cross-reactivities and real matrix complications.

4. MISG USED IN OPTICAL SENSORS

Special attention is devoted to optical biosensors. This field has been experiencing rapid growth, reflecting the increasing demand for stable, robust, and specific devices for application in areas such as food industry, diagnostics, in vivo monitoring, environmental control, and biotechnology [60]. Optical sensors can be based on various optical principles (absorbance, reflectance, luminescence, fluorescence), covering different regions of the spectra (UV, visible, IR, NIR) and allowing the measurement not only of the intensity of light, but also of other related properties, such as refractive index, scattering, diffraction and polarization. Thin films (with thickness usually <1 μm) and coatings representing a key area for the development of optical sensors require only small amounts of precursors and embedded functional molecules; exhibit fast response times and superior transparency; possess mechanical resistance; and are less susceptible to crack than monoliths (particularly when exposed to liquids). Thin films can be obtained by means of spin-coating, dip-coating, or spray-coating techniques; they are regarded as the most technologically important sol-gel configuration. Sol-gel coatings can be easily combined with optical fibers or planar waveguides, providing intrinsic evanescent wave sensors. Furthermore, thin films have great potential for miniaturization and also permit the possibility of preparing multilayer configurations. Thus, sol-gel thin films are now widely recognized as a promising strategy for the effective and low-cost mass production of reversible, robust, and portable optical chemical sensors.

MISG for DDT using both a noncovalent and a covalent approach was prepared [12]. A nonpolar porous sol-gel network was created through the use of the bridged polysilsesquioxane, his(trimethoxysilylethyl)benzene (BTEB), as the principal sol-gel component. A covalent imprinting strategy was employed by generating a sacrificial spacer through the reaction of two 3-isocyanatopropyltriethoxysilanes with one of two different template molecules: 4,4′-ethylenedianiline (EDA) or 4,4′-ethylidenebisphenol (EBP). After formation of the sol-gel, the bonds linking the spacer template to the matrix were cleaved in a manner that generated a pocket of the appropriate size bordered by amine groups that could aid in the binding of DDT through weak hydrogen bonding interactions. An environmentally sensitive fluorescent probe, 7-nitrobenz-2-oxa-1,3-diazole (NBD), located adjacent to the DDT binding site, was used to transduce the binding of analyte (Fig. 13.19). EDA-imprinted sol-gels, deposited as films on glass microscope slides, can quantitatively detect DDT in water to a LOD of 50 ppt with a response time of <60 s.

Cross-linked films consisting of the acrylamide-acrylamidophenylboronic acid copolymer that are imprinted with recognition sites for β-nicotinamide adenine dinucleotide (NAD^+), β-nicotinamide adenine dinucleotide phosphate $NADP^+$, and their reduced forms (NAD(P)H), are assembled on Au-coated glass supports (Fig. 13.20). The binding of the oxidized cofactors NAD^+ or $NADP^+$ or the reduced cofactors NADH or NADPH to the respective imprinted sites results in the swelling of the polymer films through the uptake of water. Surface plasmon resonance (SPR) spectroscopy is employed to follow the binding

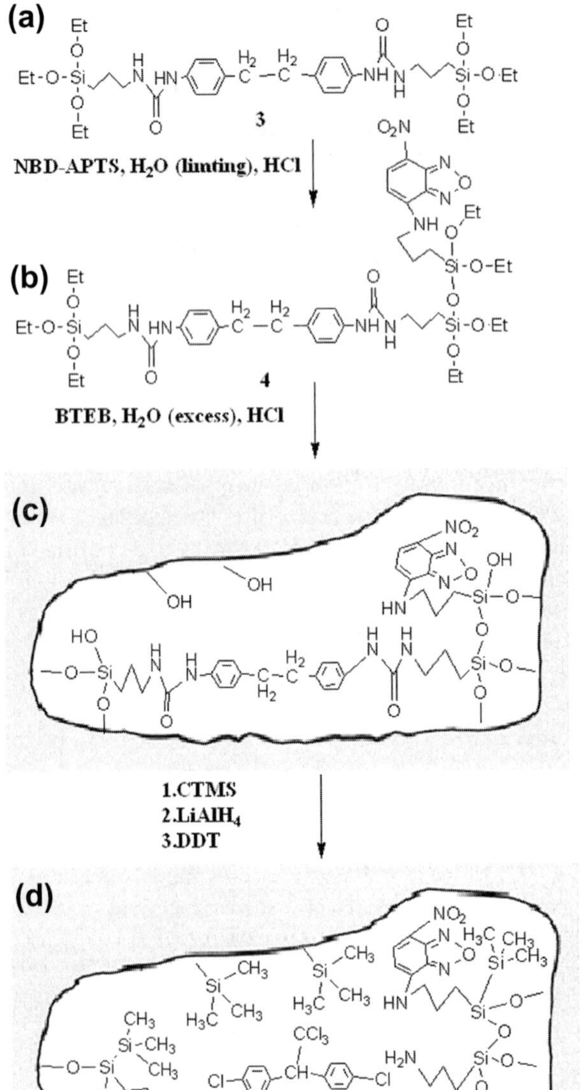

FIGURE 13.19 Schematic diagram showing the approach used to generate a molecularly imprinted fluorescence-based DDT sensor: (a) the sacrificial spacer silane (3) was reacted with an equal molar amount of NBD–APTS to link the fluorophore with the spacer silane (4); (b) a 320:1 molar ratio of BTEB and excess water was added under acid-catalyzed conditions to make the sol-gel solution; this solution was dip-coated onto substrates and thermally annealed; (c) the sol-gel was treated stepwise with CTMS to eliminate the free Si–OH groups, followed by the chemical cleavage of the urea linkages and removal of the EDA spacer; and (d) DDT can then be bound by the imprinted site and transduced by the increased quantum yield of the NBD as the local environment potentially becomes more nonpolar [12].

of the different cofactors to the respective imprinted sites. The imprinted recognition sites reveal selectivity toward the association of the imprinted cofactors. The method enables analysis of the NAD(P)$^+$ and NAD(P)H cofactors in the concentration range of 1×10^{-6} to 1×10^{-3} M [61]. The cofactor-imprinted films associated with the Au-coated glass supports serve as active interfaces for the characterization of biocatalyzed transformations that involve the cofactor-dependent enzymes.

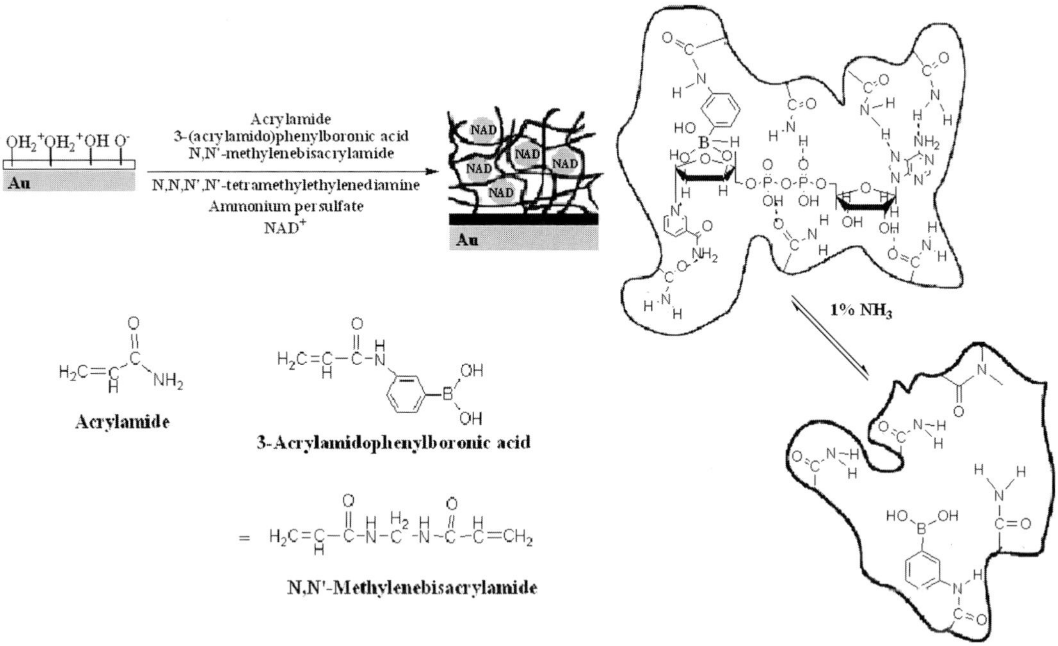

FIGURE 13.20 Schematic preparation of the NAD(P)$^+$/NAD(P)H cofactors-imprinted acrylamide-acrylamidophenylboronic acid copolymer sensing film on an Au-coated glass support. The support was primarily modified with cystamine monolayer followed by its coupling to acrylic acid. The scheme exemplifies the imprinting of NAD$^+$ [61].

A 96 well microplate for chemiluminescence (CL) detection of melamine (MM) using MISG as a sensing element was developed (Fig. 13.21). This method used 96 well microplates modified with imprinted film for preconcentration and selection. Compared with the chemical grafting method, the proposed method shows some excellent performances such as simple process and recycling utilization of 96 well microplate owing to the shedding of film in a sufficiently high temperature and dry environment. The MM-MIF was prepared via sol-gel technology with PTMOS and MTMS as functionalized organosilicon precursors. Based on the accurate and quick detection of MM, the quality of foods could be effectively monitored. In addition, the industrialization of molecularly imprinted assay kits can be implemented, and the procedures will be so simplified that they can be mastered. This method provides a number of advantages such as high sensitivity, high selectivity, small amount of chemical consumption, cost effectiveness, simple operation, and high throughput analysis for the determination of MM [62].

Films based on TiO_2 nanoparticles (NPs) have been successfully used as sensing elements in chemical sensors. TiO_2 colloidal suspensions can be obtained by spontaneous hydrolysis in acidic solutions of Ti(IV) compounds. The obtained TiO_2 NPs can be employed to build up nanostructured films. Zattoni et al. obtained TiO_2 NP dispersions by hydrolyzing potassium titanyl oxalate in the presence of a target analyte (tyrosine). They present the characterization of colloidal tyrosine-TiO_2 NP systems by flow field-flow fractionation (FlFFF) with online

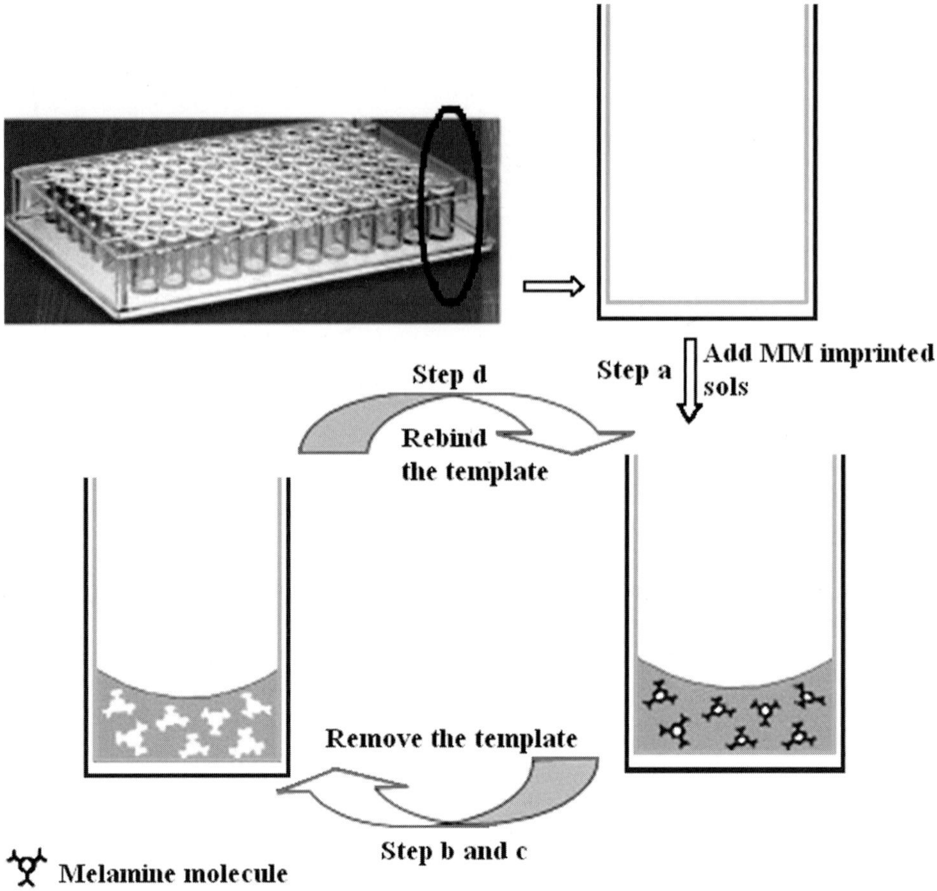

FIGURE 13.21 MISG in 96 well plate for CL detection of MM [62].

UV/Vis absorption detection and offline fluorescence analysis. Good correlation between the fluorescence intensity, which is proportional to the tyrosine uptake, and the FlFFF-based, NP mass-size frequency distribution finally confirms the existence of tyrosine-TiO_2 NP interaction [63].

In order to detect the explosive 2,4,6-trinitrotoluene (TNT), a chemical sensor was developed utilizing a planar integrated optical waveguide (IOW) attenuated total reflection spectrometry [64]. Submicron thick films of organically modified sol-gel polymers were deposited on the waveguide surface as the sensing layer (Fig. 13.22). Sol-gels were molecularly imprinted for TNT using covalently bound template molecules linked to the matrix through 1 or 2 carbamate linkages (Fig. 13.23). Upon chemical cleavage of the template and displacement of the TNT-like pendant groups from the matrix, shape-selective binding sites were created that possess a primary amine group. The amine was used to deprotonate bound TNT yielding an anionic form that absorbs visible light. Binding of TNT and subsequent

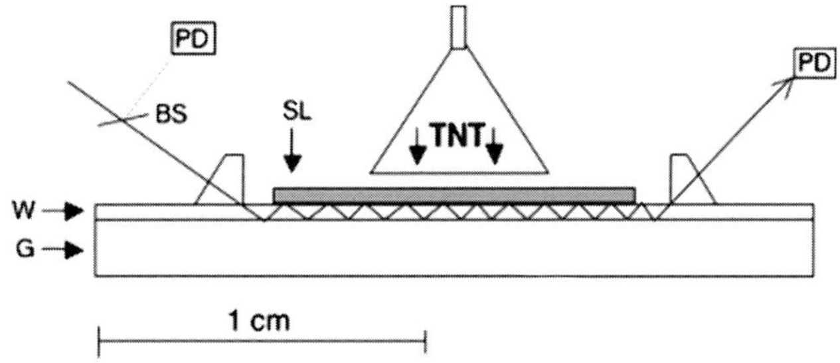

FIGURE 13.22 Schematic of the TNT chemical sensor (note the vertical dimension is not to scale). G: glass support, microscope slide 1 mm in thickness. W: waveguide layer high index SiO_2/TiO_2 glass layer formed from sol-gel film deposition, thickness 0.8–1.0 μm. SL: MISG sensing thickness 0.3–1.0 μm. BS: beamsplitter. PD: photodiode. A laser is prism coupled into the waveguide layer where the propagating beam interacts with the sensing layer upon each internal reflection. The intensity of the outcoupled light is measured [64].

conversion to the anion results in the attenuation of light propagating through the waveguide, thus creating a spectrophotometric device. Sensitivity can be achieved by taking advantage of the substantial path length provided by the use of single-mode IOWs. The LOD to gas-phase TNT was found to be 5 ppbV in ambient air at a flow rate of 40 mL min^{-1}, given a 60 s sampling time. The sensor is highly selective for TNT due to the selectivity of binding site recognition of TNT and the subsequent generation of the TNT anion. TNT detection can also be achieved by a MISG sensor integrated with surface-enhanced Raman

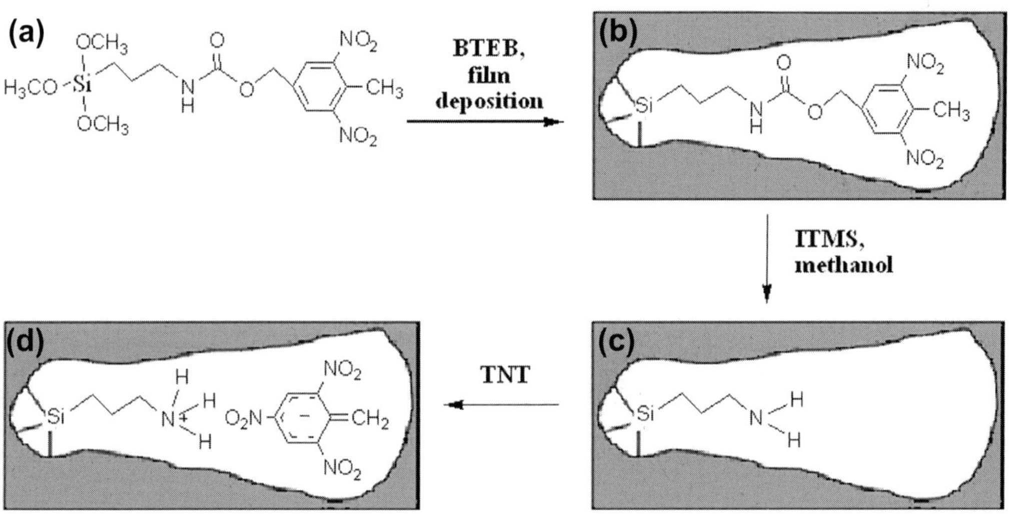

FIGURE 13.23 Molecular imprinting scheme for the TNT sol-gel sensing layer. (a) The template. (b) The template is copolymerized with BTEB by base-catalyzed hydrolysis and condensation of alkoxysilane groups. (c) The template is removed by cleaving the carbamate linkage using iodotrimethylsilane and methanol. (d) A shape-specific pocket aids in binding TNT. The amine group can TNT and deprotonate the methyl group to convert TNT into its anionic form (one resonance structure of TNT^- is shown. [64]).

scattering (SERS) [65]. Micron thick films of sol-gel-derived xerogels were deposited on a SERS-active surface as the sensing layer. Xerogels were molecularly imprinted for TNT using noncovalent interactions with the polymer matrix. Binding of the TNT within the polymer matrix results in unique SERS bands, which allow for detection and identification of the molecule in the xerogel. This MISG-SERS sensor exhibits a 3-μM LOD.

Tao et al. developed a new sensor strategy termed protein-imprinted xerogels with integrated emission sites (PIXIES) [66]. The PIXIES platform is completely self-contained, and it achieves analyte recognition without a biorecognition element (e.g., antibody). PIXIES rely on sol-gel-derived xerogels, molecular imprinting, and the selective installation of a luminescent reporter molecule directly within the molecularly imprinted sites (Fig. 13.24). In operation the templated xerogel selectively recognizes the target analyte, the analyte binds to the template site, and binding causes a change in the physicochemical properties within the template site that are sensed and reported by the luminescent probe molecule. For human interleukin-1 the PIXIES-based sensor elements exhibited the following analytical figures of merit: (1) 2 pg/mL detection limits; (2) $<$2 min response times; (3) $>$85 selectivity; (4) $<$6% relative standard deviation R.S.D. long term drift over 16 weeks of ambient storage; (5) $>$95% reversibility after more than 25 cycles; and (6) $>$85% recoveries on spiked samples.

Another simple and selective sensing system for protein recognition has been realized by the development of a quantum dots (QDs)-based MISG-coated composite for selective recognition of the template cytochrome *c* (Cyt) [67]. The MISG imprinting process resulted in an

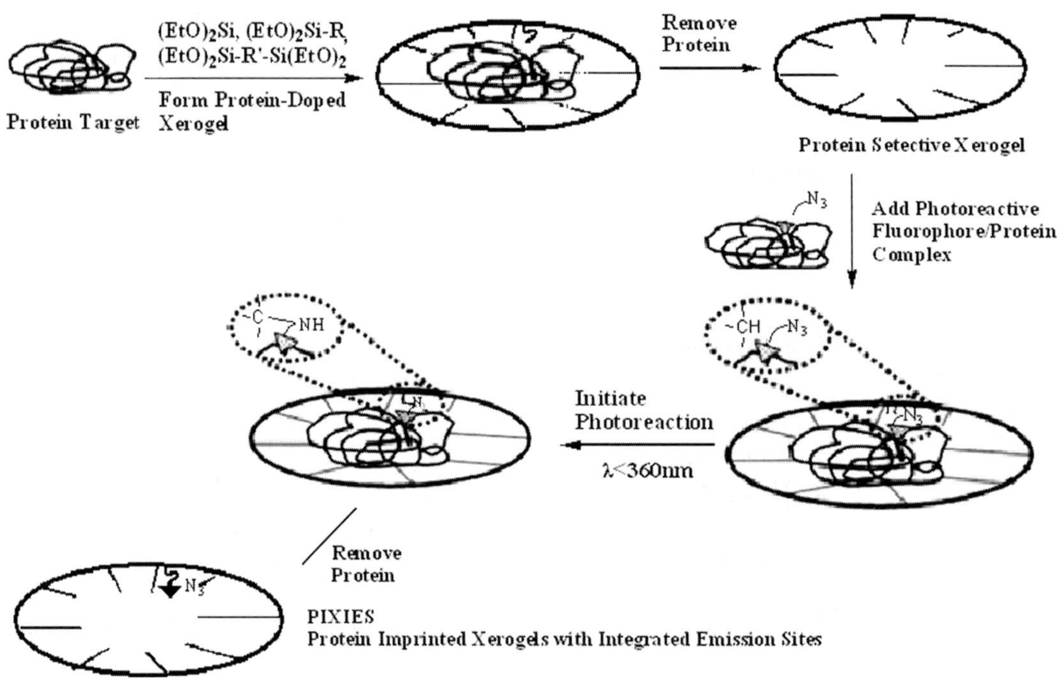

FIGURE 13.24 Outline of the generic PIXIES production protocol [66].

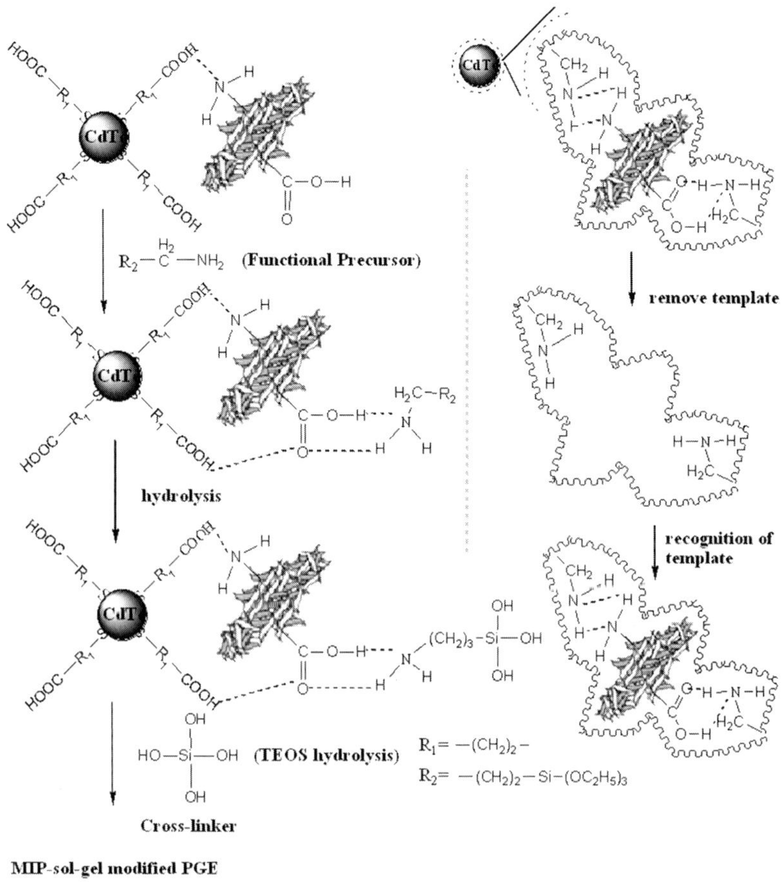

FIGURE 13.25 Preparative procedures for the fabrication of fluorescent MISG-coated CdTe QD composites [67].

increased affinity of the composites toward the corresponding template (Fig. 13.25). The fluorescence of MISG-coated QDs was stronger quenched by the template versus that of NIP-coated QDs, which indicated that the composites could recognize the corresponding template (Fig. 13.26). The results of specific experiments further exhibited the recognition ability of the composites. Under optimum conditions, the linear range for Cyt is from 0.97 μM to 24 μM, and the detection limit is 0.41 μM. The MISG-QD composites integrated the high selectivity of molecular imprinting technology and fluorescence property of QDs and could convert the specific interactions between imprinted cavities and corresponding template to the obvious changes of fluorescence signal.

There is a great demand for remote and real-time sensing techniques to detect and quantify Microcystin-LR (MCT) because cyanobacteria deteriorate the water quality and are responsible for emerging outbreaks and epidemics causing harmful diseases in humans and animals. A Fabry–Pérot sensing probe based on an optical fiber tip coated with an

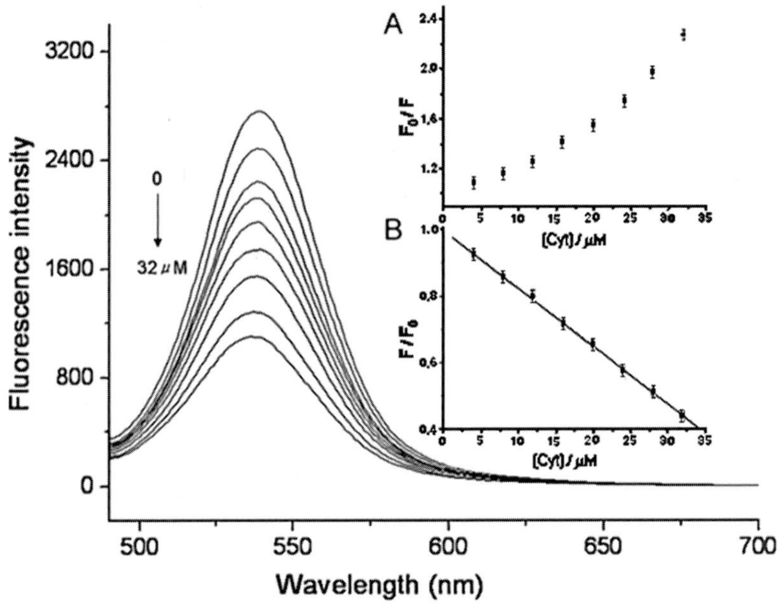

FIGURE 13.26 Fluorescence emission spectra of QDs with addition of the indicated concentrations of protein Cyt. Inset A is the Stern–Volmer curve, and inset B is the relationship between F/F_0 and C_{Cyt}. F and F_0 are the fluorescence intensities of QDs in the presence and absence of Cyt, respectively [67].

MCT selective MISG has been developed (Fig. 13.27). The proposed optical sensor shows merits such as high sensitivity and specificity, low cost, robustness, easy preparation, and preservation. The membranes were developed by imprinting MCT in a sol-gel matrix that was applied over the tip of the fiber by dip coating. The imprinting effect was obtained by curing the sol-gel membrane, prepared with APTMS, DPDMS, and TEOS, in the presence of MCT. The fiber Fabry–Pérot with a MISG sensor showed low thermal effect, thus avoiding the need of temperature control in field applications. It presented a linear response to MCT concentration within 0.3–1.4 μg L^{-1}, with a sensitivity of -12.4 ± 0.7 nm L μg^{-1} [68]. The method shows excellent selectivity for MCT against other species coexisting with the analyte in environmental waters. It was successfully applied to the determination of MCT in contaminated samples.

Taking theophylline and (1R,2S)-ephedrine as template molecules, imprinted photonic-hydrogel films are prepared by a combination of colloidal-crystal and molecular imprinting techniques [69]. This is a new approach for rapid and handy stimulant detection with high sensitivity and specificity. The key point of this approach is that the imprinted photonic polymer (IPP) consists of a three-dimensional (3D), highly ordered, and interconnected macroporous array with a thin hydrogel wall, where nanocavities complementary to analytes in shape and binding sites are distributed. This special, bicontinuous, hierarchical structure enables this polymer to report a molecular recognition event quickly, easily, sensitively, and directly without any transducers and treatments for analytes (label-free). The inherent affinity of the nanocavities, deriving from molecular imprinting, makes these sensors highly specific to analytes, even if in a competitive environment. Their sensitive and specific

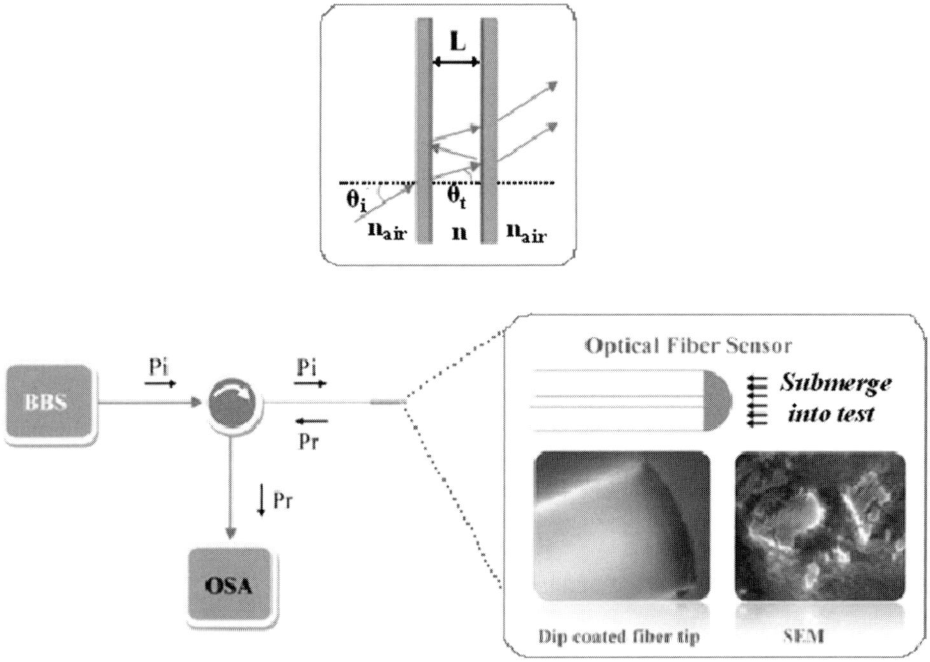

FIGURE 13.27 Experimental setup with details of the sensing head, SEM image, and scheme of the Fabry–Pérot cavity. BBS: broadband source; OSA: optical spectrum analyzer; Pi: incident power; Pr: reflected power; L: distance between mirrors; θ_i: incidence angle; θ_t: transmitted angle; and n: refractive index [68].

responses to stimulants in buffer are determined by Bragg diffractive shifts due to the lattice change of their 3D ordered macroporous arrays resulting from their preferential rebinding to the target molecules. The measurements show that the prepared hydrogel films exhibit high sensitivity in such a 0.1 fM concentration of analytes and specificity, even in a competitive urinous buffer. This method provides a rapid and handy approach for stimulant assay and drug analysis in athletic sports. We have also developed a molecularly imprinted photonic crystal hydrogel to detect an organophosphorus compound, methyl phosphonic acid (MPA), which is the degradation product of nerve agents. Monodispersed poly-methyl methacrylate (PMMA) colloidal particles (280 nm) were used to fabricate a closely packed colloidal crystal array (CCA), in which an MPA-imprinted hydrogel was infiltrated. To prepare the imprinted hydrogel, 2-hydroxyethyl-methacrylate (HEMA) and N-isopropylacrylamide (NIPA) were used as monomers, ethyleneglycol dimethacrylate (EDMA) and N,N′-methylenebisacrylamide (MBAA) as cross-linkers, and a mixture of n-octanol and acetonitrile as porogen. The diffraction intensity of MIPC decreased significantly upon the MPA adsorption with a limit of detection (LOD) of 10^{-6} M (Fig. 13.28). The MIPC provides an indirect path to detect nerve agents (Sarin, Soman, VX, and R-VX) by the monitoring the MPA released from the hydrolysis reaction of nerve agents, with LODs of 3.5×10^{-6} M, 2.5×10^{-5} M, 7.5×10^{-5} M and 7.5×10^{-5} M for Sarin, Soman, VX, and R-VX, respectively.

Nafcillin is a semisynthetic β-lactamic antibiotic employed in the treatment of serious infections caused by penicillinase-producing staphylococci. Two approaches were addressed

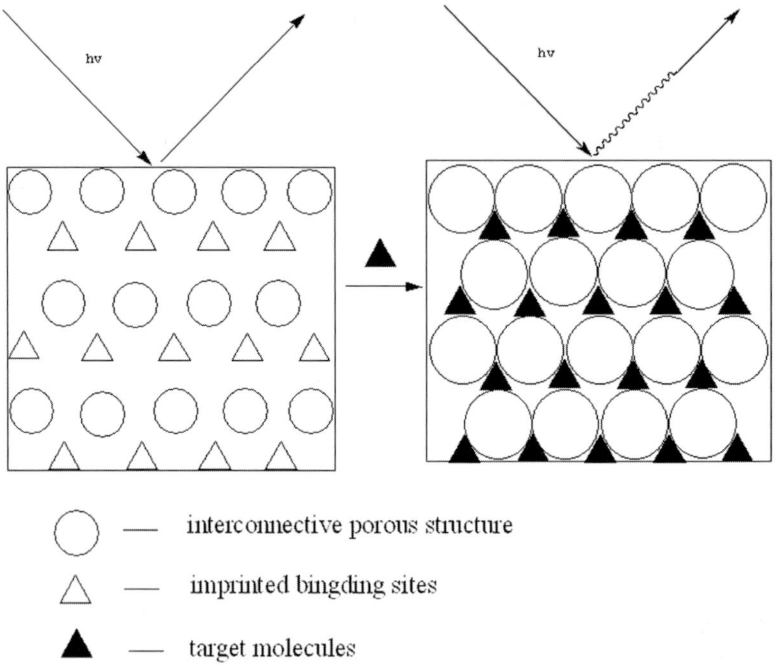

FIGURE 13.28 The mechanism of diffraction decrease of the imprinted photonic crystal hydrogel in response to the MPA absorption.

for preparation of the imprinted materials for nafcillin and the controls: as conventional monoliths, which were ground and sieved to a desired particle size for rebinding analysis, and as MISG on supporting glass slides. The specific binding sites that were created during the imprinting process were analyzed via selective room temperature phosphorescence (RTP) (sol-gel films) measurements as well as via competitive room-temperature phosphorescence ligand assay (Fig. 13.29). Results demonstrated the importance of the physical configuration of the imprinted material for minimizing nonspecific binding. The MISG films displayed enhanced specific binding characteristics with respect to the monolithic sol-gel and can be envisaged for use as recognition matrices for the screening and rapid selection of antibiotics from a combinatorial library or for the rapid control of nafcillin in biological samples [70–72].

A photoresponsive MIP, which has specific binding sites for 2,4-D through hydrogen bonding moieties, was prepared by using a photoresponsive functional monomer bearing a siloxane polymerisable group and azobenzene moieties. The binding affinity of the imprinted recognition sites was switchable by alternate irradiations with ultraviolet and visible light, suggesting that azobenzene groups located inside the binding sites could be used as chemical sensors and the *trans–cis* isomerization could regulate the affinity for the 2,4-D. In addition, the concentration of 2,4-D was able to be quantified by monitoring the *trans*-to-*cis* photoisomerization rate constant [73]. A sol-gel derived molecular-imprinted luminescent sensing material has been fabricated by a conventional sol-gel process. 2,4-D

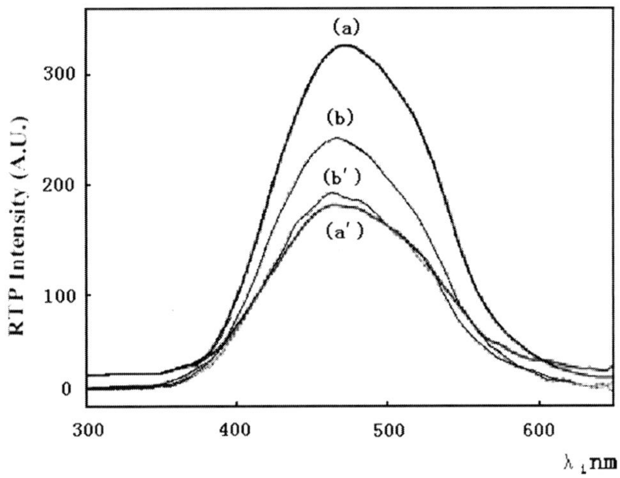

FIGURE 13.29 Room-temperature phosphorescence emission spectrum of the TPM1 imprinted sol-gel (—) before (a) and after (a′) removing the nafcillin template. Control sol-gel (- - -) before (b) and after (b′) rinsing. $\lambda_{exc} = 283$ nm, $t_d = 0.04$ ms, $t_g = 3.00$ ms.

was used as template for the imprinting process, and a tailor-made organosilane, 3-[N,N-bis(9-anthrylmethyl) amino] propyltriethoxysilane (1), was used as the functional monomer. The luminescent properties of 1 were perturbed by acid–base ion-pair formation with 2,4-D leading to suppression of photoinduced electron transfer (PET) quenching of its anthryl fluorophores. Such a PET fluorescent response was preserved after the preorganized monomer-template aggregates were cross-linked by a sol-gel process. The resultant sol-gel-derived luminescent material was found to selectively respond to 2,4-D in aqueous media at pH 7. Average binding affinity of recognition sites in the material for 2,4-D was estimated to be $(2.23 +/- 0.5) \times 10^{-6}$ M. Selectivity of the MIP was also evaluated. The present study demonstrated the feasibility of using a PET mechanism as a means of signal transduction for MISG-based luminescent sensing of nonfluorescent analytes [74].

References

[1] P.C.A. Jeronimo, A.N. Araujo, M.C.B.S.M. Montenegro, Talanta 72 (2007) 13.
[2] J. Lin, C.W. Brown, Trac-Trend. Anal. Chem. 16 (1997) 200.
[3] S. Marx, Z. Liron, Chem. Mater. 13 (2001) 3624.
[4] M.E. Diaz-Garcia, R.B. Laino, Microchim. Acta 149 (2005) 19.
[5] Y.K. Lu, X.P. Yan, Chin. J. Anal. Chem. 33 (2005) 254.
[6] E.L. Holthoff, F.V. Bright, Anal. Chim. Acta 594 (2007) 147.
[7] A. Walcarius, M.M. Collinson, Annu. Rev. Anal. Chem. 2 (2009) 121.
[8] F.H. Dickey, Proc. Natl. Acad. Sci. U S A 35 (1949) 227.
[9] M.C. Burleigh, S. Dai, E.W. Hagaman, J.S. Lin, Chem. Mater. 13 (2001) 2537.
[10] R. Makote, M.M. Collinson, Chem. Mater. 10 (1998) 2440.
[11] N. Gao, Z. Xu, F. Wang, S.j. Dong, Electroanalysis 19 (2007) 1655.
[12] A.L. Graham, C.A. Carlson, P.L. Edmiston, Anal. Chem. 74 (2002) 458.
[13] G. Jin, Y. Tang, S. Liu, S. Wang, R. Xing, Anal. Lett. 41 (2008) 1811.
[14] Z. Lin, F. Yang, X. He, Y. Zhang, J. Sep. Sci. 32 (2009) 3980.
[15] G. Jin, B. Zhang, Y. Tang, X. Zuo, S. Wang, J. Tang. Talanta 84 (2011) 644.

[16] J.E. Lofgreen, I.L. Moudrakovski, G.A. Ozin, Acs. Nano. 5 (2011) 2277.
[17] E.E. Mojica, J. Autschbach, F.V. Bright, D.S. Aga, Anal. Chim. Acta 684 (2011) 72.
[18] S. Lordel, F. Chapuis-Hugon, V. Eudes, V. Pichon, J. Chromatogr. A 1217 (2010) 6674.
[19] Z.H. Zhang, L. Liu, H. Li, S.Z. Yao, Anal. Lett. 43 (2010) 2318.
[20] Z. Lin, F. Yang, X. He, X. Zhao, Y. Zhang, J. Chromatogr. A 1216 (2009) 8612.
[21] L. Qin, X.W. He, M. Jia, W.Y. Li, Y.K. Zhang, Chem. Eur. J. 17 (2011) 1696.
[22] S. Lee, H.M. Lin, H. Chen, J. Appl. Polym. Sci. 114 (2009) 3994.
[23] C. Lin, A. Joseph, C.K. Chang, Y.C. Wang, Y.D. Lee, Anal. Chim. Acta 481 (2003) 175.
[24] H.A. Tsai, M.J. Syu, Chem. Eng. J. 168 (2011) 1369.
[25] Y.S. Chang, T.H. Ko, T.J. Hsu, M.J. Anal, Chem. 81 (2009) 2098.
[26] C.W. Hsu, M.C. Yang, J. Non-Cryst, Solid 354 (2008) 4037.
[27] S.W. Lee, I. Ichinose, T. Kunitake, Langmuir. 14 (1998) 2857.
[28] C. Pinel, P. Loisil, P. Gallezot, Adv. Mater. 9 (1997) 582.
[29] Z.P. Yang, J.L. Yan, C.J. Zhang, S.Q. Luo, Colloid. surf. B 87 (2011) 187.
[30] N. Mizutani, D.H. Yang, R. Selyanchyn, S. Korposh, S.W. Lee, T. Kunitake, Anal. Chim. Acta 694 (2011) 142.
[31] Q. Yao, Y. Zhou, J. Inorg, Organomet. Polym. Mat. 19 (2009) 466.
[32] M.R. Buchmeiser, J. Chromatogr. A 918 (2001) 233.
[33] C. Alexander, L. Davidson, W. Hayes, Tetrahedron. 59 (2003) 2025.
[34] T. Kunitake, S.W. Lee, Anal. Chim. Acta 504 (2004) 1.
[35] C.T. Chen, Y.C. Chen, Anal. Chem. 76 (2004) 1453.
[36] A. Fernandez-Gonzalez, R. Badia-Laino, M. Elena Diaz-Garcia, Microchim. Acta 172 (2010) 351.
[37] M. Azenha, B. Szefczyk, D. Loureiro, P. Kathirvel, M. Natalia, D.S. Cordeiro, A. Fernando-Silva, Langmuir 27 (2011) 5062.
[38] M. Azenha, P. Kathirvel, P. Nogueira, A. Fernando-Silva, Biosens. Bioelectron. 23 (2008) 1843.
[39] B. Pamidighantam, P.C. Lauterbur, Anal. Bioanal. Chem. 396 (2010) 1607.
[40] A. Walcarius, D. Mandler, J.A. Cox, M. Collinson, O. Lev, J. Mater, Chem. 15 (2005) 3663.
[41] W.J. Santos, M. Santhiago, I.V. Yoshida, L.T. Kubota, Anal. Chim. Acta 695 (2011) 44.
[42] D.H. Yang, M.J. Ju, A. Maeda, K. Hayashi, K. Toko, S.W. Lee, T. Kunitake, Biosens. Bioelectron. 22 (2006) 388.
[43] A. Mujahid, A. Afzal, G. Glanzing, A. Leidl, P.A. Lieberzeit, F.L. Dickert, Anal. Chim. Acta 675 (2010) 53.
[44] Z. Liu, S. Huan, J. Jiang, R. Yu, Talanta 68 (2006) 1120.
[45] Y.F. Hu, Z.H. Zhang, H.B. Zhang, L.J. Luo, S.Z. Yao, Talanta 84 (2011) 305.
[46] Z. Zhang, Y. Hu, H. Zhang, S. Yao, J. Colloid, Interface. Sci. 344 (2010) 158.
[47] Z. Zhang, Y. Hu, H. Zhang, L. Luo, S. Yao, Biosens. Bioelectron. 26 (2010) 696.
[48] U. Latif, A. Mujahid, A. Afzal, R. Sikorski, P.A. Lieberzeit, F.L. Dickert, Anal. Bioanal. Chem. 400 (2011) 2507.
[49] N. Sallacan, M. Zayats, T. Bourenko, A.B. Kharitonov, I. Willner, Anal. Chem. 74 (2002) 702.
[50] M. Lahav, A.B. Kharitonov, I. Willner, Chem. Eur. J. 7 (2001) 3992.
[51] S. Fireman-Shoresh, I. Turyan, D. Mandler, D. Avnir, S. Marx, Chiral electrochemical recognition by very thin molecularly imprinted sol-gel films, Langmuir. 21 (17) (2005) 7842–7847.
[52] M. Lahav, A.B. Kharitonov, O. Katz, T. Kunitake, I. Willner, Anal. Chem. 73 (2001) 720.
[53] S. Marx, A. Zaltsman, I. Turyan, D. Mandler, Anal. Chem. 76 (2004) 120.
[54] C.Y. Li, C.F. Wang, B. Guan, Y.Y. Zhang, S.S. Hu Sensor, Actuat. B-Chem. 107 (2005) 411.
[55] C. Li, C. Wang, C. Wang, S. Hu, Anal. Chim. Acta 545 (2005) 122.
[56] N.F. Atta, A.M. Abdel-Mageed, Talanta, 2009.
[57] A.K. Patel, P.S. Sharma, B.B. Prasad, Int. J. Pharmaceutics 371 (2009) 47.
[58] B.B. Prasad, K. Tiwari, M. Singh, P.S. Sharma, A.K. Patel, S. Srivastava, Biomed. Chromatogr. 23 (2009) 499.
[59] B.B. Prasad, R. Madhuri, M.P. Tiwari, P.S. Sharma, Sensor. Actuat. B-Chem. 146 (2010) 321.
[60] L. Basabe-Desmonts, D.N. Reinhoudt, M. Crego-Calama, Chem. Soc. Rev. 36 (2007) 993.
[61] O.A. Raitman, V.I. Chegel, A.B. Kharitonov, M. Zayats, E. Katz, I. Willner, nal. Chim. Acta 504 (2004) 101.
[62] J. Yu, C. Zhang, P. Dai, S. Ge, Anal. Chim. Acta 651 (2009) 209.
[63] A. Zattoni, P. Reschiglian, M. Montalti, N. Zaccheroni, L. Prodi, R.A. Picca, C. Malitesta, Inorganica Chim. Acta 360 (2007) 1063.
[64] N.R. Walker, M.J. Linman, M.M. Timmers, S.L. Dean, C.M. Burkett, A.J. Lloyd, J.D. Keelor, B.M. Baughman, P.L. Edmiston, Anal. Chim. Acta 593 (2007) 82.

[65] E.L. Holthoff, D.N. Stratis-Cullum, M.E. Hankus, Sensors 11 (2011) 2700.
[66] Z. Tao, E.C. Tehan, R.M. Bukowski, Y. Tang, Shugha, Anal. Chim. Acta 564 (2006) 59.
[67] W. Zhang, X.W. He, Y. Chen, W.Y. Li, Y.K. Zhang, Biosens. Bioelectron. 26 (2011) 2553.
[68] R.B. Queiros, S.O. Silva, J.P. Noronha, O. Frazao, P. Jorge, P.V. Marques, M.G. Sales, Biosens. Bioelectron. 26 (2011) 3932.
[69] X.B. Hu, G.T. Li, M.H. Li, J. Huang, Y. Li, Y.B. Gao, Y. H Zhang, Adv. Funct. Mater. 18 (2008) 575.
[70] A. Fernandez-Gonzalez, R.B. Laino, M.E. Diaz-Garcia, L. Guardia, A. Viale, J. Chromatogr. B 804 (2004) 247.
[71] L. Guardia, R. Badia, M.E. Diaz-Garcia, Biosens. Bioelectron. 21 (2006) 1822.
[72] L.R.B. Guardia, M.E. Diaz-Garcia, J. Agri. Food Chem. 55 (2007) 566.
[73] S.J. Guang, A.Z. Shi, C. Lan, B. Idriss, W. Andrew, Radiat. Phys. Chem. 80 (2010) 130.
[74] M.K. Leung, C.F. Chow, M.H.W. Lam, J. Mater, Chem. 11 (2001) 2985.

CHAPTER

14

MIP-based Sensors

Sergey Piletsky, Stas Piletsky, Iva Chianella

Cranfield Health, Cranfield University, Cranfield, Bedfordshire, UK

OUTLINE

1. INTRODUCTION—MOLECULAR IMPRINTING

Molecular imprinting is a process used for preparing affinity polymers for targets of interest. It involves formation of a complex between a given target (template) molecule and functional monomers dissolved in an appropriate solvent, which is then fixed by polymerization into a growing polymer chain. Subsequent removal of the template leaves binding sites within the polymer possessing both the correct shape and the correct orientation of functional groups to allow selective recognition of the imprint species (Fig. 14.1) [1,2]. Most often, MIPs are made of acrylic and methacrylic monomers [3], polyurethane [4], or conjugated polymers such as polyaniline and polypyrrole [5,6]. There are also examples of MIPs produced from inorganic oxides [7].

Acrylates and methacrylates remain the most popular monomers used for MIP preparation and usually produce polymers with superior binding and recognition abilities. This is mainly because of greater flexibility in selecting appropriate functionalities (more than 40,000 polymerizable compounds are commercially available) and also greater flexibility in selecting appropriate polymerization protocol. Thus acrylates and methacrylate-based MIPs can be synthesized using bulk polymerization, grafting, made in the form of membranes, micro- and nanoparticles [8]. These polymers also benefit from advanced designing approaches such as combinatorial and computational methods [9]. The synthesized polymers are robust

Molecularly Imprinted Sensors.
DOI: 10.1016/B978-0-444-56331-6.00014-1

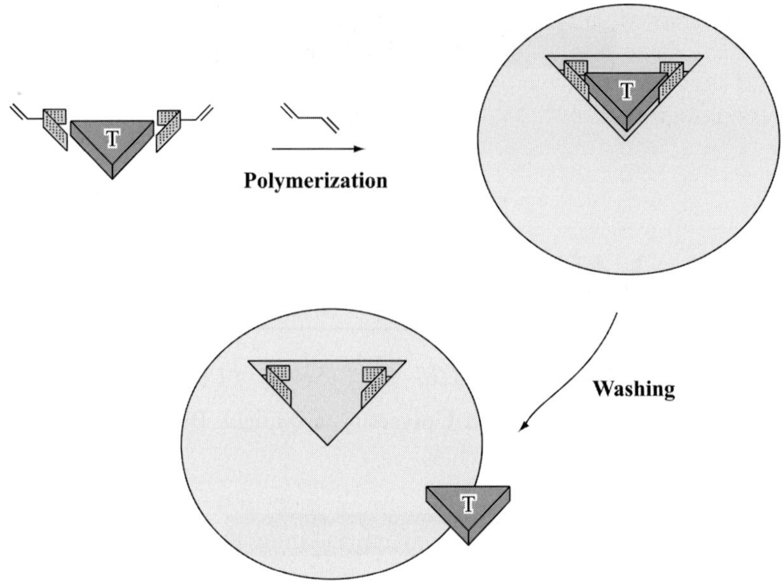

FIGURE 14.1 Scheme of molecularly imprinted polymerization.

and compatible, with a range of applications from diagnostics to separation [12]. Because of the great variety of functional monomers available, it is possible to design a MIP for virtually any target of interest. Since molecular imprinting is one of the few generic technologies available for the design of synthetic receptors, it became a focus of interest for scientists and engineers looking for a replacement of unstable biological molecules such as antibodies and enzymes in biosensors. The advantages of MIPs over natural receptor molecules are listed

TABLE 14.1 Comparison of Natural Biomolecules Used in Sensors and MIPs

Natural biomolecules	MIPs
Low stability	Stable at low/high pHs, pressure, and temperature (<140 °C)
High price of enzymes and receptors	Inexpensive and easy to prepare
Poor performance in nonaqueous media	Can work in organic solvents
Different natural biomolecules have different operational requirements (pH, ionic strength, temperature, substrate)	Polymers for different targets could operate in the same specific environment
For some important analytes natural receptors and enzymes do not exist and antibodies cannot be prepared	Polymers could be prepared for practically any compound
Poor compatibility with micromachining technology and miniaturization	Polymers are fully compatible with micromachining technology

in Table 14.1. The main advantages include superior stability, low cost, and ease of preparation. Among the disadvantages associated with this technology are: (1) high level of nonspecific binding; (2) poor performance of MIPs in aqueous environment; and (3) lack of suitable protocols for the integration of MIPs with existing sensor devices.

2. MIPs IN SENSORS

The first attempts to use bulk-imprinted polymers in sensors were made in 1992 [13]. It was shown that templates such as amino acids, nucleic acids, and cholesterol increase the current passing through membranes made of imprinted polymers [14]. Since then, more than 1200 papers focusing on MIP applications on sensors (http://www.mipdatabase.com/Database.html) have been published worldwide, which clearly indicates growing interest in this area. The prime function of MIPs exploited in sensors is, of course, selective binding. The binding of the target analyte to an MIP is registered by sensor as a change in mass (in, e.g., acoustic sensors), refractive index (in SPR sensors), or impedance (in electrochemical sensors). There are also numerous examples of sensors that rely on MIPs capable of generating optical or electrochemical signals in response to the template binding to the functional groups in the imprinting site. Very few cases exist in which MIPs are used in sensors as catalysts to produce a detectable electrochemical signal. The following examples will illustrate the design and performance of these types of MIP-based sensors in what we consider to be their most important practical applications. Because of their superior performance, only MIPs made from organic polymers will be reviewed in this work.

2.1. MIP Sensors in Gravimetric Analysis

In principle, this type of sensor is the simplest and most straightforward, which allows template detection by measuring mass of molecules readsorbed by an immobilized MIP. Gravimetric sensors such as piezoelectric devices have advantages such as relative simplicity of sensor design, easy interpretation of sensor response, and compatibility with a variety of solvents and templates. Among their disadvantages are poor performances in liquid and often low sensitivity.

Integration of MIPs with piezoelectric sensors can be performed in a variety of ways, including grafting, in-situ polymerization, by entrapment of MIP particles into polymer films deposited using, for example, spin coating, and by physical or covalent immobilization of MIP micro- and nanoparticles [15–18]. One of the first piezoelectric MIP sensors was developed by Percival and co-authors for the determination of L-menthol [19]. A MIP was cast in situ onto the surface of a gold-coated quartz crystal microbalance (QCM) electrode as a thin film. The detection of target analyte monitored as a frequency shift was possible at 0.2–1 ppm concentrations (Fig. 14.2). Several examples of application of acoustic wave MIP sensors for measuring paracetamol and nicotine in real samples (human serum and urine) were presented by Tan and co-authors [20,21]. The sensor showed high selectivity and a sensitive response to nicotine in an aqueous system. Recently, a QCM sensor for digestive proteins in saliva (amylase, lipase, and lysozyme) was developed using surface imprinting [22]. In this work, a mixed salivary protein and poly(ethylene-co-vinyl alcohol)

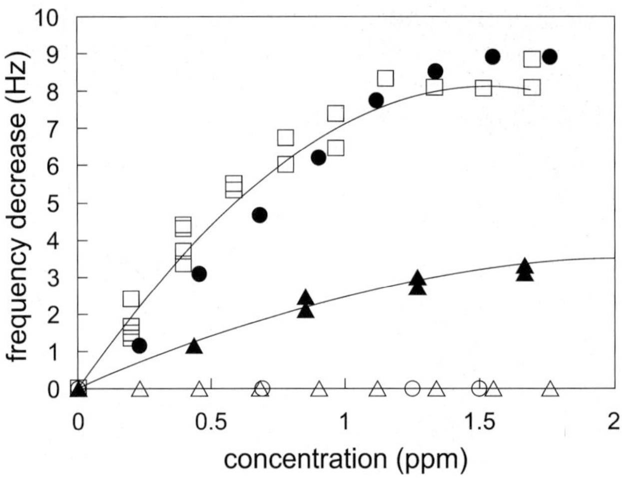

FIGURE 14.2 Response of MIP-coated quartz crystals to monoterpene analogues: L-menthol (squares), D-menthol (solid triangles), citronellol (solid circles), citronellal (triangles), and menthone (circles) *(Percival et al., 2001).*

solution was coated onto a QCM chip, and a MIP thin film formed by thermally induced phase separation. The detection limit in real samples was at the ppm level.

Polyurethane-based MIPs were applied successfully in piezoelectric sensors for the detection of benzene, toluene, ethylbenzene, and xylenes in complex matrixes such as gasoline vapors [23]. The analytes were detected at the ppt level. In another very interesting work, polyurethanes were imprinted with characteristic oils rather than with a single compound and used as sensor materials for monitoring degradation of automotive engine oils [24]. These approaches showcased an opportunity to perform analysis of complex mixtures in gas and in liquid phase.

An application of MIPs in piezoelectric sensors was expanded later for large analytes such as proteins and cells (Fig. 14.3).

In one of the first examples, the polymer surface was imprinted with yeast cells and integrated with a sensor capable of detecting yeast cells and cell membrane fragments in growth media [26]. The same group later developed a QCM sensor capable of differentiating blood subtypes [27]. In another important work, MIPs specific for the dengue viral NS1 protein were formed by polymerization of monomers onto a QCM chip surface. This MIP-QCM chip, whose results agreed with those of conventional EIA, was able to diagnose dengue infection in under an hour and exhibited good sensitivity and specificity when compared to reported monoclonal antibody-based chips for DENV2 [28]. This group later developed a QCM sensor coated with a MIP specific for epitopes of the anthrax protective antigen PA_{83} [29]. The differential response of the sensor to the protective antigen fragments was observed in the picomolar range.

An example of a piezoelectric multisensor has been developed by Sun and co-authors [30]. In this work MIPs were prepared for quinine and saccharine and used as thin-film coatings in sensor arrays for the analysis of food samples. The main advantage of the sensor was its ability to process a large volume of samples within a short time (up to 20 samples/hour). The recent development of integration of MIP nanoparticles with sensors appears very

FIGURE 14.3 Poylurethane surface imprinted with yeast cells. *(Reproduced with permission from [25].)*

promising. Thus Yaqub and co-authors have shown that it is possible to develop sensors where "plastic antibodies" compete successfully with natural molecules offering superior sensitivity and selectivity [31] (Fig. 14.4).

The main problems associated with piezoelectric or in general acoustic MIP sensors are not actually associated with MIPs, but with the transducers themselves. Acoustic sensors are in

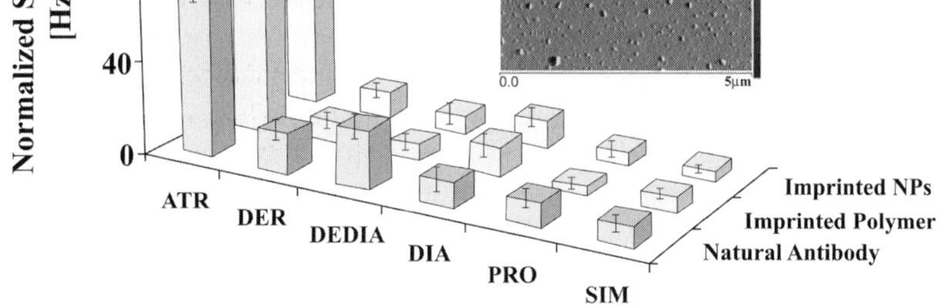

FIGURE 14.4 Sensor response of natural and atrazine-imprinted synthetic antibodies for 7 mg/L concentrations of atrazine (ATR) and its structural analogues des-ethyl atrazine (DEA), des-ethyl-des-isopropyl atrazine (DEDIA), des-isopropyl atrazine (DIA), propazine (PRO), and simazine (SIM) in water. *(Reproduced with permission from [31].)*

general less sensitive than amperometric and fluorescent sensors, although their sensitivity can be significantly improved using a surface acoustic wave mode [32]. Acoustic sensors are not particularly reliable when it comes to applications in liquid phase. The sensors' response is hugely influenced by viscosoelastic properties of the environment and affected by temperature changes [33]. The MIP-related problem in the development of sensors involves the necessity to fabricate sensors in situ, which of course would increase the price of the device. The possibility of entrapping presynthesized MIP particles in spin-coated or otherwise deposited linear polymers from the solution seems a good option that might be useful in future commercialization work.

2.2. MIP Electrochemical Sensors

Electrochemical sensors generally possess higher sensitivity than piezoelectric sensors. The first potentiometric MIP sensors based on electropolymerized materials were described by Boyle and co-authors and Vinokurov [34,5]. Here the monomers reacted with themselves and other molecules as templates forming specific polymers. Sensors for pyrrole, aromatic amines, and substituted phenols, based on polypyrrole, polyaniline, and aniline-p-aminophenol copolymers, respectively, were prepared. A similar MIP-based amperometric device was developed by Piletsky and co-authors for the detection of aniline and phenol [35]. The sensors relied on preferential adsorbance of the electroactive template by conductive MIP. Differential pulse voltammetry has been used for the detection of clenbuterol in bovine liver samples [36]. The sensor approach relied on displacement of clenbuterol specifically adsorbed by a MIP with an electroinactive analogue-isoxsuprine. Murray and co-authors developed potentiometric sensors for the analysis of lead and uranil ions in a complex media [37]. The direct detection of an "inert" template in electrochemical sensors has also been reported [38–40]. These sensors explored MIPs' ability to change conformation upon binding with the template, which leads to a change in a measurable property such as conductivity, permeability, or surface potential. Sensors show high selectivity and sensitivity in the micromolar-nanomolar range (Fig. 14.5).

A successful potentiometric sensor was later reported for what is probably the most important analyte of all time—glucose. This was based on measurement of the concentration of protons released during interaction of a metal-complexing imprinted polymer with glucose [41]. A clinically relevant measurement range of glucose concentration in plasma was reported (0 to 25 mM). Recently, a portable potentiometric MIP sensor has been developed for the detection of salbutamol in blood (Fig. 14.6). The conductometric sensor displayed a linear correlation over a broad range of salbutamol concentrations (50 to 280 nM), with a detection limit of 13.5 nM [42].

The fabrication of *Bacillus subtilis* endospore-imprinted conducting polymer films and subsequent detection of bound spores in concentration 10^2 cfu/ml via impedance measurements were reported [43]. A sensor, made of polypyrrole and a poly(3-methylthiophene) layer, was detecting dipicolinic acid and other constituents released from the core in the course of germination.

In electrochemical sensors, there is no special requirement for the thickness of polymer layer or its morphology. Most frequently, MIPs are deposited on the detector surface by in-situ UV or thermoinitiated polymerization [44,45]. A convenient way to immobilize MIP

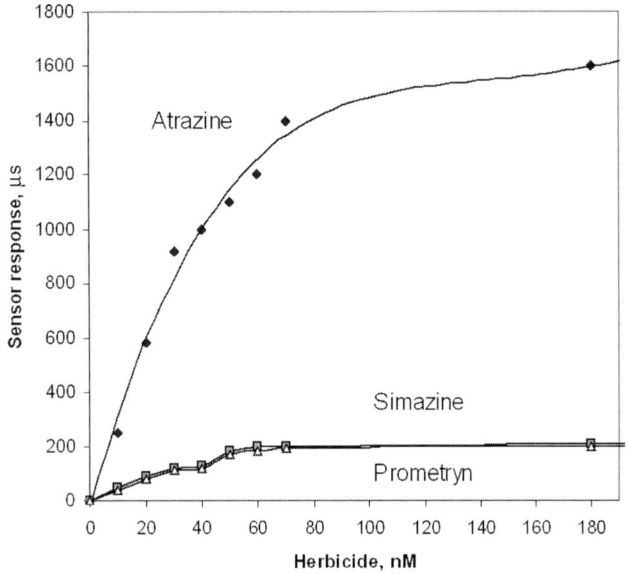

FIGURE 14.5 Sensor response to herbicides obtained with a polymer imprinted with atrazine.

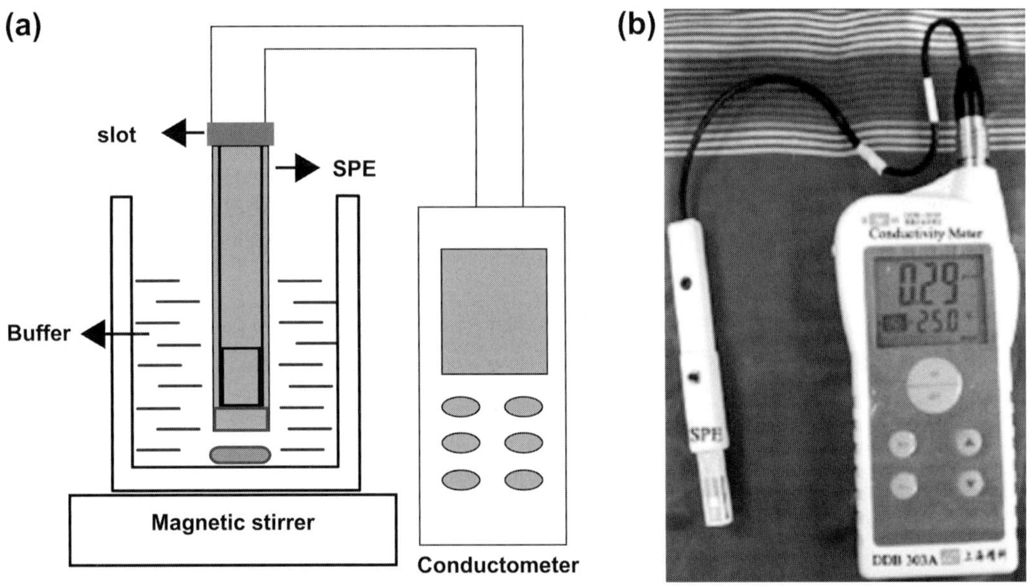

FIGURE 14.6 Setup for conductometric measurements of salbutamol with a MIP sensor (a) and photo of the portable salbutamol MIP probe (b). *(Reproduced with permission from [42].)*

particles on the surface of electrochemical sensors is to trap them into carbon paste used in screen printing or in another conductive polymer [46–48]. Using this approach, a novel sensor for explosives in soil has been developed by Alizadeh and co-authors, with a detection limit for TNT of 1.5 nM [49]. Further enhancement of the sensor response can be achieved by using conductive nanomaterials such as carbon nanotubes [50]. Other convenient ways to fabricate MIP sensors and sensor arrays involve the use of photolithography and laser deposition [51,52]. In a first example, an MIP array was created (Fig. 14.7) capable of differentiating salbuterol from the interfering analogies such as clenbuterol and terbutaline. For electrochemical detection it is essential for the polymer to be extremely thin (nm scale) or impregnated with conductive material. Recently, hybrid materials were developed, which allow grafting of the MIPs to the surface of conductive polymers that can be deposited by photolithography [53].

Significant enhancement in the sensor signal has been achieved using enzyme amplification [54]. A MIP specific for oxytetracycline was used in competitive assay format with a horseradish peroxidase labeled template assuring a detection limit for oxytetracycline as low as 0.65 nM. In an ideal sensor, the enzyme should also be replaced with a stable catalytic MIP. MIPs with catalytic properties could be made, for example, by using a transition state analogue of the reaction as a template [55]. Although immense progress has been made in the field of MIP catalysts, a comparatively small number of publications have been dedicated to catalytic sensors based on MIPs [56–58]. In one example a catalytic MIP, capable of catechol oxidation, was created by mimicking the transition stage analogue of tyrosinase [59]. The polymer was grafted to a derivative of polyaniline-bearing double bonds. The direct electron transfer was achieved between catalytic centers in the MIP and polyaniline deposited onto an electrode surface (Fig. 14.8).

Further developments in catalytic receptors based on MIPs will depend largely on success in the preparation of MIP-based catalysts that can react with water-soluble and practically important analytes.

FIGURE 14.7 An optical microscope image of the lithographic pattern created from acrylic MIPs. *(Reproduced with permission from [51].)*

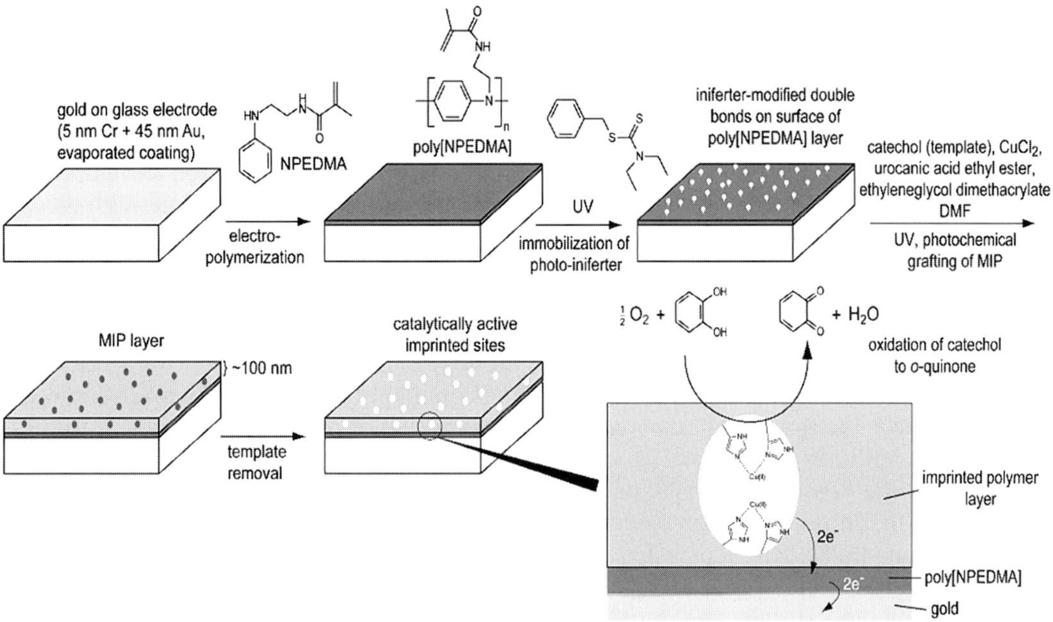

FIGURE 14.8 Preparation of the catalytic MIP-hybrid electrodes imprinted with catechol. *(Reproduced with permission from [59].)*

2.3. MIP-based Optical Sensors

Optical MIP sensors are a very popular area of research. A number of sensors have been developed for the detection of UV and visible light-absorbing and fluorescent analytes. One of the first MIP-based optical sensors specific for fluorescent dansyl-L-phenylalanine was developed by Kriz and co-authors [44]. A template, accumulated in the polymer matrix, was detected by a fiber-optic sensing device. In another example, Dickert and co-authors demonstrated the detection of fluorescent polycyclic aromatic hydrocarbons using optical sensors with imprinted polyurethanes [4]. A successful MIP sensor was produced for the detection of fluorescent aflatoxins and ochratoxins in food extracts [60]. Recently, an attempt was made to integrate several MIP sensors in arrays for the optical detection of fluorescent templates using laser pulse initiated polymerization [52]. In another example, MIPs' diffractive gratings were micropatterned onto a sensor surface using SiO_2/Si molds [61]. A serious limitation to the broad applicability of optical sensors, in particular fluorescent sensors, is the small number of practically important fluorescent templates. A solution to this problem is the development of sensors that can operate in a competitive mode by using binding competition between template and its fluorescent labeled analogue [62]. A competitive MIP sensor for the detection of chloramphenicol in blood serum was reported by Levi and co-authors [63]. The sensor included an HPLC column with template-specific MIPs. A constant flow of chloramphenicol-Methyl Red was run through the column at a concentration of 0.5 μg/ml. Free analyte displaced the adsorbed conjugate, giving a peak with an area that was proportional to chloramphenicol concentration. Unfortunately, not all analytes can be easily modified with dyes, and the modification itself can change the affinity of analyte. In principle, a specific MIP sensor can be created exploiting a competition between a nonspecific dye and template [64]. In a recent example, this approach has been used for the development of a chemiluminescent MIP sensor for terbutaline [65]. This sensor was capable of detecting a 4.0 ng/ml template in blood serum. Detection of the template in optical sensors can be done using postmodification of templates adsorbed by MIP [66]. A microfluidic disposable chip has been developed for the detection of anesthetic propofol at 655 nm after the reaction of the analyte with a color reagent (Gibbs reaction) [67]. The range of detectable analytes can also be expanded using infrared spectroscopy. Jakusch and co-authors developed an infrared evanescent wave sensor for 2,4-dichlorophenoxyacetic acid with a detection limit of 4.1 μM [68]. The infrared spectroscopy provided an additional element contributing to selectivity of sensor response.

In a broad range of applications, a template has been analyzed by measuring a refractive index using evanescent field sensor devices. The benefits in using this approach lie in the ability to detect a template that does not have a high level of optical absorbance or fluorescence. The magnitude of the signal produced by a small template interacting with MIP immobilized onto the sensor surface is usually not particularly high [69], although there are exceptions [70]. Thus a very high detection limit (pM level) was achieved in the surface plasmon resonance (SPR) MIP sensor for δ-opioid G-protein-coupled receptor agonist peptide DPDPE [71]. The key to success seems to be connected with a very thin polymer coating created on the chip surface (100 nm), in addition to the fact that higher sensor responses are generally obtained for larger target analytes such as proteins. A very sensitive SPR chip for T-2 toxin was prepared by in-situ electropolymerization of 3-aminophenylboronic-acid (3-APBA) with T-2 [72]. The sensor showed a linear response for T-2 from 2.1 fM to 33.6 fM with a detection limit of 0.1 fM. Moreover, the sensor showed very little interference from other toxins (Fig. 14.9).

The signal for small molecules can be amplified by conformational changes in the polymer induced by template–polymer interactions. Lightly cross-linked theophylline-imprinted

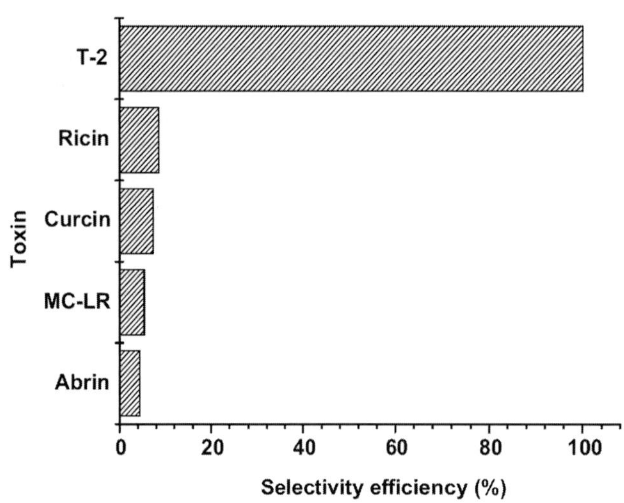

FIGURE 14.9 Selectivity of MIP for T-2 toxin with respect to Abrin [9.3 μM], MC-LR [1 μM], Curcin [14.3 μM], Ricin [23.3 μM], and T-2 toxin [3 pM]. Temperature: 25 °C. *(Reproduced with permission from [72].)*

poly*N*-(*N*-propyl)acrylamide particles were spin-coated onto a gold chip used in SPR spectroscopy [73]. In the presence of template theophylline the particles swelled, producing noticeable changes in refractive index. The detection limit in these experiments was at the μM level. A signal amplification by a factor of x100 can also be achieved by incorporating gold nanoparticles into MIP film [74]. This increase in the detection limit was crucial for the success of sensor application in complex matrices. Thus the MIP sensor developed here was capable of generating a specific response to glucose in spiked urine with a detection limit of 0.12 mg/mL. There are no special requirements to the format of MIPs integrated with fluorescent sensors, where MIP can be deposited by spin coating or by photolithography [75]. It is different, however, when MIPs are used in resonant or evanescent wave field sensors [76]. In these devices, control of MIP thickness and morphology is essential [77,78]. One way to deposit thin MIP coating onto a sensor surface is to use grafting. Surface grafting has been used successfully in construction of SPR sensors for domoic acid [79] and testosterone [70]. To achieve controlled deposition of MIPs with layer thickness 25–400 nm, Belmont and co-authors have used two approaches: spin coating followed by in-situ polymerization of monomer mixture and deposition of preformed nanoparticles [80]. From the point of view of the mass manufacturing of sensor devices, the last approach is preferred because it allows better control over all synthetic and assembling stages.

One other promising sensor approach relies on the development of responsive polymers. Template-signaling functionality can be introduced by incorporating an environment-sensitive probe into the polymer structure. Cooper and co-authors integrated a functional monomer into a cross-linked matrix, which was able to change its fluorescence properties in the presence of proton donors [81]. Polar templates were recognized by a strong quenching of fluorescent emission induced by hydrogen bonding. Dickert and co-authors showed a similar effect by using betaine dyes (phenol blue) with strong hypsochromic effects for protic solvent gas-phase analysis [82]. A fluorescent sensor for cAMP detection was constructed using an environmentally sensitive dye trans-4-[p-(N,N-dimethylamino)styryl]-N-vinylbenzylpyridinium chloride) [83]. The resulting polymer displayed both functions of template recognition and sensing (Fig. 14.10).

Ye and Mosbach reported a new type of proximity scintillation assay for (S)-propranolol. During the imprinting reaction, they covalently incorporated a scintillation monomer (4-hydroxymethyl-2,5-diphenyloxazole acrylate) into MIP microparticles [84]. This monomer was capable of transforming β-radiation from the bound tritium-labelled template into a fluorescent signal. Jenkins and co-authors developed a very sensitive lanthanide-based luminescent sensor for sarin and soman with a detection limit of 7 ppt [85]. The sensor functioned by binding selectively and reversibly the phosphonate hydrolysis product of this agent to a MIP containing a coordinatively bound Eu^{3+} ion. This binding led to the appearance of a narrow luminescence band in the 610-nm region of the Eu^{3+} spectrum, which can be monitored using a miniature spectrometer. Very promising combinations of group-specific fluorescent reporters with template-specific MIPs were described for sugars [86], carboxylic acids [87], and primary amines [88]. In these cases anthracene and thioacetale reporting groups were modified in order to introduce polymerizable and recognition functionalities. The approach involves the transformation of nonspecific interactions of monomers, such as a boronic acid derivative (Fig. 14.11) with cis-dioles, into specific interactions. This is done by incorporating these monomers into specific binding sites created by imprinting.

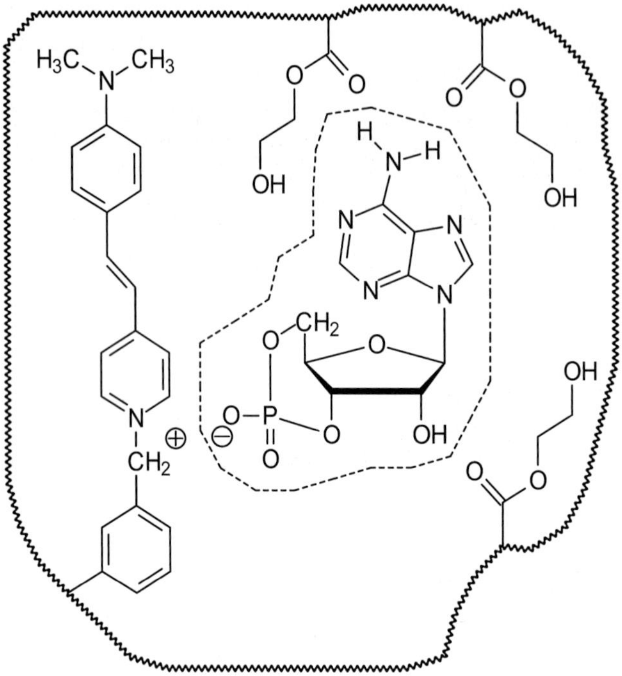

FIGURE 14.10 Schematic representation of the polymer binding site for cAMP with a signaling trans-4-[p-(N,N-dimethylamino)styryl]-N-vinylbenzylpyridinium monomer *(Turkewitsch et al., 1998).*

MIPs were integrated with quantum dots producing fluorescent sensors capable of responding to the template by quenching or enhancing photoluminescence emission [89]. In this example, MIPs specific for caffeine were cast into thin films using CdSe/ZnS core-shell derivatized with 4-vinyl pyridine. The emission intensity increased several-fold when the template reacted with recognition sites of the cross-linked polymeric matrix. An integrated MIP-quantum dots sensor was fabricated and successfully utilized as a recognition element for cytochrome c with the detection limit of 0.41 μM [90]. The fluorescence of MIP-coated quantum dots was strongly quenched by the template (Fig. 14.12).

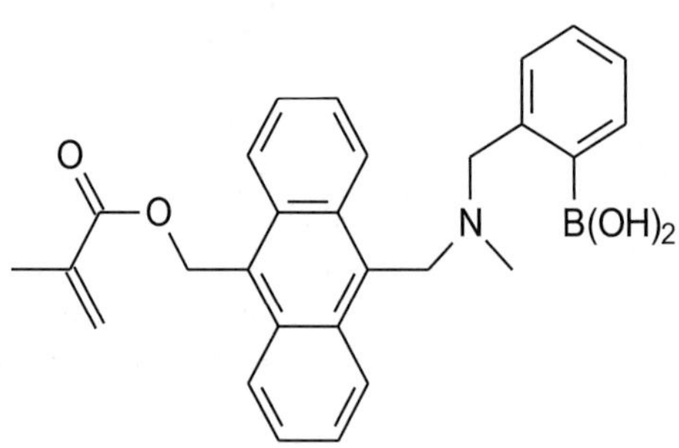

FIGURE 14.11 Structure of the fluorescent monomer used in MIPs for sugar derivatives. *(Adapted with permission from [86].)*

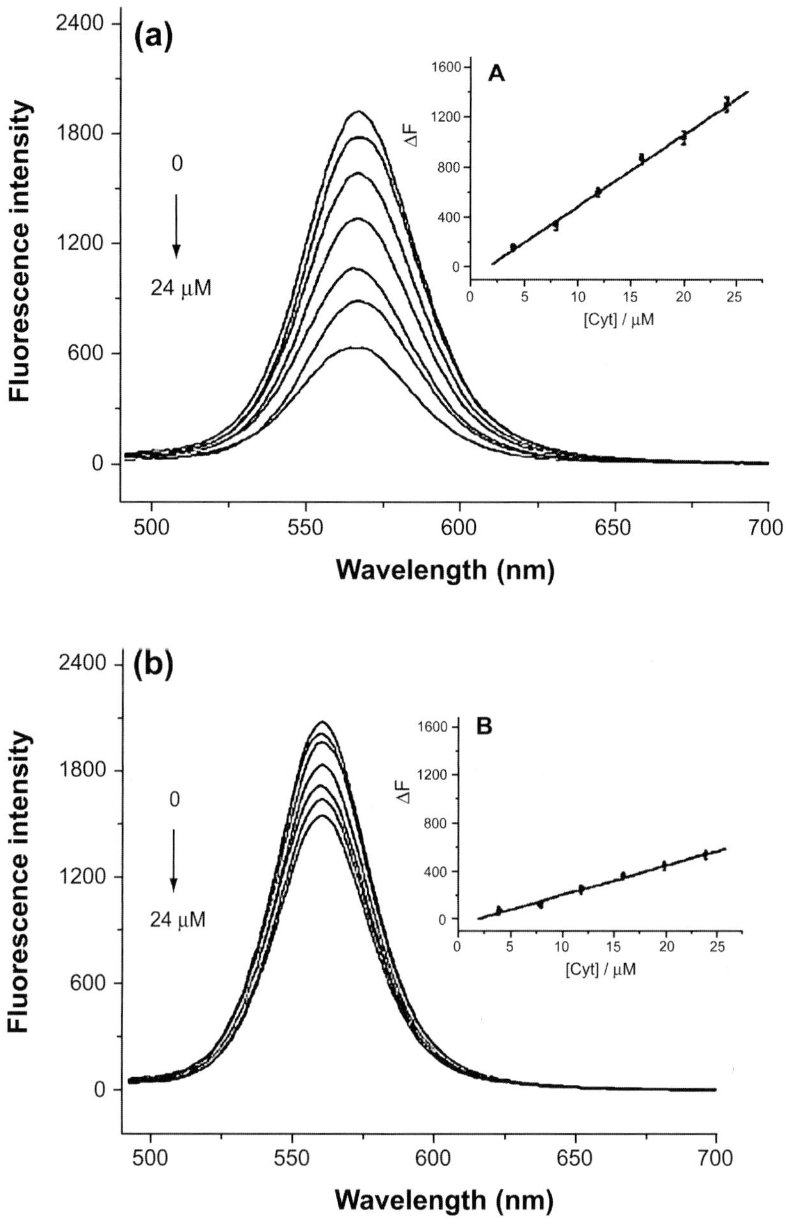

FIGURE 14.12 Fluorescence emission spectra of (a) MIP- and (b) NIP-coated quantum dots with addition of the indicated concentration of protein Cyt in PBS (10 mM, pH 7.0)—Tri (0.5%) solution. *(Reproduced with permission from [90].)*

Optical MIP sensors are a versatile technology that can be adapted to environmental, food, and clinical analyses. In our view, the main problem associated with optical MIP sensors is interference from matrix components in biological samples such as blood plasma, saliva, and urine [91].

3. NICHE AREAS FOR APPLICATION OF MIP SENSORS

Three particular properties make commercial application of MIP sensors attractive: (1) polymers are highly stable and can be autoclaved; (2) they are fully compatible with microfabrication technology; and (3) they involve low cost of the materials and easy processes of polymer preparation in comparison with natural and other artificial receptor systems. Despite the fact that many successful examples of the development of MIP-based electrochemical sensors exist, the current level of commercial activity related to their development and marketing remains low. The remaining problems to be resolved continue to be poor performance of MIPs in water, high level of nonspecific binding, and difficulty integrating MIPs with transducers. With further advances in technology MIP sensors will find clear niches in the sensor market. The most promising areas of MIP sensor applications are medical diagnostics and environmental analyses. One very important niche for MIP sensors is the rapid detection of chemical and biological warfare agents under battlefield and civil conditions. At the time of writing we are aware of two examples where MIP sensors are commercialized or very close to commercialization. One example is the optical sensor for the detection of aflatoxins and ochratoxins developed by Toximet Ltd. (http://www.toximet.com/index.html), which detects fluorescent toxins absorbed onto the MIP cartridge [60]. The second one covers the set of optical kits produced by Raptor Detection Technologies (http://www.raptordetectiontech.com/?q=about-us), which detect explosives by producing noticeable color changes in MIP-coated wipes. Several other companies are also working on the development of MIP sensors, although no product has yet been presented to the public. Among these companies are Sphere Medical (http://www.spheremedical.com/) and Semorex (http://www.semorex.com/index.html). Turning exciting science into practical products is a long and hard path, but it is clear that MIPs have come of age. Serious attempts are now being made to realize their huge potential benefits in the form of tangible products that will lead to generation of wealth and improvements in the quality of life.

References

[1] G. Wulff, Angew. Chem. Int. Ed. Engl. 34 (1995) 1812.
[2] A.G. Mayes, K. Mosbach, TrAC-Trends Anal. Chem. 16 (1997) 321.
[3] C. Alexander, H.S. Andersson, L.I. Andersson, R.J. Ansell, N. Kirsch, I.A. Nicholls, J. O'Mahony, M.J. Whitcombe, J. Mol. Recognit. 19 (2006) 106.
[4] F.L. Dickert, M. Tortschanoff, W.E. Bulst, G. Fischerauer, Anal. Chem. 71 (1999) 4559.
[5] I.A. Vinokurov, Sensor. Actuat. B-Chem. 10 (1992) 31.
[6] C. Malitesta, I. Losito, P.G. Zambonin, Anal. Chem. 71 (1999) 1366.
[7] M. Lahav, A.B. Kharitonov, O. Katz, T. Kunitake, I. Willner, Anal. Chem. 73 (2001) 720.
[8] A. Poma, A.P.F. Turner, S.A. Piletsky, Trends Biotechnol. 28 (2010) 629.
[9] F. Lanza, B. Sellergren, Anal. Chem. 71 (1999) 2092.

[10] T. Takeuchi, D. Fukuma, J. Matsui, Anal. Chem. 71 (1999) 285.
[11] I. Chianella, M. Lotierzo, S.A. Piletsky, I.E. Tothill, B. Chen, K. Karim, A.P.F. Tuner, Anal. Chem. 74 (2002) 1288.
[12] S.A. Piletsky, S. Alcock, A.P.F. Turner, Trends Biotechnol. 19 (2001) 9.
[13] S.A. Piletsky, I.A. Butovich, V.P. Kukhar, Zh Anal. Khim. 47 (1992) 1681.
[14] S.A. Piletsky, E.V. Piletskaya, T.L. Panasyuk, A.V. Elskaya, R. Levi, I. Karube, G. Wulff, Macromolecules 31 (1998) 2137.
[15] K. Das, J. Penelle, V.M. Rotello, Langmuir. 19 (2003) 3921.
[16] Y. Fu, H.O. Finklea, Anal. Chem. 75 (2003) 5387.
[17] L. Ye, P.A.G. Cormack, K. Mosbach, Anal. Chim. Acta 435 (2001) 187.
[18] F. Horemans, J. Alenus, E. Bongaers, A. Weustenraed, R. Thoelen, J. Duchateau, L. Lutsen, D. Vanderzande, P. Wagner, T.J. Cleij, Sensor. Actuat. B-Chem. 148 (2010) 392.
[19] C.J. Percival, S. Stanley, M. Galle, A. Braithwaite, M.I. Newton, G. McHale, W. Hayes, Anal. Chem. 73 (2001) 4225.
[20] Y.G. Tan, Z.L. Zhou, P. Wang, L.H. Nie, S.Z. Yao, Talanta 55 (2001) 337.
[21] Y.G. Tan, J. Yin, C.D. Liang, H. Peng, L.H. Nie, S.Z. Yao, Bioelectrochemistry 53 (2001) 141.
[22] M.H. Lee, J.L. Thomas, H.Y. Tseng, W.C. Lin, B.D. Liu, H.Y. Lin, ACS Appl. Mater. Interf. 3 (2011) 3064.
[23] R.A. Iglesias, F. Tsow, R. Wang, E.S. Forzani, N.J. Tao, Anal. Chem. 81 (2009) 8930.
[24] F.L. Dickert, P. Forth, P.A. Lieberzeit, G. Voigt, Fresenius' J. Anal. Chem. 366 (2000) 802.
[25] F.L. Dickert, P.A. Lieberzeit, O. Hayden, in: Molecular Imprinting of Polymers, Landes Bioscience, Georgetown, US, 2006.
[26] F.L. Dickert, O. Hayden, Anal. Chem. 74 (2002) 1302.
[27] A. Seifner, P. Lieberzeit, C. Jungbauer, F.L. Dickert, Anal. Chim. Acta 651 (2009) 215.
[28] D.F. Tai, C.Y. Lin, T.Z. Wu, J.H. Huang, P.Y. Shu, Clin. Chem. 52 (2006) 1486.
[29] D.F. Tai, M.H. Jhang, G.Y. Chen, S.C. Wang, K.H. Lu, Y.D. Lee, H.T. Liu, Anal. Chem. 82 (2010) 2290.
[30] H. Sun, Z.H. Mo, J.T.S. Choy, D.R. Zhu, Y.S. Fung, Sensor. Actuat. B-Chem. 131 (2008) 148.
[31] S. Yaqub, U. Latif, F.L. Dickert, Sensor Actuat. B-Chem. 160 (2011) 227.
[32] W. Wang, S.T. He, S.Z. Li, M.H. Liu, Y. Pan, Sensor. Actuat. B-Chem. 125 (2007) 422.
[33] W. Wang, S.T. He, Sensor. Actuat. B-Chem. 138 (2009) 432.
[34] A. Boyle, E.M. Genies, M. Lapkowski, Synthetic Met. 28 (1989) 769.
[35] S.A. Piletskii, Y.I. Kurys, A.E. Rachkov, A.V. Elskaya, Russ. J. Electrochem. 30 (1994) 990.
[36] A. Pizzariello, M. Stred'ansky, S. Stred'anska, S. Miertus, Sensor. Actuat. B-Chem. 76 (2001) 286.
[37] G.M. Murray, A.L. Jenkins, A. Bzhelyansky, O.M. Uy, Johns Hopkins APL Techn. Dig. 18 (1997) 464.
[38] E. Hedborg, F. Winquist, I. Lundström, L.I. Andersson, K. Mosbach, Sensor. Actuat. A-Phys. 37-8 (1993) 796.
[39] T.A. Sergeyeva, S.A. Piletsky, A.A. Brovko, E.A. Slinchenko, L.M. Sergeeva, A.V. Elskaya, Anal. Chim. Acta 392 (1999) 105.
[40] Y. Yoshimi, R. Ohdaira, C. Iiyama, K. Sakai, Sensor. Actuat. B-Chem. 73 (2001) 49.
[41] G.H. Chen, Z.B. Guan, C.T. Chen, L.T. Fu, V. Sundaresan, F.H. Arnold, Nat. Biotechnol. 15 (1997) 354.
[42] C.Y. Chai, G.Y. Liu, F. Li, X.F. Liu, B. Yao, L. Wang, Anal. Chim. Acta 675 (2010) 185.
[43] A. Namvar, K. Warriner, Biosens. Bioelectron. 22 (2007) 2018.
[44] D. Kriz, O. Ramström, A. Svensson, K. Mosbach, Anal. Chem. 67 (1995) 2142.
[45] S.P. Pogorelova, M. Zayats, T. Bourenko, A.B. Kharitonov, O. Lioubashevski, E. Katz, I. Willner, Anal. Chem. 75 (2003) 509.
[46] S. Kröger, A.P.F. Turner, K. Mosbach, K. Haupt, Anal. Chem. 71 (1999) 3698.
[47] E. Mazzotta, R.A. Picca, C. Malitesta, S.A. Piletsky, E.V. Piletska, Biosens. Bioelectron. 23 (2008) 1152.
[48] B.B. Prasad, R. Madhuri, M.P. Tiwari, P.S. Sharma, Biosens. Bioelectron. 25 (2010) 2140.
[49] T. Alizadeh, M. Zare, M.R. Ganjali, P. Norouzi, B. Tavana, Biosens. Bioelectron. 25 (2010) 1166.
[50] B.B. Prasad, R. Madhuri, M.P. Tiwari, P.S. Sharma, Electrochim. Acta 55 (2010) 9146.
[51] H.C. Huang, S.Y. Huang, C.I. Lin, Y.D. Lee, Anal. Chim. Acta 582 (2007) 137.
[52] O.Y.F. Henry, S.A. Piletsky, D.C. Cullen, Biosens. Bioelectron. 23 (2008) 1769.
[53] D. Lakshmi, M.J. Whitcombe, F. Davis, I. Chianella, E.V. Piletska, A. Guerreiro, S. Subrahmanyam, P.S. Brito, S.A. Fowler, S.A. Piletsky, Chem. Comm. 19 (2009) 2759.
[54] J.P. Li, F.Y. Jiang, X.P. Wei, Anal. Chem. 82 (2010) 6074.
[55] A. Strikovsky, J. Hradil, G. Wulff, Reac. Func. Polym. 54 (2003) 49.

[56] K. Sode, S. Ohta, Y. Yanai, T. Yamazaki, Biosens. Bioelectron. 18 (2003) 1485.
[57] K. Lettau, A. Warsinke, A. Laschewski, K. Mosbach, E. Yilmaz, F.W. Scheller, Chem. Mater. 16 (2004) 2745.
[58] J.d.R.M. Neto, W.d.J.R. Santos, P.R. Lima, S.M.C.N. Tanaka, A.A. Tanaka, L.T. Kubota, Sensor. Actuat. B-Chem. 152 (2011) 220.
[59] D. Lakshmi, A. Bossi, M.J. Whitcombe, I. Chianella, S.A. Fowler, S. Subrahmanyam, E.V. Piletska, S.A. Piletsky, Anal. Chem. 81 (2009) 3576.
[60] E. Piletska, K. Karim, R. Coker, S. Piletsky, J. Chromatogr. A 1217 (2010) 2543.
[61] C.A. Barrios, C. Zhenhe, F. Navarro-Villoslada, D. López-Romero, M.C. Moreno-Bondi, Biosens. Bioelectron. 26 (2011) 2801.
[62] K. Haupt, A.G. Mayes, K. Mosbach, Anal. Chem. 70 (1998) 3936.
[63] R. Levi, S. McNiven, S.A. Piletsky, S.H. Cheong, K. Yano, I. Karube, Anal. Chem. 69 (1997) 2017.
[64] S.A. Piletsky, E. Terpetschnig, H.S. Andersson, I.A. Nicholls, O.S. Wolfbeis, Fresenius' J. Anal. Chem. 364 (1999) 512.
[65] D.Y. He, Z.J. Zhang, H.J. Zhou, Y. Huang, Talanta 69 (2006) 1215.
[66] L. Wang, Z. Zhang, Talanta 76 (2008) 768.
[67] C.C. Hong, P.H. Chang, C.C. Lin, C.L. Hong, Biosens. Bioelectron. 25 (2010) 2058.
[68] M. Jakusch, M. Janotta, B. Mizaikoff, K. Mosbach, K. Haupt, Anal. Chem. 71 (1999) 4786.
[69] O.A. Raitman, V.I. Chegel, A.B. Kharitonov, M. Zayats, E. Katz, I. Willner, Anal. Chim. Acta 504 (2004) 101.
[70] Q.Q. Wei, T.X. Wei, Chin. Chem. Lett. 22 (2011) 721.
[71] S. Devanathan, Z. Salamon, A. Nagar, S. Narang, D. Schleich, P. Darman, V. Hruby, G. Tollin, Anal. Chem. 77 (2005) 2569.
[72] G. Gupta, A.S.B. Bhaskar, B.K. Tripathi, P. Pandey, M. Boopathi, P.V.L. Rao, B. Singh, R. Vijayaraghavan, Biosens. Bioelectron. 26 (2011) 2534.
[73] B.K. Lavine, D.J. Westover, N. Kaval, N. Mirjankar, L. Oxenford, G.K. Mwangi, Talanta 72 (2007) 1042.
[74] J. Wang, S. Banerji, N. Menegazzo, W. Peng, Q.J. Zou, K.S. Booksh, Talanta 86 (2011) 133.
[75] G.P. González, P.F. Hernando, J.S.D. Alegría, Anal. Chim. Acta 638 (2009) 209.
[76] P.J.R. Roche, S.M. Ng, R. Narayanaswamy, N. Goddard, K.M. Page, Sensor. Actuat. B-Chem. 139 (2009) 22.
[77] K. Reimhult, K. Yoshimatsu, K. Risveden, S. Chen, L. Ye, A. Krozer, Biosens. Bioelectron. 23 (2008) 1908.
[78] G. Sener, L. Uzun, R.I. Say, A. Denizli, Sensor. Actuat. B-Chem. 160 (2011) 791.
[79] M. Lotierzo, O.Y.F. Henry, S. Piletsky, I. Tothill, D. Cullen, M. Kania, B. Hock, A.P.F. Turner, Biosens. Bioelectron. 20 (2004) 145.
[80] A.S. Belmont, S. Jaeger, D. Knopp, R. Niessner, G. Gauglitz, K. Haupt, Biosens. Bioelectron. 22 (2007) 3267.
[81] M.E. Cooper, B.P. Hoag, D.L. Gin, Abstr. Pap. Am. Chem. S 213 (1997) 115.
[82] F.L. Dickert, U. Geiger, P. Lieberzeitv, U. Reutner, Sensor. Actuat. B-Chem. 70 (2000) 263.
[83] P. Turkewitsch, B. Wandelt, G.D. Darling, W.S. Powell, Anal. Chem. 70 (1998) 2025.
[84] K. Ye, J. Mosbach, Am. Chem. Soc. 123 (2001) 2901.
[85] A.L. Jenkins, O.M. Uy, G.M. Murray, Anal. Chem. 71 (1999) 373.
[86] W. Wang, S. Gao, B. Wang, Org. Lett. 1 (1999) 1209.
[87] H. Zhang, W. Verboom, D.N. Reinhoud, Tetrahedr. Lett. 42 (2001) 4413.
[88] S. Subrahmanyam, S.A. Piletsky, E.V. Piletska, B.N. Chen, R. Day, A.P.F. Turner, Adv. Mater. 12 (2000) 722.
[89] C.I. Lin, A.K. Joseph, C.K. Chang, Y.D. Lee, Biosens. Bioelectron. 20 (2004) 127.
[90] W. Zhang, X.W. He, Y. Chen, W.Y. Li, Y.K. Zhang, Biosens. Bioelectron. 26 (2011) 2553.
[91] J.C.C. Yu, E.P.C. Lai, Reac. Func. Polym. 63 (2005) 171.

Index

Note: Page numbers followed by *f* indicate figures, *t* indicate tables and *b* indicate boxes.

D

E

F

K

L

N

Q

R

S

T

U

V

W

X

Y

Z